Genetic Engineering: Techniques and Applications

Genetic Engineering: Techniques and Applications

Editor: Enrique Preston

www.callistoreference.com

Callisto Reference,
118-35 Queens Blvd., Suite 400,
Forest Hills, NY 11375, USA

Visit us on the World Wide Web at:
www.callistoreference.com

ISBN: 978-1-63239-870-3 (Hardback)

The publisher's policy is to use permanent paper from mills that operate a sustainable forestry policy. Furthermore, the publisher ensures that the text paper and cover boards used have met acceptable environmental accreditation standards.

Printed in the United States of America.

Cataloging-in-Publication Data

Genetic engineering : techniques and applications / edited by Enrique Preston.
    p. cm.
Includes bibliographical references and index.
ISBN 978-1-63239-870-3
1. Genetic engineering. 2. Transgenic organisms. 3. Genetic recombination. 4. Biotechnology.
I. Preston, Enrique.
QH442 .G45 2017
660.65--dc23

Table of Contents

Preface

Genetic engineering is the alteration of genome using biotechnology. It is a compilation of technologies used to modify the genetic makeup of cells, including the transfer of genes within and across species boundaries to produce improved or novel organisms. This book elucidates new techniques and their applications in a multidisciplinary approach keeping the focus on genetic engineering. Also included in this book is a detailed explanation of the various concepts and applications of this discipline in different fields like medicine, manufacturing, gene therapy, etc. The various studies that are constantly contributing towards advancing technologies and evolution of this field are examined in detail. Researchers and students in this field will be assisted by this book.

This book unites the global concepts and researches in an organized manner for a comprehensive understanding of the subject. It is a ripe text for all researchers, students, scientists or anyone else who is interested in acquiring a better knowledge of this dynamic field.

I extend my sincere thanks to the contributors for such eloquent research chapters. Finally, I thank my family for being a source of support and help.

Editor

The Rose (*Rosa hybrida*) NAC Transcription Factor 3 Gene, *RhNAC3*, Involved in ABA Signaling Pathway Both in Rose and *Arabidopsis*

Guimei Jiang[1❾], **Xinqiang Jiang**[2❾], **Peitao Lü**[1], **Jitao Liu**[1], **Junping Gao**[1], **Changqing Zhang**[1*]

1 Department of Ornamental Horticulture, College of Agriculture and Biotechnology, China Agricultural University, Beijing, PR China, **2** College of Landscape Architecture and Forestry, Qingdao Agricultural University, Qingdao, PR China

Abstract

Plant transcription factors involved in stress responses are generally classified by their involvement in either the abscisic acid (ABA)-dependent or the ABA-independent regulatory pathways. A stress-associated NAC gene from rose (*Rosa hybrida*), *RhNAC3*, was previously found to increase dehydration tolerance in both rose and *Arabidopsis*. However, the regulatory mechanism involved in RhNAC3 action is still not fully understood. In this study, we isolated and analyzed the upstream regulatory sequence of *RhNAC3* and found many stress-related *cis*-elements to be present in the promoter, with five ABA-responsive element (ABRE) motifs being of particular interest. Characterization of *Arabidopsis thaliana* plants transformed with the putative *RhNAC3* promoter sequence fused to the β-glucuronidase (GUS) reporter gene revealed that *RhNAC3* is expressed at high basal levels in leaf guard cells and in vascular tissues. Moreover, the ABRE motifs in the *RhNAC3* promoter were observed to have a cumulative effect on the transcriptional activity of this gene both in the presence and absence of exogenous ABA. Overexpression of *RhNAC3* in *A. thaliana* resulted in ABA hypersensitivity during seed germination and promoted leaf closure after ABA or drought treatments. Additionally, the expression of 11 ABA-responsive genes was induced to a greater degree by dehydration in the transgenic plants overexpressing *RhNAC3* than control lines transformed with the vector alone. Further analysis revealed that all these genes contain NAC binding *cis*-elements in their promoter regions, and RhNAC3 was found to partially bind to these putative NAC recognition sites. We further found that of 219 *A. thaliana* genes previously shown by microarray analysis to be regulated by heterologous overexpression *RhNAC3*, 85 are responsive to ABA. In rose, the expression of genes downstream of the ABA-signaling pathways was also repressed in *RhNAC3*-silenced petals. Taken together, we propose that the rose RhNAC3 protein could mediate ABA signaling both in rose and in *A. thaliana*.

Editor: Ji-Hong Liu, Key Laboratory of Horticultural Plant Biology (MOE), China

Funding: This work was supported by the National Natural Science Foundation of China (Grant No. 31372096) and Beijing Nova Program (Grant No. 2009B51). The funders had no role in study design, data collection and analysis, decision to publish, or preparation of the manuscript.

Competing Interests: The authors have declared that no competing interests exist.

* Email: chqzhang@cau.edu.cn

❾ These authors contributed equally to this work.

Introduction

Drought, or dehydration, is one of the major limiting factors for plant growth, development, and productivity and plants have evolved a range of physiological, biochemical and molecular responses to promote drought stress tolerance [1]. One such response to drought stress is the production of the plant hormone abscisic acid (ABA), which mediates numerous downstream responses, including stomatal closure, thereby restricting water loss. Using genomic and transcriptomic analyses, the products of drought-inducible genes, including those regulated by ABA, have been classified into two groups: structural proteins and regulatory proteins [2], such as transcription factors (TFs).

Drought or dehydration-induced TFs have been isolated from many plant species and demonstrated to be involved in drought tolerance, such as *DREB2A*, *DREB2C*, *AREB1*, and *WRKY57* from *A. thaliana* [3,4], *OsbZIP46* from *Oryza sativa* [5],

TaMYB30-B from *Triticum aestivum* [6], *RhNAC2* from *Rosa hybrida* [7], and *ThbZIP1* from *Tamarix hispida* [8]. TFs involved in stress responses are typically classified as being involved either in ABA-dependent or the ABA-independent regulatory pathways [9] and structure and sequence analyses of the promoters of ABA-dependent TF genes have identified many stress-related *cis*-elements. These include the G-box (CACGTG, a MYC recognition site), the dehydration-responsive element/C-repeat (DRE/CRT) and the ABA-responsive element [10,11]. Among these, the conserved ABA-responsive *cis*-element, PyACGTG/TGC, also named ABA-responsive element (ABRE), is a signature sequence for genes involved in the ABA signaling pathway [12,13], and is important for promoter activity under osmotic stress conditions, such as those resulting from dehydration and high salinity [12,14]. The functions of ABA-dependent TFs have also been investigated through overexpression in many plant species [11], which has been found to result in hypersensitivity to ABA during seed germination [5,15], constitutive stomatal closure [16] and severely inhibited root growth [17]. Microarray analysis further revealed that constitutive overexpression of ABA-related TF genes, such as *ABO3* and *MYB96*, generally increases the expression of downstream ABA-responsive genes, and results in drought tolerance [18,19].

Another class of ABA-related TFs are NAC (*NAM*, *ATAF*1 and 2, and *CUC2*) proteins, plant-specific transcriptional regulators that contain conserved N-terminal NAC domains and divergent C-terminal regions [20]. NAC TFs play important roles in regulating numerous aspects of growth and development, including cell division, and senescence, as well as responses to environmental stress stimuli [21]. Many are involved in ABA mediated signaling during their response to abiotic stresses, such as *A. thaliana* ANAC019 and ANAC055, soybean GmNAC011 and GmNAC020 and rice OsNAC5 [22,23,24], and their overexpression can result in enhanced ABA sensitivity at both the germination and post-germination developmental stages [25].

Different members of the NAC family have also been shown to be responsive to dehydration in rose petals [7]. Of these, *RhNAC2* has been found to promote petal cell expansion, in association with the regulation of cell wall-related genes [7], while *RhNAC3* regulates osmotic stress-related genes when exposed to drought stress [26]. However, nothing has been reported to date regarding the mechanism by which RhNAC3 participates in the ABA regulatory pathway. In this study, we characterized the upstream regulatory sequence of *RhNAC3* and found that ABREs in the *RhNAC3* promoter are needed for gene activity in both the presence and absence of exogenous ABA. Transgenic *A. thaliana* overexpressing *RhNAC3* showed enhanced ABA sensitivity during seed germination and during stomatal closure, and ABA-responsive genes were also up-regulated under dehydration conditions in both rose and the *A. thaliana* overexpressing *RhNAC* lines. These data indicate that ectopically expressed RhNAC3 enhances ABA sensitivity in *A. thaliana* and is involved in an ABA-dependent signaling pathway, at least some components of which are likely conserved between rose and *A. thaliana*.

Materials and Methods

Regulatory *cis*-element analysis

The upstream regulatory sequence of *RhNAC3* was isolated using PCR-based genome walking method [27]. (We state clearly that no specific permissions were required for these locations/ activities and confirm that the field studies did not involve endangered or protected species). And the primers used are listed in Table S1. All amplified fragments were sub-cloned into the pGEM T-Easy Vector (Promega, Madison, WI, USA) and

transformed into *Escherichia coli DH5a* cells after sequencing. The position of the translation start site was designated "0". The *cis*-acting elements were analyzed and annotated using two software programs from the Plant *Cis*-acting Regulatory DNA Elements (PLACE) [28] (http://www.dna.affrc.go.jp/PLACE/) and Plant *Cis*-acting Regulatory Elements (PlantCARE) [29] (http://bioinformatics.psb.ugent.be/webtools/plantcare/html/) software programs. For NAC-binding site analysis of RhNAC3 upregulated genes in *A. thaliana*, a 1,000 bp regulatory sequence upstream of the genes was searched and analyzed by TAIR Loci Upstream Seq –1,000 bp of 'Sequence Bulk Download and Analysis' at www.arabidopsis.org.

Construction of plant expression vectors and *Arabidopsis* transformation

The 977 bp upstream regulatory sequence of *RhNAC3* was amplified with 5′ ACC*AAGCTT*CATTCTACTTGTCCAAAT-CTGAACCTC 3′ and 5′ GCT*CTAGA*CCGTATCAGAGA-GATGAAACAGGAA 3′ (Table S1) and the product digested with *Hind* III and *Xba*I, and inserted into the pBI121 binary vector. The resulting Pro$_{RhNAC3}$:GUS plasmid was introduced into the *Agrobacterium tumefaciens* strain *GV3101* and transformed into *A. thaliana* (Columbia) by the floral dip method [30]. Ten independent lines of kanamycin-resistant transgenic plants were obtained. The homozygous T3 generation seeds of the transgenic lines were used for subsequent experiments.

Histochemical staining and quantitative GUS activity assay

Histochemical staining for GUS activity was performed as described by Li *et al.* (2009) [31]. Plant samples exposed to different treatments were immersed in GUS staining buffer (0.5 mM 5-bromo-4-chloro-3-indoly-β-D-GlcA, 0.5 M NaH$_2$PO$_4$, pH 7.0, 1 mM EDTA, 0.5 mM potassium ferricyanide and 0.5 mM potassium ferrocyanide). After staining at 37°C for 3–10 h, the samples were immersed in 95% (v/v) ethanol at 37°C to remove chlorophyll. For histochemical analysis of the Pro$_{RhNAC3}$: GUS transgenic *A. thaliana* plants in response to ABA treatment, 9-day-old seedlings were grown on MS medium supplemented with 100 µM ABA for 4 days, before being sampled for histochemical GUS staining. The GUS staining patterns were examined under a microscope (BX51; Olympus) and analyzed using Photoshop CS6 software (Adobe, McLean, VA). Quantitative assays of GUS activity were performed as described by Jefferson *et al.* (1987) [32]. All experiments were performed three times to give three independent biological replicates.

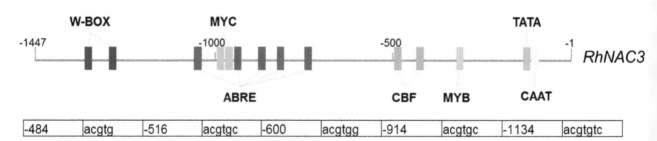

Figure 1. Schematic representation of the *RhNAC3* promoter. The major stress-related *cis*-acting elements in the 1447 bp promoter of *RhNAC3* are shown. The position and putative sequences of ABRE elements are listed.

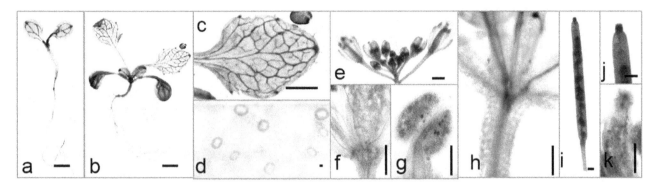

Figure 2. Histochemical analysis of *RhNAC3* expression in *A. thaliana*. The *GUS* gene driven by the *RhNAC3* promoter was expressed in 5-day-old *A. thaliana* seedlings (**a**), 14 day-old seedlings (**b**), young leaves (**c**), stomata of young leaves (**d**), flowers (**e**), petals (**f**), stigma (**g**), apical stems of inflorescences (**h**), mature siliques (**i** and **j**; the latter is magnified); and immature seeds (**k**), Scale bars = 1 mm.

Construction of the truncated *RhNAC3* promoter-GUS fusion and transient expression assays

Three truncated *RhNAC3* promoter fragments, N0 (−1447 to −160 bp), N1 (−707 to −160 bp) and N2 (−377 to −160 bp) were amplified from rose genomic DNA and cloned into a modified pUC19 plasmid containing the GUS reporter gene, as described by Dai *et al.* (2012) [7]. To mutate the ABRE cis-element, we replaced the ACGT of the ABRE core sequence with TTTA using overlap PCR methods [33]. Mutation fragments of N0 (mN0, five ABREs mutated) and N1 (mN1, three ABREs mutated) were amplified, and cloned into the modified pUC19 as described for N0 and N1. *A. thaliana* mesophyll protoplasts were transformed with the resulting vectors: N0, mN0, N1, mN1 and N2, and an empty (normal) vector control (NC). For the ABA treatment experiments, *Arabidopsis* mesophyll protoplasts harboring the different constructs (N0, N1, N2 and NC) were exposed to 10 μM ABA (Sigma, St. Louis, MO). GUS activity was measured in protoplast extracts after 24 h of incubation with ABA. Isolation of *A. thaliana* mesophyll protoplasts, transformation of protoplasts and GUS activity assays were carried out as previously described [32,34]. The primers are listed in Table S1, and the experiments were performed in triplicate.

Seed germination assay and root growth measurements

The *RhNAC3*-overexpressing plants (overexpressor OE#3, OE#6 and OE#12) had previously been generated [26], and wild type (WT) and vector (VC) plants were used as controls. Approximately 50 seeds were plated onto solid MS medium supplemented with either 0, 0.2, 0.4 or 0.8 μM ABA. After vernalization at 4°C for 3 days, the seeds were moved to a temperature controlled room at 23±1°C under long-day conditions (16 h light/8 h dark cycle), with a light intensity of 80–100 μmol/m²/s and 40–60% relative humility. The rates of radicle emergence and cotyledon greening were measured after 7 days. All experiments were performed in triplicate. To measure seedling root growth, 5-day old seedlings of OE#3, OE#6 and OE#12 were transferred to plates of MS medium containing 5, 10, or 30 μM ABA respectively, and grown vertically. After growth for 10 d, primary root length and lateral root number was measured and analyzed using the Image J software (http://rsbweb.nih.gov/ij/). The WT and VC plants were used as controls.

Stomatal aperture measurements

For ABA-induced stomatal closure, mature leaves from light-grown 3-week-old control and *RhNAC3* transgenic plants were detached and incubated in stomatal opening solution (10 mM KCl, 100 μM CaCl₂ and 10 mM MES, pH 6.1) for 2 h at 22°C [16], before being transferred to fresh stomatal opening solution containing 0 μM or 10 μM ABA. Stomata on abaxial surfaces were photographed through a light microscope (BX51; Olympus), and the stomatal aperture (the ratio of width to length) was measured (n = 20). For drought-induced stomatal closure, 3-week-old seedlings of WT, VC and *RhNAC3*-overexpressing lines were grown for 10 d without water. Plants grown under normal conditions were used as control. Leaves in the same position on the plant were sampled, and the stomata on the leaf abaxial surfaces were immediately photographed. Stomatal aperture was measured (n = 20) and all experiments were repeated three times.

Quantitative reverse transcription PCR analysis

The detached leaves of 3-week-old vector (VC) and *RhNAC3* overexpressor (OE#3, 6 and 12) *A. thaliana* plants were dehydrated for 3 h at 23–25°C, 40–50% relative humidity, and 100 μmol m⁻²s⁻¹ light intensity, then sampled for quantitative reverse transcription polymerase chain reaction (qRT-PCR) analysis. Total RNAs were isolated from the leaf samples using the Trizol agent (Invitrogen, Carlsbad, CA). DNase-treated RNA (1 μg) was used for first-strand cDNA synthesis (Invitrogen, Carlsbad, CA) and the cDNA (2 μL) was used as the template in a 20 μL qRT-PCR using a qPCR Kit (Kapa Biosystems, Woburn, MA). The *A. thaliana* Actin2 gene (GenBank accession no. NM_112764) was used as an internal control. The 11 selected genes and gene specific primers used for the qRT-PCR analysis are listed in Table S1. Each qRT-PCR evaluation was performed with three biological replicates.

For qRT-PCR analysis of downstream genes of *RhNAC3* action in rose, nine putative ABA signaling and downstream rose genes from the ABA-signaling pathways were selected from our rose transcriptome databases [7] (Table S2). The rose cDNAs from Tobacco Rattle Virus (TRV) and *RhNAC3*-silenced petals were obtained in our previous study [26]. *RhUbi1* (accession no. JK622648) was used as the internal control. The gene specific primers for qRT-PCR are listed in Table S1. Each qRT-PCR analysis was performed with three biological replicates.

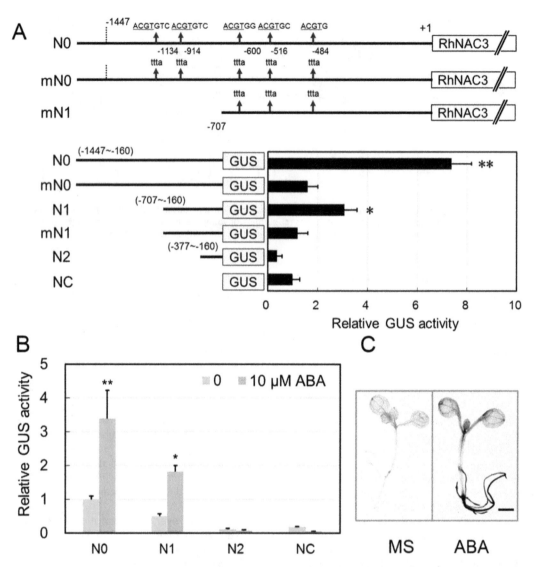

Figure 3. Deletion analysis and ABA dose dependent response of *RhNAC3* promoter activity. A, Assays of GUS activity in *A. thaliana* protoplasts containing RhNAC3 promoter deletion and ABRE mutation constructs. The numbers on top represent the positions of the ABRE *cis*-elements and mutations of ABREs in the *RhNAC3* promoter region. Relative GUS activity in transient expression experiments using five different constructs (N0, mN0, N1, mN1 and N2) and the vector control (NC) is shown at the bottom. GUS activity was determined after 24 h of incubation. Error bars represent standard error (*n* = 3). **: *P* < 0.01, *: *P* < 0.05, *t* test. **B,** Effects of exogenous ABA on GUS activity in *A. thaliana* protoplasts containing *RhNAC3* promoter deletion constructs. The truncated *RhNAC3* promoter constructs (N0, N1 and N2) and vector control (NC) were transformed into *A. thaliana* protoplasts, which were then exposed to 0 and 10 μM exogenous ABA. GUS activity in protoplast extracts was measured after 24 h of incubation with ABA. Error bars represent standard error (*n* = 5). **C,** Histochemical analysis of *RhNAC3* promoter::GUS expression in response to ABA. 9-day-old transgenic seedlings were grown on MS medium only or MS medium plus ABA (transferred to MS medium plus 100 μM ABA for 4 days) before being subjected to histochemical GUS staining. Scale bar = 1 mm.

Electrophoretic mobility-shift assay

The electrophoretic mobility-shift assay (EMSA) was performed according as previously described [26] with minor modifications. To construct the GST-RhNAC3 fusion protein, the N-terminal of RhNAC3 (RhNAC3^{N1-162}) was amplified by PCR (primers are listed in Table S1) and the PCR product was ligated into the pGEX-2T vector (Pharmacia LKB Biotechnology, Piscataway, NJ) via the *Bam*HI and *Sac*I sites and the recombinant vector was expressed in *Escherichia coli* BL21 cells. The fusion protein was induced by 0.2 mM isopropyl β-D-1-thiogalactopyranoside (IPTG), and the cells *E. coli* incubated at 28°C for a further 6 h. The recombinant protein was purified by GST-agarose affinity chromatography (GE Healthcare, http://www. gehealthcare.com/). Biotin-labeled DNA fragments used in the EMSA contain one or two putative NAC binding sequences [21]. The probes were incubated with the fusion protein at room temperature for 25 min in binding buffer (10× concentration 100 mM Tris, 500 mM KCl, 10 mM dithiothreitol; pH 7.5). Each 20 μL binding reaction contained 0.2 pmol biotin probe and 2 μg fusion protein, and 1 μg Poly (dI•dC) was added to the reaction to minimize nonspecific interactions. The reaction products were analyzed using 5% native polyacrylamide gel electrophoresis and 0.5× Tris-borate/EDTA buffer. After electrophoresis, the DNA fragments on the gel were transferred to a nitrocellulose membrane using 0.5× Tris-borate/EDTA at 380 mA (~100 V for 30 min at 4°C. After UV cross-linking, the membrane wa

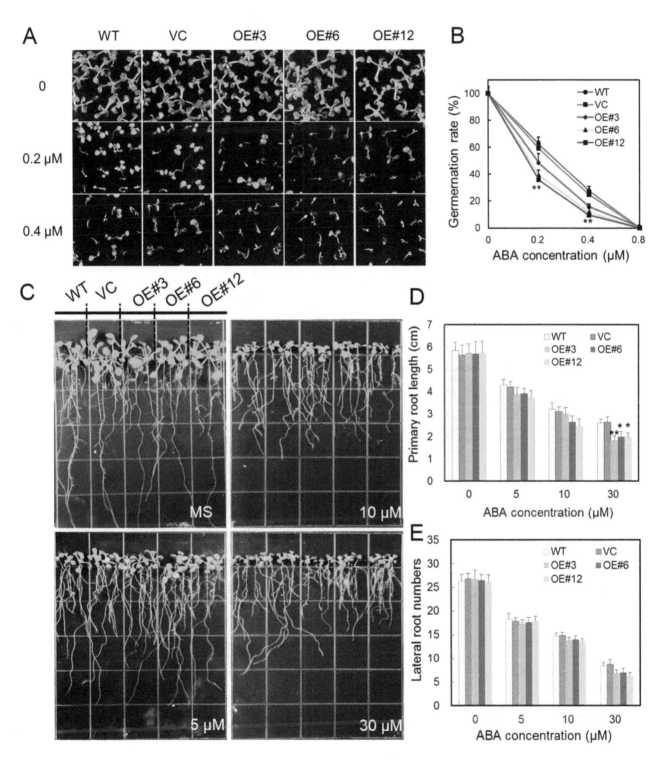

Figure 4. Effect of ABA concentration on seed germination and root growth in WT and *RhNAC3* overexpressing *A. thaliana* plants. A,
Seed germination phenotypes. The homozygous T3 seeds of *RhNAC3*-overexpressing lines (OE#3, OE#6 and OE#12), wild type (WT) and vector plants were plated on MS supplemented with 0, 0.2 or 0.4 μM ABA. Images were obtained 7 days after planting. **B,** Seed germination rates. The germination rates were measured 7 days after planting. Error bars represent standard error (*n* = 3). **: *P*<0.01, *: *P*<0.05, *t* test. **C,** Root growth phenotypes. Five-day-old seedlings of WT, vector only and three *RhNAC3*-overexpressing *A. thaliana* lines (OE#3, #6 and #12) were transferred to MS plates supplemented with 0, 5, 10 and 30 μM ABA. Root phenotypes were visualized 10 days after planting. **D,** Primary root length analysis. **: *P*< 0.01, *: *P*<0.05, *t* test. **E,** Lateral root number analysis. Both primary root length and lateral root number were measured after 10 days of growth. Three independent experiments were performed using 15 plants in each experiment in D and E. Error bars represent standard error (*n* = 3).

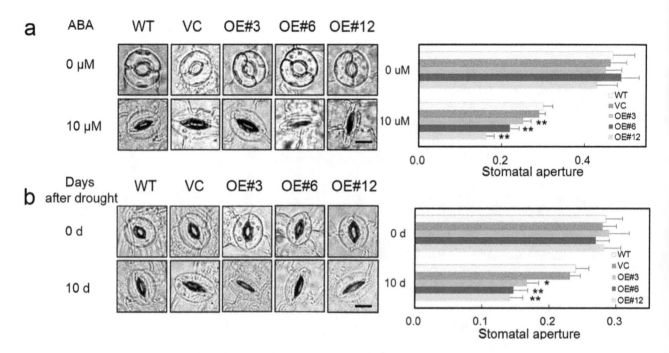

Figure 5. Stomatal aperture of the *RhNAC3* overexpressing *A. thaliana* plants in response to ABA and drought treatments. A, Stomatal aperture in response to ABA. Mature leaves from three-week old wild type (WT), vector control and independent *RhNAC3*-overexpressing plants (OE#3, OE#6 and OE#12) were treated with a stomatal opening solution for 2 h (0 μM) and incubated with 10 μM ABA for 2 h (10 μM). Stomata on the abaxial surfaces were imaged by light microscopy. Stomatal aperture (the ratio of width to length) was quantified using at least 20 guard cells from each sample. *Bar* 10 μm. **B,** Stomatal aperture of *RhNAC3* overexpressing lines in response to drought stress. Three-week old seedlings of WT, vector control and independent *RhNAC3*-overexpressing plants (OE#3, OE#6 and OE#12) were subjected to 10 days without water. Plants grown under normal well watered conditions were used as a control. The leaves were harvested and the stomata on the leaf abaxial surfaces were immediately photographed. Stomatal apertures were then quantified (n = 20). *Bar* 10 μm. ******: $P<0.01$, *****: $P<0.05$, t test.

transferred to conjugate/blocking buffer by mixing 16.75 μL stabilized streptavidin-horseradish peroxidase conjugate with 5 mL blocking buffer. After washing, biotin-labeled DNA was detected by chemiluminescence according to the manufacturer's protocol (Pierce, http://www.piercenet.com/).

Results

Structure and sequence analysis of the *RhNAC3* promoter

To elucidate the regulation of *RhNAC3* transcription, a 1,447 bp fragment corresponding to the sequence immediately upstream of its translational start site (TSS) (GenBank accession number: KJ000025) was isolated by PCR-based genome walking. Subsequent sequence analysis of this region revealed a number of putative *cis*-elements, including elements associated with ABA, cold, pathogen and wounding responses (Figure 1, Figure S1 and Table S2). The TATA box (TTATTT) was found −104 bp upstream of the TSS and a CAAT-box sequence (CAAT) was found 2 bp downstream of the TATA-box sequence. Five ABRE-related sequence motifs (ACGTG), located at positions −1140 to

Table 1. Analysis of putative NAC binding *cis*-elements in the promoter regions of genes downstream from *RhNAC3* action.

Gene name	NAC protein binding sites			
	CACG	CGTG	GTGC	CATGTG
RD29A	4	1	1	
RD29B	2	4	2	
RD20	7	4	2	2
RD26	5	6	3	
COR47	1	1		1
COR15A	4	1		
KIN2	6	3		
ABI1	2			
ABI3	1	2	1	
ABF4	3	1		
ABA3	4	4		

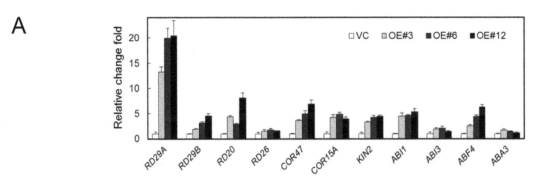

B

P1: *RD29A* pro (-313~-347bp), AT**GTGC**CGTTTGTTATAATAAACAGCCA**CACG**AC
P2: *RD20* pro (-623~-650bp), GGTTCCAGC**CACGTGCATGTG**ACCACAT
P3: *COR47* pro (-913~-952bp),TC**CATGTG**TGGGCCCGTTCCCACCTCTTTTAATA**CGTG**TC
P4: *COR15A* pro (-14~-36bp), AATAAATA**CACGTG**AAGGAAATG

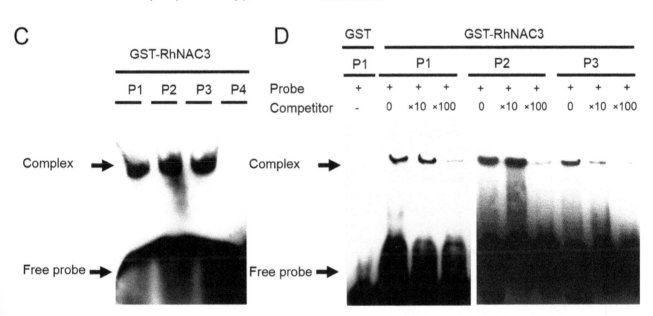

Figure 6. RhNAC3 binding to the regulatory sequences of ABA-related *A. thaliana* genes. A, ABA-related gene expression in *RhNAC3* overexpressing *A. thaliana* lines The aerial parts of light-grown, 3-week old vector control and three independent *RhNAC3* overexpressing *A. thaliana* lines (OE#3, OE#6 and OE#12) were dehydrated for 3 h and sampled (23–25°C, 40–50% relative humidity). The expression patterns of 11 ABA-responsive genes were analyzed by qPCR and the data represents the fold induction of each gene by dehydration relative to the control treatment. Mean values from three independent biological replicates were normalized to the levels of the internal control gene *Actin2*. **B,** Sequences and positions of putative RhNAC3 binding elements used in an electrophoretic mobility shift assay (EMSA). Probes were derived from the regulatory sequences of 4 selected ABA-responsive *A. thaliana* genes. Underlined letters indicate the core sequences of NAC protein targeted promoters. The sense strands of the oligonucleotide probes corresponding to the predicted RhNAC3 binding sites are shown. **C,** Interaction between GST-RhNAC3N$^{1-162}$ and biotin-labeled probes indicated in (B). **D,** DNA-binding specificity for RhNAC3 with interacting probes. The arrows indicate the positions of protein/DNA complexes and the free probes. Purified protein (2 μg) was incubated with 0.2 pmol of the biotin probe. GST incubated with the P1 probe was used as a control, and a 10 or 100 fold excess of the unlabeled P1, P2, or P3 probes was used for competitive binding.

−1136, −918 to −914, −605 to −601, −521 to −517 and −488 to −484, were identified, which we hypothesized might be involved in an ABA response. Other potential regulatory elements that were found include two CBF sequences (RYCGAC) at positions −493 to −488 and −404 to −399, and a Myb-type TF recognition sequence (GGATA) at position −291 to −287. There are also two Myc-type TF recognition sequences (CACATG) at positions −985 to −980 and −951 to −946, and two W-box sequences (TGACT) at positions −1300 to −1296 and −1231 to −1227, relative to the TSS. Collectively, the presence of these *cis*-elements suggests that *RhNAC3* may play a role in responses to a

variety of stresses, such as cold, pathogen challenge or wounding, particularly via the ABA-dependent pathway.

Activity of the *RhNAC3* promoter in Pro$_{RhNAC3}$:*GUS* transgenic *A. thaliana* lines

We next examined the spatial expression pattern of the *RhNAC3* gene in 10 independent *A. thaliana* lines (Pro$_{RhNAC3}$:-*GUS*) that had been transformed with a construct containing the *RhNAC3* promoter fused to the *GUS* reporter gene. The T3 generation homozygotes of six lines were selected to analyze by GUS staining. Histochemical GUS staining revealed that Pro-$_{RhNAC3}$:*GUS* was expressed almost throughout the entire plant

Table 2. Up-regulated genes involved in the ABA response in *A. thaliana* lines overexpressing *RhNAC3*.

Affy ID[a]	Description[b]	AGI Code[c]	Fold change[d]	P-value[e]
	Signal transduction (21)			
267069_at	Calmodulin (CAM)-binding protein of 25 kDa	At2g41010	6.04	0.025
258947_at	Calcium-binding EF-hand family protein	At3g01830	2.58	0.045
255844_at	Protein kinase family protein / peptidoglycan-binding LysM domain-containing protein	At2g33580	2.48	0
255503_at	Concanavalin A-like lectin protein kinase family protein	At4g02420	2.39	0.045
251054_at	Lectin receptor kinase a4.3	At5g01540	2.39	0.02
261662_at	MAP kinase kinase 7	At1g18350	2.19	0.019
266371_at	Calcium-binding EF-hand family protein	At2g41410	2.14	0.006
257751_at	MAP kinase substrate 1	At3g18690	2.09	0.011
266037_at	Protein kinase superfamily protein	At2g05940	2	0.001
	Transcriptional regulation (32)			
261648_at	Salt tolerance zinc finger	At1g27730	10.1	0.011
257022_at	Zinc-finger protein 2	At3g19580	7.56	0.004
248448_at	Integrase-type DNA-binding superfamily protein	At5g51190	4.41	0.04
257053_at	Ethylene responsive element binding factor 4	At3g15210	3.69	0.022
266719_at	Circadian clock associated 1	At2g46830	3.44	0.049
246932_at	Integrase-type DNA-binding superfamily protein	At5g25190	3.04	0.044
252859_at	Integrase-type DNA-binding superfamily protein	At4g39780	2.76	0.032
266656_at	Zinc finger C-x8-C-x5-C-x3-H type family protein	At2g25900	2.41	0.018
258436_at	RING/U-box superfamily protein	At3g16720	2.4	0.001
245051_at	WRKY DNA-binding protein 15	At2g23320	2.34	0.003
259626_at	Basic region/leucine zipper motif 60	At1g42990	2.26	0.007
256426_at	RING/FYVE/PHD zinc finger superfamily protein	At1g33420	2.05	0.002
252009_at	A20/AN1-like zinc finger family protein	At3g52800	2.01	0.047
	Stress responsive (19)			
247708_at	Zinc finger (C3HC4-type RING finger) family protein	At5g59550	4.75	0.034
257763_s_at	Receptor like protein 38	At3g23110	3.9	0.001
262911_s_at	HSP20-like chaperones superfamily protein	At1g59860	2.45	0.03
262383_at	Toll-Interleukin-Resistance (TIR) domain-containing protein	At1g72940	2.34	0.043
259105_at	Rubber elongation factor protein (REF)	At3g05500	2.22	0.044
253046_at	Cytochrome P450, family 81, subfamily D, polypeptide 8	At4g37370	2.2	0
	Enzymes and metabolism (55)			
254975_at	2-oxoglutarate (2OG) and Fe(II)-dependent oxygenase superfamily protein	At4g10500	7.62	0.003
256933_at	Bifunctional inhibitor/lipid-transfer protein/seed storage 2S albumin superfamily protein	At3g22600	4.07	0.005
266993_at	Major facilitator superfamily protein	At2g39210	3.58	0.005
263852_at	Nudix hydrolase homolog 6	At2g04450	3.52	0.021
252908_at	Glycolipid transfer protein (GLTP) family protein	At4g39670	3.45	0
248330_at	NAD(P)-binding Rossmann-fold superfamily protein	At5g52810	3.33	0.003
248970_at	Solute:sodium symporters; urea transmembrane transporters	At5g45380	3.12	0.003
255630_at	C2 calcium/lipid-binding plant phosphoribosyltransferase family protein	At4g00700	2.94	0.048
253332_at	Peroxidase superfamily protein	At4g33420	2.86	0.028
249188_at	HXXXD-type acyl-transferase family protein	At5g42830	2.78	0.006
266761_at	NAD(P)-binding Rossmann-fold superfamily protein	At2g47130	2.73	0.001
252098_at	Eukaryotic aspartyl protease family protein	At3g51330	2.67	0.019
253806_at	RING membrane-anchor 2	At4g28270	2.61	0.037
249910_at	Arogenate dehydratase 2	At5g22630	2.58	0.013
245035_at	Acireductone dioxygenase 3	At2g26400	2.5	0.018
251422_at	Preprotein translocase Sec, Sec61-beta subunit protein	At3g60540	2.48	0.032
267337_at	HXXXD-type acyl-transferase family protein	At2g39980	2.38	0.02

Table 2. Cont.

Affy ID[a]	Description[b]	AGI Code[c]	Fold change[d]	P-value[e]
	Signal transduction (21)			
247604_at	COBRA-like protein 5 precursor	At5g60950	2.38	0.05
262237_at	Thioesterase superfamily protein	At1g48320	2.32	0
264843_at	2-oxoglutarate (2OG) and Fe(II)-dependent oxygenase superfamily protein	At1g03400	2.31	0.005
267300_at	UDP-Glycosyltransferase superfamily protein	At2g30140	2.24	0.006
253238_at	O-Glycosyl hydrolases family 17 protein	At4g34480	2.23	0.019
	Cell expansion related (4)			
247866_at	Xyloglucan endotransglucosylase/hydrolase 25	At5g57550	2.29	0.04
248263_at	Plant invertase/pectin methylesterase inhibitor superfamily	At5g53370	2.12	0.024
	Others (61)			
244966_at	Polyketide cyclase/dehydrase and lipid transport superfamily protein	At1g02470	5.58	0.041
256337_at	Serine-type endopeptidase inhibitors	At1g72060	4.84	0.008
257264_at	Receptor-like protein kinase-related family protein	At3g22060	3.77	0.015
258792_at	Glycine-rich protein	At3g04640	3.73	0.031
247193_at	MATE efflux family protein	At5g65380	3.2	0.049
250942_at	Legume lectin family protein	At5g03350	3.18	0.003
254832_at	Bifunctional inhibitor/lipid-transfer protein/seed storage 2S albumin superfamily protein	At4g12490	3.13	0.026
266097_at	SOUL heme-binding family protein	At2g37970	3.1	0.033
246289_at	VQ motif-containing protein	At3g56880	3.02	0.019
246495_at	Unknown protein	At5g16200	2.87	0.038
257690_at	SAUR-like auxin-responsive protein family	At3g12830	2.83	0.009
249769_at	Sigma factor E	At5g24120	2.75	0.027
263948_at	Late embryogenesis abundant (LEA) hydroxyproline-rich glycoprotein family	At2g35980	2.46	0.016
259502_at	Galactose oxidase/kelch repeat superfamily protein	At1g15670	2.32	0.002
248592_at	hydroxyproline-rich glycoprotein family protein	At5g49280	2.21	0.025
259410_at	Regulator of Vps4 activity in the MVB pathway protein	At1g13340	2.2	0.002
266247_at	Cysteine/Histidine-rich C1 domain family protein	At2g27660	2.15	0.019
252053_at	Syntaxin of plants 122	At3g52400	2.13	0.039
259507_at	P-loop containing nucleoside triphosphate hydrolases superfamily protein	At1g43910	2.11	0.03
264951_at	Target of Myb protein 1	At1g76970	2.11	0
262703_at	SAUR-like auxin-responsive protein family	At1g16510	2.1	0.025
258501_at	Glycine-rich protein	At3g06780	2.08	0.02
262571_at	Protein of unknown function (DUF1644)	At1g15430	2.08	0.024
251859_at	Proteophosphoglycan-related	At3g54680	2.03	0.02
	Unknown (27)			
253859_at	unknown protein	At4g27657	8.71	0.017
256891_at	unknown protein	At3g19030	3.98	0.031
260656_at	unknown protein	At1g19380	3.95	0.022
266017_at	unknown protein	At2g18690	3.53	0.008
265276_at	unknown protein	At2g28400	3.05	0.011
258188_at	unknown protein	At3g17800	2.57	0.021
258275_at	unknown protein	At3g15760	2.47	0.031
266259_at	unknown protein	At2g27830	2.19	0.009
252057_at	unknown protein	At3g52480	2.04	0.03

Genes derived from the *RhNAC3* up-regulated genes identified by the ATH1 microarray analysis in our previous study [26], classified to be responsive to ABA treatment according to the AtGenExpress global stress expression dataset [34].
[a] Affymetrix identification codes for the probes.
[b] Description as given by the Munich Information Center for Protein Sequences (MIPS) database.
[c] Represents a hyperlink to TAIR (www.arabidopsis.org) for more information.
[d] The ratio of three independent transgenic lines compared with the ratio of vector control plants. Genes expressed in RhNAC3 overexpressing transgenic plants with an up-regulation ratio higher than 2.0 are shown.
[e] Indicates one-way ANOVA of the differences in mean transcript expression levels between the transgenic and vector control plants at the 0.05 significance level.

A

Clone ID[a]	Accesion number	Description[b]
RU25535	JK622599	Responsive-to-dessication protein 29
RU07831	——	Responsive-to-dessication protein 28
RU01455	JK622958	Responsive-to-dessication protein 21
RU06450	JK620694	Cold regulated protein 47
RU04740	——	Kinesin motor protein (kin2)
RU22946	JK620897	ABI1 (ABA insensitive 1)
RU24499	JK620250	ABI2 (ABA insensitive 2)
RU03861	JK618281	ABF4 (ABRE binding factor 4)
RU26868	——	ABA3/ATABA3/LOS5/SIR3 (ABA deficient 3)

B

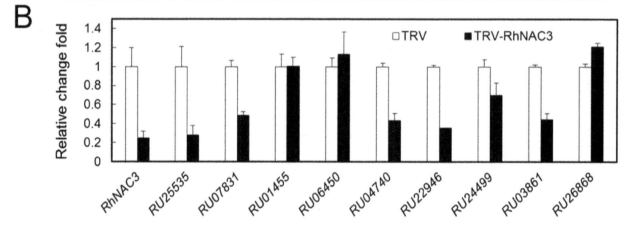

C

P1: *RU25535*-pro (-231~-201), 5'- GATTC**CACG**TCGCCTTTAAGC**CACGTG**GCGG -3'

P2: *RU04740*-pro (-444~-422), 5'- CCACGACGAC**CACGT**TCCCTTG -3'

P3: *RU03861*-pro (-543~-511), 5'- GAGCGGCCGCTTT**CGTG**GTTTCAATGGCTTAAA-3'

D   **E**

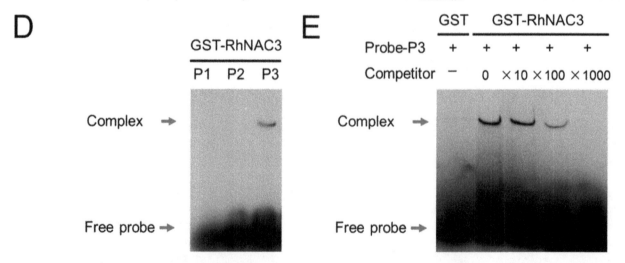

Figure 7. ABA-related gene expression in *RhNAC3*-silenced rose petals. A, The putative ABA signaling and downstream rose genes from the ABA-signaling pathway in rose. a, The clone ID from the rose transcriptome database [7]. b, Description of the *A. thaliana* homolog given by The Arabidopsis Information Resource (TAIR, http://www.arabidopsis.org). **B**, qRT-PCR analysis of *RhNAC3*-silenced rose petals. The rose cDNAs of TRV and *RhNAC3*-silenced (TRV-*RhNAC3*) petals were described in our previous report [26]. Data represent the fold change of each gene by TRV-*RhNAC3* relative to the TRV control. *RhUbi1* was used as the internal control. Error bars indicate SE (*n* = 3). **C**, Sequences and positions of putative RhNAC3 binding elements used for the EMSA. Probes were derived from the regulatory sequence of three selected ABA-related rose genes. Underlined letters indicate the core sequences of putative NAC protein-binding sites. The sense strands of oligonucleotide probes corresponding to the predicted RhNAC3 binding sites are shown. **D**, DNA-binding specificity for RhNAC3 with the probes indicated in **C**. The arrows indicate the positions of protein/

DNA complexes and the free probes, respectively. Purified protein (2 μg) was incubated with 0.2 pmol of biotin probe. **E**, DNA-binding specificity for RhNAC3 with *RU03861*. The *RU03861* (P3) probe incubated with GST was used as a control, and a 10, 100, and 1000 fold excess of the unlabeled P3 was used for competitive binding.

during the seedling stage of development (Figure 2a and b), and particularly strong staining was observed in the vascular system and leaf stomatal guard cells (Figure 2c and d). In addition, GUS staining was detected in the flower petals (Figure 2e and f), stigma (Figure 2g) and apical stem of the inflorescence (Figure 2h), while in mature siliques, staining was primarily localized to the stigma and immature seeds (Figure 2i, j and k). We conclude that the *RhNAC3* gene was expressed ubiquitously in a number of different plant tissues, with higher basal expression levels in leaf guard cells and areas of the vascular system.

ABREs are important for *RhNAC3* promoter activity

To assess the potential role of ABREs in the transcriptional activity of the *RhNAC3* promoter, we made three truncated promoter fragments containing five (N0), three (N1) or no (N2) ABREs, and two fragments, mN0 and mN1 with replacement of ACGT by TTTA in ABRE core sequence. These fragments were fused to the *GUS* reporter gene in the plant transient expression vector pUC19 (Figure 3A, top). The resulting constructs, as well as the vector control (NC), were transformed into *A. thaliana* protoplasts and relative GUS activity was measured. We observed that extracts from the protoplasts transformed with constructs containing a higher number of ABRE copies had higher GUS activity than control transformant extracts. Specifically, the N0 and N1 construct extracts had 7.2-fold and 3.1-fold greater GUS activity, respectively, than the NC extract, while constructs with ABRE mutations (mN0 and mN1) conferred only slight GUS activity and minimal activity was detected in the N2 construct extract (Figure 3A, bottom). We also investigated the transcriptional activity of three truncated *RhNAC3* promoter fragments in *A. thaliana* protoplasts exposed to exogenous ABA. Transcription of the *RhNAC3* promoter (N0 and N1) was found to be induced with a higher GUS activity for the N0 fragment (five ABREs) than for N1 (three ABREs), while no difference was seen for N2 (no ABRE) when ABA was added (Figure 3B). The potential role of the ABREs in the *RhNAC3* promoter in ABA induced transcription was also evaluated using the Pro$_{RhNAC3}$:*GUS* transgenic *A. thaliana* lines. Seedlings of thee lines had stronger GUS staining after treatment with ABA than those without ABA treatment (Figure 3C). Taken together these data indicate that ABREs in the *RhNAC3* promoter have a cumulative effect on the transcription activity of *RhNAC3* both in the presence and absence of ABA, and that ABA significantly induces *RhNAC3* transcription.

A. thaliana plants overexpressing *RhNAC3* show hypersensitivity to ABA during germination

Our previous study showed that three representative *RhNAC3*-overexpressing *A. thaliana* lines (OE#3, OE#6 and OE#12) had enhanced drought tolerance, with a higher water-retaining ability [26]. To understand the roles of *RhNAC3* in the ABA signaling pathway, we investigated the seed germination rates of *RhNAC3* overexpressing lines following ABA treatment. More than 98% of the seeds sown on control MS medium germinated well, while the germination rate of both the control and transgenic seeds decreased when grown for 7 days on MS medium plus ABA. In the presence of 0.2 μM ABA, the germination rates of WT and Vector (VC) seeds were 63% and 59%, while the rates for OE#3, OE#6 and OE#12 were 49%, 40% and 36%, respectively. A higher concentration of ABA (0.4 μM) resulted in lower germina-

tion rates for both the control and *RhNAC3*-overexpressing transgenic plants, and the latter showed a greater decrease (Figure 4A and 4B). We also compared the effects of ABA on the root architectures of the *RhNAC3*-overexpressing and control plants. Primary root growth of *RhNAC3*-overexpressors was inhibited more by 30 μM ABA treatment than that of the control plants, while no significant differences in the number of lateral roots was observed (Figure 4C–E). We conclude from these results that *RhNAC3* overexpression in *A. thaliana* results in ABA hypersensitivity at the seed germination stage.

RhNAC3 participated positively in ABA- and drought-induced stomatal closure

Since the leaves of *RhNAC3*-overexpressing *A. thaliana* plants have greater water-retaining capacity than those of WT plants [26], we examined ABA-dependent stomatal movement phenotypes. Expanded leaves of 3-week-old plants (12 h day/12 h night) were submerged in stomatal opening solution, treated with 10 μM ABA for 2 h, and then the stomatal apertures of the guard cells were measured in the focal planes of the outer edge in epidermal strips. None of the plants exhibited altered stomatal movement in the absence of ABA, and most of the guard cells examined were fully opened. No obvious difference in stomatal aperture (the ratio of width to length) were observed between the controls (WT and VC) and *RhNAC3* overexpressors (OE#3, OE#6 and OE#12). However, in the presence of 10 μM ABA, stomatal closure in the leaves of *RhNAC3* overexpressing transgenic plants was substantially enhanced compared with that of control plant leaves. The stomatal aperture ratios in the OE#3, OE#6, and OE#12 lines were approximately 0.16, 0.22 and 0.25, respectively, compared with 0.30 and 0.29 for the WT and VC plants (Figure 5A).

We also investigated stomatal movement in the leaves of *RhNAC3*-overexpressing *A. thaliana* plants grown under drought conditions. Water was withheld for 10 days from three-week-old plants that had previously been grown under normal conditions, after which the stomatal apertures of the expanded leaves were measured. Under normal growth conditions (0 days), both the controls (WT and VC) and overexpressor lines (OE#3, OE#6 and OE#12) showed no obvious difference, with stomatal apertures of 0.29, 0.28, 0.29, 0.27 and 0.28, respectively. However, after 10 days of exposure to drought conditions, the stomatal aperture ratios had decreased to 0.17, 0.15 and 0.12 in OE#3, OE#6 and OE#12 lines, respectively, but only 0.24 in WT and 0.23 in VC plants (Figure 5B). These results indicate that RhNAC3 is involved in stress responses in an ABA-dependent manner.

RhNAC3 activates ABA-responsive gene expression in rose and *Arabidopsis*

Given that the expression of RhNAC3 showed an association with both ABA sensitivity and drought tolerance, we investigated the expression profiles of drought-induced ABA-responsive genes. Eleven representative genes were selected, including ABA and stress-induced downstream marker genes (*RD29A*, *RD29B*, *RD20*, *RD26*, *COR47*, *COR15A*, and *KIN2*), the ABA-responsive protein phosphatase 2C gene (*ABI1*), the B3-domain transcription factor ABA-insensitive 3 (*ABI3*), the ABA-activated basic Leu zipper TF gene *ABF4*, and the ABA-biosynthesis gene

ABA3 (Table 1 and Figure 6). Under dehydrating conditions, *RD26*, *ABI3* and *ABA3* showed a slightly higher level of expression in *RhNAC3* overexpressing plants than in VC plants, but all the other tested genes showed more than a two-fold greater expression (Figure 6A). The promoters of these genes were then screened for putative NAC binding *cis*-elements and indeed they were identified in all the tested genes (Table 1). We then used an electrophoretic mobility shift assay (EMSA) to investigate whether the RhNAC3 protein directly binds to 4 of the selected genes: *RD29A*, *RD20*, *COR47* and *COR15A*. We observed that RhNAC3 bound to the putative NAC recognition sites of *RD29A*, *RD20* and *COR47*, but no binding signal was detected for *COR15A*, which may therefore be an indirect target of RhNAC3 following its overexpression in *A. thaliana* (Figure 6B–D). In our previous study, we used the ATH1 microarray to identify 219 *RhNAC3*-up-regulated genes [26] and in this current study, we further analyzed the response to ABA of these genes using the AtGenExpress global stress expression dataset [35]. In total, 85 of the 219 genes were found to be ABA responsive, including those encoding proteins involved in signal transduction (e.g. calmodulin-binding protein) and TFs (e.g. zinc-finger protein) (Table 2). In rose, we found 9 putative signaling and downstream genes of the ABA-signaling pathway from our rose transcriptome databases [7] (Figure 7A). qRT-PCR analysis revealed that the expression levels of 6 of these genes were substantially repressed in *RhNAC3*-silenced rose petals (with a fold change <0.8) (Figure 7B). Three ABA-responsive genes with putative NAC binding *cis*-elements (*RU25535*, *RU04740* and *RU03861*) were selected for RhNAC3 binding assays (Table S3). EMSA revealed that RhNAC3 could bind to the promoter region of *RU03861*, a rose homolog of *ABF4*, whereas no binding signals were detected for *RU25535* and *RU04740* (Figure 7C–E), which may therefore be indirect targets of RhNAC3 in rose. Collectively, these data suggest that RhNAC3 positively activates ABA-responsive gene expression and is involved in the ABA signaling pathway in rose and *A. thaliana*.

Discussion

RhNAC3 is involved in the ABA-dependent signaling pathway

Plants respond and adapt to drought stresses through a broad range of molecular and biochemical processes that result in cellular physiological changes [1]. Many drought-inducible genes with various functions, including a number of TFs that regulate stress-inducible gene expression, have been identified by molecular and genomic analyses of *A. thaliana*, rice and other plants, [9,18,23]. These TFs have been classified as being involved in one of two signal transduction pathways: ABA-independent or ABA-dependent [2,36]. ABA-dependent gene induction is controlled by at least five different classes of TFs at the transcriptional level: AREB (bZIPs), NACs, MYB/MYCs, AZF/STZs and DREB1D [2,37]. Many ABA-inducible genes contain a conserved ABRE motif in their promoter regions [38], which functions as a *cis*-element in ABA-regulated gene expression. In a previous study, we determined that *RhNAC3* expression is induced by exogenous ABA [26], implying RhNAC3 is involved in the ABA signaling pathway during stress responses. In this current study, we further analyzed the *RhNAC3* promoter, which was found to contain five ABRE motifs, as well as other stress-responsive elements (Figure 1, Figure S1 and Table S2). Promoter activity was detected ubiquitously in a number of tissues upon transformation into *A. thaliana*, and was highest in leaf guard cells and some areas of the vascular system (Figure 2). In rose, *RhNAC3* expression was also

detected in sepals, petals, gynoecia, stamens and receptacles [26]. ABREs were found to be important for the activity of the *RhNAC3* promoter, and the multiple copies had a cumulative effect on transcriptional activity in both the presence and absence of exogenous ABA in *A. thaliana* protoplasts (Figure 3). It has been shown that multiple ABRE elements can collectively confer ABA responsiveness to a minimal promoter, whereas a single copy of ABRE is insufficient for the full ABA response of *AREB1* and *AREB2*, two basic leucine zipper TFs [14]. ABREs are also regarded as one of the major types of *cis*-acting elements in the promoter regions of stress-inducible genes during osmotic stress-responsive transcriptional regulation [12,39]. This is consistent with our previous findings that *RhNAC3* confers dehydration tolerance to rose petals, mainly through the regulation of osmotic adjustment-associated genes [26].

RhNAC3 overexpression in *A. thaliana* enhances ABA sensitivity

Many ABA-responsive TFs have been isolated and characterized from different plant species, including *A. thaliana* [16], rice [5] soybean [22], maize [40] and *Citrus reshni* [41]. Overexpression of these ABA-responsive TF genes has been reported to result in a range of phenotypic changes, including dwarfing [42,43], ABA hypersensitivity [44], lateral root formation [22] and stomatal closure [45]. In our study, *RhNAC3* overexpression in *A. thaliana* lead to ABA hypersensitivity during seedling germination and primary root growth (Figure 4), and promoted stomatal closure after exogenous ABA or drought treatments (Figure 5). We note that this differs from the effects of the soybean ABA-inducible gene *GmNAC20*, which promotes lateral root formation enhances salt and freezing tolerance when overexpressed in transgenic *Arabidopsis* [22].

RhNAC3 enhanced ABA-responsive gene expression in rose and *A. thaliana*

ABA-inducible TFs involved in ABA signaling pathways typically up-regulate ABA-responsive genes or stress-responsive genes [13,18] and such downstream genes have studied using qRT-PCR [16], cDNA microarrays [13] and other transcriptomic analyses [23]. Among these genes, AREB TFs play a primary role in the ABA-dependent signaling pathway [13,46], while other TFs (NAC, MYB/MYC and AZF/STZ TFs) can play additional direct or indirect regulatory roles. The diverse *cis*-elements in the promoter regions of these TFs suggest additional potential mechanisms of transcriptional regulation for ABA-signaling downstream genes as a consequence of abiotic stresses [47]. In the current study, 11 representative ABA-induced genes were investigated, all of which were found to be up-regulated in *RhNAC3* overexpressing *A. thaliana* plants (Figure 6A). Further analysis showed that NAC binding *cis*-elements were present in the upstream regulatory sequences of these genes (Table 1) and that RhNAC3 was able to bind to the putative NAC recognition sites of some of the tested genes (Figure 6B–D). We conclude that RhNAC3 may therefore directly or indirectly regulate their expression at the transcriptional level. These genes were selected based on their involvement in the ABA-dependent signaling pathway and the fact that their overexpression in *A. thaliana* has been shown to result in an increased ABA sensitivity [25,48]. In a previous microarray study we found that the expression of 219 genes was up-regulated in *RhNAC3* overexpressing plants [26], of which 85 responded to ABA treatment in the current study (Table 2). These results suggest that these genes may contribute to ABA sensitivity in the *RhNAC3* overexpressing *A. thaliana* plants an idea that is supported by previous experimental evidence. For example, *CYP81D8* expression has been suggested to be regulated by the ABA-dependent pathway under osmotic stress conditions [49], and loss-of-function mutations of *atrdufs* (At5g59550) resulted in hyposensitivity to ABA and reduced tolerance to drought stress [50]. Moreover, the expression of downstream genes in the ABA-signaling pathway was also repressed in *RhNAC3*-silenced rose petals (Figure 7). In our previous study, RhNAC3 was shown to bind to the promoter of *RU23063*, a rose homolog of *ABI2* [26], which encodes a protein phosphatase 2C involved in ABA signal transduction [51]. Here RhNAC3 was observed to bind to the promoter region of the ABA-responsive rose gene *ABF4*, which encodes the ABRE binding factor 4 (Figure 7C–E). Taken together, the data suggest that *RhNAC3* regulated genes that were responsive to osmotic stress, are also involved in the ABA-dependent signaling pathway in both rose and *A. thaliana*.

In conclusion, we found that: (1) ABRE elements in the *RhNAC3* promoter were necessary for, and had a cumulative effect on, its transcription activity in both the presence and absence of exogenous ABA; (2) *RhNAC3*-overexpressing *A. thaliana* lines showed ABA hypersensitivity during seed germination and constitutive leaf stomatal closure under ABA or drought treatment; and (3) RhNAC3 up-regulated the expression level of ABA-responsive genes, which were responsive to osmotic stress. These findings provide new evidence that RhNAC3 is a positive mediator of ABA signaling in the regulation of drought stress tolerance in rose and at least some components of the associated signaling pathways are conserved between rose and *A. thaliana*.

References

1. Lawlor DW (2013) Genetic engineering to improve plant performance under drought: physiological evaluation of achievements, limitations, and possibilities. J Exp Bot 64: 83–108.
2. Shinozaki K, Yamaguchi-Shinozaki K (2007) Gene networks involved in drought stress response and tolerance. J Exp Bot 58: 221–227.
3. Lee S, Kang J, Park HJ, Kim MD, Bae MS, et al. (2010) DREB2C interacts with ABF2, a bZIP protein regulating abscisic acid-responsive gene expression, and its overexpression affects abscisic acid sensitivity. Plant Physiol 153: 716–727.
4. Sakuma Y, Maruyama K, Osakabe Y, Qin F, Seki M, et al. (2006) Functional analysis of an *Arabidopsis* transcription factor, DREB2A, involved in drought-responsive gene expression. Plant Cell 18: 1292–1309.
5. Tang N, Zhang H, Li X, Xiao J, Xiong L (2012) Constitutive activation of transcription factor OsbZIP46 improves drought tolerance in Rice. Plant Physiol 158: 1755–1768.
6. Zhang L, Zhao G, Xia C, Jia J, Liu X, et al. (2012) A wheat R2R3-MYB gene, TaMYB30-B, improves drought stress tolerance in transgenic *Arabidopsis*. J Exp Bot 63: 5873–5885.
7. Dai F, Zhang C, Jiang X, Kang M, Yin X, et al. (2012) *RhNAC2* and *RhEXPA4* are involved in the regulation of dehydration tolerance during the expansion of rose petals. Plant Physiol 160: 2064–2082.
8. Ji X, Liu G, Liu Y, Zheng L, Nie X, et al. (2013) The bZIP protein from *Tamarix hispida*, ThbZIP1, is ACGT elements binding factor that enhances abiotic stress signaling in transgenic *Arabidopsis*. BMC Plant Biol 13: 151.
9. Lata C, Yadav A, Prasad M (2012) Role of plant transcription factors in abiotic stress tolerance. abiotic stress response in plants-Physiological, Biochemical and Genetic Perspectives: 269–296.
10. Abe H, Urao T, Ito T, Seki M, Shinozaki K, et al. (2003) *Arabidopsis* AtMYC2 (bHLH) and AtMYB2 (MYB) function as transcriptional activators in abscisic acid signaling. Plant Cell 15: 63–78.
11. Cramer GR, Urano K, Delrot S, Pezzotti M, Shinozaki K (2011) Effects of abiotic stress on plants: a systems biology perspective. BMC Plant Biol 11: 163.
12. Kim JS, Mizoi J, Yoshida T, Fujita Y, Nakajima J, et al. (2011) An ABRE promoter sequence is involved in osmotic stress-responsive expression of the *DREB2A* gene, which encodes a transcription factor regulating drought-inducible genes in *Arabidopsis*. Plant Cell Physiol. 52: 2136–2146.
13. Yoshida T, Fujita Y, Sayama H, Kidokoro S, Maruyama K, et al. (2010) AREB1, AREB2, and ABF3 are master transcription factors that cooperatively regulate ABRE-dependent ABA signaling involved in drought stress tolerance and require ABA for full activation. Plant J 61: 672–685.
14. Uno Y, Furihata T, Abe H, Yoshida R, Shinozaki K, et al. (2000) *Arabidopsis* basic leucine zipper transcription factors involved in an abscisic acid-dependent

signal transduction pathway under drought and high-salinity conditions. Proc Natl Acad Sci U S A 97: 11632–11637.
15. Cho SK, Ryu MY, Seo DH, Kang BG, Kim WT (2011) The *Arabidopsis* RING E3 ubiquitin ligase AtAIRP2 plays combinatory roles with AtAIRP1 in abscisic acid-mediated drought stress responses. Plant Physiol 157: 2240–2257.
16. Ryu MY, Cho SK, Kim WT (2010) The *Arabidopsis* C3H2C3-type RING E3 ubiquitin ligase AtAIRP1 is a positive regulator of an abscisic acid-dependent response to drought stress. Plant Physiol 154: 1983–1997.
17. Fujita Y, Fujita M, Satoh R, Maruyama K, Parvez MM, et al. (2005) AREB1 is a transcription activator of novel ABRE-dependent ABA signaling that enhances drought stress tolerance in *Arabidopsis*. Plant Cell 17: 3470–3488.
18. Ren X, Chen Z, Liu Y, Zhang H, Zhang M, et al. (2010) ABO3, a WRKY transcription factor, mediates plant responses to abscisic acid and drought tolerance in *Arabidopsis*. Plant J 63: 417–429.
19. Seo PJ, Xiang F, Qiao M, Park J, Lee YN, et al. (2009) The MYB96 transcription factor mediates abscisic acid signaling during drought stress response in *Arabidopsis*. Plant Physiol 151: 275–289.
20. Puranik S, Sahu PP, Srivastava PS, Prasad M (2012) NAC proteins: regulation and role in stress tolerance. Trends Plant Sci 17: 369–381.
21. Olsen AN, Ernst HA, Leggio LL, Skriver K (2005) NAC transcription factors: structurally distinct, functionally diverse. Trends in plant science 10: 79–87.
22. Hao Y, Wei W, Song Q, Chen H, Zhang Y, et al. (2011) Soybean NAC transcription factors promote abiotic stress tolerance and lateral root formation in transgenic plants. Plant J 68: 302–313.
23. Jeong JS, Kim YS, Redillas MC, Jang G, Jung H, et al. (2013) *OsNAC5* overexpression enlarges root diameter in rice plants leading to enhanced drought tolerance and increased grain yield in the field. Plant Biotechnol. J 11: 101–114.
24. Tran LSP, Nakashima K, Sakuma Y, Simpson SD, Fujita Y, et al. (2004) Isolation and functional analysis of *Arabidopsis* stress-inducible NAC transcription factors that bind to a drought-responsive *cis*-element in the *early responsive to dehydration stress 1* promoter. Plant Cell 16: 2481–2498.
25. Fujita M, Fujita Y, Maruyama K, Seki M, Hiratsu K, et al. (2004) A dehydration-induced NAC protein, RD26, is involved in a novel ABA-dependent stress-signaling pathway. Plant J 39: 863–876.
26. Jiang X, Zhang C, Lü P, Jiang G, Liu X, et al. (2014) RhNAC3, a stress-associated NAC transcription factor, has a role in dehydration tolerance through regulating osmotic stress-related genes in rose petals. Plant Biotechnol. J 12: 38–48.
27. Liu Y, Chen Y (2007) High-efficiency thermal asymmetric interlaced PCR for amplification of unknown flanking sequences. BioTechniques 43: 649–656.
28. Higo K, Ugawa Y, Iwamoto M, Korenaga T (1999) Plant cis-acting regulatory DNA elements (PLACE) database: 1999. Nucleic acids research, 27: 297–300.

Supporting Information

Figure S1　The promoter sequence of the *RhNAC3* gene. A cumulative result of the 1447 bp promoter sequence showing the positions of important putative cis-acting elements deduced from PlantCARE and PLACE database. The regulatory elements identified by the programs are colorful boxed with appropriate annotations. TATA: TATA box, CAAT: CAAT box, ABRE: ABRE element, CBF: cold binding factor, MYB: MYB binding site, MYC: MYC binding site, WBOX: WRKY binding site.

Table S1　Primer sequences used in this study.

Table S2　*cis*-elements of the upstream regulatory region of *RhNAC3*.

Table S3　Upstream regulatory region of three ABA-related rose genes.

Acknowledgments

We thank PlantScribe (www.plantscribe.com) for editing this manuscript.

Author Contributions

Conceived and designed the experiments: CZ JG XJ. Performed the experiments: GJ XJ PL JL. Analyzed the data: CZ XJ. Contributed reagents/materials/analysis tools: CZ GJ. Contributed to the writing of the manuscript: CZ XJ.

29. Lescot M, Déhais P, Thijs G, Marchal K, Moreau Y, et al. (2002) PlantCARE, a database of plant cis-acting regulatory elements and a portal to tools for in silico analysis of promoter sequences. Nucleic acids research, 30: 325–327.

30. Clough SJ, Bent AF (1998) Floral dip: a simplified method for Agrobacterium-mediated transformation of *Arabidopsis thaliana*. Plant J 16: 735–743.

31. Li Y, Wu Z, Ma N, Gao J (2009) Regulation of the rose *Rh-PIP2;1* promoter by hormones and abiotic stresses in *Arabidopsis*. Plant Cell Rep 28: 185–196.

32. Jefferson RA, Kavanagh TA, Bevan MW (1987) GUS fusions: β-glucuronidase as a sensitive and versatile gene fusion marker in higher plants. EMBO J. 6: 3901–3907.

33. Urban A, Neukirchen S, Jaeger KE (1997) A rapid and efficient method for site-directed mutagenesis using one-step overlap extension PCR. Nuclei Acids Res 25: 2227–2228.

34. Sheen J (2001) Signal transduction in maize and *Arabidopsis* mesophyll protoplasts. Plant Physiol 127: 1466–1475.

35. Kilian J, Whitehead D, Horak J, Wanke D, Weinl S, et al (2007). The AtGenExpress global stress expression data set: protocols, evaluation and model data analysis of UV-B light, drought and cold stress responses. Plant J 50: 347–363.

36. Yamaguchi-Shinozaki K, Shinozaki K (2005) Organization of *cis*-acting regulatory elements in osmotic- and cold-stress-responsive promoters. Trends Plant Sci 10: 88–94.

37. Nakashima K, Yamaguchi-Shinozaki K (2013) ABA signaling in stress-response and seed development. Plant Cell Rep 32: 959–970.

38. Bonetta D, McCourt P (1998) Genetic analysis of ABA signal transduction pathways. Trends Plant Sci 3: 231–235.

39. Fujita Y, Fujita M, Shinozaki K, Yamaguchi-Shinozaki K (2011) ABA-mediated transcriptional regulation in response to osmotic stress in plants. J Plant Res 124: 509–525.

40. Yan F, Deng W, Wang X, Yang C, Li Z (2012) Maize (*Zea mays* L.) homologue of *ABA-insensitive* (*ABI*) *5* gene plays a negative regulatory role in abiotic stresses response. Plant Growth Regul 68: 383–393.

41. Xian L, Sun P, Hu S, Wu J, Liu JH (2014) Molecular cloning and characterization of *CrNCED1*, a gene encoding 9-cis-epoxycarotenoid dioxy-genase in *Citrus reshni*, with functions in tolerance to multiple abiotic stresses. Planta 239: 61–77.

42. Vannini C, Locatelli F, Bracale M, Magnani E, Marsoni M, et al. (2004) Overexpression of the rice *Osmyb4* gene increases chilling and freezing tolerance of *Arabidopsis thaliana* plants. Plant J 37: 115–127.

43. Cominelli E, Sala T, Calvi D, Gusmaroli G, Tonelli C (2008) Over-expression of the *Arabidopsis AtMYB41* gene alters cell expansion and leaf surface permeability. Plant J 53: 53–64.

44. Lu G, Gao C, Zheng X, Han B (2009) Identification of OsbZIP72 as a positive regulator of ABA response and drought tolerance in rice. Planta 229: 605–615.

45. Jung C, Seo JS, Han SW, Koo YJ, Kim CH, et al. (2008) Overexpression of AtMYB44 enhances stomatal closure to confer abiotic stress tolerance in transgenic Arabidopsis. Plant Physiol 146: 623–635.

46. Fujita Y, Yoshida T, Yamaguchi-Shinozaki K (2013) Pivotal role of the AREB/ABF-SnRK2 pathway in ABRE-mediated transcription in response to osmotic stress in plants. Physiol Plant 147: 15–27.

47. Ithal N, Reddy AR (2004) Rice flavonoid pathway genes, *OsDfr* and *OsAns*, are induced by dehydration, high salt and ABA, and contain stress responsive promoter elements that interact with the transcription activator, OsC1-MYB. Plant Sci 166: 1505–1513.

48. Yamaguchi-Shinozaki K, Shinozaki K (1994) A novel *cis*-acting element in an *Arabidopsis* gene is involved in responsiveness to drought, low-temperature, or high-salt stress. Plant Cell 6: 251–264.

49. Narusaka M, Seki M, Umezawa T, Ishida J, Nakajima M, et al. (2004) Crosstalk in the responses to abiotic and biotic stresses in *Arabidopsis*: analysis of gene expression in cytochrome P450 gene superfamily by cDNA microarray. Plant Mol Biol 55: 327–342.

50. Kim SJ, Ryu MY, Kim WT (2012) Suppression of *Arabidopsis* RING-DUF1117 E3 ubiquitin ligases, AtRDUF1 and AtRDUF2, reduces tolerance to ABA-mediated drought stress. Biochem Biophys Res Commun 420: 141–147.

51. Rodriguez PL, Benning, G Grill E (1998) ABI2, a second protein phosphatase 2C involved in abscisic acid signal transduction in Arabidopsis, FEBS Lett 421: 185–190.

2

Cloning of *Gossypium hirsutum* Sucrose Non-Fermenting 1-Related Protein Kinase 2 Gene (*GhSnRK2*) and Its Overexpression in Transgenic *Arabidopsis* Escalates Drought and Low Temperature Tolerance

Babatunde Bello, Xueyan Zhang, Chuanliang Liu, Zhaoen Yang, Zuoren Yang, Qianhua Wang, Ge Zhao, Fuguang Li*

State Key Laboratory of Cotton Biology, Cotton Research Institute, Chinese Academy of Agricultural Sciences, Beijing, China

Abstract

The molecular mechanisms of stress tolerance and the use of modern genetics approaches for the improvement of drought stress tolerance have been major focuses of plant molecular biologists. In the present study, we cloned the *Gossypium hirsutum* sucrose non-fermenting 1-related protein kinase 2 (*GhSnRK2*) gene and investigated its functions in transgenic Arabidopsis. We further elucidated the function of this gene in transgenic cotton using virus-induced gene silencing (VIGS) techniques. We hypothesized that *GhSnRK2* participates in the stress signaling pathway and elucidated its role in enhancing stress tolerance in plants via various stress-related pathways and stress-responsive genes. We determined that the subcellular localization of the *GhSnRK2*-green fluorescent protein (GFP) was localized in the nuclei and cytoplasm. In contrast to wild-type plants, transgenic plants overexpressing *GhSnRK2* exhibited increased tolerance to drought, cold, abscisic acid and salt stresses, suggesting that *GhSnRK2* acts as a positive regulator in response to cold and drought stresses. Plants overexpressing *GhSnRK2* displayed evidence of reduced water loss, turgor regulation, elevated relative water content, biomass, and proline accumulation. qRT-PCR analysis of *GhSnRK2* expression suggested that this gene may function in diverse tissues. Under normal and stress conditions, the expression levels of stress-inducible genes, such as *AtRD29A*, *AtRD29B*, *AtP5CS1*, *AtABI3*, *AtCBF1*, and *AtABI5*, were increased in the *GhSnRK2*-overexpressing plants compared to the wild-type plants. *GhSnRK2* gene silencing alleviated drought tolerance in cotton plants, indicating that VIGS technique can certainly be used as an effective means to examine gene function by knocking down the expression of distinctly expressed genes. The results of this study suggested that the *GhSnRK2* gene, when incorporated into Arabidopsis, functions in positive responses to drought stress and in low temperature tolerance.

Editor: Girdhar K. Pandey, University of Delhi South Campus, India

Funding: This work was supported by Major Program of Joint Funds (Sinkiang) of the National Natural Science Foundation of China (Grant No. U1303282). The funders had no role in study design, data collection and analysis, decision to publish, or preparation of the manuscript.

Competing Interests: The authors have declared that no competing interests exist.

* Email: aylifug@163.com

Introduction

Plants have developed complex signaling pathways in response to various environmental stresses, such as salt, drought, and cold, and have acquired metabolic functions and developmental methods to survive changing environmental conditions [1]. Improving crop resistance to drought stress would be the most valuable means to improve agricultural productivity and to reduce crop loss caused by environmental stress. As a result, understanding the mechanisms of drought tolerance and developing drought-resistant crop plants have been major targets of plant molecular biologists and biotechnologists. Low-temperature constraints have been progressively overcome by the recognition of cold-tolerant genes for applications in transgenic plants. Transgenic approach has reveal many possibilities to improve cold stress in plants by incorporating or deleting genes that regulate a particular characteristic [2]. These approaches also provide unique opportunities to improve the genetic quality of plants via the development of particular crop varieties that exhibit enhanced resistance to biotic and cold stresses and improved nutritional quality. The plant response to salt stress typically results in osmotic alterations, which play a major role in ensuring osmotic balance in plant cells. During plant stress responses, the regulation of gene expression involves both universal and unique changes in the transcript levels of certain plant genes [3]. Plants directly or indirectly respond to stresses by initiating signal transduction pathways. Various abiotic stresses result in both general and specific effects on plant growth and development. For example, drought limits plant growth due to difficulties in maintaining turgor pressure, photosynthetic decline, osmotic stress-imposed constraints on plant processes and interference with nutrient availability as the soil dries [4].

Protein kinases and phosphatases are major elements of stress signals which are transmitted to different cellular regions via specialized signaling pathways. Some of the protein kinases involved in stress signal transduction in plants, such as mitogen-activated protein kinases (*MAPKs*) [5,6,7,8], glycogen synthase kinase 3 (*GSK3*) [9,10], and S6 kinase (*S6K*) [11], are similar among all eukaryotic organisms, whereas others, including calcium-dependent protein kinases (*CDPKs*) [12,13,14] and *SNF1*-related kinases (*SnRKs*), are plant-specific.

SnRK2 is an important stress-related protein kinase in plants that has been implicated in stress and abscisic acid-mediated signaling pathways [15]. Previously, it was reported that *SRK2C/SnRK2.8*, a subclass II member, was strongly activated by drought stress and that plants overexpressing *SRK2C* exhibited improved drought tolerance as a result of the up-regulation of many stress-inducible genes [16]. *SnRK2s* have a molecular weight of approximately 40 kDa and are monomeric serine/threonine protein kinases [17,18]. Based on phylogenetic analysis, three groups of *SnRK2* family members have been identified. Group 1 consists of kinases not activated by ABA, group 2 consists of kinases not activated or activated very weakly by ABA, and group 3 consists of kinases strongly activated by ABA. The amino acid sequences of all *SnRK2s* can be separated into two regions, the highly conserved N-terminal kinase domain and the regulatory C-terminal domain, which contains stretches of acidic amino acids. Furthermore, the C-terminal domain consists of two subdomains, Domain I and Domain II. Domain I is characteristic of all *SnRK2* family members and is required for activation by osmotic stress. Based on our investigation, we found that the expression of *GhSnRK2* was induced by PEG. PEG was quite commonly used in physiological experiments to induce controlled drought stress. In the present study, we generated a gene construct containing *GhSnRK2* driven by the constitutive cauliflower mosaic virus (CaMV) 35S promoter and transformed this construct into *Arabidopsis* in order to investigate the functional analysis of *GhSnRK2* gene. We monitored the activities of this gene with respect to drought and low temperature tolerance in transgenic plants. We further elucidated the function of this gene in transgenic cotton using virus-induced gene silencing (VIGS) techniques. We showed that *GhSnRK2*, the cotton *SnRK2* gene, is involved in multi-stress responses. Our results promote the analysis of gene function in *G. hirsutum* to facilitate the exploitation of desirable genes from this species. The elucidation of *GhSnRK2* gene function contributes to our understanding of the mechanism by which this plant adapts to abiotic stress and provides a valuable gene resource for plant breeders.

Materials and Methods

Cloning of theGhSnRK2 gene

Total RNA was extracted from CCRI24 cotton tissues using Trizol (Sigma-Aldrich) according to the manufacturer's instructions. Reverse transcription (RT) was performed using total RNA extracted from seedlings oligo(dT)16 primer, and SuperScript II reverse transcriptase (Promega). The RT product was used in PCRs to amplify the predicted *GhSnRK2* open reading frame using primer star polymerase (enzymes). The cDNA regions of *GhSnRK2* were cloned into the T-simple vector. All of the clones were confirmed via sequencing. Primers specific to the sequence of *GhSnRK2* were designed, synthesized, and used to clone the *SnRK2* gene. For plant transformation, *GhSnRK2* cDNAs were introduced into the modified pCAMBIA2301 plant transformation vector under the control of the CaMV 35S promoter.

Localization of the *GhSnRK2*-GFP fusion protein

The method of [19] was adopted to perform the subcellular localization assay. The *GhSnRK2* coding sequence was cloned and ligated into the XbaI and SpeI sites of the PCAMBIA2301-GFP vector to generate PCAMBIA2301-*GhSnRK2*-GFP, which expressed the *GhSnRK2*-GFP fusion protein under the control of the CaMV35S promoter. The construct was used for transient transformation of onion. Onion epidermal peels were bombarded with DNA-coated gold particles, and GFP expression was visualized 24 h later. Transformed onion cells were observed under a confocal microscope.

Effect of polyethylene glycol (PEG) treatment on *GhSnRK2* gene expression

The effect of PEG on the expression level of the *GhSnRK2* gene was evaluated via qRT-PCR. The root of a three-week-old upland cotton plant was submerged in 10% PEG solution, and samples were collected from the root at one hour intervals for 6 hours. RNA was extracted from the samples, and RT was performed as described above to generate cDNA for qRT-PCR analysis.

Arabidopsis transformation and screening of transgenic plants

pCAMBIA2301 carrying *GhSnRK2* was introduced into *Agrobacterium tumefaciens* strain GV3101. The transgenic *Arabidopsis* plants were generated using the flower dipping method [20], and transgenic plants were selected based on their growth in 0.8% agar containing half-strength MS salts and kanamycin. The transformants were transferred to soil and allowed to set seed. The T3 generation of the transgenic plants was used for all experiments.

Plant materials and growth conditions

Arabidopsis thaliana seeds and ecotype Col-0 wild-type, mutant, and transgenic seeds were surface-sterilized using bleach (5% Sodium hypochlorite) and 0.1% Triton X-100. After cold exposure at 4°C for 2 d, the seeds were germinated and cultured in plates containing $0.5 \times$ MS medium, 0.8% agar, and 1% sucrose under continuous light at 22°C. For plants grown in soil, 7-day-old seedlings were transferred from the MS plates to soil and cultured under 80–100 μmol m^{-2}s^{-1} photoperiodic cycles of 16 h light and 8 h dark at 22°C in a growth chamber under fluorescent light using one cool-light and one warm-light tube, each of which was suspended several inches above the plants. Mature seeds obtained from transformed plants (T$_0$ generation) were cultured in MS medium containing the antibiotic kanamycin in order to screen for transformed plants. The seeds were allowed to grow in the medium for two weeks before transfer to soil for growth continuation. Cotton seeds were planted in moist soil and grown in a growth chamber with a 14 h photoperiod at a 20/30°C night/day temperature cycle, with a light intensity of 400 μmolm^{-2}s^{-2} and at 60% relative humidity.

Drought tolerance assay

Drought stress tolerance was measured by transferring 2-week-old plants cultured in Petri dishes to pots (10-cm diameter) filled with a 1:1 (v/v) vermiculite:perlite mixture. The seedlings were cultured for 2 weeks with constant watering before the water was withheld. After 9 days without water, all of the pots were re-watered simultaneously, and plant re-growth was scored 4 days later. Thirty-five plants for each individual transgenic line were used in each repeated experiment, and one representative image is shown. Plants were scored as survivors if there were healthy green young leaves after re-watering treatment. The survival rate was calculated as the ratio of number of surviving plants to the total number of treated plants in the flower pot. Representative transgenic and WT plants of were photographed before and after drought treatment. Each stress assay was repeated at least three times. The water use efficiency (WUE) of the plants was calculated as the amount of water needed to maintain the weight of each experimental pot containing the plants. To determine the soil water content (SWC), the pots were immediately weighed after saturation with water prior to the initiation of drought stress and then periodically weighed during the drought stress period. The SWC was calculated as (final fresh weight-dry weight)/(initial weight–dry weight)x100. The plants were exposed to water stress by reducing the water content to 25 to 30% of the SWC. For plant tugor pressure assay, drought stressed plants were removed from the growth chamber and transferred to a dark laboratory cupboard for 4 h before turgor measurements using the pressure chamber [21]. To measure the drought tolerance of GhSnRK2 gene silenced and non-silenced plants, water was withheld for 5 days, approximately two weeks post-inoculation, and the survival rate was recorded as the percentage of plants that survived after re-watering for 3 days. Each of the ten treated groups consisted of five plants.

Freezing and cold stress treatment

Transgenic plants overexpressing the GhSnRK2 gene and wild-type Arabidopsis plants were cultured in pots filled with a 1:1 mixture of perlite and vermiculite under a long-day photoperiod at 22°C. Freezing stress was performed by transferring 4-week-old plants to a chamber in which the temperature was adjusted to − 4°C and −8°C for 10 h, respectively, and then returned to the typical standard growth conditions. For cold stress, 4-week-old plants were transferred to 4°C for three weeks. The plants were analyzed after recovery for 7 days under normal growth conditions. The survival rate (%) under freezing and cold conditions was calculated as the percentage of clearly green plants after returning to normal conditions, and plants exhibiting clear sign of wilting were denoted as dead. Thirty-five plants for each individual line were used in each repeated experiment, and one representative image is shown.

Transpirational water loss assay

The leaves of mutant and wild-type seedlings at the rosette stage were detached and placed in a weighing boat, and the changes in fresh weight over time were monitored using an electronic balance. The rate of water loss was calculated as the loss in fresh weight of the samples. Six plants of each transgenic and WT line were analyzed in this assay, and this experiment was replicated three times [22].

Relative water content

The relative water content of the leaves was measured according to the method of [23]. Fully expanded leaves were cut from the plants, and the fresh weight (FW) was recorded immediately. Then, the fresh portions were immersed in distilled water for 4 h and the turgid weight (TW) was recorded. Finally, the dry weight (DW) was recorded after drying for 48 h at 80°C in an oven. The RWC was calculated according to the following formula: RWC (%) = (FW−DW)/(TW−DW)×100.

Measurement of stomatal closure in response to ABA treatment

Stomatal closure assays were conducted as described by [24]. Leaves from GhSnRK2 transgenic and WT plants were immersed in a solution containing 50 μM CaCl$_2$, 10 mM KCl, and 10 mM 2-(N-morpholino) ethanesulfonic acid (MES)-Tris, pH 6.15, and were exposed to light for 2 h. Subsequently, various concentrations of ABA were added to the solution. The stomatal apertures were measured after 2 h of ABA treatment.

Determination of proline content

The free proline content was measured using the method described by [25]. Leaf segments were homogenized in 3% sulfosalicylic acid, and the homogenates were centrifuged at 3000 g for 20 min. Mixtures containing 2 ml of sample supernatant, 2 ml of acetic acid, and 2 ml of 2.5% acid ninhydrin solution were boiled for 30 min, and the absorbance at 520 nm (A520) was measured.

Chlorophyll content assay

The determination of chlorophyll content was performed according to the method of [26]. Extracts were obtained from 0.1 g (fresh weight) leaf samples from 4-week-old plants and were homogenized in 1 ml of 80% acetone to quantify the chlorophyll content via spectrophotometric analysis.

Seedling growth in response to ABA and NaCl treatment

The sensitivity of seedling germination to ABA and NaCl was assessed on MS agar plates containing various concentrations of ABA and NaCl solution [27]. Seedlings from WT and transgenic plants were placed on MS agar plates supplemented with distilled water or different concentrations of ABA or NaCl and were placed in the growth chamber at 22°C.

Germination assay

To evaluate seed germination, the method of [27] was employed. Seeds from wild-type and transgenic plants were surface-sterilized using sodium hypochlorite and were placed on Murashige and Skoog (MS) solid medium containing various concentrations of ABA (0 μm, 0.3 μm, or 0.5 μm). For the evaluation of seed germination under salt stress, the MS medium was supplemented with various concentrations of NaCl (0 mM, 50 mM, or 100 mM). The percentage of germinating seeds was recorded after 7 days.

Biomass accumulation

Plants were harvested for biomass measurements after 4 weeks of germination. The fresh weight of each individual plant shoot and root was measured immediately after harvesting. For fresh weight biomass, the fresh weight of the root and the shoot was measured immediately after harvesting, and the dry weight was recorded in (g) after drying in an oven to a constant weight at 70° for 48 h.

qRT-PCR

Total RNA was extracted from Arabidopsis seedlings using Trizol reagent (Takara) and was treated with RNase-free DNase (Promega). cDNA was synthesized using a Promega kit according to the manufacturer's protocol. qPCR was performed using a SYBR green PCR master mix kit. After PCR, the data was quantified using the comparative CT method (2$^{-ΔΔCT}$ method) based on the CT values [28]. AtACT2 (gene accession At3G18780) and cotton histone3 (gene accession AF024716) were used as internal controls, and the relative expression level of each target gene was quantified. Each RT-PCR measurement was performed in triplicate. All RT-PCR experiments were reproduced at least three times using independent cDNA samples.

Silencing construct development and cotton transformation

The method described by [29] was employed to construct the VIGS vector. Specific fragments (approximately 300 bp) were amplified via PCR and cloned into the T-simple vector according to the manufacturer's specifications (Promega). The resulting clones were sequenced. The plasmids were digested using the restriction enzymes Xba1 and BamH1 and were subsequently ligated into the VIGS vector (*PYL156:(pTRV-RNA2)-GhSnRK2*). The resulting constructs were named according to the putative function of the stress-related gene. For transformation, *Agrobacterium* (GV3101) carrying *GhSnRK2* derivatives were cultured at 28°C in LB medium containing appropriate antibiotics. Cells were harvested from cultures grown overnight, re-suspended in inoculation buffer (200 μM acetosyringone, 10 mM MES, pH 5.5), and incubated for 2 h at room temperature in a shaker. *Agrobacterium* strains containing the *GhSnRK2* derivative *PYL156:(pTRV-RNA2)-GhSnRk2*, *PYL192:(pTRV-RNA1)*, *PYL156:(pTRV-RNA2)-GrCLA1*(positive control), or*PYL156:(pTRV-RNA2)* vector (OD 1.5) were mixed at a 1 : 1 ratio in 5 mM MES buffer (pH 5.5) and inoculated into *Gossypium hirsutum* leaves using a needleless syringe. The inoculated plants were maintained in a green house at 23±2°C for effective viral infection and spreading.

Experimental design and statistical analysis

The data obtained were subjected to statistical analysis and were expressed as the means and standard error (SE) of at least three replicates. All experimental data are presented as the means of at least three independent replicates, and the analysis for significance was performed using Student's T-test.

The sequences of primers used in this study are listed in Table S2.

Results

Phylogenetic analysis, subcellular localization, and expression pattern of *GhSnRK2* in cotton plant

The phylogenetic tree constructed using the full length amino acid sequence of selected *SnRK2* genes to analyze the evolutionary relationship between *GhSnRK2* and other *SnRK2* family genes is shown in the neighbor-joining tree developed based on an alignment of the complete protein sequences. The bootstrap values are shown on the branches. *GhSnRK2* clustered with the known stress-related genes *AtSnRK2.1*, AED91326.1; *AtSnRK2.4*, AEE28666.1; *AtSnRK2.5*, AED97781.1; *AtSnRK2.10*, AEE33751.1; and *Oryza sativa* (*RK1*), ABB89146.1. *SnRK2.10* and *SnRK2.4* are closely related to *GhSnRK2*, suggesting that *GhSnRK2* belongs to the *SnRK2* family (Figure 1A). Alignment of the *GhSnRK2* amino acid sequence with that of other *SnRK2s* revealed that the *GhSnRK2* protein is highly similar to other *SnRK2s*; the relatively conserved motif is underlined. The deduced amino acid sequence displays relatively high homology with the monocot *SnRK2* family members *Oryza sativa* (*RK1*), ABB89146 and with the dicot species *AtSnRK2.10*, AEE33751.1 (Figure S1A). The *GhSnRK2*-GFP fusion protein driven by the CaMV 35S promoter was transiently expressed in onion epidermal cells, and the green fluorescent *GhSnRK2* protein

signals were localized to the nuclei and the cytoplasm, whereas GFP alone was detected throughout the cell (Figure1B). qRT-PCR was performed to quantify the expression level of the *GhSnRK2* gene. Treatment of 3-week-old CCRI24 upland cotton cultivar plants with 10% PEG for different periods induced the expression of the *GhSnRK2* gene for 3 h, after which the expression level declined (n = 3) (Figure 1C). To investigate the distribution of *GhSnRK2* in different tissues, samples from root (RT), stem (ST), cauline leaves (CL), rosette leaves (RL) and flowers (FL) were analyzed. *GhSnRK2* gene expression was detected in all of the examined tissues in distinct expression patterns and was highest in the root and lowest in the flower, suggesting that the *GhSnRK2* gene may most actively function in the root (Figure 1D). Cotton *histone*3 (gene accession AF024716) was used as internal control to normalize expression data. The values are presented as the means of three replicates, and the error bars denote the SE. Different letters denote the means ± standard deviation displaying significant difference at P≤0.05.

Plant transformation vector and the expression pattern of the *GhSnRK2* gene in the transgenic lines

The schematic representation of the 35S-*GhSnRK2* construct used for *Arabidopsis* transformation (Figure 2A). The full-length *GhSnRK2* cDNA was introduced into the *Arabidopsis* genome via the floral dip method using *Agrobacterium* GV3101; the XbaI–BamHI region of *GhSnRK2* was inserted between the CaMV 35S promoter and the nopaline synthase gene terminator in the pCAMBIA2301 vector to generate the recombinant plasmid *35S-GhSnRK2-NOST*. The neomycin phosphotransferase II (*NPTII*) gene, driven by the nopaline synthase gene promoter (NOSP), was carried by the *pCAMBIA2301* vector for transgenic cell selection in kanamycin-containing LB medium. *GhSnRK2* in transgenic *Arabidopsis* plants was confirmed by PCR analysis. Genomic DNA from the first generation of the plants (T1) was extracted and used as a template for gene-specific primers. Of the 43 plants evaluated, 35 were positive for *GhSnRK2* amplification. No product was formed by the amplification reaction using untransformed plant DNA as the template (Figure S2B). At T1 generation, the ratio of dead to surviving plants was approximately 1:3 in kanamycin-containing LB medium. However, at T3 generation, all of the plants became homozygous, as their survival rate was 100% in kanamycin-containing LB medium (Figure S2A, Table S1). The values are expressed as the mean germination rate (%) of approximately 200 seeds. qRT-PCR analysis was performed to evaluate the differences in transgene expression between all of the independently generated transgenic lines. Various transgene expression patterns were detected in the transgenic lines overexpressing the *GhSnRK2* gene, but no expression was detected in the WT line, indicating that the *GhSnRK2* genes were successfully introduced into the transformed plants (Figure 2B). The values are presented as the means of three experimental replicates; the error bars indicate the standard deviations. *AtACT2* (gene accession At3G18780) was used as internal control to normalize expression data. Three transgenic lines (L1, L2, and L4) were selected for further investigation based on their seed availability and relative expression pattern.

Overexpression of *GhSnRK2* in transgenic plants improves their tolerance to drought and cold stresses

To investigate the roles of *GhSnRK2* in stress response pathways, the responses of the three selected transgenic *GhSnRK2*-overexpressing *Arabidopsis* lines to drought and low temperature stresses were analyzed. Constitutive overexpression of

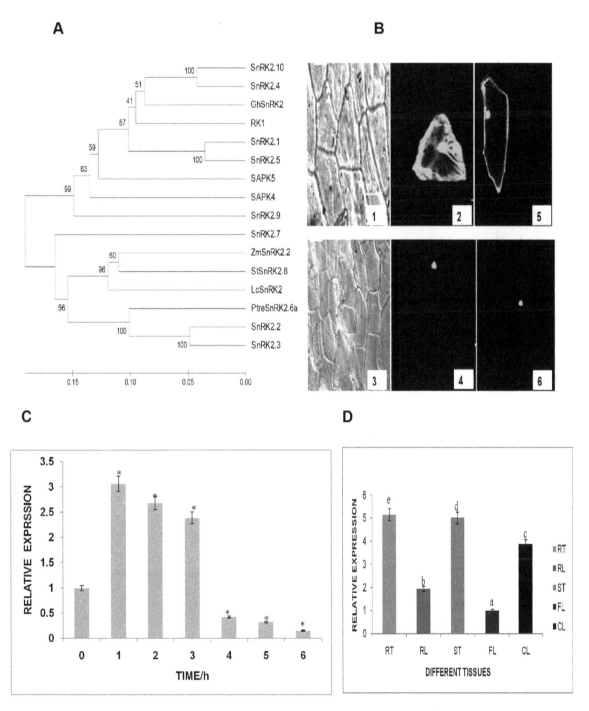

Figure 1. Phylogenetic analysis, subcellular localization, and expression pattern of *GhSnRK2* **in cotton plant.** (A) A phylogenetic tree of *GhSnRK2* and other *SnRK2* proteins from different plants was constructed using the neighbor-joining method with MEGA 5. The sequences used for analysis are listed by accession number: *Litchi chinensis* (*LcSnRK2*), AFX72761.1; *Arabidopsis thaliana* (*AtSNRK2.2*), CP002686.1; *Arabidopsis thaliana* (*AtSnRK2.3*), AED98274.1; *AtSnRK2.1*, AED91326.1; *AtSnRK2.9*, AEC07398.1; *AtSnRK2.10*, AEE33751.1; *AtSnRK2.4*, AEE28666.1; *AtSnRK2.5*, AED97781.1; *AtSnRK2.7*, AEE87152.1; *Populus tremula* (*PtreSnRK2.6a*), AGW51610.1; *Zea mays* (*ZmSnRK2.2*), NM_001137717.1; *Solanum tuberosum* (*StSnRK2.8*), AFR68945.1; *Oryza sativa* (*RK1*), ABB89146.1; *Oryza sativa* (*SAPK4*), BAD18000.1; *Sorghum bicolor* (*SAPK4*), AGM39623.1; and *Zea mays* (*SAPK5*), ACG42286.1. The bootstrap values are shown on the tree branches. (B) Subcellular localization of the *GhSnRK2-GFP* protein. (2 and 5) GFP alone; (4 and 6) *GhSnRK2-GFP* in onion epidermal cells; (1 and 3) corresponding bright-field images. (C) The expression pattern of the *GhSnRK2* gene in cotton plants subjected to 10% PEG stress. The gene expression data were normalized to that of the cotton histone 3 gene. The values are presented as the means of three experimental replicates. The vertical axis represents the relative expression level. The values from 1 to 6 indicate the time (h) of PEG treatment. Asterisk denotes a significant difference (P<0.05) compared with the control (0 h). (D) Relative expression levels of the *GhSnRK2* gene in various cotton plant tissues. Samples from root (RT) stem (ST), cauline leaves (CL), rosettes leaves (RL) and flowers (FL) were analyzed. The vertical axis represents the relative expression level. The letters denote significant differences (P<0.05) based on Duncan's multiple range tests. The cotton histone 3 gene was used as an internal control for normalization of the gene expression data.

A

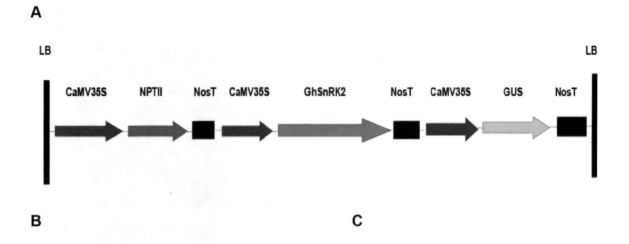

B   **C**

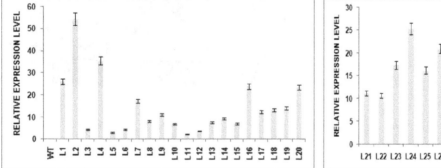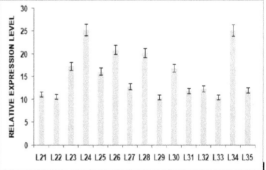

Figure 2. Plant transformation vector and the expression pattern of the *GhSnRK2* gene in the transgenic lines. (A) Schematic representation of the T-DNA region of the binary vector *pCAMBIA2301-GhSnRK2*. (B) Expression pattern of the *GhSnRK2* gene in the transgenic plants. Various upregulated expression patterns of the *GhSnRK2* gene in transgenic lines were detected, as indicated by the vertical axis. The values are presented as the means of three experimental replicates; the error bars indicate the standard deviations. The *AtACT2* gene was used as an internal control for normalization of gene expression.

the 35S-*GhSnRK2* gene resulted in an increase in the drought tolerance of the transgenic plants. After withholding water for 9 days, some of the WT plants grew slowly, wilted severely and died, whereas few of the *GhSnRK2* transgenic plants wilted. The recovery rate of the transgenic plants was higher than that of the WT plants after re-watering. Indeed, the survival rate was significantly different between the *GhSnRK2* transgenic and wild-type plants: L2 (100%), L4 (94.4%), L1 (88.9%), and WT (4.4%); Student's T-test (n = 3) (p<0.05) (Figure 3A, B). These results suggested that the *GhSnRK2* gene is involved in the response to drought tolerance.

Under low temperature stress conditions, the survival rate of the transgenic lines was significantly higher than that of the WT line. Most of the transgenic plants were intact and recovered from this stress, whereas most WT plants were found to have died and could not recover after transferring them to the normal growth conditions. The survival rates at 4°C were: WT (16.67%), L2 (80.55%), L1 (77.78%), and L4 (78.33%). The survival rates at −4°C were: WT (5.56%), L1 (54.44%), L2 (58.33%), and L4 (56.11%). At −8°C, only 3.3% of the WT plants survived, whereas the survival rates of transgenic lines ranged between 30% and 33.33% after 7 days under normal growth conditions (Figure 3C, D). However, the survival rate under different low temperatures

clearly indicates the difference between the transgenic and WT plants, suggesting that the *GhSnRK2* overexpression alleviates low temperature stress in transgenic Arabidopsis. Our findings revealed that the water use efficiency was higher in *GhSnRK2*-overexpressing plants compared to the corresponding WT plants, suggesting that *GhSnRK2*-overexpressing plants exhibit greater photosynthetic potential during drought stress treatment (Figure 3E). Normal turgor pressure which is regulated by the amount of water in the plant's cells is required for healthy growth of a plant and is a powerful determinant of the plant's drought tolerance. We determined the turgor of the *GhSnRK2* transgenic plant to understand the role of turgor pressure in cellular signaling during water deficit condition. Turgor values of *GhSnRK2* transgenic plant was notably higher compared to the corresponding WT plant and this may be attributable to osmotic adjustment, like stomatal closure, and the maintenance of water content which prevent plants from desiccation and turgor loss. Student's T-test (n = 3) (p<0.05) (Figure 3F).

Biochemical and physiological assays of plants overexpressing the *GhSnRK2* gene

The rate of water loss and the RWC were investigated to further understand the tolerance of *GhSnRK2*-overexpressing plants to

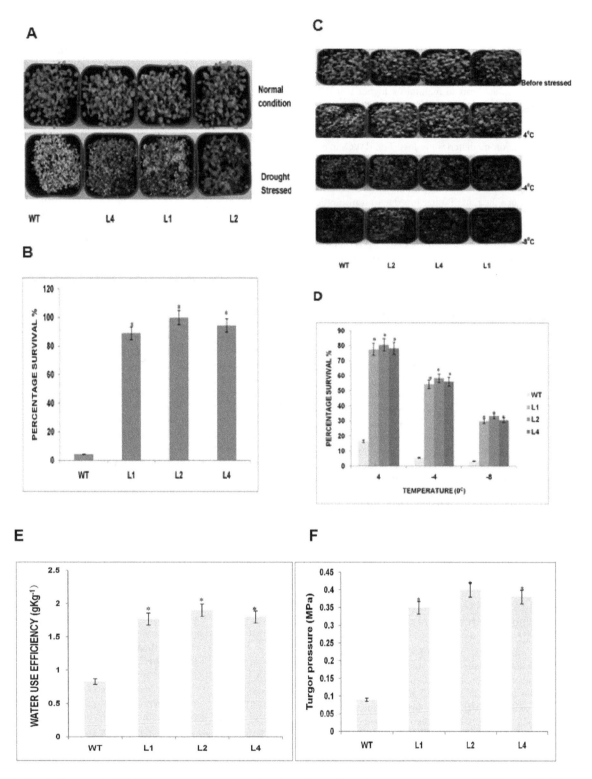

Figure 3. Survival rates of *GhSnRK2* transgenic plants under drought and low temperature stresses. (**A**) WT and *GhSnRK2* transgenic plants before and after drought stress. (**B**) Seedlings were cultured for 2 weeks with constant watering before the withholding of water. After 9 days without water, all of the plants were irrigated, and plant re-growth was scored 4 days later. The plants were scored as survivors if there were healthy green young leaves after re-watering. The survival rate was calculated as the ratio of the number of surviving plants to the total number of treated plants in the flower pot. Asterisk denotes a significant difference (P<0.05). (**C**) WT and *GhSnRK2* transgenic plants before and after low temperature stress. (**D**) Survival rates under low temperature stress conditions. The survival rate after transferring 4-week-old *GhSnRK2* transgenic and WT plants to a low temperature chamber at 4°C, −4°C or −8°C for 10 h, followed by returning the plants to normal growth conditions. Clearly green plants after returning to the normal growth condition were scored as survivors, and plants exhibiting clear signs of wilting were denoted as dead. The mean survival rates of the WT line were compared with those of the transgenic lines using Student's T-test. (**E**) The water use efficiency of the WT and *GhSnRK2* transgenic plants. Asterisk denotes a significant difference (P<0.05).

water stress via the maintenance of a higher RWC and a reduced rate of water loss. Our results revealed that due to their smaller stomata aperture, the transgenic *Arabidopsis* plants overexpressing the *GhSnRK2* gene lost water more slowly than the WT plant in the same period under normal conditions, which may underlie the capacity of the transgenic plants to maintain a higher leaf RWC and to tolerate water stress conditions; "tstat">t Critical two-tail (Figure 4A). The *GhSnRK2*-overexpressing plants exhibited a higher RWC than the WT plants, which is most likely due to the reduction in transpiration-mediated water loss. The RWCs in each line were L2(74.56%), L4(72.09%), L1 (70.04%), and WT (53.98%); Student's T-test (n = 3) (p<0.05) (Figure 4B). It is known that during the plant response to abiotic stress, stomata typically close to reduce the rate of water loss due to transpiration. Moreover, abscisic acid (ABA) regulates the stomatal aperture under water-deficient conditions. We found that the stomata apertures of the *GhSnRK2* transgenic plants in response to exogenous ABA treatment were smaller than those of the WT plants, indicating the protective effect of the *GhSnRK2* protein in response to ABA. At 0 μM ABA, the stomata aperture of the WT plants was observed slightly wider than that of the *GhSnRK2* transgenic plants. However, when the concentration of ABA was increased to 1 μM, the aperture size of the transgenic plants was significantly reduced. A similar pattern was detected at 5 μM ABA (n = 5) (Figure 4C, D). Free proline is an osmoprotective molecule that accumulates under stress conditions. Our results revealed that the proline contents of the *GhSnRK2* transgenic lines were significantly higher than those of the WT line, indicating that the transgenic plants accumulate a higher amount of proline than the WT plants. Moreover, the upregulation of Arabidopsis pyrroline-5-carboxylate synthetase 1, a proline biosynthesis gene, in *GhSnRK2*-overexpressing plants suggested that proline production may be improved by *GhSnRK2* via the regulation of proline biosynthesis genes; Student's T-test (n = 3) (p<0.05) (Figure 4E). Greater leaf chlorophyll content at all stages of plant development has been associated with improved transpiration efficiency under drought stress, and this trait may indicate the presence of drought avoidance mechanisms. Based on our results, the chlorophyll content of the *GhSnRK2* transgenic lines was higher than that of the WT line, suggesting that *GhSnRK2*-overexpressing plants exhibit greater photosynthetic potential (p<0.05) (Figure 4F).

Overexpression of *GhSnRK* in transgenic plants enhances seedling growth in response to NaCl and exogenous ABA treatment

To investigate the effect of *GhSnRK2* overexpression in response to NaCl and exogenous ABA treatment, a root growth experiment was conducted to elucidate the physiological differences between the transgenic and WT plants. The transgenic plants overexpressing *GhSnRK2* exhibited enhanced seedling growth under NaCl and ABA stress treatments, suggesting that *GhSnRK2* may be involved in the oxidative stress response pathway. Seedlings of WT and *GhSnRK2* transgenic lines grew normally in 0 mM NaCl. In 100 mM and 150 mMNaCl, the transgenic seedlings formed longer roots and displayed significantly larger growth than the WT seedlings. When the NaCl concentration was increased to 200 mM, the growth of the WT seedlings was completely inhibited, and the color of these seedlings was found to have turned brownish, whereas the transgenic seedlings remained green and continued to grow, although at a slower rate (n = 3) (Figure 5A, B).

The seedlings from the WT and *GhSnRK2*-overexpressing lines cultured under normal conditions displayed no significant difference. However, when various concentrations of ABA

(0.5 μM, 1 μM, or 2 μM) were introduced to the MS medium and the plants were allowed to grow vertically, the root growth of the *GhSnRK2*-overexpressing lines, although slightly inhibited, displayed greater elongation than the WT line, the root growth of which was severely inhibited (n = 3). These results indicated that *GhSnRK2*-overexpressing plants were more tolerant to ABA than the WT plants (Figure 5C, D).

GhSnRK2 overexpression in transgenic plants enhances seed germination in response to NaCl and exogenous ABA treatment

We investigated the germination of *GhSnRK2*-overexpressing plants under exogenous ABA and salt stresses to determine whether *GhSnRK2* is involved in stress response pathways. The germination rates of WT and *GhSnRK2*-overexpressing seeds were similar at 0 mM NaCl and ABA. However, when different concentrations of ABA or NaCl were introduced, the germination of the WT and *GhSnRK2*-overexpressing seeds was inhibited. When 50 mM NaCl was introduced to the "MS" medium, the germination rate was 47% for the WT line and 85%, 89%, and 86% for the L1, L2, and L4 lines, respectively. At100 mM NaCl, the germination rates were: WT (20.5%), L1(52%), L2 (56%), and L4 (54%)(Figure 6A, B). These results suggested that the *GhSnRK2*-overexpressing plants were more tolerant to salt stress than the corresponding WT plants. ABA plays a prominent role in the regulation of germinating and post-germinating growth arrest and mediates the adaptation of the plant to stress. To investigate this hypothesis, we measured the response of *GhSnRK2*-overexpressing plants to different concentrations of ABA during the germination stage. At 0.3 μm ABA the seed germination rates were WT (45%), L1 (69%), L2 (73%), and L4 (70%), respectively. When the concentration of ABA was increased to 0.5 μm, the germination rate of the WT line was clearly reduced (27%), whereas the germination rate of the transgenic lines ranged from 43% to 48% (Figure 6A, C). In summary, our findings revealed that the *GhSnRK2*-overexpressing mutants were more tolerant to NaCl and exogenous ABA stresses than the corresponding WT plants. Student's t-test was used to compare the mean germination rates. Asterisks denote a significant difference compared to the control (P≤0.05).

We further investigated the biomass accumulation of *GhSnRK2* transgenic and WT plants. The transgenic plants exhibited highly significant increases in both the fresh and dry weight biomasses compared with the corresponding WT plants (Figure 6D, E). These increases were all statistically significant; suggesting that overexpression of the *GhSnRk2* gene increases biomass accumulation in plants. The increased biomass accumulation may be attributable to the increased chlorophyll content in *GhSnRK2* overexpressing plant, as chlorophyll in plant cells carries out the bulk of energy fixation in the process of photosynthesis and is probably the most-often used estimator of plant biomass.

Expression analysis of stress-responsive marker genes

To further elucidate the biological function and molecular mechanisms of the *GhSnRK2* gene, we determined the transcript levels of several stress-associated genes in *GhSnRK2*-overexpressing lines. Seedlings from 10-day-old *GhSnRK2* transgenic and WT *Arabidopsis* lines cultured in MS medium were in the presence or absence of NaCl (250 mM) for 2 h. qRT-PCR was performed as described in the Materials and Methods section. Our findings revealed that *GhSnRK2* regulates the expression levels of ABA and stress-responsive marker genes, suggesting that *GhSnRK2* may positively affect ABA signaling and plant stress

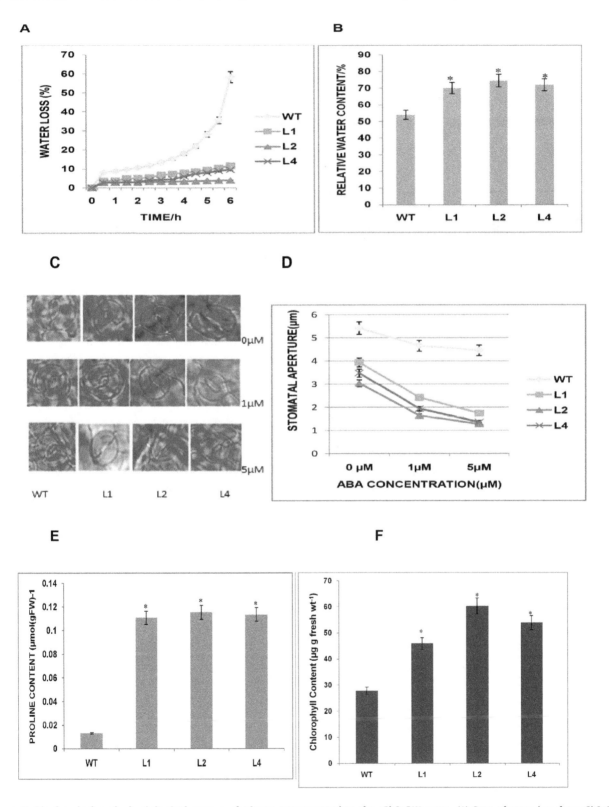

Figure 4. Biochemical and physiological assays of plants overexpressing the *GhSnRK2* gene. (A) Rate of water loss from *GhSnRK2* transgenic plants. Six plants of each transgenic and WT line were analyzed. Three biological replicates produced similar results. Asterisk denotes a significant difference (P<0.05). (B) The RWC of *GhSnRK2* transgenic plants. (C) Stomata aperture size of WT and *GhSnRK2* transgenic plants treated with different concentrations of ABA (D) The stomata apertures were measured after 2 h of treatment with different concentrations of ABA, and the mean values of the WT and transgenic lines at each ABA concentration were compared using Student's T-test (P<0.05). Asterisk denotes a significant difference (P<0.05). (E) Proline accumulation in WT and *GhSnRK2* transgenic plants. The proline content of *GhSnRK2* transgenic lines was consistently higher than that of the WT line. Student's T-test revealed a significant difference (p<0.05) between the transgenic and WT lines (n = 3). (F) The chlorophyll content of the WT and *GhSnRK2* transgenic plants. The mean values were compared using Student's "t-test". Asterisk denotes a significant difference (P<0.05).

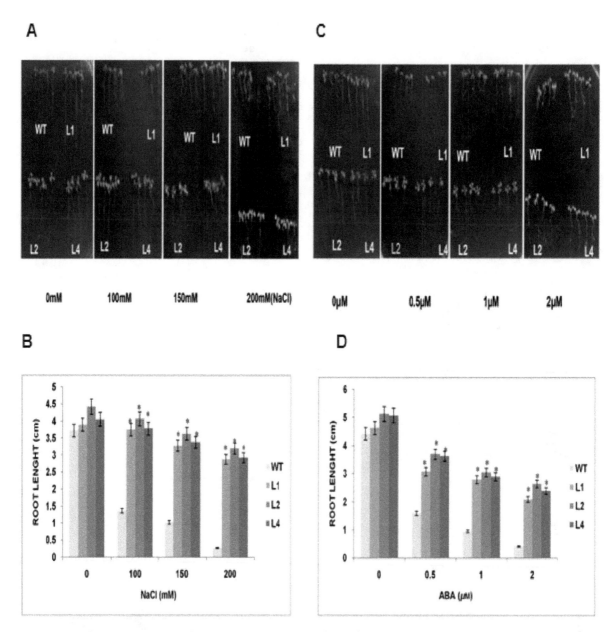

Figure 5. Seedling growth of the WT and *GhSnRK2*-overexpressing lines in response to NaCl and ABA treatment. (A) WT and *GhSnRK2* transgenic plants subjected to different concentrations of NaCl. (B) Approximately one-week-old seedlings were transferred to 1/2MS medium supplemented with different concentrations of NaCl; the root length was measured after 7 days. Each of the three biological replicates consisted of 16 plants. The mean values were compared using Student's T-test (p<0.05). (C) The root growth of the WT and *GhSnRK2* transgenic plants treated with different concentrations of ABA. (D) The plants were cultured vertically on MS-containing medium treated with different concentrations of ABA (0.5 *μM*, 1 *μM*, or 2 *μM*); the root length was measured after 7 days. Each of the three biological replicates consisted of 16 plants. Student's T-test was used to compare the mean values of the WT and transgenic lines. Asterisk denotes a significant difference (P<0.05).

responses. The expression levels of C-repeat binding factor-1 (*AtCBF1*; accession: ABV27087), delta-1-pyrroline-5-carboxylate-synthase-1 (*AtP5CS1*; accession:NP001189715), protein abscisic acid-insensitive-5 (*AtABI5*; accession:NP565840), protein abscisic acid-insensitive-3 (*AtABI3*; accession:CAA48241), desiccation-responsive genes (*Atrd29B*; accession: BAA02375 and *Atrd29A*; accession:BAA02376) in the *GhSnRK2*-overexpressing lines were significantly higher than those of the WT lines (p≤0.05), revealing that *GhSnRK2* is actively involved in stress signaling pathways (Figure 7). Three experimental replicates produced similar results (n = 3); Student's T-test.

VIGS efficiency and the transcript level of *GhSnRK2* in gene silenced plants

We employed the VIGS technique to further dissect the function of the *GhSnRK2* gene in transgenic cotton. The TRV VIGS vectors were modified based on a pTRV1 containing RNA-dependent RNA polymerase (RdRp), movement protein (MP), a 16 kDa cysteine-rich protein (16K), CaMV 35S promoters (2X35S) and a NOS terminator (NOSt) T-DNA vector. pTRV2 contains the coat protein (CP), multiple cloning sites (MCSs), CaMV 35S promoters (2X35S) and a NOS terminator (NOSt) T-DNA vector. Both vectors contain Rz, which is designated as a

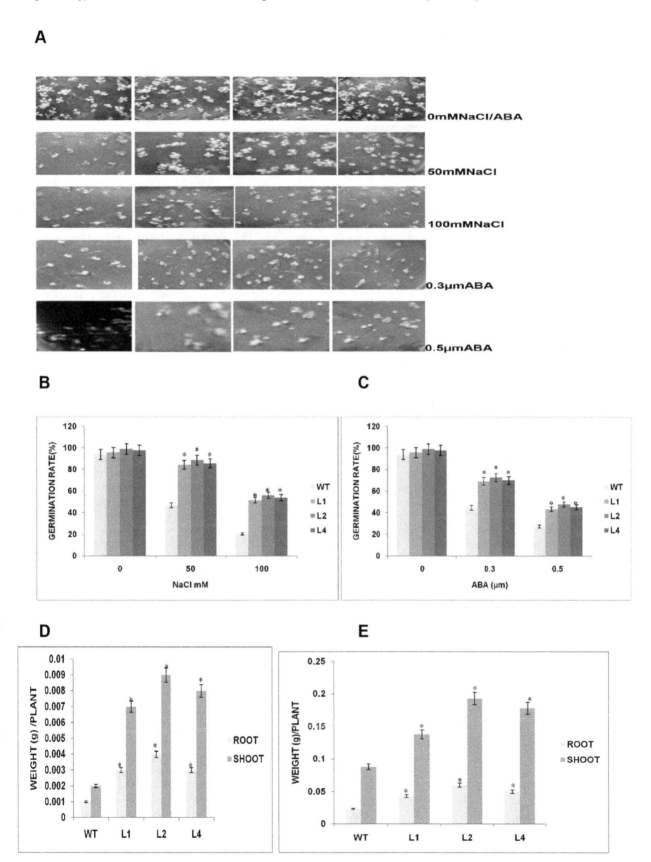

Figure 6. Seed germination of the WT and *GhSnRK2* **plants subjected to NaCl and exogenous ABA treatment and biomass accumulation of these plants.** (A) Seed germination frequency of the WT and *GhSnRK2* transgenic plants cultured on MS medium supplemented with different concentrations of NaCl (50 mM or 100 mM) or ABA (0 μM, 0.3 μM, or 0.5 μM). (B) The survival rate of the WT and *GhSnRK2* transgenic plants cultured in MS medium containing 50 mM or100 mMNaCl. The mean values were compared using Student's T-test (p<0.05). (C) The

germination rate in MS medium supplemented with 0.3 µM or 0.5 µM ABA. The values are presented as the mean germination rates (%) of approximately 200 seeds. Asterisk denotes a significant difference (P<0.05). (D) Biomass accumulation of the GhSnRK2 transgenic and WT plants. For dry weight biomass, the dry weight in the roots and shoots was recorded after drying in an oven to a constant weight at 70°C for 48 h. (E) Fresh weight biomass of GhSnRK2 transgenic plants and corresponding WT plants. The fresh weight of the roots and shoots was measured immediately after harvesting. Each of the three biological replicates consisted of 12 plants. Student's T-test was performed. Asterisk denotes a significant difference (P<0.05).

self-cleaving ribozyme, LB (left border) and RB (right border) of the T-DNA (Figure S1B). After inoculating *Agrobacterium* strains containing *GhSnRK2* derivatives (PYL156: (pTRV-RNA2)-*GhSnRk2*, PYL192:(pTRV-RNA1), PYL156:(pTRV-RNA2)-*GhCLA1* (positive control), or PYL156:(pTRV-RNA2) vector) into *Gossypium hirsutum* leaves, the phenotype of both silenced and non-silenced plants was monitored daily for gene silencing efficiency. Loss of normal green coloration in plant leaves and albino phenotype was detected in PYL156:(pTRV-RNA2)-*GrCLA1* silenced plants, which served as the positive control indicating gene silencing efficiency. The *GrCLA1* gene represents the ideal visual marker for gene silencing efficiency due to its involvement in chloroplast development, as its loss of function results in an albino phenotype in true leaves (Figure 8A). The transcript levels of *GhSnRK2* in gene silenced and non-silenced plants were analyzed via qRT-PCR. The transcript level of *GhSnRK2* in the gene silenced two plant cultivars were downregulated compared to the non-silence plants inoculated with empty vector and WT plants, which exhibited a higher expression level. The downregulation of *GhSnRK2* indicated that the gene was successfully knocked down in the silenced plants (Figure 8B). qRT-PCR analysis of the distribution of the TRV-construct in the silenced plants revealed that its infection results in gene-specific transcript degradation, that pTRV-*GhSnRK2* functions in diverse tissues, and that pTRV-*GhSnRK2* is abundantly expressed in the root. The roots of gene silenced plants exhibited further down regulated expression levels compared to the stems and the leaves. The TRV-construct displays increased infectivity and meristem invasion, both of which are key requirements for efficient VIGS-based functional characterization of genes in root tissues (Figure 8C).

Physiological assay and responses of GhSnRK2 gene silenced plants to various stresses

To further elucidate the function of the *GhSnRK2* gene in stress tolerance, we investigated the response of *GhSnRK2* gene silenced plants to various stresses. Our findings validated the importance of this technique for stress tolerance studies. Under water-deficient conditions, the silenced plants wilted and drooped, regardless of the cultivar (Figure 9A). The phenotype of the non-silenced plants inoculated with empty vector (TRV) was indistinguishable from that of the WT plants. The two cotton cultivars (CRI99668 and CRI409) inoculated with the target gene displayed similar phenotypes and symptoms. The WT and control vector-treated plant survival rates were 80% and 77.3%, respectively, whereas the survival rate of the two gene silenced cotton cultivars (CRI99668 and CRI409) were 34.66% and 38.67%, respectively (Figure 9B). Student's T-test revealed a significant difference (p<0.05) between the survival rate of the gene silenced and non-silenced plants. The rate of water loss from the detached leaves was determined in both the gene silenced and non-silenced plants (Figure 9C). After monitoring the water loss for 6-h, we found that the *GhSnRK2* gene silenced plants inoculated lost more water than the non-silenced plants inoculated with empty vector and the WT plants. The rate of water loss was slightly higher in the CRI99668 cultivars than in the CRI409 cultivars. Under similar

conditions, the gene silenced plants, regardless of the cultivar, exhibited a substantially reduced water content compared with the non-silenced plants inoculated with empty vector and the WT plants (Figure 9D). Oxidative stress, including that mediated by salinity, alters the physiological and morphological responses of plants. To investigate the effect of salt stress on *GhSnRK2* gene silenced plants, the gene silenced and non-silenced plants were treated with 150 mM NaCl for 7 days, and the effect of NaCl was measured following this stress. Our findings revealed that the chlorophyll content of the non-silenced plants was higher than that of the gene silenced plants, indicating that the *GhSnRK2* gene may be involved in the oxidative stress response (Figure 9E). Growth retardation was detected in the *GhSnRK2* silenced plants, and the effect of salt accumulation in the plant cells was detected as elevated blisters, which were visibly detectable on the leaf surface of *GhSnRK2* gene silenced plants (Figure 9F).

Discussion

The ability of plants to withstand water shortage while sustaining proper physiological activities can be associated with drought tolerance. In this study, we generated a gene construct containing *GhSnRK2* driven by the CaMV 35S promoter and functionally characterized it via gene overexpression in Arabidopsis. We further elucidated the function of this gene in transgenic cotton plants using VIGS techniques. We found that the amino acid sequence of *GhSnRK2* is highly similar to that of other *SnRK2* proteins, and a neighbor-joining tree developed based on an alignment of the complete protein sequences revealed that *GhSnRK2* clustered with known stress-related genes from other plants. The conserved motif detected demonstrated that *GhSnRK2* is a functional protein. The conserved domain (CD) region in these proteins plays a crucial role in protein interactions, DNA binding, enzyme activity, and other important cellular processes. Furthermore, we found that *GhSnRK2* was localized to the cytoplasm and the nucleus, suggesting that *GhSnRK2* might perform diverse functions in cotton plants. Our findings revealed that *GhSnRK2* is abundantly expressed in the root and is widely distributed throughout plant tissues, suggesting that it may function in diverse tissues. Transgenic *Arabidopsis* plants overexpressing *GhSnRK2* exhibited enhanced tolerance to various abiotic stresses. Consistent with previous studies, *SnRK2* genes have been implicated in the response to multi-environmental stresses [16,30,31]. The improved drought tolerance of plants overexpressing the *GhSnRK2* gene may be due to the abundance of gene expression in roots. In addition, protein kinases that are specifically found in roots, such as *SnRK2C*, may perform certain important functions in root tissues as a sensor of water and nutrients in soil. The root tip plays a vital role in the response to water and nutrient detection via appropriate signal transduction [32]. Turgor regulation of the *GhSnRK2* transgenic plant under water deficit condition may be attributable to the dynamic process of cell wall adjustment of the transgenic plant due to their enhanced drought-tolerant. Plants can maintain turgor by solute accumulation, i.e. by osmotic adjustment, and possibly by elastic adjustment of their cell walls [33].

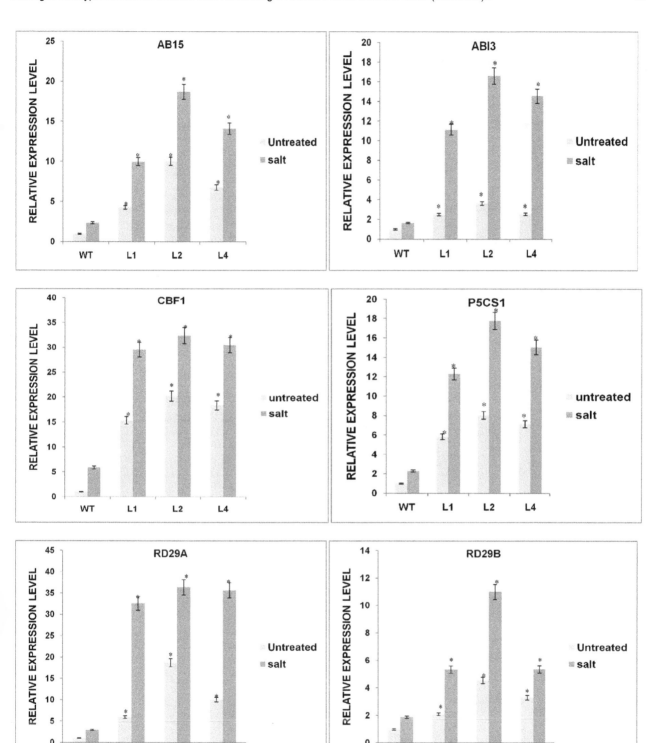

Figure 7. Expression analysis of stress-responsive marker genes. The relative transcript levels of the stress-responsive genes *AtABI5, AtABI3, AtP5CS1, AtRD29A, AtCBF1,* and *AtRD29B* in the *GhSnRK2*-overexpressing and WT lines. qRTPCR was performed for gene expression analysis. The vertical axis displays the expression pattern. Three biological replicates produced similar results.

Based on our results, the transgenic plants overexpressing *GhSnRK2* exhibited enhanced low temperature tolerance compared to WT plants. This result may be attributable to the resistance of the membrane system of the transgenic plant cells to cold stress. These findings corroborated the results of [30], who reported enhanced multi-stress tolerance in *Arabidopsis* plants overexpressing *TaSnRK2.8*. Numerous studies have demonstrated that the membrane systems of the cell are the primary site of low temperature contusion in plants [34,35]. Indeed, it is well confirmed that low temperature-mediated membrane injury

A

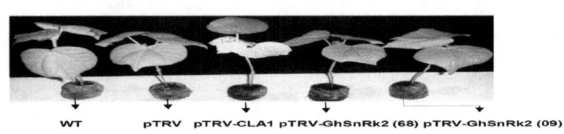

B

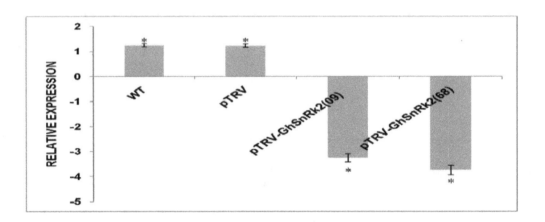

C

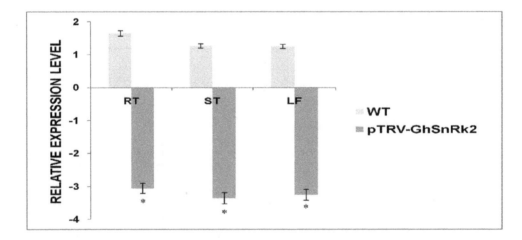

Figure 8. Silencing efficiency and transcript level of the *GhSnRK2* gene silenced plants. (A) Phenotype of gene silenced and non-silenced plants; 14 dpi. Wild-type (WT), negative control Empty vector (pTRV), positive control (pTRV-GrCLA1), gene silenced CRI409 cotton cultivars [pTRV-GhSnRk2 (09)], and gene silenced CRI99668 cotton cultivars [pTRV-*GhSnRK2* (68)]. (B) The silencing efficiency as determined by the expression pattern of WT, vector control (pTRV) and *GhSnRK2* gene silenced plants based on qRT-PCR. (C) The distribution of the TRV- construct in the gene silenced plants. Samples from the root, the stem and the leaves of gene silenced and non-silenced plants were analyzed via qRT-PCR. The values are presented as the means of three biological replicates. Asterisk denotes a significant difference (P<0.05).

occurs primarily as a result of acute dehydration associated with low temperature [35,36].

Our findings revealed the accumulation of compatible osmolytes, such as free proline, in the *GhSnRK2*-overexpressing plants; therefore, we speculate that the *GhSnRK2* transgenic plants exhibit increased stress tolerance by regulating downstream gene expression and accumulating a larger amount of compatible osmolytes, which may account for the increase in stress tolerance

A

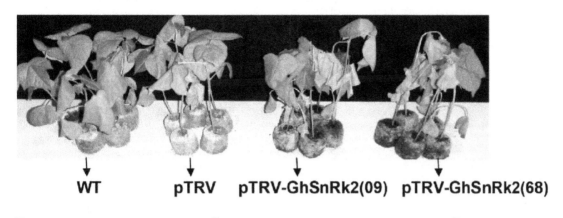

B   **C**   **D**

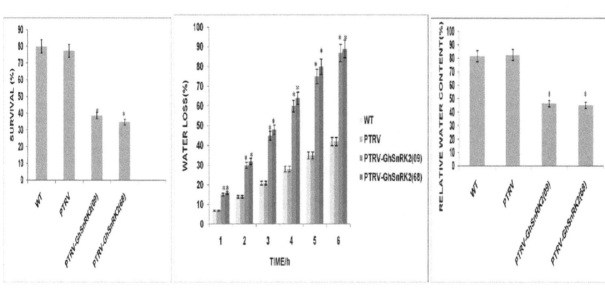

E   **F**

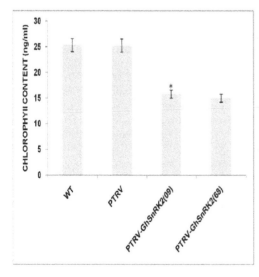

Figure 9. Physiological assay of *GhSnRK2* gene silenced plants. (A) Phenotype of drought stressed plants. Water was withheld from *GhSnRK2* gene silenced and non-silenced plants for 5 days approximately two weeks post-inoculation. Each of the ten treated groups consisted of five plants. A photograph of each group representative was captured. (B) The survival rate was determined by withholding water from *GhSnRK2* gene silenced and non-silenced plants for 5 days approximately two weeks post-inoculation, and the survival rate was recorded as the percentage of plants that survived after re-watering for 3 days. (C) The relative water loss was determined two weeks after inoculation. The reduction in the fresh weight from the initial weight was determined at the indicated time and represented as the percentage of water loss. The experiment was replicated three times. (D) The mean RWC was analyzed after immersing the fresh detached leaves in distilled water for 4 h and oven drying at 80°C for 48 h. The values are presented as the means of three biological replicates. Asterisk denotes a significant difference (P<0.05). (E) The change in the chlorophyll content of *GhSnRK2* gene silenced and non-silenced plants under salt stress. (F) The phenotype of *GhSnRK2* gene silenced and non-silenced plants under salt stress.

of *GhSnRK2* plants. Proline is an osmoprotective molecule that accumulates in response to water stress and salinity [37,38]. Proline is considered as a typical physiological parameter for evaluating abiotic stress tolerance and resistance in plants. Many plants reduce their cellular osmotic potential via the accumulation of intracellular organic osmolytes, such as proline, to maintain a stable intracellular environment when subjected to external environmental stresses [39,40,41].

We detected decreased stomatal apertures in the *GhSnRK2* transgenic plants compared with the corresponding WT plants. When exposed to adverse environmental conditions, the stomata must close for the plant to survive [42]. Stomatal closure reduces water loss, particularly in plants that have been exposed to water stress conditions caused by high solute concentrations in the nutrient medium, such that minimizing the rate of water loss due to transpiration reduces the accumulation of toxic ions [43]. The reduced rate of water loss from *GhSnRK2* transgenic plants may be due to the decrease in the size of the stomata aperture. ABA is involved in stomata movement, enhanced drought tolerance and plant osmoregulation. Under water stress conditions, ABA is released into leaves, inducing the release of potassium salts from the guard cells, resulting in stomatal closure. Plant guard cells modulate the opening and closure of the stomata in response to phytohormones and various environmental signals, such as light and temperature, thereby regulating gas exchange for photosynthesis and water status via transpiration [44]. The *SnRK2* gene *SRK2E*, or *SnRK2.6*, plays a significant role in stomatal closure in *Arabidopsis* leaves [45,46].

The *GhSnRK2* gene was upregulated after treatment with 10% PEG, which may suggest that the *GhSnRK2* gene sufficiently confers drought tolerance, as one strategy that can be employed to mimic the effect of drought on plants is treatment with PEG. PEG was quite commonly used in physiological experiments to induce controlled drought stress.

We speculate that the *GhSnRK2*-overexpressing plants were more tolerant to NaCl than the WT plants, which may be due to the improved accumulation of Na+ in the vacuoles, increased sequestration of Na+ into the vacuoles, improved cellular Na+ exclusion, or their greater capacity for vacuolar osmoregulation. High salt can result in oxidative damage to cell membranes. The ability to avert such damage is consistent with the degree of tolerance exhibited by the plant [47]. Under salt stress, this damage is less extensive in transgenic plants overexpressing the *GhSnRK2* gene than in WT plants because *GhSnRK2* overexpression can tolerate the accumulation of Na+ in the vacuoles, thus preventing the toxicity of excess Na+. Hypersensitivity to ABA treatment during seed germination and early seedling development is typically followed by improved drought tolerance [48,49]. Our finding reveals that the transgenic plants overexpressing *GhSnRK2* exhibited enhanced seedling growth under ABA stress treatment. This may be attributable to ABA-independent signaling pathway involving SnRK2 in root growth. This finding reveals the complexity in ABA-signaling, as different signaling components may function in different cells or tissues in ABA signal

transduction. *SnRK2.2* and *SnRK2.3* double mutant showed strong ABA-insensitive phenotypes in seed germination and root growth inhibition.[50]. The decreased stomatal apertures detected in the *GhSnRK2* transgenic plants may be attributable to the role of *SnRK2* in the regulation of ABA-induced activation of plasma membrane anion channels in guard cells and concomitant response to their closure. These results suggest that *SnRK2* functions in ABA signaling of stomatal closure. *SRK2E* knockout mutant lost the ABA-dependent stomatal closure. SRK2E affects ABA-signaling in stomatal closure, but not in germination stage [51]. ABA and PYR/PYLs inhibit protein phosphatase2C (PP2C), which in turn relieves the repression of positive factors, such as SnRK2s [52].

We analyzed the expression of the stress-inducible marker genes *RD29A*, *RD29B*, *P5CS1*, *ABI3*, *CBF1*, and *ABI5* in *GhSnRK2*-overexpressing plants, and our results revealed increases in the transcript levels of these marker genes. Consistent with a previous study, the upregulation of stress response genes, such as *P5CS1*, can contribute to enhanced salt stress tolerance in plants [53]. *ABI3* and *ABI5* activation is necessary to sustain the germination of seedlings during intense drought stress [54]. *ABI5* and *ABI3* regulate numerous ABA responses, such as osmotic water permeability of the plasma membrane, stomatal closure, drought-induced resistance, germination, and inhibition of vegetative growth. We speculated that *GhSnRK2* functions in the transcriptional regulation of ABA-inducible genes in seedlings, suggesting that *SnRK2* protein kinases are involved in several processes of ABA signal transduction, including transcriptional as well as post-transcriptional regulation pathways. Therefore, we suggest that stress signaling pathways may be involved in *GhSnRK2*-mediated stress tolerance.

Our findings revealed the downregulation of *GhSnRK2* in gene silenced plants, indicating that this gene was knocked down in the gene silenced plants. TRV spread vigorously throughout the entire plant, including meristem tissue, and the symptoms induced by TRV are not severe compared with those induced by other viruses [55]. The albino phenotype observed in the *GhCLA1* gene silenced plants in this study is an indicator of effective VIGS function in inoculated plants, which is consistent with previous findings [56]. The results of this study revealed that *GhSnRK2* gene silencing in *Gossypium hirsutum* greatly reduced its tolerance to drought stress, corroborating the findings of [57], who reported that silencing of the function of the *SpMPK1*, *SpMPK2*, and *SpMPK3* genes in tomato plants alleviates their tolerance to drought stress. The decrease in the chlorophyll content of silenced plants in response to salt stress indicates that *GhSnRK2* may be involved in oxidative stress. High-concentration salt stress inhibits plant biochemical processes. Decreased chlorophyll content in salt-stressed pumpkin plants was detected by [58]. The stunted growth observed in salt-stressed *GhSnRK2* gene silenced plants corroborate the findings of [59], who reported that the typical effect of salt stress in plants is growth retardation as a result of cell elongation inhibition.

Conclusions

Despite the significant innovations that have been made in elucidating the genetic mechanisms underlying drought tolerance, considerable challenges remain. The present study revealed that the genetic manipulation of the *GhSnRK2* gene from cotton using a transgenic technique results in enhanced drought and cold stress tolerance in plants. Thus, *GhSnRK2* is hypothesized to participate in the stress signaling pathway, and therefore, overexpression of *GhSnRK2* may alleviate abiotic stress by regulating stress-responsive genes. Moreover, the findings in this study have helped to elucidate that *GhSnRK2* enhances stress tolerance in plants by affecting various stress-related pathways. Thus, the *GhSnRK2* gene represents a candidate gene for future research of abiotic stress signaling pathways and the genetic modification of novel *Gossypium hirsutum* varieties.

Supporting Information

Figure S1 Multiple sequence alignment of *GhSnRK2* and closely related *SnRK2s* from other plants species and VIGS construct. (A) Alignment of *GhSnRK2* and closely related *SnRK2s*. The relatively conserved motif is underlined. The deduced amino acid sequence displays relatively high homology with the monocot *SnRK2* family members *Oryza sativa* (*RK1*), ABB89146 and with the dicot species *AtSnRK2.10*, AEE33751.1. (B) Virus-induced gene silencing construct. The TRV VIGS vectors were modified based on a pTRV1 containing RNA-dependent RNA polymerase (RdRp), movement protein (MP), a 16 kDa cysteine-rich protein (16K), CaMV 35S promoters (2X35S) and a NOS terminator (NOSt) T-DNA vector. pTRV2 contains the coat protein (CP), multiple cloning sites (MCSs), CaMV 35S promoters (2X35S) and a NOS terminator (NOSt) T-DNA vector. Both vectors contain Rz, which is designated as a self-cleaving ribozyme, LB (left border) and RB (right border) of the T-DNA.

Figure S2 Survival rate in kanamycin-containing medium and PCR confirmation of *GhSnRK2* gene expression in transgenic Arabidopsis. (A) The survival rate of WT and *GhSnRK2* transgenic plants in MS medium supplemented with the antibiotic kanamycin. Photograph of a representative plant was captured after 9 days of germination in kanamycin-containing medium. (B) Confirmation of *GhSnRK2* gene in transgenic Arabidopsis. Genomic DNA from the first generation of the plants (T1) was extracted and used as a template for gene-specific primers. Lane 1: DNA molecular marker III standard; lane 2: negative control.

Table S1 Survival (%) of *GhSnRK2* transgenic plants in kanamycin-containing medium. At T1 generation, the ratio of dead to surviving plant was approximately 1:3 in kanamycin-containing LB medium. The survival rate at T3 generation was 100% in kanamycin-containing LB medium. The values are expressed as the mean germination rate (%) of approximately 200 seeds.

Table S2 The lists of primers sequences used in this study.

Acknowledgments

Our appreciation goes to the staff and students of the State Key Laboratory of Cotton Biology, Chinese Academy of Agricultural Sciences, Cotton Research Institute, Anyang, China, for their support during the course of this research. We also thank Jiahe Wu for VIGS technical support and suggestions during manuscript preparation.

Author Contributions

Conceived and designed the experiments: BB FL XZ Zhaoen Yang CL QW GZ Zuoren Yang. Performed the experiments: BB FL XZ Zhaoen Yang CL QW GZ Zuoren Yang. Analyzed the data: BB FL XZ Zhaoen Yang CL QW GZ Zuoren Yang. Contributed reagents/materials/analysis tools: BB FL XZ Zhaoen Yang CL QW GZ Zuoren Yang. Wrote the paper: BB FL XZ Zhaoen Yang CL QW GZ Zuoren Yang.

References

1. Genoud T, Metraux JP (1999) Crosstalk in plant cell signaling: structure and function of the genetic network. Trends in Plant Science 4: 503–507.
2. Kumar N, Bhatt RP (2006) Transgenics: An emerging approach for cold tolerance to enhance vegetables production in high altitude areas. Indian J. Crop Sci 1: 8–12.
3. Shinozaki K, Yamaguchi-Shinozaki K (2000) Molecular responses to dehydration and low temperature: differences and cross-talk between two stress signaling pathways. Curr Opin Plant Biol 3: 217–223
4. Guo-Tao H, Shi-Liang M, Li-Ping B, Li Z, Hui M, et al. (2000) The role of root border cells in plant defense. Trends Plant Sci 5: 128–133
5. Cho K, Agrawal GK, Jwa NS, Kubo A, Rakwal R (2009) Rice OsSIPK and its orthologs: a "central master switch" for stress responses. Biochem Biophys Res Commun 379: 649–653.
6. Mishra NS, Tuteja R, Tuteja N (2006) Signaling through MAP kinase networks in plants. Arch Biochem Biophys 452: 55–68.
7. Pitzschke A, Schikora A, Hirt H (2009) MAPK cascade signalling networks in plant defence. Curr Opin Plant Biol 12: 421–426.
8. Rodriguez MC, Petersen M, Mundy J (2010) Mitogen-activated protein kinase signaling in plants. Annu Rev Plant Biol 61: 621–649.
9. Jonak C, Hirt H (2002) Glycogen synthase kinase 3/SHAGGY-like kinases in plants: an emerging family with novel functions. Trends Plant Sci 7: 457–461.
10. Koh S, Lee SC, Kim MK, Koh JH, Lee S, et al. (2007) T-DNA tagged knockout mutation of rice OsGSK1, an orthologue of Arabidopsis BIN2, with enhanced tolerance to various abiotic stresses. Plant Mol Biol 65: 453–466.
11. Mahfouz MM, Kim S, Delauney AJ, Verma DP (2006) Arabidopsis TARGET OF RAPAMYCIN interacts with RAPTOR, which regulates the activity of S6 kinase in response to osmotic stress signals. Plant Cell 18: 477–490.
12. Das R, Pandey GK (2010) Expressional analysis and role of calcium regulated kinases in abiotic stress signaling. Curr Genomics 11: 2–13.
13. Hrabak EM, Chan CWM, Gribskov M, Harper JF, Choi JH, et al. (2003) The Arabidopsis CDPK-SnRK superfamily of protein kinases. Plant Physiol 132: 666–680.
14. Wurzinger B, Mair A, Pfister B, Teige M (2011) Cross-talk of calcium-dependent protein kinase and MAP kinase signaling. Plant Signal Behav 6: 8–12.
15. Coello P, Hirano E, Hey SJ, Muttucumaru N, Martinez-Barajas E, et al. (2012) Evidence that abscisic acid promotes degradation of SNF1-related protein kinase (SnRK) 1 in wheat and activation of a putative calcium-dependent SnRK2. J Exp Bot 63: 913–924
16. Umezawa T, Yoshida R, Maruyama K, Yamaguchi-Shinozaki K, Shinozaki K (2004) SRK2C, a SNF1-related protein kinase 2,improves drought tolerance by controlling stress-responsive gene expression in Arabidopsis thaliana. Proc Natl Acad Sci. USA 101: 17306–173011.
17. Hardie DG (1999) Plant protein serine/threonine kinases: classification and functions. Annu Rev Plant Physiol Plant Mol Biol 50: 97–131.
18. Mikołajczyk M, Awotunde OS, Muszyńska G, Klessi DF, Dobrowolska G (2000) Osmotic stress induces rapid activation of a salicylic acid-induced protein kinase and a homolog of protein kinase ASK1 in tobacco cells. Plant cell 12: 165–178
19. Wang X, Xu W, Xu Y, Chong K, Xu Z, et al (2004) Wheat RAN1, a nuclear small G protein, is involved in regulation of cell division in yeast. Plant Sci 167: 1183–1190.
20. Clough SJ, Bent AF (1998) Floral dip: a simplified method for Agrobacterium-mediated transformation of Arabidopsis thaliana. Plant J 16(6):735–743
21. Tyree MT, Hammel HT (1972) The measurement of the turgor pressure and the water relations of plants by the pressure-bomb technique. J Exp Bot 23: 267–282
22. Duan J, Zhang M, Zhang H, Xiong H, Liu P, et al. (2012) OsMIOX, a myoinositol oxygenase gene, improves drought tolerance through scavenging of reactive oxygen species in rice (Oryza sativa L.). Plant Sci 196: 143–151.

23. Parida AK, Dasgaonkar VS, Phalak MS, Umalkar GV, Aurangabadkar LP (2007) Alterations in photosynthetic pigments, protein, and osmotic components in cotton genotypes subjected to short-term drought stress followed by recovery. Plant Biotechnology Reports 1: 37–48.

24. Pei ZM, Kuchitsu K, Ward JM, Schwarz M, Schroeder JI (1997 Mar) Differential abscisic acid regulation of guard cell slow anion channels in Arabidopsis wild-type and abi1 and abi2 mutants. Plant cell 9(3):409–423.

25. Bates L, Waldren RP, Teare ID (1973) Rapid determination of free proline for water stress studies. Plant and Soil 39: 205–207

26. Arnon DI (1949) Copper enzymes in isolated chloroplasts, polyphenoxidase in beta vulgaris. Plant physiology 24: 1–15.

27. Xiong L, Ishitani M, Lee H, and Zhu JK (2001b) The *Arabidopsis LOS5/ABA3* locus encodes a molybdenum cofactor sulfurase and modulates cold and osmotic stress responsive gene expression. Plant cell 13: 2063–2083.

28. Livak KJ, Schmittgen TD (2001) Analysis of Relative Gene Expression Data Using Real-Time Quantitative PCR and the 2DDCT Method. METHODS 25: 402–408

29. Gao X, Britt RC, Shan L, He P (2011) Agrobacterium-Mediated Virus-Induced Gene Silencing Assay In Cotton. J. Vis. Exp 54:e2938

30. Kobayashi Y, Murata M, Minami H, Yamamoto S, Kagaya Y (2005) Abscisic acid-activated SNRK2 protein kinases function in the gene-regulation pathway of ABA signal transduction by phosphorylating ABA response element binding factors. Plant J 44: 939–949.

31. Zhang H, Mao X, Wang C, Jing R (2010) Overexpression of a common wheat gene *TaSnRK2.8* enhances tolerance to drought, salt and low temperature in *Arabidopsis*. PLoS ONE 5: e16041.

32. Hawes MC, Gunawardena U, Miyasaka S, Zhao X (2000) The role of root border cells in plant defense. Trends Plant Sci 5: 128–133.

33. Dainty J (1976) Water relations of plant cells. *In* AP Gottingen, MH Zimmermann, eds, Encyclopedia of Plant Physiology, Vol 2. New Series, Part A. Springer-Verlag, Berlin, pp 12–35.

34. Levitt J (1980) Response of plant to Environmental Stress, water, radiation, salt and other stresses. Academic press, New York.

35. Steponkus PL (1984) Role of the plasma membrane in freezing injury and cold acclimation. Annu. Rev. Plant Physiol 35: 543–584.

36. Steponkus PL, Uemura M, Webb MS (1993) A contrast of the cryostability of the plasma membrane of winter rye and spring oat-two species that widely differ in their freezing tolerance and plasma membrane lipid composition. In: Steponkus P L, editor. Adv. Low-Temperature Biol. Vol. 2. London: JAI Press pp. 211–312.

37. Claussen W (2005) Proline as a measure of stress in tomato plants. Plant Sci 168: 241–248.

38. Younis ME, Hasaneen MNA, Tourky MNS (2009) Plant growth, metabolism and adaptation in relation to stress conditions. XXIV. Salinity biofertility interactive effects on proline, glycine and various antioxidants in Lactuca sativa. Plant Omics J 2: 197–205.

39. Zhu JK (2002) Salt and drought stress signal transduction in plants. Annu Rev Plant Biol 53: 247–273

40. Granier C, Tardieu F (1999) Water deficit and spatial pattern of leaf development variability in responses can be simulated using a simple model of leaf development. Plant Physiol 119: 609–620.

41. Wang ZQ, Yuan YZ, Ou JQ, Lin QH, Zhang CF (2007) Glutamine synthetase and glutamate dehydrogenase contribute differentially to proline accumulation in leaves of wheat (*Triticum aestivum*) seedlings exposed to different salinity. Plant Physiol 164: 695–701.

42. Pareek A, Sopory SK, Bohnert HJ, Govindjee EDS (2010) Abiotic Stress Adaptation in Plants: Physiological, Molecular and Genomic Foundation, Springer, Dordrecht, pp. 283e305.

43. Everard JD, Gucci R, Kahn JA, Flore WH (1994). Gas exchange and carbon partitioning in the leaves of celery (Apium graveolens L.) at various levels of root zone salinity, Plant Physiol 106: 281e292.

44. Schroeder JI, Kwak JM, Allen GJ (2001) Guard cell abscisic acid signalling and engineering drought hardiness in plants. Nature 410: 327–330.

45. Yoshida R, Hobo T, Ichimura K, Mizoguchi T, Takahashi F (2002) ABA-activated SnRK2 protein kinase is required for dehydration stress signaling in Arabidopsis Plant Cell Physiol 43: 1473–1483.

46. Mustilli AC, Merlot S, Vavasseur A, Fenzi F, Giraudat J (2002) Arabidopsis OST1 Protein Kinase Mediates the Regulation of Stomatal Aperture by Abscisic Acid and Acts Upstream of Reactive Oxygen Species Production. Plant Cell 14: 3089–3099.

47. Chinnusamy V, Jagendorf A, Zhu JK (2005) Understanding and improving salt tolerance in plants, Crop Sci 45: 437e448.

48. Hu H, Dai M, Yao J, Xiao B, Li X, et al. (2006) Overexpressing a NAM, ATAF, and CUC (NAC) transcription factor enhances drought resistance and salt tolerance in rice. Proc Natl Acad Sci USA 103: 12987–12992.

49. Ko JH, Yang SH, Han KH (2006) Upregulation of an Arabidopsis RING-H2 gene, *XERICO*, confers drought tolerance through increased abscisic acid biosynthesis. Plant J 47 343–355.

50. Hiroaki F, Paul E, Jian-Kang Z (2007) Identification of Two Protein Kinases Required for Abscisic Acid Regulation of Seed Germination, Root Growth, and Gene Expression in *Arabidopsis*. The Plant Cell 19 (2):485–494.

51. Riichiro Y, Tokunori H, Kazuya I, Tsuyoshi M, Fuminori T, et al. (2002) ABA-Activated SnRK2 Protein Kinase is Required for Dehydration Stress Signaling in *Arabidopsis*. *Plant Cell Physiol* 43 (12): 1473–1483.

52. Sang-Youl P, Pauline F, Noriyuki N, Davin RJ, Hiroaki F, et al. (2009) Abscisic Acid Inhibits Type 2C Protein Phosphatases via the PYR/PYL Family of START Proteins. Science 324: 1068

53. Kavi PB, Zonglie H, Cuo-Hua M, Chein-An AH, DeshPal SV (1995) Overexpression of A1-Pyrroline-5-Carboxylate Synthetase lncreases Proline Production and Confers Osmotolerance in Transgenic Plants Plant Physiol 108: 1387–1394

54. Lopez-Molina L, Mongrand S, McLachlin DT, Chait BT, Chua NH (2002) ABI5 acts downstream of ABI3 to execute an ABA-dependent growth arrest during germination. Plant J 32: 317–328.

55. Ratcliff F, Martin-Hernandez AM, Baulcombe DC (2001) Tobacco rattle virus as a vector for analysis of gene functions by silencing. Plant J 25: 237–245

56. Liu Y, Schiff M, Dinesh-Kumar S (2002) Virus-induced gene silencing in tomato. Plant J 31: 777–786.

57. Cui L, Jian-Min Y, Yun-Zhou L, Zhen-Cai Z, Qiao-Li W, et al. (2013) Silencing the *SpMPK1, SpMPK2*, and *SpMPK3* Genes in Tomato Reduces Abscisic Acid—Mediated Drought Tolerance Int. J. Mol. Sci 14: 21983–21996

58. Senay S, Fikret Y, Sebnem K, Sebnem E (2011) The effect of salt stress on growth, chlorophyll content, lipid peroxidation and antioxidative enzymes of pumpkin seedling. African Journal of Agricultural Research 6(21): 4920–4924

59. Yasar F, Ellialtioglu S, Yildiz K (2008) Effect of salt stress on antioxidant defense systems, lipid peroxidation, and chlorophyll content in green bean. Russian J. Plant Physiol 55(6): 782–786.

3

A *Malus* Crabapple Chalcone Synthase Gene, *McCHS*, Regulates Red Petal Color and Flavonoid Biosynthesis

Deqiang Tai[1,3,9], **Ji Tian**[1,2,9], **Jie Zhang**[1,2,9], **Tingting Song**[1,2], **Yuncong Yao**[1,2*]

1 Department of Plant Science and Technology, Beijing University of Agriculture, Beijing, China, 2 Key Laboratory of New Technology in Agricultural Application of Beijing, Beijing University of Agriculture, Beijing, China, 3 College of Horticulture, Shanxi Agricultural University, Taigu, Shanxi, China

Abstract

Chalcone synthase is a key and often rate-limiting enzyme in the biosynthesis of anthocyanin pigments that accumulate in plant organs such as flowers and fruits, but the relationship between *CHS* expression and the petal coloration level in different cultivars is still unclear. In this study, three typical crabapple cultivars were chosen based on different petal colors and coloration patterns. The two extreme color cultivars, 'Royalty' and 'Flame', have dark red and white petals respectively, while the intermediate cultivar 'Radiant' has pink petals. We detected the flavonoids accumulation and the expression levels of *McCHS* during petals expansion process in different cultivars. The results showed *McCHS* have their special expression patterns in each tested cultivars, and is responsible for the red coloration and color variation in crabapple petals, especially for color fade process in 'Radiant'. Furthermore, tobacco plants constitutively expressing *McCHS* displayed a higher anthocyanins accumulation and a deeper red petal color compared with control untransformed lines. Moreover, the expression levels of several anthocyanin biosynthetic genes were higher in the transgenic *McCHS* overexpressing tobacco lines than in the control plants. A close relationship was observed between the expression of *McCHS* and the transcription factors McMYB4 and McMYB5 during petals development in different crabapple cultivars, suggesting that the expression of *McCHS* was regulated by these transcription factors. We conclude that the endogenous *McCHS* gene is a critical factor in the regulation of anthocyanin biosynthesis during petal coloration in *Malus* crabapple.

Editor: Xianlong Zhang, National Key Laboratory of Crop Genetic Improvement, China

Funding: Financial support was provided by the Project of Construction of Innovative Teams and Teacher Career Development for Universities and Colleges Under Beijing Municipality (IDHT20140509), the National Natural Science Foundation of China (31301762) and the National Modern Agricultural Science City Achievement for People Service Technology Demonstration Project (Z121100007412003). The funders had no role in study design, data collection and analysis, decision to publish, or preparation of the manuscript.

Competing Interests: The authors have declared that no competing interests exist.

* Email: yaoyc_20@126.com

9 These authors contributed equally to this work.

Introduction

The plant phenylpropanoid biosynthetic pathway leads to the formation of numerous compounds that are involved in diverse physiological and biochemical processes [1]. Some well-studied examples of these compounds include anthocyanins, flavonols and proanthocyanidins of the flavonoid family, which play a central role in the pigmentation of plant organs, seed germination, UV-B protection and defense against pathogens and biotic stresses [2–8]. Previous studies have focused on anthocyanin biosynthesis in *Arabidopsis thaliana* [9], *Petunia hybrida* [10], *Zea mays* [11] and *Malus domestica* [12–13], and anthocyanin biosynthetic genes have been characterized that are regulated by three classes of transcription factors (TFs): MYB, basic helix-loop-helix (bHLH) and WD40 proteins [14–16]. The reaction catalyzed by chalcone synthase (CHS) is thought to be the key regulatory step in the synthesis of flavonoids by catalyzing the condensation of one molecule of 4-coumaroyl-CoA with three molecules of malonyl-CoA to form naringenin chalcone, a major pigment of many flowers, leaves and fruits [17–19]. Indeed, chalcones provide the structural precursors for a broad range of flavonoids, flavonols,

flavanones, anthocyanin glycosides and other derived compounds (Figure 1). Consequently, there has been much interest in *CHS* and its involvement in many aspects of plant physiology and biochemistry.

The first *CHS* gene was cloned from parsley (*Petroselinum crispum*) in 1983 [20] and, since then, numerous *CHS* genes have been isolated, mostly from monocots and dicots, including the legume soybean (*Glycine max*), alfalfa (*Medicago sativa*), pea (*Pisum sativum*), *A. thaliana*, barley (*Hordeum vulgare*), corn (*Zea mays*), grape (*Vitis vinifera*) and others [21–27]. CHS protein sequences are highly conserved among different plants [28], with amino acid homologies of approximately 80–90% [29], and molecular evolution analysis of *CHS* genes has shown them to be ubiquitous in plants, including early land plants and algae of the Charophyceae [30].

CHS has been well studied in the context of the synthesis and accumulation of anthocyanin pigments and several reports have described the effects of altering *CHS* expression in transgenic plants. For example, expression of an antisense *CHS* gene in petunia resulted in flowers that were pale colored, or even white, due to an inhibition of anthocyanin production, and plant fertility

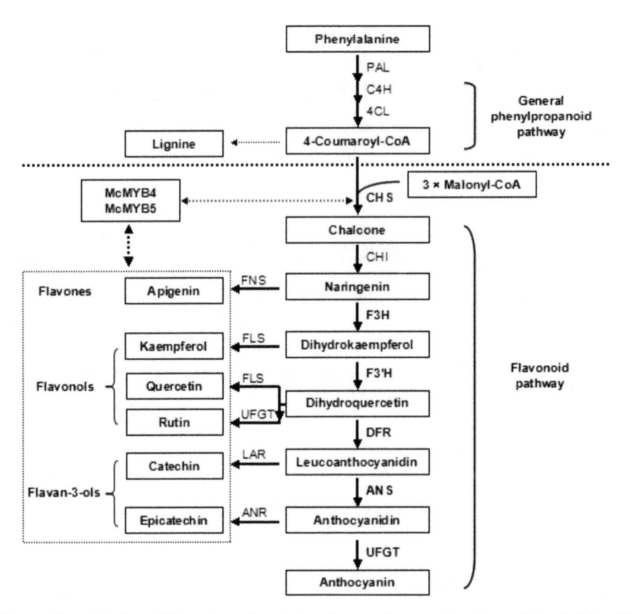

Figure 1. Scheme of the flavonoid biosynthetic pathway in plants. Genes encoding enzymes for each step are indicated as follows: *PAL*, *phenylalanine ammonia-lyase*; *C4H*, *cinnamate 4-hydroxylase*; *4CL*, *4-coumarate-CoA ligase*; *CHS*, *chalcone synthase*; *CHI*, *chalcone isomerase*; *F3H*, *flavanone 3-hydroxylase*; *F3'H*, *flavonoid 3'-hydroxylase*; *DFR*, *dihydro-flavonol 4-reductase*; *FNS*, *flavonol synthase*; *FLS*, *flavonol synthase*; *LAR*, *leucoanthocyanidin reductase*; *ANS*, *anthocyanidin synthase*; *UFGT*, *UDP-glucose: flavonoid-3-O-glycosyltransferase*.

was also affected [17,31]. Similarly, fruits from *CHS*-silenced strawberry (*Fragaria* × *ananassa*) lines were reported to have a lighter pink/orange coloration and a significant decrease in the levels of all flavonols, proanthocyanidins and anthocyanins [32]. As another example, fruits from apple (*Malus domestica*) *CHS*-RNAi knockout lines were shown to have no detectable anthocyanins accumulation and substantially reduced levels of dihydrochalcones and flavonoids [33].

Malus crabapple has one of the most economically important sets of ornamental apple germplasm resources. This represents a valuable source of research material, since crabapple exhibits excellent stress resistance and is useful for investigating the mechanism of plant pigmentation, due to the diversity of color in its leaves, flowers and fruits as a consequence of anthocyanins accumulation [34]. In this current study, we compared three crabapple cultivars: two extreme color cultivars, 'Royalty' and

'Flame' with dark red and white petals respectively, and an intermediate cultivar, 'Radiant', with pink petals [35]. To understand better the function of *McCHS* involved in the biosynthesis of anthocyanins and coloration level in the petals of these cultivars, we compared the expression of *McCHS* and other flavonoid biosynthesis pathway structural genes during petal expansion in these three typical cultivars by quantitative real-time PCR (qRT-PCR). Meanwhile, HPLC analysis provided an insight into the accumulation of anthocyanins and other flavonoids compounds in these different crabapple cultivars. The results suggest that the expression of *McCHS* in petals is well controlled, in a tissue- and developmentally specific fashion. We also overexpressed the *McCHS* gene in tobacco to evaluate its activity and the consequences of its expression. To sum up, *McCHS* expression is associated with petal coloration and the expression level of *McCHS* determined the coloration level of petals in different crabapple cultivars. Meanwhile, the expression level of this gene may be regulated by multiple MYB transcription factors.

Results

Petal phenotypes of three *Malus* crabapple cultivars, 'Royalty', 'Radiant' and 'Flame'

The petal phenotypes of three *Malus* crabapple cultivars, 'Royalty', 'Radiant' and 'Flame', are shown in Figure 2A. 'Royalty' petals at stage I were dark red and during flowering the color became more vivid and bright, ultimately reaching maximum color strength at stage IV of full bloom. At the later stage V, after full bloom, the pigmentation faded and dulled (Figure 2B). The petals of 'Flame' at stage I showed an obvious light red color, while from stage II to stage V, the color gradually became fainter and eventually the petals turned white (Figure 2D). Interestingly, the petals of 'Radiant' were red at stage I, and the color was gradually becoming pink during petal expansion (Figure 2C). Overall, 'Royalty' has the most red and vivid flowers, while the 'Flame' petals are almost white during petal development, and 'Radiant' petals have a significantly color fade process during petal expansion. Next, it is determined how the anthocyanins color the three typical cultivars.

Quantification and identification of flavonoid composition during petal development in three *Malus* crabapple cultivars

To gain insight into the *Malus* crabapple petal flavonoid composition and its variation among cultivars, the levels of two anthocyanins (cyanidin and pelargonidin) and five other flavonoids were measured in the petals of the three cultivars (Figure 2E). Consistent with the petal color observations in the red cultivar 'Royalty', the abundance of anthocyanins was significantly higher than the other two cultivars, and a gradually decrease in the anthocyanins occurred in petal development, while anthocyanins were only detected in white cultiver 'Flame' petals at the first stage. Quercetin and apigenin showed the same spatial and temporal accumulation patterns, and the abundance of these compounds also decreased during petal development in all three varieties. Rutin had a similar profile to quercetin and apigenin and decreased substantially to undetectable levels except for in 'Flame' at stage II. In contrast, the catechin content of 'Flame' petal was much higher than petals of 'Royalty' and 'Radiant', and catechin was only found at the first stage of development in 'Royalty' and 'Radiant' petals. As the precursor for proanthocyanidin, the accumulation of catechin suggests that proanthocyanidin is the main flavonoids compound in colorless tissues and compared with other color-petal cultivars, the proanthocyanidin biosynthetic pathway is primary flavonoids biosynthetic branch pathway in 'Flame'. In addition, the levels of kaempferol in 'Royalty' petals were almost same with those in the other two varieties in the first stage and decreased with the development of petals, the kaempferol contents in 'Radiant' and 'Flame' were barely detectable from the second stage to the fifth stage (Figure 2E).

Characterization of the *Malus* crabapple McCHS gene

Based on homology with *Malus* × *domestica* sequences and related sequences in the Genome Database for Rosaceae (http://www.rosaceae.org), a gene sequence for the full length *CHS* (*McCHS*) was amplified from total RNA of *Malus* crabapple 'Royalty' leaves by RT-PCR and RACE. Specifically, the *McCHS* cDNA (accession no. FJ599763) is 1,529 bp in length and is predicted to encode a protein of 389 amino acids with a high level of homology to with *CHS* genes from a range of species (Figure 3A), including *Malus domestica* (MdCHS2; 99% identity), *Malus domestica* (MdCHS1; 99% identity), *Malus domestica* (MdCHS320; 99% identity), *Pyrus bretschneideri* (PbCHS; 98% identity), *Malus domestica* (MdCHS46; 94% identity) and so on. It appears that chalcone synthase protein sequences have undergone little sequence diversification in the species examined and that they are highly similar in length in different plant species. Other than the putative *CHS* gene sequences, there are also related sequences that are more divergent from characterized *CHS* genes and although these genes show high sequence similarity to a *Fragaria ananassa CHS* (FrCHS; AB201756), e.g. *Rubus idaeus* (RiPKS-5; EF694718; 97% identity), their functions have yet to be determined. Collectively they are described as polyketide synthase (*PKS*) genes, a more general description for the *CHS* and *CHS-Like* gene family (Figure 3B).

We performed qRT-PCR to determine the expression of *McCHS* during petal development of the three crabapple cultivars (Figure 3C). Compared with that in colorless cultivars, the relative expression level of *McCHS* in 'Royalty' was much higher, whereas *McCHS* transcripts were barely detectable in 'Flame' petals, except at stage IV. In 'Royalty' and 'Radiant' the transcript levels decreased gradually among the petal development. The results showed the expression level of *McCHS* was consistent with the variation of petal color and anthocyanins accumulation in different crabapple cultivars, and the expression intensity of *McCHS* induced the petal pigmentation level in crabapple. In addition, the color variation in 'Radiant' petals was induced by the down-regulation of *McCHS* transcriptional level.

The expressions of *McCHS* in leaves and fruits at maturity were also investigated in the three cultivars. The phenotypes of leaf, fruit flesh and fruit pericarp are shown in Figure S1. When comparing leaves, the highest expression of *McCHS* was in 'Radiant', with a value more than twice that of 'Royalty' and four times that of 'Flame' (Figure 4A). *McCHS* transcripts were most abundant in the flesh and pericarp of 'Flame' fruits, but they were almost undetectable in the pericarp of 'Royalty' and 'Radiant' (Figure 4A). In fruit flesh, the expression profile of *McCHS* showed an opposite trend from that of leaves (Figure 4A). These results suggest that the expression of *McCHS* have different patterns in different tissues/organs among these three cultivars, so we deduced that the expression of *McCHS* may be also involved in other flavonoids compounds biosynthesis. As an important factor to determine the leaf color, the chlorophyll contents in leaves from the three cultivars were measured at five development stages, the results suggested that the chlorophyll content was higher in ever-red-leaf 'Royalty' leaves than spring-red-leaf 'Radiant' and ever-green-leaf 'Flame', and gradually decreased with the development of leaves in crabapple. So the red leaf color of crabapple is depended on the accumulation of anthocyanin and expression of *McCHS*, not the chlorophyll content (Figure 4B).

RNA expression profiles of flavonoid biosynthetic genes in *Malus* crabapple petals

To further confirm how *CHS* regulated the other anthocyanin biosynthetic genes to affect anthocyanin biosynthesis, we analyzed the expression of anthocyanin biosynthetic genes that are located downstream of *McCHS* (Figure 5). The expression levels of the downstream genes *McF3H*, *McF3'H*, *McDFR*, *McANS* and *McUFGT* were almost all higher in 'Royalty' and 'Radiant' petals than in 'Flame' petals at all developmental stages, with an exception being *McUFGT* at stage II. To some extent, the expression patterns of *McF3H*, *McF3'H* and *McDFR* showed a similar pattern among the three *Malus* crabapple cultivars and was closely related to flavonol levels (Figure 2E and 5). While the

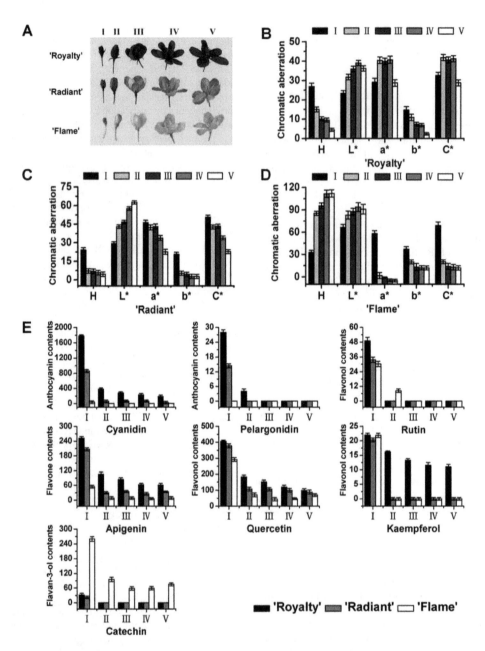

Figure 2. Flower developmental series of three *Malus* crabapple cultivars. (A) Typical flower phenotypes of *Malus* crabapple 'Royalty', 'Radiant' and 'Flame' cultivars through development. Five stages of each cultivar are shown. (B) Color changes in 'Royalty' petals. (C) Color changes in 'Radiant' petals. (D) Color changes in 'Flame' petals. (E) Content of flavonoids and anthocyanin in 'Royalty', 'Radiant' and 'Flame' petals. A spectrophotometric colorimeter was used to measure dynamic changes in petals, and HPLC was used to analyze the flavonoids and total anthocyanin contents. Five stages were tested in this study: (I) 6 days before full bloom; (II) 3 days before full bloom; (III) 1 day before full bloom; (IV) full bloom; and (V) 3 days after full bloom. Error bars indicate the standard error of the mean ± SE of three replicate measurements.

expression of *McF3H*, *McF3'H* and *McDFR* in 'Radiant' and 'Flame' petals was similar to *McCHS*, but not in 'Royalty'. Most importantly, *McCHS* showed the same spatial and temporal expression pattern as *McANS* and *McUFGT* in all three cultivars. Taken together, the results of transcriptional level suggested *McCHS* gene regulate the flavonoid biosynthetic genes to color the plants.

Quantitative PCR expression analysis of putative transcription factor genes in *Malus* crabapple petals

Transcriptional control of flavonoid biosynthesis is one of the best characterized regulatory systems in plants, and involves

integrating both developmental and various biotic and abiotic stress signals and the promoters of flavonoid biosynthetic genes via the control of TFs [37–38]. Previously, we showed that anthocyanin biosynthetic genes are regulated by three classes of TFs [14–16]. To explore the transcriptional regulatory mechanisms of *McCHS*, we investigated the expression profiles of nine MYB TFs during the five petal developmental stages (Figure 6). For all three crabapple cultivars, the expressions of *McMYB1*, *McMYB2* and *McMYB14* were barely detected in petals, except at stage II. The expression levels of *McMYB3* and *McMYB7* showed a very similar profile in petals from the same cultivar and almost

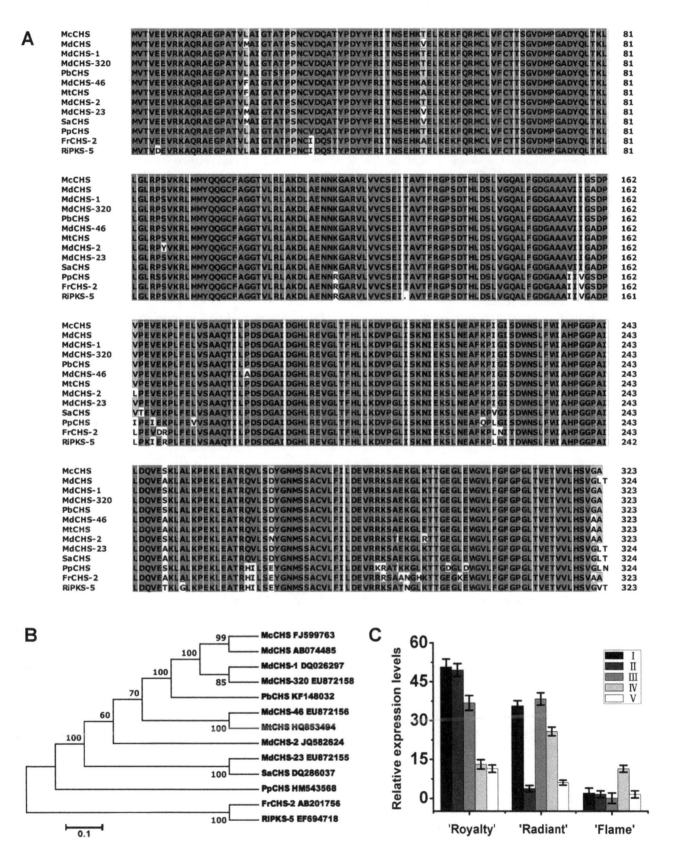

Figure 3. Sequence characteristics of McCHS and relationships with other CHS proteins. (A) Protein sequence alignment of McCHS and other known anthocyanin biosynthetic proteins from other plant species. Identical residues are shown in black, conserved residues in dark grey and similar residues in light grey. (B) Phylogenetic relationships between McCHS and CHS sequences from other species involved in anthocyanin biosynthesis. Phylogenetic and molecular evolutionary analysis was conducted using MEGA version 5.10 (using the minimum evolution phylogeny test and 1000 bootstrap replicates). (C) Relative expression profile of *McCHS* at different developmental stages of the petals of three *Malus* crabapple

cultivars: 'Royalty', 'Radiant' and 'Flame'. Five stages were tested in this study: (I) 6 days before full bloom; (II) 3 days before full bloom; (III) 1 day before full bloom; (IV) full bloom; and (V) 3 days after full bloom. Error bars represent the standard error of the mean ± SE of three replicate reactions. All real-time PCR reactions were normalized using the Ct value corresponding to a *Malus* crabapple 18S ribosomal RNA gene (DQ341382). The GenBank accession numbers of the proteins are as follows: McCHS (*Malus domestica*; FJ599763), MdCHS (*Malus domestica*; AB074485), MdCHS-1 (*Malus domestica*; DQ026297), MdCHS-320 (*Malus domestica*; EU872158), PbCHS (*Pyrus bretschneideri*; KF148032), MdCHS-46 (*Malus domestica*; EU872156), MtCHS (*Malus toringoides*; HQ853494), MdCHS-2 (*Malus toringoides*; JQ582624), MdCHS-23 (*Malus toringoides*; EU872155), SaCHS (*Sorbus aucuparia*; DQ286037), PpCHS (*Prunus persica*; HM543568), FrCHS-2 (*Fragaria ananassa*; AB201756), RiPKS-5 (*Rubus idaeus*; EF694718).

increaed with the development of crabapple petals. Furthermore, *McMYB4* expression displayed a similar pattern to that of *McMYB5* in petals from the same cultivars and decreased at the early development stages and increased at later development stages of crabapple petals. The profiles of *McMYB6* and *McMYB10* transcript accumulation suggested more specialized expression patterns between 'Royalty', 'Radiant' and 'Flame' petals. In 'Royalty' and 'Flame', the expressions of *McMYB6* and *McMYB10* were increased at the early development stages and decreased at later development stages of crabapple petals, but the transcriptions of *McMYB6* and *McMYB10* were increased with the development of petals in 'Radiant'.

McCHS showed decreased expression in 'Royalty' petals during development, while the expression of *McMYB4* increased over the same period, revealing a negative correlation between the *McMYB4* and genes in the flavonoid biosynthetic pathway, and the anthocyanins accumulation was negatively regulated by *McMYB4*. In 'Radiant' petals, the expression of *McMYB5* was almost consistent with the expression pattern of *McCHS* and the contents of anthocyanins, flavonol (kaempferol, quercetin and rutin) and apigenin. In 'Flame' petals, the expression of *McCHS*

also showed a positive correlation with *McMYB5* expression (except for the second petal development stage). Meanwhile, the expression trend of TFs and *McCHS* were consistent with the accumulation of kaempferol, quercetin, rutin, apigenin and catechin, which were the main flavonoids compounds in white petal cultivar 'Flame'. These data are consistent with *McMYB4*, *McMYB5* that may be involved in the regulation of *McCHS* expression during petal development in these cultivars, and these results suggested that due to different accmulation of flavonoids compounds, the *McCHS* gene have different biosynthesis functions and regulated by different transcription factors.

Overexpression *McCHS* in tobacco plants

To investigate the temporal and spatial expression of *McCHS* in plants, transgenic tobacco plants expressing *McCHS* under the control of a constitutive CaMV35S promoter were generated, and two independent T2 lines were characterized. Compared to wild type, tobacco plants transformed with *35S::McCHS* developed a more intense pigmentation in the petals, especially in *McCHS-ox-1* line (Figure 7A). The lightness value L*, hue value a*, hue value b* of the *McCHS*-overexpressing tobacco petals were also

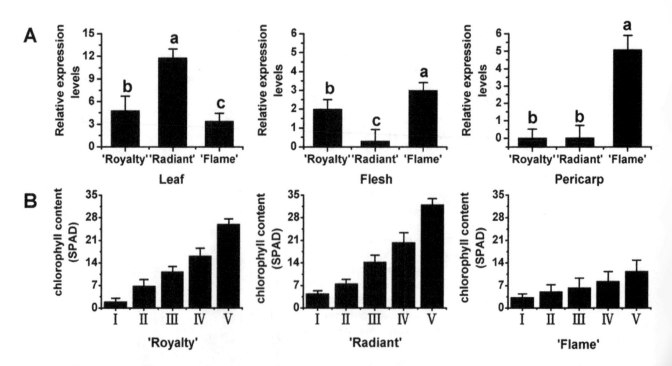

Figure 4. Relative expression profile of *McCHS* in different organs or tissues and chlorophyll content in leaves of the three *Malus* crabapple cultivars: 'Royalty', 'Radiant' and 'Flame'. (A) Relative expression profile of *McCHS* in different organs or tissues. (B) The Chlorophyll contents in leaves of the three crabapple cultivars. All real time-PCR reactions were normalized using the Ct value corresponding to a *Malus* crabapple 18S ribosomal RNA gene (DQ341382). Similar material was collected at the same time from the three cultivars. Leaves were all collected at the time of petal stage IV (Figure 2) and the fruits were collected at a fully ripe stage. The Chlorophyll contents of leaves on the tree in vivo were measured by Chlorophyll Content Meter (CL-01, Hansatech, UK). Five Phyllotaxy was measured according to the order of the leaves, respectively. Error bars correspond to the standard error of the mean ± SE of three replicate reactions. Different letters above the bars indicate significantly different values (*P*<0.05) calculated using one-way analysis of variance (ANOVA) followed by a Duncan's multiple range test.

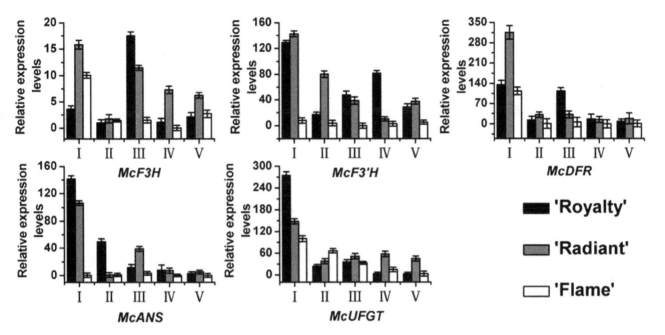

Figure 5. Relative expression profile of downstream genes of _McCHS_ during flower developmental stages. Real-time PCR was used to analyze the expression patterns of _McCHS, McF3H, McF3'H, McDFR, McANS_ and _McUFGT_ in petals of _Malus_ cultivars 'Royalty' and 'Radiant', 'Flame'. All real time-PCR reactions were normalized using the Ct value corresponding to a _Malus_ crabapple 18S ribosomal RNA gene (DQ341382). Stages referred to on the x axis are: (I) 6 days before full bloom; (II) 3 days before full bloom; (III) 1 day before full bloom; (IV) full bloom; and (V) 3 days after full bloom. Error bars correspond to the standard error of the mean ± SE of three replicate reactions.

measured. The 'H' value (hue angle) decreased to one-third of that of control plant petals and the 'a*' values were 4.0 and 5.0-fold higher than in control plant petals, indicative of a stronger red color (Figure 7B). In addition, we confirmed by microscopy that more anthocyanins accumulated in the petal cells of the transgenic tobacco lines (Figure 7C).

To confirm the biochemical activity of _McCHS_, anthocyanins content and the expression level of the _McCHS_ gene were detected in the transgenic tobacco petals, (Figure 7D). The anthocyanins content of petals was approximately 3-fold greater in the _McCHS-ox-1_ line and 2.5-fold greater in the _McCHS-ox-3_ line compared to control petals. These results are consistent with the visual flower color phenotypes.

The qRT-PCR results confirmed a massive elevation of _McCHS_ transcript levels in the transgenic lines, and the absence of expression in the control plants, as expected. The accumulation of anthocyanins was consistent with the relative expression profile of _McCHS_ in _McCHS_-overexpressing tobacco petals. Moreover, the over-expression of _McCHS_ in tobacco significantly promoted the expression of the downstream endogenous genes _NtF3H, NtF3'H, NtDFR, NtANS_, and _NtUFGT_ (Figure 7E). To further explore the molecular regulation mechnism in transgenic tobacco, we tested the expression of several endogenous tobacco anthocyanin regulation factors. Interestingly, the results showed the massive accumulated _McCHS_ can alter the expressions of these transcription factors. The expression levels of _NtAn1a, NtTTG1, MYB305, NtMYC2a_ and _NtMYC2b_ were increased, and _NtAn1b_ and _NtTTG2_ were decreased by overexpressed _McCHS_ (Figure 8).

Discussion

CHS, which catalyzes the first committed step in anthocyanin biosynthesis, plays a central role and provides a common chalcone precursor for the production of all intermediates and final products of the flavonoid biosynthetic pathway. Many studies have analyzed the activity of _CHS_ in different plants, including _P. hybrida_ [17], _Antirrhinum majus_ [39], maize [40], _Arabidopsis_ [41], strawberry and tomato [32,42] and apple [33]. The first characterized _chs_ mutant, the light white mutant of _A. majus_, whose single _CHS_ gene was knocked out, was reported to lack anthocyanins and UV-absorbing flavonoids [39]. This mutant established a relationship between CHS activity and the white-flowered phenotype. In maize, recessive mutations in both _CHS_ genes of maize, _C2 and Whp_, results in a white pollen phenotype and male sterility [26,40]. Moreover, down-regulation of _MdCHS_ was reported to lead to major changes in plant development, resulting in plants with shortened internode lengths and leaf areas, and a greatly reduced growth rate, as well as a loss of anthocyanins, tannins and phenylpropanoid-related coloration of the stem and fruit skin [33]. But the important role of _CHS_ in petal coloration level in different cultivars is still unknown.

In this current study, we evaluated the expression of _McCHS_, associated downstream genes and several MYB TFs in three different _Malus_ crabapples cultivars, and determined that _McCHS_ shows different expression patterns in various tissues and organs. A positive correlation is identified between the expression of _McCHS_ and the accumulation of anthocyanins in petals, but not in other organs, indicated that the the petal coloration level is determined by the expression level of _McCHS_ specially, the more expression of _McCHS_ has more red color in crabapple petals. Meanwhile, the expression of _McCHS_ is responsible for the color fade process during petal expansion in color variation cultivar. The different expression patterns of this gene in accumulation during leaves, petals and fruits development in different cultivars, and different tissues/organs, which may be explained by diversity flavonoids compounds accumulation during leaves, petals and fruits development in different cultivars, and is regulated by various organ-specific transcription factors. In nectarine, _CHS_ expression was

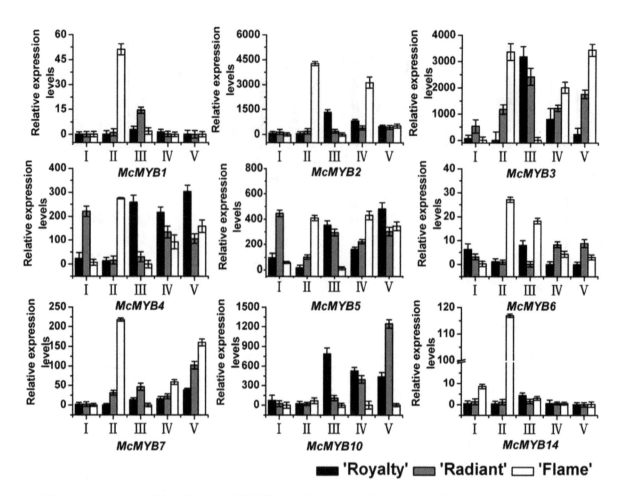

Figure 6. Relative expression profiles of potential MYB flavonoid activators during flower development. Real-time PCR was used to analyze the transcripts in petals of *Malus* cultivars 'Royalty' and 'Radiant', 'Flame', and all real time-PCR reactions were normalized using the Ct value corresponding to a *Malus* crabapple 18S ribosomal RNA gene (DQ341382). Stages referred to on the x axis are: (I) 6 days before full bloom; (II) 3 days before full bloom; (III) 1 day before full bloom; (IV) full bloom; and (V) 3 days after full bloom. Error bars correspond to the standard error of the mean ± SE of three replicate reactions.

reported that the expression in the skin was much higher than that in the fruit flesh [36]. The same result was found in apple, the expression of *CHS* was much higher in the fruit skin than in flesh and green leaves [43]. To sum up, *CHS* expression was higher in red tissues/organs than in colorless tissues/organs. Several reports showed that *ANS* [44–46] and *UFGT* [44,47–49] were two key anthocyanin biosynthetic pathway genes. A positive correlation between the expression of *McCHS* and its downstream genes *McANS, McUFGT* was observed. So this is additional evidence that *McCHS* gene is involved in anthocyanin biosynthesis. Furthermore, the expression patterns of the downstream genes *McF3H, McF3'H* and *McDFR* shows similar pattern to *McCHS* in 'Radiant' and 'Flame' petals. This may suggest that the transcript abundance of *McCHS* may influence the expression of other anthocyanin biosynthetic genes, and thus regulate the production of anthocyanin compounds, or they may be regulated by the same TFs.

NtAn1a, NtAn1b, NtTTG1, NtTTG2, MYB305, NtMYC2a and *NtMYC2b* have been proved to play an important role in tobacco anthocyanin biosynthesis pathway [50–55]. In our results, overexpressed *McCHS* can alter the expression of endogenous tobacco anthocyanin regulation factors. So we presume that massive expression of *McCHS* may be as a feedback signal to

activate or inhibit the anthocyanin-related transcription factors, resulting in enhanced anthocyanins accumulation in tobacco.

The expression of *CHS* is not only responsible for the anthocyanins biosynthesis, but also for the flavonol and other flavonoids compounds biosynthesis in plants. So the transcription level of *McCHS* is different in various crabapple cultivars, and may be regulated by several MYB transcription factors. We identify a close relationship between *McCHS* and *McMYB4*, *McMYB5*. The three cultivars show individual patterns of correlated expression and some of the results support the conclusions resulting from previously published data. MYB4 functions as a transcription repressor which involved in the regulation of diversity secondary metabolism, such as anthocyanins, lignin, flavonol, proanthocyanidin, and as a target gene that the transcript level of *CHS* was regulated by MYB4 in *A. thaliana*, turnip, Kiwifruit, *Pinus taeda* and so on [56–60]. The transcript levels of *MYB5* and all proanthocyanidin-specific genes were previously shown to be down-regulated, while anthocyanin-specific gene expression increased, leading to a switch from proanthocyanidin biosynthesis to anthocyanin biosynthesis in the mature phase of grape berries [61–64]. Transient expression assays have shown that VvMYB5a and VvMYB5b could activate several grapevine flavonoid pathway genes and affect the

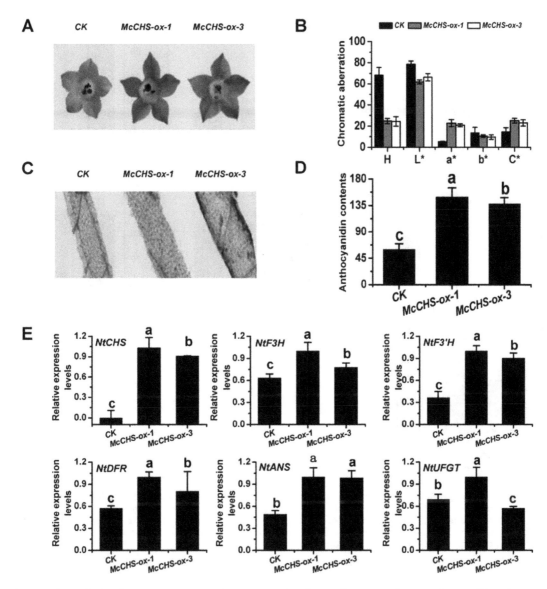

Figure 7. Phenotypic analysis of *McCHS* overexpressing transgenic tobacco flowers and expression profiles of target genes. (A) Typical flower phenotypes of control lines (*CK*) and *McCHS-ox-1 and McCHS-ox-3* tobacco plants overexpressing *McCHS*. (B) Petal color of control plants (*CK*) and *McCHS-ox-1* and *McCHS-ox-3*. (C) Microscopic observation of the transgenic tobacco petals. (D) Content of total anthocyanin in petals of control lines (*CK*) and *McCHS-ox-1* and *McCHS-ox-3*. (E) Relative expression profiles of endogenous anthocyanin biosynthesis genes in transgenic tobacco flowers. A spectrophotometric colorimeter was used to measure color changes in petals and HPLC was used to analyze the total anthocyanin content in the petals of transgenic tobacco. Real-time PCR was used to assess the expression of target genes (*NtCHS, NtF3H, NtF3'H, NtDFR, NtANS, NtUFGT*) in *McCHS-ox* tobacco plants. *CK* refers to wild type tobacco. All real time-PCR reactions were normalized using the Ct value corresponding to the *NtActin* gene (GQ339768). Error bars correspond to the standard error of the mean ± SE of three replicate reactions. Analysis of relative expression levels of control and *McCHS-ox* lines and different letters above the bars indicate significantly different values (*P*<0.05) calculated using one-way analysis of variance (ANOVA) followed by a Duncan's multiple range test.

biosynthesis of anthocyanins, flavonols, tannins and lignins in reproductive organs when overexpressed in tobacco (*Nicotiana tabacum*), including *CHS* gene [65–66]. Meanwhile, the promoter of *McCHS* has several MYB-binding cis-elements and maybe regulated by MYB TFs [67]. All of these informations support our observation in crabapple that *McCHS* can be regulated by these transcription factors in different crabapple cultivars due to the various flavonoids compounds accumulation (Figure 2E, Figure 3C and Figure 6).

To confirm that the accumulation of flavonoids was influenced by expression of the *McCHS* gene, we over-expressed *McCHS* under the control of a constitutive CaMV35S promoter in tobacco

and observed an increase in anthocyanins accumulation, as well as increased expression of downstream endogenous tobacco anthocyanin biosynthesis genes. This may indicate that massive expression of *McCHS* may regulate the downstream genes to promote anthocyanin accumulation. It has also been shown that overexpression of *CHS* can lead to an accumulation of flavonoids. Such as, overexpression of a *CHS* gene in *Linum usitatissimum*, tomato and potato (*Solanum tuberosum*) resulted in an increase in total phenolic compounds in *Linum usitatissimum* [68], an accumulation of naringenin in the tomato fruit flesh and an accumulation of anthocyanins in potato tubers, respectively [69]. Besides overexpression, silencing of *CHS* gene expression in

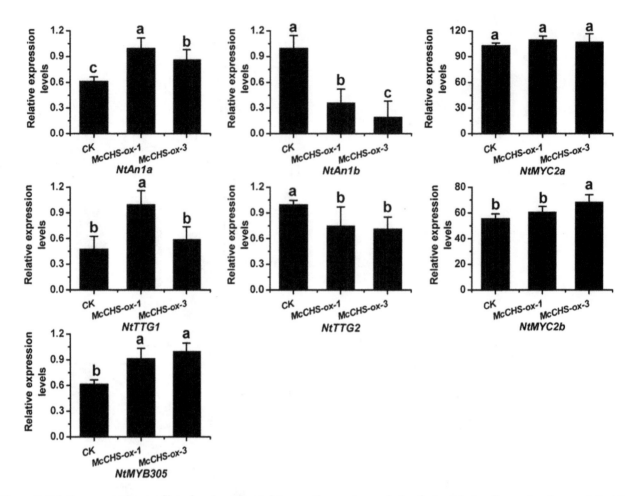

Figure 8. Relative expression profiles of endogenous tobacco anthocyanin regulation factors during flower development. Real-time PCR was used to analyze the transcripts in transgenic tobacco flowers, and all real time-PCR reactions were normalized using the Ct value corresponding to the *NtActin* gene (GQ339768). Error bars correspond to the standard error of the mean ± SE of three replicate reactions. Analysis of relative expression levels of control and *McCHS-ox* lines and different letters above the bars indicate significantly different values ($P<0.05$) calculated using one-way analysis of variance (ANOVA) followed by a Duncan's multiple range test.

several plants has been shown to result in reduced anthocyanin levels in flowers. *CHS* gene silencing in tobacco [70], chrysanthemum [71], *P. hybrida* [17,72], carnation [73], Gutterson rose [73], lisianthus [74] and gentian [75] resulted in plants with very pale pink, or entirely white flowers due to reduced anthocyanin levels, which again is congruent with the results reported here for *Malus* crabapple.

The expression of *CHS* is regulated by several factors, such as environmental conditions, the variation in the conditions in plants (such as pH), transcription factors and so on [12,76–79]. Recently, the report showed that the different methylation levels of *OgCHS* promoter altered the expression of this gene in different *Oncidium* orchid cultivars [19]. Therefore, epigenetic modification is a new sight that it can affect the anthocyanin biosynthetic gene expression. Future study is required on the possible regulation mechanism controlling *McCHS* expression in crabapple.

Conclusions

In this study, we have shown evidences that the expression level of *McCHS* is consistent with the variation of petal color and anthocyanins accumulation in different crabapple cultivars, and the expression intensity of *McCHS* determined the petal pigmentation level in crabapple specially. On the other hand,

the expression of *McCHS* is responsible for the color fade process during petal expansion in color variation cultivar. The regulation mechanism of *McCHS* expression is complicated, and several transcription factors might be involved in regulating *McCHS* expression in crabapple petals. Key future questions that need to be addressed include but not limited to the transcription regulation and epigenetic modification of *McCHS* expression. Undoubtedly the work described in this report will trigger a series of exciting future research projects to elucidate the mechanisms governing plant coloration.

Materials and Methods

Plant materials

The petals of *Malus* crabapple *cv*. 'Royalty', 'Radiant' and 'Flame' were collected at different development stages (I, six days before full bloom; II, three days before full bloom; III, one day before full bloom; IV, full bloom V, three days after full bloom). These trees were grown in the Crabapple germplasm of Beijing University of Agriculture (Changping District, Beijing, China) received standard horticultural practices, and disease and insect control. Control and T2 transgenic tobacco (*Nicotiana benthamiana*) plants were grown in a greenhouse. Flowers were collected for studies at the full-bloom stage (IV). The leaves of *Malus* crabapple *cv.* 'Royalty', 'Radiant' and 'Flame' were collected at the same time with petals at the full bloom stage. Fruits were collected at their fully mature stage in October 2012. At sampling time all the plant materials were immediately frozen in liquid nitrogen and stored at -80°C until RNA or phenolic compounds were extracted.

Color analysis

For flower color evaluation, color components of the CIE *L**a*b** coordinate, namely lightness and hue, were measured immediately after flowers were picked with a hand spectrophotometer (Konica Minolta CR-400, Minolta, Japan, Tokyo, Japan). The 'L*' value represents brightness and darkness, the 'a*' value represents greeness and redness as the value increases from negative to positive, and 'b*' represents blueness and yellowness. The 'C*' value represents chroma, calculated according to C* = $1/2(a*2+b*2)$. The hue angle (H) was calculated according to the following equation: H = arctan (b*/a*) [80]. Three areas of the petal adaxial surface were subjected to color measurement. Values were obtained and averaged from three replicate petals at 5 stages from three different flowers. For leaf color evaluation, the Chlorophyll contents of leaves on the tree in vivo were measured by Chlorophyll Content Meter (CL-01, Hansatech, UK). The same part of leaf adaxial surface was subjected to color measurement. Five Phyllotaxy was measured according to the order of the leaves, respectively. Values were obtained and averaged from three replicate leaves of five Phyllotaxy from three different cultivars.

Identification and quantification of anthocyanins and flavonols

Pigments were quantified by high-performance liquid chromatography (HPLC). One hundred milligrams of petals were homogenized with a mortar and pestle in liquid nitrogen, then 1 ml of extraction solution (methanol: water: formic acid: trifluoroacetic acid [70:27:2:1, v/v]) was added and the mixture was stored overnight at 4°C in the dark. The mixture was then centrifuged at 4°C at 12,000×g for 15 min, and the supernatants were filtered through a 0.22 μm Millipore filter (Billerica, MA, USA). Anthocyanins were identified and quantified using an HPLC1100-DAD system (Agilent Technologies, Waldbronn, Germany). Detection was performed at 520 nm for anthocyanins, 350 nm for flavonoids. A NUCLEODURH C18 column (250 mm64.66 mm) (Pretech Instruments, Sollentuna, Sweden) operating at 25°C was used for separation and compounds were eluted in a mobile phase consisting of solvent A, (trifluoroacetic acid: formic acid: water [0.1:2:97.9]) and solvent B (trifluoroacetic acid: formic acid: acetonitrile: water [0.1:2:35:62.9]) at a flow rate of 0.8 ml min^{-1}. The elution program followed the procedure described by Wu and Prior [81] with some modifications. Solvent B was initially 30% and increased linearly in steps to 35% at 5 min, 40% at 10 min, 50% at 30 min, 55% at 50 min, 60% at 70 min, 30% at 80 min. HPLC analysis was performed as described in Ohno [82].

RNA extraction and Quantitative real-time PCR analysis

Total RNA was extracted from flowers, leaves and fruit flesh and pericarp using an RNA Extract Kit (Aidlab, Beijing, China), according to the manufacturer's instructions. DNase I (TaKara, Japan) was added to remove genomic DNA and the samples were subjected to cDNA synthesis using the Access RT-PCR System (Promega, USA) according to the manufacturer's instructions. The expression levels of *McCHS*, *McF3H*, *McF3'H*, *McDFR*, *McANS*, *McUFGT*, *McMYB1*, *McMYB2*, *McMYB3*, *McMYB4*, *McMYB5*, *McMYB6*, *McMYB7*, *McMYB10*, *McMYB14* were analyzed using quantitative real-time PCR (qRT-PCR) with SYBR Green qPCR Mix (Takara, Japan) and a Bio-Rad CFX96 Real-Time PCR System (BIO-RAD, USA), according to the manufacturers' instructions. The expression levels of *NtCHS*, *NtF3H*, *NtF3'H*, *NtDFR*, *NtANS* and *NtUFGT* were analyzed using the same techniques. Primers are designed by NCBI Primer BLAST and listed in Table S1. *Malus* ribosomal 18S rRNA gene (DQ341382) was used as a reference gene for the target genes *McCHS*, *McF3H*, *McF3'H*, *McDFR*, *McANS*, and *McUFGT*, *McMYB1*, *McMYB2*, *McMYB3*, *McMYB4*, *McMYB5*, *McMYB6*, *McMYB7*, *McMYB10*, *McMYB14*; *NtActin* (GQ339768) was used as a reference gene for the target genes *NtCHS*, *NtF3H*, *NtF3'H*, *NtDFR*, *NtANS*, *NtUFGT*. The resulting cDNA samples for reverse transcription were serially diluted (1, 1/10, 1/100, 1/1000, 1/10000). Real-time RT-PCR analysis was carried out in a total volume of 20 μl, containing 9 μl of 2×SYBR Green qPCR Mix (Takara, Japan), 0.1 μM of each specific primer, and 100 ng of template cDNA. The reaction mixtures were heated to 95°C for 30 s, followed by 39 cycles at 95°C for 10 s, 59°C for 15 s, and 72°C for 30 s. A melting curve was generated for each sample at the end of each run to ensure the purity of the amplified products. Standard dilution curves were performed for each gene fragment, and all data were normalized to the level of the reference gene transcript. Primers for real-time experiments were designed using primer premier v5.0 software with forward and reverse primers corresponding to two different exons.

Phylogeny and sequence alignment

Protein consensus sequences were determined from the coding sequences of *CHS* and aligned with translated genome reference sequences and published CHS protein sequences from other species with DNAMAN 5.2.2. A phylogenetic tree was produced using MEGA version 5.10 [83] based on the coding sequence alignment of chalcone synthase-like genes, using a minimum evolution phylogeny test and 1,000 bootstrap replicates.

Expression vector construction and tobacco transformation

The entire *McCHS* coding sequence was amplified by PCR with the *McCHS-F* and *McCHS-R* primers, using cDNA from 'Royalty' petals as a template. The primers for *McCHS* amplification contained *Spe*I and *Kpn*I restriction enzyme sites, which were used to clone the coding region of the *McCHS* gene into the pBI121 vector. This was then used to transform *Agrobacterium* strain LBA4404 which in turn was used to transform tobacco (*N. benthamiana*) Wisconsin 38 using the leaf disk method [84]. Transgenic plants were selected based on kanamycin resistance. T2 progeny from the transgenic plants was used for further analysis and compared to wild-type non-transformed lines grown under the same conditions. All the primers used are listed in Table S1.

Statistical analysis

Statistical analysis and graphing was carried out using the OriginPro 8 statistical software (OriginLab Corporation, USA). Microsoft Office PowerPoint 2003 was used for artwork. Error bars represents mean ± SE of three replicate reactions.

Acknowledgments

We thank the Fruit Tree Key Laboratory at the Beijing University of Agriculture. We also thank the Beijing Key Laboratory for Agricultural Application and New Technique for providing experimental resources. We are grateful to all technicians in the BUA Crabapple Germplasm Resource Garden and thank PlantScribe (www.plantscribe.com) for carefully editing this paper.

Author Contributions

Conceived and designed the experiments: JT YY. Performed the experiments: DT JZ TS. Analyzed the data: JZ JT YY. Contributed reagents/materials/analysis tools: YY TS. Wrote the paper: DT JT YY.

References

1. Toni M . Kutchan (2005) A role for intra- and intercellular translocation in natural product biosynthesis. Curr Opin Plant Biol 8: 292–300.
2. Debeaujon I, Nesi N, Perez P, Devic M, Grandjean O, et al. (2003) Proanthocyanidin-accumulating cells in *Arabidopsis Testa*: Regulation of differentiation and role in seed development. Plant Cell 15: 2514–2531.
3. Price SF, Breen PJ, Valladao M, Watson BT (1995) Cluster sun exposure and quercetin in Pinot noir grapes and wine. Am J Enol Vitic 46: 187–194.
4. Gronquist M, Bezzerides A, Attygalle A, Meinwald J, Eisner M, et al. (2001) Attractive and defensive functions of the ultraviolet pigments of a flower (*Hypericum calycinum*). PNAS 98: 13745–13750.
5. Downey MO, Harvey JS, Robinson SP (2004) The effect of bunch shading on berry development and flavonoid accumulation in Shiraz grapes. Aust J Grape Wine R 10: 55–73.
6. Albert NW, Lewis DH, Zhang HB, Irving IJ, Jameson PE, et al. (2009) Light-induced vegetative anthocyanin pigmentation in petunia. J Exp Bot 60: 2191–2202.
7. Czemmel S, Stracke R, Weisshaar B, Cordon N, Harris NN, et al. (2009) The grapevine R2R3-MYB transcription factor VvMYBF1 regulates flavonol synthesis in developing grape berries. Plant Physiol 151: 1513–1530.
8. Punyasiri PA, Abeysinghe IS, Kumar V, Treutter D, Duy D, et al. (2004) Flavonoid biosynthesis in the tea plant *Camellia sinensis*: properties of enzymes of the prominent epicatechin and catechin pathways. Arch Biochem Biophys 431: 22–30.
9. Burbulis LE, Shirley BW (1999) Interactions among enzymes of the *Arabidopsis* flavonoid biosynthetic pathway. PNAS 96: 12929–12934.
10. Quattrocchio F, Wing J, Van der Woude K, Souer E, de Vetten N, et al. (1999) Molecular analysis of the *anthocyanin2* gene of petunia and its role in the evolution of flower color. Plant Cell 11: 1433–1444.
11. Winkel-Shirley B (2001) Flavonoid biosynthesis: A colorful model for genetics, biochemistry, cell biology, and biotechnology. Plant Physiol 126: 485–493.
12. Chagné D, Wang KL, Espley RV, Volz RK, How NM, et al. (2013) An ancient duplication of apple MYB transcription factors is responsible for novel red fruit-flesh phenotypes. Plant Physiol 161: 225–239.
13. Feng FJ, Li MJ, Ma FW, Cheng LL (2013) Phenylpropanoid metabolites and expression of key genes involved in anthocyanin biosynthesis in the shaded peel of apple fruit in response to sun exposure. Plant Physiol Bioch 69: 54–61.
14. Hichri L, Barrieu F, Bogs J, Kappel C, Delrot S, et al. (2011) Recent advances in the transcriptional regulation of the flavonoid biosynthetic pathway. J Exp Bot 62: 2465–2483.
15. Bai YH, Pattanaik S, Patra B, Werkman RJ, Xie HC, et al. (2011) Flavonoid-related basic helix-loop-helix regulators, NtAn1a and NtAn1b, of tobacco have originated from two ancestors and are functionally active. Planta 234: 363–375.
16. Zhao L, Gao LP, Wang HX, Chen XT, Wang YS, et al. (2013) The R2R3-MYB, bHLH, WD40, and related transcription factors in flavonoid biosynthesis. Funct Integr Genomic 13: 75–98.
17. Napoli C, Lemieux C, Jorgensen R (1990) Introduction of a chimeric chalcone synthase gene into petunia results in reversible co-suppression of homologous genes *in trans*. Plant Cell 2: 279–289.
18. Li SJ, Deng XM, Mao HZ, Hong Y (2005) Enhanced anthocyanin synthesis in foliage plant *caladium biocolor*. Plant Cell Rep 23: 716–720.
19. Liu XJ, Chuang YN, Chiou CY, Chin DC, Shen FQ, et al. (2012) Methylation effect on chalcone synthase gene expression determines anthocyanin pigmentation in floral tissues of two *Oncidium* orchid cultivars. Planta 236: 401–409.
20. Reimold U, Kroger M, Kreuzaler F, Hahlbrock K (1983) Coding and 3′ noncoding nucleotide sequence of chalcone synthase messenger RNA and assignment of amino acid sequence of the enzyme. EMBO J 2: 1801–1806.
21. Akada S, Kung SD, Dube SK (1991) The nucleotide sequence of gene 1 of the soybean chalcone synthase multigene family. Plant Mol Biol 16: 751–752.
22. Junghans H, Dalkin K, Dixon RA (1993) Stress responses in alfalfa (*Medicago sativa* L.). 15. Characterization and expression patterns of members of a subset of the chalcone synthase multigene family. Plant Mol Biol 22: 239–253.
23. Harker CL, Ellis THN, Coen ES (1990) Identification and genetic regulation of the chalcone synthase multigene family in pea. Plant Cell 2: 185–194.
24. Feinbaum RL, Ausubet FM (1988) Transcriptional regulation of the *Arabidopsis thaliana* chalcone synthase gene. Mol Cell Biol 8: 1985–1992.
25. Rohde W, Dorr S, Salamini F, Becker D (1991) Structure of a chalcone synthase gene from *Hordeum vulgare*. Plant Mol Biol 16: 1103–1106.
26. Franken P, Klosgen UN, Weydemann U, Drouard LM, Saedler H, et al. (1991) The duplicated chalcone synthase genes C2 and *Whp* (*white pollen*) of *Zea mays* are independently regulated; evidence for translationalcontrol of *Whp* expression by the anthocyanin intensifying gene in. EMBO J 10: 2605–2612.
27. Sparvoli F, Martin C, Scienza A, Gavazzi G, Tonelli C (1994) Cloning and molecular analysis of structural genes involved in flavonoid and stilbene biosynthesis in grape (*Vitis vinifera* L.) Plant Mol Bio 24: 743–755.
28. Goff SA, Ricke D, Lan TH, Presting G, Wang RL, et al. (2002) A draft sequence of the rice genome (*Oryza sativa* L. ssp. *japonica*). Science 296: 92–100.
29. Beerhues L, Wiermann R (1988) Chalcone synthases from spinach (*Spinacia oleracea* L.). Planta 173: 544–553.
30. Schroder J (1997) A family of plant-specific polyketide synthases: facts and predictions. Trends Plant Sci 10: 373–378.
31. Blokland R, Geest N, Mol JNM, Kooter JM (1994) Transgene-mediated suppression of chalcone synthase expression in *Petunia hybrida* results from an increase in RNA turnover. Plant J 6: 861–877.
32. Lunkenbein S, Coiner H, De Vos CHR, Schaart JG, Boone MJ, et al. (2006) Molecular characterization of a stable antisense chalcone synthase phenotype in strawberry (*Fragaria ×ananassa*). J Agr Food Chem 54: 2145–2153.
33. Dare AP, Tomes S, Jones M, McGhie TK, Stevenson DE, et al. (2013) Phenotypic changes associated with RNA interference silencing of chalcone synthase in apple (*Malus ×domestica*). Plant J 74: 398–410.
34. Wang YS, Gao LP, Shan Y, Liu YJ, Tian YW, et al. (2012) Influence of shade on flavonoid biosynthesis in tea (*Camellia sinensis* (L.) O. Kuntze). Sci Hortic 141: 7–16.
35. Shen HX, Zhang J, Yao YC, Tian J, Song TT, et al. (2012) Isolation and expression of McF3H gene in the leaves of crabapple. Acta Physiol Plant 34: 1353–1361.
36. Ravaglia D, Espley RV, Henry-Kirk RA, Andreotti C, Ziosi V, et al. (2013) Transcriptional regulation of flavonoid biosynthesis in nectarine (*Prunus persica*) by a set of R2R3 MYB transcription factors. BMC Plant Biol 68: 1471–1485.
37. Grotewold E (2006) The genetics and biochemistry of floral pigments. Annu Rev Plant Biol, 57: 761–780.
38. Czemmel S, Heppel SC, Bogs J (2012) R2R3 MYB transcription factors: key regulators of the flavonoid biosynthetic pathway in grapevine. Protoplasma 249: 109–118.
39. Kuckuck H (1936) Uber vier neue Serien multipler Allele bei *Antirrhinum majus*. Mol Genet Genomic 71: 429–440.
40. MO YY, Nagel C, Taylor LP (1992) Biochemical complementation of chalcone synthase mutants defines a role for flavonols in functional pollen. PNAS 89: 7213–7217.
41. Shirley BW, Kubasek WL, Storz G, Bruggemann E, Koornneef M, et al. (1995) Analysis of *Arabidopsis* mutants deficient in flavonoid biosynthesis. Plant J 8: 659–671.
42. Schijlen EGWM, De Vos CHR, Martens S, Jonker HH, Rosin FM, et al. (2007) RNA interference silencing of chalcone synthase, the first step in the flavonoid biosynthesis pathway, leads to parthenocarpic tomato fruits. Plant Physiol 144: 1520–1530.
43. Espley RV, Hellens RP, Putterill J, Stevenson DE, Ammal SK, et al. (2007) Red colouration in apple fruit is due to the activity of the MYB transcription factor, MdMYB10. Plant J 2007 49: 414–427.
44. Wei YZ, Hu FC, Hu GB, Li XJ, Huang XM, et al. (2011) Differential expression of anthocyanin biosynthetic genes in relation to anthocyanin accumulation in the pericarp of *Litchi Chinensis* Sonn. PLoS One 6: e19455.
45. Zhang B, Hu Z, Zhang Y, Li Y, Zhou S, et al. (2012) A putative functional MYB transcription factor induced by low temperature regulates anthocyanin biosynthesis in purple kale (*Brassica Oleracea* var. *acephala* f. tricolor). Plant Cell Rep 31: 281–289.

46. Debes MA, Arias ME, Grellet-Bournonville CF, Wulff AF, Martínez-Zamora MG, et al. (2011) White-fruited *Duchesnea indica* (*Rosaceae*) is impaired in *ANS* gene expression. Am J Bot 98: 2077–2083.

47. Kobayashi S, Lshimaru M, Ding CK, Yakushiji H, Goto N (2001) Comparison of UDP-glucose: Flavonoid 3-O-glucosyltransferase (UFGT) gene sequences between white grapes (*Vitsi vinifera*) and their spots with red skin. Plant Sci J 160: 543–550.

48. Kobayashi S, Ishimaru M, Hiraoka K, Honda C (2002) Myb-related genes of the Kyoho grape (*Vitis labruscana*) regulate anthocyanin biosynthesis. Planta 215: 924–933.

49. Walker AR, Lee E, Bogs J, Debra AJ, David M, et al. (2007) White grapes arose through the mutation of two similar and adjacent regulatory genes. Plant J 49: 772–785.

50. Bai YH, Pattanaik S, Patra B, Werkman JR, Xie CH, et al. (2011) Flavonoid-related basic helix-loop-helix regulators, NtAn1a and NtAn1b, of tobacco have originated from two ancestors and are functionally active. Planta 234: 363–375.

51. Wang Y, Liu R, Chen L, Wang Y, Liang Y, et al. (2009) Nicotiana tabacum TTG1 contributes to ParA1-induced signalling and cell death in leaf trichomes. J Cell Sci 122: 2673–2685.

52. Li B, Gao R, Cui R, Lu B, Li X, et al. (2012) Tobacco TTG2 suppresses resistance to pathogens by sequestering NPR1 from the nucleus. J Cell Sci 125: 4913–4922.

53. Zhu Q, Li B, Mu S, Han B, Cui R, et al. (2013) TTG2-regulated development is related to expression of putative auxin response factor genes in tobacco. BMC Genomics 14: 806–821.

54. Wang WJ, Liu GS, Niu HX, Timko MP, Zhang HB (2014) The F-box protein COI1 functions upstream of MYB305 to regulate primary carbohydrate metabolism in tobacco (*Nicotiana tabacum* L. cv. TN90). J Exp Bot 65: 2147–2160.

55. Zhang HB, Bokowieca MT, Rushton PJ, Han SC, Timko MP (2011) NtMYC2b Form Nuclear Complexes with the NtJAZ1 Repressor and Regulate Multiple Jasmonate-Inducible Steps in Nicotine Biosynthesis. Mol Plant 5: 73–84.

56. Jin HL, Cominelli E, Bailey P, Parr A, Mehrtens F, et al. (2000) Transcriptional repression by AtMYB4 controls production of UV-protecting sunscreens in *Arabidopsis*. EMBO J 22: 6150–6161.

57. Schenke D, Bottcher C, Scheel D (2011) Crosstalk between abiotic ultraviolet-B stress and biotic (flg22) stress signalling in *Arabidopsis* prevents flavonol accumulation in favor of pathogen defence compound production. Plant cell environ 11: 1849–1864.

58. Patzlaff A, McInnis S, Courtenay A, Surman C, Newman LJ, et al. (2003) Characterisation of a pine MYB that regulates lignification. Plant J 36: 743–754.

59. Akagi T, Ikegami A, Tsujimoto T, Kobayashi S, Sato A, et al. (2009) DkMyb4 is a MYB transcription factor involved in proanthocyanidin biosynthesis in persimmon fruit. Plant Physiol 151: 2028–2045.

60. Wang Y, Zhou B, Sun M, Li YH, Kawabata S (2012) UV-A light induces anthocyanin biosynthesis in a manner distinct from synergistic Blue + UV-B light and UV-A/Blue light responses in different parts of the hypocotyls in turnip seedlings. Plant Cell Physiol 53: 1470–1480.

61. Yamamoto NG, Wan GH, Masaki K, Kobayashi S (2002) Structure and transcription of three chalcone synthase genes of grapevine (*Vitis vinifera*). Plant Sci 162: 867–872.

62. Bogs J, Downey MO, Harvey JS, Ashton AR, Tanner GJ, et al. (2005) Proanthocyanidin synthesis and expression of genes encoding leucoanthocyanidin reductase and anthocyanidin reductase in developing grape berries and grapevine leaves. Plant Physiol 139: 652–663.

63. Bogs J, Jaffé FW, Takos AM, Walker AR, Robinson SP (2007) The grapevine transcription factor VvMYBPA1 regulates proanthocyanidin synthesis during fruit development. Plant Physiol 143: 1347–1361.

64. Cutanda-Perez MC, Ageorges A, Gomez C, Vialet S, Terrier N, et al. (2009) Ectopic expression of *VlmybA1* in grapevine activates a narrow set of genes involved in anthocyanin synthesis and transport. Plant Mol Biol 69: 633–648.

65. Deluc L, Barrieu F, Marchive C, Lauvergeat V, Decendit A, et al. (2006) Characterization of a grapevine R2R3-MYB transcription factor that regulates the phenylpropanoid pathway. Plant Physiol 140: 499–511.

66. Deluc L, Bogs J, Walker AR, Ferrier T, Decendit A, et al. (2008) The Transcription factor VvMYB5b contributes to the regulation of anthocyanin and proanthocyanidin biosynthesis in developing grape berries. Plant Physiol 147: 2041–2053.

67. Tian J, Shen HX, Zhang J, Song T, Yao YC (2011) Characteristics of chalcone synthase promoters from different leaf-color *Malus* crabapple cultivars. Sci Hortic 129: 449–458.

68. Davies KM (2007) Genetic modification of plant metabolism for human health benefits. Mutat Res-fund Mol M 622: 122–137.

69. Schijlen EGWM, De Vos CHR, Tunen AJV, Bovy AG (2004) Modification of flavonoid biosynthesis in crop plants. Phytochemistry 65: 2631–2648

70. Wang CK, Chen PY, Wang HM, To KY (2006) Cosuppression of tobacco chalcone synthase using Petunia chalcone synthase construct results in white flowers. Bot Stud 47: 71–82.

71. Courtney-Gutterson N, Napoli C, Lemieux C, Morgan A, Firoozabady E, et al. (1994) Modification of flower color in florist's chrysanthemum: Production of a white-flowering variety through molecular genetics. Nat Biotechnol 12: 268–271.

72. Van der Krol AR, Mur LA, Beld M, Mol JNM, Stuitje AR (1990) Flavonoid genes in petunia: addition of a limited number of gene copies may lead to a suppression of gene expression. Plant Cell 2: 291–299.

73. Gutterson N (1995) Anthocyanin biosynthetic genes and their application to flower color modification through sense suppression. Hortscience 30: 964–966.

74. Deroles SC, Bradley JM, Schwinn KE, Markham KR, Bloor S, et al. (1998) An antisense chalcone synthase cDNA leads to novel colour patterns in lisianthus (*Eustoma grandiflorum*) flowers. Mol Breeding 4: 59–66.

75. Nakatsuka T, Mishiba K, Abe Y, Kubota A, Kakizaki Y, et al. (2008) Flower color modification of gentian plants by RNAi-mediated gene silencing. Plant biotechnol J 25: 61–68.

76. Wang K, Micheletti D, Palmer J, VOLZ R, Lozano L, et al. (2011) High temperature reduces apple fruit color via modulation of the anthocyanin regulatory complex. Plant Cell Environ 34: 1176–1190.

77. Crifò T, Petrone G, Lo Cicero L, Lo Piero AR (2012) Short cold storage enhances the anthocyanin contents and level of transcripts related to their biosynthesis in blood oranges. J Agric Food Chem 60: 476–481.

78. Lo Piero AR, Puglisi I, Rapisarda P, Petrone G (2005) Anthocyanins accumulation and related gene expression in red orange fruit induced by low temperature storage. J Agric Food Chem 53: 9083–9088.

79. Zhang Y, Zhang J, Song T, Li J, Tian J, et al. (2014) Low medium pH value enhances anthocyanin accumulation in *Malus* crabapple leaves. PLoS One 9: e97904.

80. McGuire RG (1992) Reporting of objective colour measurements. Hortscience 27: 1254–1255.

81. Wu X, Prior RL (2005) Identification and characterization of anthocyanins by high-performance liquid chromatography-electrospray ionization-tandem mass spectrometry in common foods in the United States: Vegetables, nuts, and grains. J Agr Food Chem 53: 3101–3113.

82. Ohno S, Hosokawa M, Hoshino A, Kitamura Y, Morita Y, et al. (2011) A bHLH transcription factor, DvIVS, is involved in regulation of anthocyanin synthesis in dahlia (*Dahlia variabilis*). J Exp Bot 62: 5105–5116.

83. Tamura K, Peterson D, Peterson N, Stecher G, Nei M, et al. (2011) MEGA5, molecular evolutionary genetics analysis using maximum likelihood, evolutionary distance, and maximum parsimony methods. Mol Biol Evol 28: 2731–2739.

84. Horsch RB, Fry J, Hoffmann N, Neidermeyer J, Rogers SG, et al. (1989) Leaf disc transformation. Plant Mol Biol 5: 63–71.

4

In COS Cells Vpu Can Both Stabilize Tetherin Expression and Counteract Its Antiviral Activity

Abdul A. Waheed*, Nishani D. Kuruppu, Kathryn L. Felton, Darren D'Souza, Eric O. Freed

Virus-Cell Interaction Section, HIV Drug Resistance Program, NCI-Frederick, Frederick, Maryland, United States of America

Abstract

The interferon-inducible cellular protein tetherin (CD317/BST-2) inhibits the release of a broad range of enveloped viruses. The HIV-1 accessory protein Vpu enhances virus particle release by counteracting this host restriction factor. While the antagonism of human tetherin by Vpu has been associated with both proteasomal and lysosomal degradation, the link between Vpu-mediated tetherin degradation and the ability of Vpu to counteract the antiviral activity of tetherin remains poorly understood. Here, we show that human tetherin is expressed at low levels in African green monkey kidney (COS) cells. However, Vpu markedly increases tetherin expression in this cell line, apparently by sequestering it in an internal compartment that bears lysosomal markers. This stabilization of tetherin by Vpu requires the transmembrane sequence of human tetherin. Although Vpu stabilizes human tetherin in COS cells, it still counteracts the ability of tetherin to suppress virus release. The enhancement of virus release by Vpu in COS cells is associated with a modest reduction in cell-surface tetherin expression, even though the overall expression of tetherin is higher in the presence of Vpu. This study demonstrates that COS cells provide a model system in which Vpu-mediated enhancement of HIV-1 release is uncoupled from Vpu-mediated tetherin degradation.

Editor: Michael Schindler, Helmholtz Zentrum Muenchen - German Research Center for Environmental Health, Germany

Funding: Research in the Freed lab is supported by the Intramural Research Program of the NIH, National Cancer Institute, and by the Intramural AIDS targeted Antiviral Program. The funders had no role in study design, data collection and analysis, decision to publish, or preparation of the manuscript.

Competing Interests: The authors have declared that no competing interests exist.

* Email: waheedab@mail.nih.gov

Introduction

Mammalian cells have evolved a variety of strategies to prevent virus replication. These include constitutive or inducible expression of a number of restriction factors that interfere with different stages of the virus replication cycle. Many of these restriction factors are induced by type-I interferon (IFN) as a component of the innate immune system. Host cell restriction factors target the incoming virus, act at the level of transcription, or disrupt late stages of the replication cycle. Tetherin was identified as an IFNα-inducible restriction factor that tethers mature viral particles to the infected cell surface [1,2]. While the physiological function of tetherin is not clearly understood, it is expressed constitutively in terminally differentiated B cells, monocytes, primary bone marrow stromal cells, and plasmacytoid dendritic cells [3–6]. Tetherin is a type-II integral membrane protein with an unusual topology: it bears an N-terminal cytoplasmic domain, followed by a trans-membrane (TM) domain, a coiled-coil, and a putative C-terminal glycosylphosphatidylinositol (GPI) anchor [7]. Membrane anchors at both N- and C-terminal regions of tetherin are required for antiviral activity [1,8,9]. Tetherin restricts the release of a broad spectrum of enveloped viruses, including not only HIV-1 but also other retroviruses, filoviruses, arenaviruses, and herpesviruses [10–12].

Lentiviruses have developed several distinct strategies for evading the antiviral activity of tetherin. HIV-2 and some strains of simian immunodeficiency virus (e.g., SIVtan) express an Env glycoprotein that acts as a tetherin antagonist by inducing its sequestration in an intracellular compartment that bears markers for the trans-Golgi network (TGN) [13–15]. Serra-Moreno et al. reported that a Nef-deleted strain of SIV adapts to overcome rhesus tetherin by acquiring changes in the cytoplasmic tail of Env [16]. Other strains of SIV antagonize simian but not human tetherin through their Nef proteins [17–19]. The herpes simplex virus 1 (HSV-1) glycoprotein M, the Env proteins of equine infectious anemia virus (EIAV), human endogenous retrovirus type K (HERV-K), and feline immunodeficiency virus (FIV), and the chikungunya virus non-structural protein 1 (nsP1) antagonize tetherin restriction [20–24]. HIV-1 Vpu counteracts human, chimpanzee, and gorilla tetherin but is relatively inactive against tetherin from other non-human primates or from non-primate species (e.g., the mouse) [17,18,25–27]. Nonetheless, Shengai et al. reported that Vpu from some simian-human immunodeficiency virus (SHIV) chimeras is capable of antagonizing macaque tetherin [28]. SIVcpz, the chimpanzee precursor to HIV-1 [29], encodes a Nef protein that is able to counteract chimpanzee but not human tetherin. Following transfer to humans, group M HIV-1 (the main pathogenic group of HIV-1 responsible for the AIDS epidemic) acquired the ability to antagonize human tetherin through its Vpu protein [18]. In contrast, Vpu from the less-pathogenic HIV-1 group O strains has limited capacity to downregulate tetherin [18]. Thus, the ability of Vpu to counteract human tetherin may have played a significant role in the current AIDS pandemic.

The mechanism by which HIV-1 Vpu counteracts the antiviral activity of human tetherin remains to be elucidated (for reviews, [30,31]). A number of diverse findings have been reported. Upon Vpu expression, previous studies have observed reduced cell-surface tetherin levels with no effect on overall expression [2,32] or a reduction in overall tetherin levels via lysosomal [33,34] or proteasomal [27,35,36] degradation pathways. Immuno-electron microscopy analysis indicated that Vpu shifts the localization of tetherin from the plasma membrane (PM) to early and recycling endosomes [37]. Several studies have demonstrated Vpu-mediated antagonism of tetherin in the absence of significantly reduced expression [1,38,39]. It has also been reported that Vpu induces the sequestration of tetherin in a perinuclear compartment, consistent with Vpu interfering with the trafficking of tetherin from the TGN to the PM [40]. Vpu has been reported to disrupt both the transport of newly synthesized tetherin to the PM and the recycling of internalized tetherin back to the cell surface without affecting rates of tetherin internalization [40–42]. Rollason et al. reported that Vpu translocates endogenous tetherin out of lipid rafts and induces its internalization into endosomes and degradation in lysosomes [43].

While the main function of tetherin is to restrict HIV-1 release, it may also affect cell-cell virus transfer and particle infectivity. Data have been published supporting a role for tetherin in enhancing [44] or restricting [45–47] cell-cell spread of HIV-1. It has also been reported that tetherin expression can impair specific HIV-1 particle infectivity [48], whereas others observed a similar reduction in particle release and infectivity for Vpu(−) compared with Vpu(+) virus produced in (tetherin-expressing) HeLa cells [37]. Recently, it was reported that anti-tetherin activity of Vpu protects HIV-1-infected cells from antibody-dependent cell-mediated cytotoxicity (ADCC) [49–51].

In this study, we show that Vpu-mediated degradation of tetherin occurs by both proteasomal and lysosomal pathways, and that the absence of Vpu antagonism of non-human tetherin is associated with a lack of tetherin degradation. We observe that human tetherin is poorly expressed in agm kidney (COS) cells and that Vpu markedly enhances expression of human tetherin in this cell line. The stabilization of tetherin by Vpu is specific to human tetherin, and requires the tetherin TM sequence implicated in Vpu–tetherin binding. Although the cellular expression of tetherin is enhanced by Vpu in COS cells, its surface expression and virus-restricting capacity are reduced by Vpu. These observations confirm that Vpu-mediated degradation of tetherin and antagonism of tetherin restriction are two separable functions of Vpu [38,39] and demonstrate that the consequences of the Vpu–tetherin interaction are strongly cell type dependent.

Materials and Methods

Plasmids, antibodies, and chemicals

The full-length infectious HIV-1 molecular clone pNL4-3 and the Vpu-defective counterparts pNL4-3delVpu and pNL4-3Udel were described previously [53–55]. The codon-optimized plasmid pcDNA-Vphu was used for expressing the Vpu protein [56]. pNL4-3delVpu, pNL4-3Udel, and pcDNA-Vphu were kindly provided by K. Strebel. The HA epitope-tagged tetherin expression vector, the deletion mutant derivatives delCT and delGPI, and the mouse, rh, agm, and chimeric tetherin constructs [hu(rhTM)] or agm [hu(agmTM)] [1,25] were generously provided by P. Bieniasz. The fluorescently-tagged Gag expression plasmid containing a monomeric Eos fluorescent protein (mEosFP)

was constructed by inserting the mEosFP coding region between the MA and CA domains of Gag using a strategy similar to that reported by Hubner et al. to construct Gag-iGFP [57]. Briefly, a synthetic viral protease cleavage sequence, SQNYPIVQ, followed by a polylinker containing BstBI and PacI restriction sites was introduced between the MA and CA domains of the HIV-1 Gag expression vector pNL4-3ΔPolΔEnv, in which the *pol* and *env* genes are deleted. A large portion of *pol* (nt 2429–4377) was removed by restriction digestion with BciI and NsiI, and a portion of *env* (nt 6530–7611) was deleted by restriction enzymes NsiI and BgiII. The fragments were ligated after treating with T4 DNA polymerase to create blunt ends. The mEosFP coding region was amplified from the pmEosFP2-C1 plasmid [58] (a kind gift from J. Lippincott-Schwartz) flanked by AclI (5′) and BsiEI (3′) restriction sites. This fragment was inserted into the BstBI and PacI restriction sites of the synthetic cleavage sequence to generate pNL4-3Gag-imEosFPΔPolΔEnv. Anti-HA and anti-tubulin antibodies, and concanamycin A, were purchased from Sigma (St. Louis, MO). Anti-Vpu [59], anti-human tetherin [39], and anti-HIV-1 Ig were obtained from the NIH AIDS Research and Reference Reagent Program. Anti-TGN46 was purchased from AbD Serotec (MorphoSys), anti-CD63 from Santa Cruz Biotechnology, and anti-LAMP-1 from BD Biosciences. MG132 was obtained from A.G. Scientific Inc, N-acetyl-leu-leu-norleucinal (ALLN) from Calbiochem, and bafilomycin from Tocris Bioscience.

Cell culture, transfection and preparation of virus stocks

293T, COS, and Vero cells were obtained from American Type Culture Collection and maintained in Dulbecco's modified Eagle's medium (DMEM) containing 10% fetal bovine serum (FBS). One day after plating, the cells were transfected with the indicated plasmid DNA in the absence or presence of WT or mutant tetherin expression vectors using Lipofectamine 2000 (Invitrogen Corp. Carlsbad, CA) according to the manufacturer's recommendations. Cells and virus were harvested 24 h posttransfection.

Western blotting analysis

For immunoblot analyses, cells were washed with PBS, and then lysed in a buffer containing 50 mM Tris-HCl (pH 7.4), 150 mM NaCl, 1 mM EDTA, 0.5% Triton X-100, and protease inhibitor cocktail (Roche). Proteins were denatured by boiling in sample buffer and subjected to SDS-PAGE, transferred to PVDF membrane, and incubated with appropriate antibodies as described in the text. Membranes were then incubated with HRP-conjugated secondary antibodies, and chemiluminescence signal was detected by using Western Lightning Chemiluminescence Reagent Plus (PerkinElmer, Wellesley, MA). Quantification of the intensities of the protein bands was performed by using an Alpha Innotech Fluorchem SP imager (Alpha Innotech Inc., CA).

Pulse-chase analysis of tetherin

293T and COS cells were transfected with the vector expressing HA-tagged human tetherin in the absence and presence of pcDNA-Vphu. One day post-transfections cells were pulse-labeled for 30 min with [35S]Met-Cys, after which the labeling medium was replaced with DMEM containing 10% FBS and cultured for 1, 2, or 4 h. Cells were harvested at each time point and solubilized in NP40 containing lysis buffer and immunoprecipitated with anti-HA antibodies and analyzed by SDS-PAGE followed by fluorography [60].

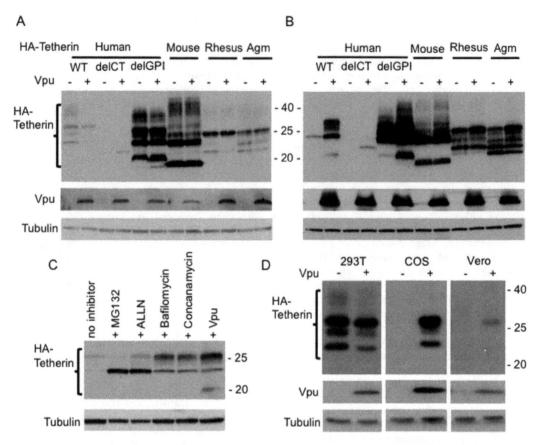

Figure 1. Vpu induces the degradation of human tetherin in 293T cells but stabilizes its expression in COS cells. 293T (A) or COS (B) cells were transfected with vectors expressing HA-tagged human, mouse, rhesus, and agm tetherins without or with Vpu expression vector at a tetherin:Vpu DNA ratio of 1:5. Total transfected DNA was held constant with empty vector. Truncated human tetherins that lack the cytoplasmic tail (delCT) or putative GPI anchor (delGPI) were also used to identify the regions of tetherin required for Vpu-mediated degradation. One day posttransfection, cells were lysed and subjected to western blot analysis with the indicated antibodies. Vpu decreased tetherin expression by 3.1 fold in 293T cells whereas it increased tetherin expression by ~5-fold in COS cells; levels of the delCT mutant were increased ~3-fold in COS cells and 1.6 fold in 293T cells by Vpu. (C) COS cells were transfected with HA-tagged human tetherin expression vector and cell lysates were subjected to western blot analysis with anti-HA or anti-tubulin antibodies. Vpu was also co-expressed to compare the pattern of tetherin expression. Molecular mass markers are shown on the right (in kDa). In this experiment, proteasomal inhibitors (MG132 and ALLN) increased the expression of the ~23 kDa tetherin species by 3.2 fold, whereas lysosomal inhibitors (bafilomycin and concanamycin) rescued the expression of the ~26 kDa tetherin species by 2.6 fold and the ~23 kDa tetherin species by 1.7 fold compared to the no inhibitor control. Co-expression of Vpu increased the expression of the ~26 kDa tetherin species by 5.2 fold and the ~23 kDa tetherin species by 1.7 fold. (D) 293T, COS, and Vero cells were transfected with HA-tagged human tetherin expression vector in the absence and presence of Vpu expression vector, and one-day posttransfection cells were lysed and subjected to Western blot analysis as above. Vpu reduced the expression of tetherin by 2.4 fold in 293T cells, whereas the levels were increased by ~20-fold in COS cells and ~4-fold in Vero cells. Similar results were obtained in an independent experiment.

Virus release assays

293T cells were cotransfected with pNL4-3 or pNL4-3delVpu and human or agm tetherin plasmids in the absence or presence pcDNA-Vphu. One day posttransfection, cells were metabolically labeled for 2 h with [35S]Met-Cys and virions were pelleted in an ultracentrifuge. Viral proteins in cell and virus lysates were immunoprecipitated with HIV-Ig and analyzed by SDS-PAGE followed by fluorography [60]. The virus release efficiency was calculated as the amount of virion-associated Gag as a fraction of total (cell plus virion-associated) Gag.

Flow cytometry

293T and COS cell lines were transfected with human tetherin in the absence or presence of Vpu. Twenty-four hours after transfection, cells were harvested by adding a solution of 1 mM EDTA in PBS and washed twice in ice-cold 1% BSA-PBS. The cells were then incubated with polyclonal anti-tetherin antiserum

or normal rabbit serum in 1% BSA-PBS for 1 h at 4°C. The cells were then washed three times in 1% BSA-PBS, and stained with Alexa-488-conjugated anti-rabbit IgG secondary antibody in 1% BSA-PBS for 1 h at 4°C. The cells were fixed in 1% paraformaldehyde after washing three times and analyzed with a Becton Dickinson FACS Calibur flow cytometer.

Immunofluorescence microscopy

For microscopy studies, 293T and COS cells cultured in chamber slides were transfected with pNL4-3Gag-imEosFPΔPol-ΔEnv and human tetherin expression vector in the absence or presence of Vpu plasmid. 12, 18, or 24 h posttransfection, cells were rinsed with PBS and fixed in 3.7% paraformaldehyde in PBS for 30 min. The cells were rinsed with PBS, permeabilized with methanol at −20°C for 4 min, washed in PBS and incubated with 0.1 M glycine-PBS for 10 min to quench the remaining aldehyde residues. After blocking with 3% BSA-PBS for 30 min, cells were

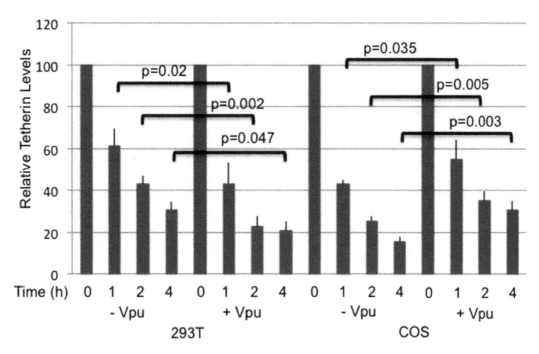

Figure 2. Pulse-chase analysis of tetherin. 293T and COS cells were transfected with HA-tagged human tetherin in the absence and presence of Vpu, 24 h post-transfection cells were labeled with [35S]Met/Cys for 30 min and then chased for 1, 2, or 4 h. Cells were lysed and immunoprecipitated with anti-HA antibodies and analyzed by SDS-PAGE followed by fluorography. Data shown are means ± SD from 4–6 independent experiments.

incubated with primary antibodies (specific for tetherin, HA, Vpu, TGN46, CD63, or LAMP-1) appropriately diluted in 3% BSA-PBS for 1 h. The cells were washed with PBS three times and then incubated with secondary antibody conjugated with either Texas Red or Alexa-488 diluted in 3% BSA-PBS. In co-staining experiments in which both antibodies were raised in mouse (HA,

LAMP-1, CD63), the primary antibodies were directly labeled with either Zenon Alexa Flour 488 or Zenon Alexa Flour 594 mouse IgG1 labeling kit (Invitrogen). After washing with PBS three times, cells were mounted with Vectashield mounting media with DAPI (Vector Laboratories) and examined with a Delta-Vision RT deconvolution microscope. Colocalization was quan-

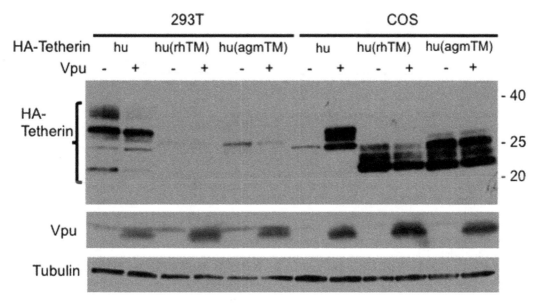

Figure 3. The TM domain of human tetherin is required for its rapid degradation in COS cells. 293T and COS cells were transfected with human (hu) or chimeric tetherins that carry the TM sequences from rh or agm tetherin [hu(rhTM) and hu(agmTM), respectively in the absence or presence of Vpu (1:5 DNA ratio). One day posttransfection, cell lysates were subjected to western blot analysis with the indicated antibodies. Vpu decreased tetherin expression by 3.0-fold in 293T cells, whereas the levels were increased by 4.5-fold in COS cells. Molecular mass markers are shown on the right (in kDa).

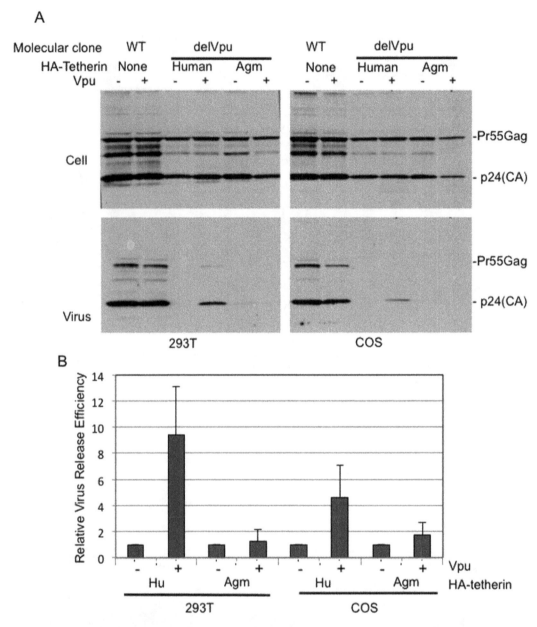

Figure 4. Vpu counteracts the virus release inhibition mediated by tetherin in COS cells. 293T or COS cells were co-transfected with WT pNL4-3 or Vpu-defective (pNL4-3/delVpu) molecular clones and human or agm tetherin expression vectors in the absence or presence of Vpu expression plasmid (15:1:5 NL4-3:tetherin:Vpu DNA ratio). (A) One day posttransfection, virus-containing supernatants were harvested and virions were pelleted by ultracentrifugation. Cell and virus lysates were subjected to western blot analysis with HIV-Ig. (B) One day posttransfection with pNL4-3/delVpu without (−Vpu) or with (+Vpu) Vpu expression plasmid and either human or agm tetherin expression vector, cells were metabolically labeled for 2 h, HIV proteins from cell and virus lysates were immunoprecipitated with HIV-Ig and analyzed by SDS-PAGE followed by fluorography. Relative virus release efficiency was calculated as the amount of virion-associated p24 relative to total Gag (cell + virion), normalized to 1 for release in the absence of Vpu. Data shown are means ± SD from four independent experiments.

tified by Pearson correlation coefficient (R) values using the softWoRx colocalization module. For surface staining of tetherin, cells were incubated with human tetherin antibody prior to fixation with formaldehyde.

Results

Vpu-mediated tetherin degradation in 293T cells is prevented by proteasome inhibitors and to a lesser extent by lysosomal inhibitors

To investigate the mechanism by which Vpu counteracts the antiviral activity of tetherin, we expressed HA-tagged tetherin in the absence or presence of Vpu in 293T cells, in which endogenous tetherin expression in undetectable [1]. As shown in Fig. S1A, steady-state levels of tetherin were significantly reduced

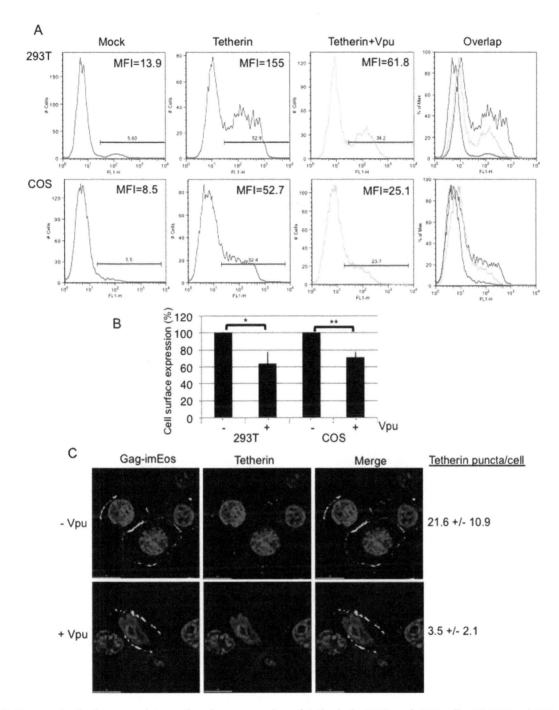

Figure 5. Vpu modestly down-regulates cell-surface expression of tetherin in 293T and COS cells. (A) 293T and COS cells were transfected with vectors expressing human tetherin alone or with the Vpu expression plasmid (1:5 DNA ratio). Twenty-four h posttransfection, cells were stained with anti-tetherin Ab and cell surface expression was analyzed by flow cytometry and mean fluorescent intensity (MFI) of tetherin-positive cells was determined. (B) Percent of cells positive for tetherin expression in the absence (−) or presence (+) of Vpu. Data from three experiments are shown, ±SD. P values were calculated using the two-tailed unpaired t-test. * = 0.02, ** = 0.002. (C) COS cells were cotransfected with pNL4-3Gag-imEosFPΔPolΔEnv and human tetherin plasmid without or with Vpu (15:1:5 NL4-3:tetherin:Vpu DNA ratio). Twenty-four h post-transfection, cell surface tetherin was stained with anti-tetherin Ab prior to fixation with formaldehyde and images were acquired with a Delta-Vision RT deconvolution microscope. 10–15 cells were scored to quantify cell-surface tetherin expression. Scale bar represents 15 μm.

by Vpu expressed from a CMV-driven expression vector. A similar reduction in tetherin levels was observed when Vpu was expressed from the full-length HIV-1 molecular clone pNL4-3 (Fig. S1A). No reduction in tetherin levels was observed when 293T cells were cotransfected with HA-tetherin and the Vpu-

defective molecular clones, delVpu and Udel (Fig. S1A). These results demonstrate, consistent with some previous reports [27,32,34,36], that Vpu reduces the steady-state levels of tetherin.

Both proteasomal and lysosomal routes have been proposed for Vpu-mediated tetherin degradation [27,33–36]. To investigate

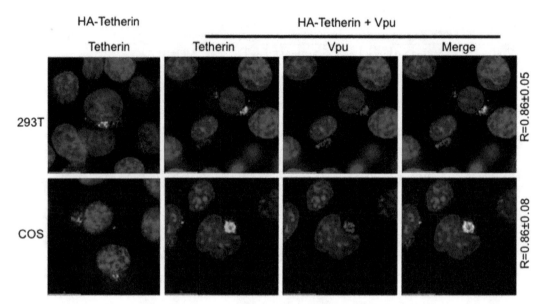

Figure 6. Vpu induces the sequestration of tetherin in COS cells. 293T and COS cells were transfected with vector expressing HA-tagged tetherin alone or in combination with Vpu expression vector (1:5 DNA ratio). Twenty-four h post-transfection, cells were fixed with 4% formaldehyde, permeabilized with methanol and stained with anti-HA (green), anti-Vpu (red), or DAPI (blue) and were analyzed with a Delta-Vision RT deconvolution microscope. Numbers represent the Pearson correlation coefficient (R) ± SD from 10–15 cells. Scale bars, 15 μm.

through which pathway tetherin undergoes degradation, we used proteasomal (MG132 and ALLN) or lysosomal (bafilomycin and concanamycin) inhibitors to treat 293T cells expressing tetherin with or without Vpu. We observed that treating Vpu and tetherin-expressing cells with proteasomal inhibitors markedly enhanced the expression of the putative non-glycosylated and singly glycosylated forms (~23–26 kDa), and the higher molecular weight (~30–36 kDa), fully glycosylated forms, of tetherin (Fig. S1B). Treating tetherin-expressing cells with lysosomal inhibitors markedly increased the expression of the ~26 kDa form, whereas the highly glycosylated forms of tetherin (~30–36 kDa) were not recovered significantly (Fig. S1B). These results suggest that the Vpu-mediated degradation of tetherin in 293T cells proceeds via both proteasomal and lysosomal routes, and that different tetherin species appear to be degraded predominantly by one or the other pathway.

Tetherin from mouse, rhesus macaque and African green monkey are not degraded by Vpu

Recent studies have demonstrated that tetherins from mouse, rhesus macaque (rh), and African green monkey (agm) inhibit HIV-1 release, but are not counteracted by Vpu [17,25–27]. To investigate whether the absence of Vpu antagonism is due to the inability of Vpu to induce the degradation of these non-human tetherins, we expressed mouse, rh and agm tetherin in 293T cells in the absence or presence of Vpu and analyzed tetherin expression by western blot. As shown above, the expression of human tetherin is reduced considerably in the presence of Vpu. In contrast, and consistent with previous reports [17,26,27], Vpu has no effect on the expression levels of mouse, rh, or agm tetherin (Fig. 1A). Next, we investigated whether Vpu could also degrade human tetherin mutants that do not inhibit HIV-1 release. We expressed inactive human tetherin mutants that lack the cytoplasmic tail (delCT) or the putative GPI anchor (delGPI). We observed that Vpu has only minor effects on the expression of these mutants, demonstrating that Vpu-mediated degradation of human tetherin requires both the CT and the putative GPI anchor (Fig. 1A).

Vpu stabilizes human tetherin in COS cells

To investigate whether the Vpu-mediated degradation of human tetherin is cell type-specific, we performed experiments similar to those described above in COS cells, an agm kidney cell line. We have chosen COS cells as they are widely used in cellular and molecular biology research [61] and are often used in HIV assembly studies [62–65]. We observed that HA-tagged human tetherin is poorly expressed in COS cells, but, surprisingly, its expression levels are markedly increased upon coexpression with Vpu (Fig. 1B). The deletion mutant delCT is also poorly expressed in COS cells but shows a modest increase in the presence of Vpu. A smaller increase in the expression of the delCT mutant was also observed in 293T cells in the presence of Vpu; the reason for this increase is not clear. The delGPI mutant and mouse, rh, and agm tetherin were highly expressed with or without Vpu (Fig. 1B). To investigate whether the lower expression of human tetherin in COS cells was due to its rapid turnover, we treated COS cells expressing human tetherin with proteasomal (MG132 and ALLN) or lysosomal (bafilomycin and concanamycin) inhibitors. As shown in Fig. 1C, proteasomal inhibitors rescued the expression of the ~23 kDa tetherin species, whereas lysosomal inhibitors rescued the expression of the ~26 kDa tetherin species, and, to a lesser extent, the ~23 kDa species. The stabilization pattern obtained in the presence of Vpu most closely resembled that observed in the presence of lysosomal inhibitors (+Vpu lane, Fig. 1C). These results demonstrate that human tetherin undergoes degradation in COS cells by both proteasomal and lysosomal pathways, with perhaps a greater contribution from the lysosomal pathway. To investigate whether the stabilization of tetherin by Vpu is specific to COS cells we tested in Vero cells, another cell line derived from agm kidney and used in HIV-1 assembly studies [66,67]. 293T, COS, and Vero cells were transfected with the vector expressing human tetherin without or with Vpu expression vector. Although transfection efficiency is lower in Vero cells, tetherin expression was stabilized by Vpu in Vero cells as in COS cells (Fig. 1D). In 293T cells, as observed before Vpu reduced tetherin expression. To investigate the rate of tetherin turnover in 293T and COS cells

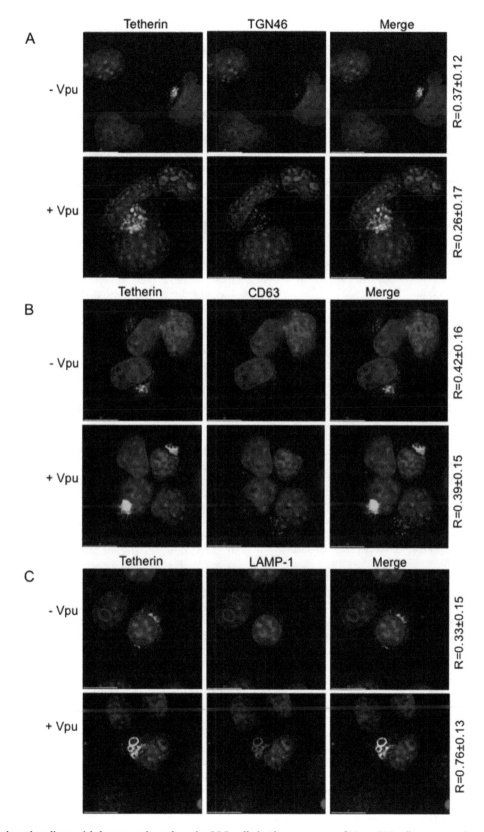

Figure 7. Tetherin colocalizes with lysosomal markers in COS cells in the presence of Vpu. COS cells were transfected with HA-tagged tetherin expression vector alone or in combination with Vpu expression vector (1:5 DNA ratio). Twenty-four h post-transfection, cells were fixed and permeabilized as in Fig. 6 and stained with anti-HA (green), DAPI (blue), and (A) the TGN marker TGN46 (red), (B) the multivesicular body marker CD63 (red), or (C) the lysosomal marker LAMP-1 (red). Images were acquired with a DeltaVision RT deconvolution microscope. Numbers represent the Pearson correlation coefficient (R) ± SD from 10–20 cells. Scale bars, 15 μm.

we performed pulse-chase analysis. 293T and COS cells were transfected in parallel with tetherin expression vector in the absence and presence of Vpu, pulse-labeled for 30 min and chased for 1, 2, or 4 h. In 293T cells Vpu significantly reduced tetherin levels within 1 h, whereas in COS cells the levels significantly increased compared to those of in the absence of Vpu (Fig. 2).

The low expression of human tetherin in COS cells depends upon the TM domain

To investigate a possible role for the TM domain of human tetherin in Vpu-mediated stabilization in COS cells, we expressed chimeric human tetherins that carry the TM sequences from rh or agm tetherin [hu(rhTM) and hu(agmTM)] [25]. We observed that these chimeric tetherins are well expressed in COS cells and their expression is not up-regulated by Vpu (Fig. 3). In contrast, these chimeric tetherins are poorly expressed in 293T cells. These results indicate that the TM domain of tetherin influences its expression levels in a cell type-dependent manner and suggest that Vpu binding to the TM domain of human tetherin protects it from rapid degradation in COS cells.

Vpu counteracts virus release inhibition mediated by human tetherin in COS cells

To examine the impact of tetherin stabilization on the ability of Vpu to counteract tetherin-mediated restriction of virus release, we expressed NL4-3delVpu with human or agm tetherin in the absence or presence of Vpu and measured virus release efficiency by radio-immunoprecipitation analysis. 293T cells were examined in parallel. As reported [1,68], the release efficiency of delVpu virus is comparable to that of the WT in both 293T and COS cells, as these cell lines are deficient for tetherin expression. In contrast, in the presence of human or agm tetherin, HIV-1 release is severely inhibited in both of these cell lines (Fig. 4A and Fig. S2). Vpu co-expression rescued HIV-1 release by ~10-fold in 293T cells (Fig. 4B). In COS cells, although Vpu increased the steady-state levels of tetherin (Fig. 1B) it was still able to rescue HIV-1 release by ~5-fold. As expected [26], Vpu showed minimal effect on virus release in the presence of agm tetherin independent of cell line (Fig. 4B and Fig. S2).

Cell-surface expression of human tetherin is modestly reduced by Vpu in COS cells

The results presented above raise the possibility that although Vpu increases overall human tetherin expression in COS cells it reduces tetherin levels at sites of virus assembly. We therefore examined the effect of Vpu expression on cell-surface levels of tetherin by flow cytometry using a polyclonal antiserum specific for the ectodomain of tetherin [39]. Again, 293T cells were included for comparison. The mean fluorescent intensity (MFI) of tetherin was reduced by 2.9-fold in 293T and 2.7-fold in COS cells by coexpression of Vpu (Fig. 5A). Further, Vpu coexpression reduced the number of cells expressing detectable levels of tetherin on average by ~36% in 293T and ~29% in COS cells (Fig. 5B). As an independent means of measuring cell-surface tetherin expression, we analyzed PM localization of tetherin by immuno-fluorescence and also measured colocalization between Gag and tetherin by co-expressing Gag containing a monomeric, photo-activatable fluorescent protein (mEos) inserted between the MA and CA domains (Gag-imEos). As previously reported [10], in the presence of Gag, tetherin showed a punctate staining on the cell surface that partially overlapped with the Gag localization pattern. The number of puncta was scored and expressed as the average number of puncta per cell. As shown in Fig. 5C, the number of

tetherin puncta (counted from 10–15 cells) on the cell surface (red) of COS cells is markedly higher in the absence of Vpu (21.6 ± 10.9) than in its presence (3.5 ± 2.1). Together, these results indicate that although the expression levels of human tetherin in COS cells are upregulated by Vpu the amount of tetherin on the cell surface is reduced in the presence of Vpu.

Vpu co-expression results in intracellular sequestration of human tetherin in COS cells

We next investigated the localization of tetherin in Vpu-expressing COS cells. Human tetherin was expressed in 293T and COS cells in the absence or presence of Vpu, and transfected cells were stained with anti-HA and anti-Vpu antibodies. As shown in Fig. 6, in 293T cells human tetherin is localized both on the cell surface and in intracellular compartments. Vpu colocalizes with human tetherin to a significant extent (Pearson correlation coefficient $R = 0.86 \pm 0.05$). In COS cells, the Vpu–tetherin colocalization values are essentially the same ($R = 0.86 \pm 0.08$) as those in 293T cells. However, in COS cells, Vpu sequesters human tetherin into intracellular vesicular compartments in the cytosol (Fig. 6). To investigate time course of Vpu-induced subcellular distribution of tetherin in 293T and COS cells we fixed the cells at time points 12, 18, and 24 h post-transfection and examined tetherin localization. In COS cells Vpu induced the formation of intracellular vesicular compartments; these are seen as early as 12 h and continuously grow at 18 h and form vesicular structures at 24 post-transfection (Fig. S3B). In 293T cells Vpu decreases tetherin levels as early as 18 h post-transfection (Fig. S3A).

Tetherin colocalizes with lysosomal markers in COS cells in the presence of Vpu

We next characterized the compartments in which tetherin is sequestered by Vpu in COS cells. Cells expressing tetherin in the absence or presence of Vpu were stained for tetherin and the TGN marker TGN46 or the late endosome marker CD63. Co-localization values for tetherin and TGN46 were 0.37 ± 0.12 and 0.26 ± 0.17 in the absence and presence of Vpu, respectively (Fig. 7A). The R-values for tetherin and CD63 were 0.42 ± 0.16 and 0.39 ± 0.15 in the absence and presence of Vpu, respectively (Fig. 7B). These relatively low colocalization values suggest that tetherin is not sequestered in the TGN or in late endosomes by Vpu in COS cells. We then stained with the lysosomal marker LAMP-1 (lysosomal-associated membrane protein 1). The coloca-lization values were significantly higher in the presence of Vpu (0.76 ± 0.13) than in its absence (0.33 ± 0.15) (Fig. 7C). These results indicate that in COS cells Vpu sequesters tetherin in a compartment that is positive for lysosomal markers. As shown earlier (Fig. 1C), treating COS cells with lysosomal inhibitors in the absence of Vpu resulted in the stabilization of tetherin to a similar extent as observed in the presence of Vpu. We therefore investigated whether adding lysosomal inhibitors to tetherin-transfected COS cells leads to tetherin accumulation in LAMP-1-positive compartments. Indeed, treating COS cells with con-canamycin or bafilomycin results in the accumulation of tetherin in LAMP-1-positive compartments even in the absence of Vpu (Fig. S4). In the presence of Vpu, lysosomal inhibitors have little or no additional effect on the co-localization of tetherin with LAMP-1 (Fig. S4). The fact that in COS cells Vpu induces the accumulation of tetherin in a LAMP-1-positive compartment without inducing tetherin degradation suggests that the LAMP-1-positive compartment in which tetherin accumulates in COS cells represents an aberrant, non-functional lysosomal compartment.

Discussion

In this study, we analyzed the relationship between Vpu-mediated degradation of tetherin and enhancement of virus release. Our results demonstrate that Vpu degrades tetherin in 293T cells by both proteasomal and lysosomal pathways, and that reduced levels of tetherin are associated with Vpu-mediated enhancement of virus release in this cell line (Fig. 4). In contrast, we find that in COS and Vero cells, human tetherin is poorly expressed in the absence of Vpu but its expression is stabilized in the presence of Vpu. Nonetheless, despite the fact that human tetherin levels are higher in the presence of Vpu, virus release is still enhanced by Vpu in COS cells. These results add to a growing body of evidence indicating that Vpu can overcome the antiviral activity of tetherin without inducing its degradation. For example, Strebel and co-workers reported that during the course of productive infection in T cells Vpu enhances HIV-1 release without reducing tetherin's cell surface or intracellular expression [39]. Goffinet et al. observed that Vpu is able to induce a small increase in the expression of a tetherin mutant (K18, 21A) in 293T cells while still antagonizing its antiviral activity [38]. In our study, although Vpu enhanced the total expression of human tetherin in COS cells, the cell-surface expression was modestly reduced by Vpu. These results are consistent with observations from others that Vpu-induced reductions in cell-surface tetherin expression correlate with loss of antiviral activity, but is uncoupled from total cellular levels of tetherin [27,32,35,38,40].

The molecular determinants in human tetherin for Vpu antagonism are reported to reside in the TM domain, with specific residues apparently regulating direct Vpu–tetherin interactions [25,26,33,35,40,52,69,70]. The inability of Vpu to antagonize non-human tetherin correlates with a lack of interaction with Vpu [17,26,27]. The interaction of Vpu with human tetherin likely plays a role in Vpu-mediated tetherin stabilization in COS cells, as Vpu has no significant effect on the expression of chimeric tetherins that carry the TM domain from rh or agm tetherin. Thus, we speculate that Vpu binding to human tetherin protects it from rapid degradation in COS cells. Moreover, the TM-chimeric tetherins were well expressed in COS cells but poorly expressed in 293T cells, indicating that the TM domain of tetherin regulates its cell-type dependent stability.

We observed that in COS cells Vpu sequesters human tetherin in a compartment that bears markers for lysosomes (e.g., LAMP-1) but shows little overlap with multivesicular body (CD63) or TGN (TGN46) markers. Earlier studies showed sequestration of tetherin in the TGN induced by Vpu [14,71], HIV-2 Env [14,15], and SIV Env [13]. Tetherin accumulation in perinuclear patches was also observed to be induced by Lassa virus Z protein, but the nature of these compartments was not characterized [11]. It is intriguing that the increased colocalization of tetherin with LAMP-1 that we observe in COS cells in the presence of Vpu results in the stabilization rather than the degradation of tetherin. We speculate that Vpu expression in COS cells results in the generation of an aberrant lysosomal compartment, as suggested by the swollen, ring-like morphology of these structures (Fig. 7). Accumulation of tetherin in a LAMP-1-positive compartment in COS cells treated with lysosomal inhibitors even in the absence of Vpu is consistent with the hypothesis that Vpu induces formation of aberrant lysosomal compartments in which tetherin accumulates. This could provide HIV-1 with a Vpu-dependent mechanism for disrupting lysosomal function, thereby preventing the degradation of virally encoded proteins. While this hypothesis will require further testing, the results of this study demonstrate that COS cells provide a system for the molecular dissection of Vpu and tetherin activities.

Supporting Information

Figure S1 Vpu mediates proteasomal and lysosomal degradation of tetherin. (A) 293T cells were transfected with empty vector alone or HA-tagged tetherin expression plasmid without or with combination of Vpu or the indicated proviral expression plasmid (1:5 DNA ratio). One day posttransfection, cells were lysed and subjected to SDS-PAGE and subsequent western blot analysis with anti-HA (top panel) or anti-tubulin (lower panel) antibodies. Levels of tetherin expression were reduced in the presence of Vpu by approximately three-fold in this and repeat experiments, as determined by band quantification with an Alpha Innotech Fluorchem SP imager. Vpu-defective molecular clones, delVpu and Udel did not reduce tetherin levels. Molecular mass markers are shown on the right (in kDa). (B) 293T cells were transfected with HA-tagged tetherin expression vector alone (left panel; −Vpu) or in combination with the Vpu expression plasmid (right panel; +Vpu). One day posttransfection, cells were treated with proteasomal and lysosomal inhibitors at the following concentrations: MG132 (25 μM), ALLN (25 μM), bafilomycin (0.15 μM) and concanamycin (0.5 μM). Levels of HA-tagged tetherin and tubulin were determined by western blotting with anti-HA and anti-tubulin antibodies, respectively. Molecular mass markers are shown on the right (in kDa). Treating Vpu and tetherin-expressing cells with proteasomal inhibitors enhanced the expression of tetherin approximately six-fold. Lysosomal inhibitors increased expression of the ~26 kDa band by approximately four-fold, whereas the highly glycosylated forms of tetherin (~30–36 kDa) were not recovered significantly.

Figure S2 293T or COS cells were co-transfected with molecular clones (pNL4-3 or pNL4-3/delVpu) and human or agm tetherin expression vectors in the absence or presence of Vpu expression plasmid as in Fig. 4A. One day posttransfection, virus supernatants were collected and RT activity was measured. Data shown are means ± SD from two independent experiments.

Figure S3 293T and COS cells were transfected with vectors expressing human tetherin alone or with the Vpu expression plasmid (1:5 DNA ratio) and fixed after 12, 18, and 24 h post-transfection with 4% formaldehyde. Cells were then permeabilized and stained with anti-tetherin Ab as in Fig. 6. Images shown are representative from 8–10 cells. Scale bars, 15 μm.

Figure S4 COS cells were transfected with HA-tagged tetherin expression vector in the absence or presence of Vpu (1:5 DNA ratio). One day post-transfection, cells were treated with lysosomal inhibitors concanamycin (0.5 μM) and bafilomycin (0.15 μM), fixed, permeabilized as in Fig. 6 and stained with anti-HA (green), DAPI (blue), and LAMP-1 (red). Numbers represent the Pearson correlation coefficient (R) ± SD from 10–12 cells. Scale bars, 15 μm.

Acknowledgments

We thank members of the Freed laboratory for helpful discussion and critical review of the manuscript. The plasmid pNL4-3Gag-imEosFPΔ-PolΔEnv was constructed and generously provided by K. Waki. We thank

K. Strebel, P. Bieniasz, and J. Lippincott-Schwartz for their generous gifts of plasmids, and K. Lee and V. KewalRamani for assistance with flow cytometry. Anti-HIV-1 Ig, anti-Vpu and anti-human tetherin antisera were obtained from the NIH AIDS Research and Reference Reagent Program.

Author Contributions

Conceived and designed the experiments: AW EF. Performed the experiments: AW NK KF DD. Analyzed the data: AW EF. Contributed to the writing of the manuscript: AW EF.

References

1. Neil SJ, Zang T, Bieniasz PD (2008) Tetherin inhibits retrovirus release and is antagonized by HIV-1 Vpu. Nature 451: 425–430.

2. Van Damme N, Goff D, Katsura C, Jorgenson RL, Mitchell R, et al. (2008) The interferon-induced protein BST-2 restricts HIV-1 release and is downregulated from the cell surface by the viral Vpu protein. Cell Host Microbe 3: 245–252.

3. Goto T, Kennel SJ, Abe M, Takishita M, Kosaka M, et al. (1994) A novel membrane antigen selectively expressed on terminally differentiated human B cells. Blood 84: 1922–1930.

4. Ishikawa J, Kaisho T, Tomizawa H, Lee BO, Kobune Y, et al. (1995) Molecular cloning and chromosomal mapping of a bone marrow stromal cell surface gene, BST2, that may be involved in pre-B-cell growth. Genomics 26: 527–534.

5. Yang WK, Kiggans JO, Yang DM, Ou CY, Tennant RW, et al. (1980) Synthesis and circularization of N- and B-tropic retroviral DNA Fv-1 permissive and restrictive mouse cells. Proc Natl Acad Sci U S A 77: 2994–2998.

6. Blasius AL, Giurisato E, Cella M, Schreiber RD, Shaw AS, et al. (2006) Bone marrow stromal cell antigen 2 is a specific marker of type I IFN-producing cells in the naive mouse, but a promiscuous cell surface antigen following IFN stimulation. J Immunol 177: 3260–3265.

7. Kupzig S, Korolchuk V, Rollason R, Sugden A, Wilde A, et al. (2003) Bst-2/HM1.24 is a raft-associated apical membrane protein with an unusual topology. Traffic 4: 694–709.

8. Perez-Caballero D, Zang T, Ebrahimi A, McNatt MW, Gregory DA, et al. (2009) Tetherin inhibits HIV-1 release by directly tethering virions to cells. Cell 139: 499–511.

9. Andrew AJ, Kao S, Strebel K (2011) C-terminal hydrophobic region in human bone marrow stromal cell antigen 2 (BST-2)/tetherin protein functions as second transmembrane motif. J Biol Chem 286: 39967–39981.

10. Jouvenet N, Neil SJ, Zhadina M, Zang T, Kratovac Z, et al. (2009) Broad-spectrum inhibition of retroviral and filoviral particle release by tetherin. J Virol 83: 1837–1844.

11. Sakuma T, Noda T, Urata S, Kawaoka Y, Yasuda J (2009) Inhibition of Lassa and Marburg virus production by tetherin. J Virol 83: 2382–2385.

12. Mansouri M, Viswanathan K, Douglas JL, Hines J, Gustin J, et al. (2009) Molecular mechanism of BST2/tetherin downregulation by K5/MIR2 of Kaposi's sarcoma-associated herpesvirus. J Virol 83: 9672–9681.

13. Gupta RK, Mlcochova P, Pelchen-Matthews A, Petit SJ, Mattiuzzo G, et al. (2009) Simian immunodeficiency virus envelope glycoprotein counteracts tetherin/BST-2/CD317 by intracellular sequestration. Proc Natl Acad Sci U S A 106: 20889–20894.

14. Hauser H, Lopez LA, Yang SJ, Oldenburg JE, Exline CM, et al. (2010) HIV-1 Vpu and HIV-2 Env counteract BST-2/tetherin by sequestration in a perinuclear compartment. Retrovirology 7: 51.

15. Le Tortorec A, Neil SJ (2009) Antagonism to and intracellular sequestration of human tetherin by the human immunodeficiency virus type 2 envelope glycoprotein. J Virol 83: 11966–11978.

16. Serra-Moreno R, Jia B, Breed M, Alvarez X, Evans DT (2011) Compensatory changes in the cytoplasmic tail of gp41 confer resistance to tetherin/BST-2 in a pathogenic nef-deleted SIV. Cell Host Microbe 9: 46–57.

17. Jia B, Serra-Moreno R, Neidermyer W, Rahmberg A, Mackey J, et al. (2009) Species-specific activity of SIV Nef and HIV-1 Vpu in overcoming restriction by tetherin/BST2. PLoS Pathog 5: e1000429.

18. Sauter D, Schindler M, Specht A, Landford WN, Munch J, et al. (2009) Tetherin-driven adaptation of Vpu and Nef function and the evolution of pandemic and nonpandemic HIV-1 strains. Cell Host Microbe 6: 409–421.

19. Zhang F, Wilson SJ, Landford WC, Virgen B, Gregory D, et al. (2009) Nef proteins from simian immunodeficiency viruses are tetherin antagonists. Cell Host Microbe 6: 54–67.

20. Blondeau C, Pelchen-Matthews A, Mlcochova P, Marsh M, Milne RS, et al. (2013) Tetherin restricts herpes simplex virus 1 and is antagonized by glycoprotein M. J Virol 87: 13124–13133.

21. Yin X, Hu Z, Gu Q, Wu X, Zheng YH, et al. (2014) Equine tetherin blocks retrovirus release and its activity is antagonized by equine infectious anemia virus envelope protein. J Virol 88: 1259–1270.

22. Morrison JH, Guevara RB, Marcano AC, Saenz DT, Fadel HJ, et al. (2014) Feline immunodeficiency virus envelope glycoproteins antagonize tetherin through a distinctive mechanism that requires virion incorporation. J Virol 88: 3255–3272.

23. Jones PH, Maric M, Madison MN, Maury W, Roller RJ, et al. (2013) BST-2/tetherin-mediated restriction of chikungunya (CHIKV) VLP budding is counteracted by CHIKV non-structural protein 1 (nsP1). Virology 438: 37–49.

24. Lemaitre C, Harper F, Pierron G, Heidmann T, Dewannieux M (2014) The HERV-K human endogenous retrovirus envelope protein antagonises Tetherin antiviral activity. J Virol.

25. McNatt MW, Zang T, Hatziioannou T, Bartlett M, Fofana IB, et al. (2009) Species-specific activity of HIV-1 Vpu and positive selection of tetherin transmembrane domain variants. PLoS Pathog 5: e1000300.

26. Rong L, Zhang J, Lu J, Pan Q, Lorgeoux RP, et al. (2009) The transmembrane domain of BST-2 determines its sensitivity to down-modulation by human immunodeficiency virus type 1 Vpu. J Virol 83: 7536–7546.

27. Goffinet C, Allespach I, Homann S, Tervo HM, Habermann A, et al. (2009) HIV-1 antagonism of CD317 is species specific and involves Vpu-mediated proteasomal degradation of the restriction factor. Cell Host Microbe 5: 285–297.

28. Shingai M, Yoshida T, Martin MA, Strebel K (2011) Some human immunodeficiency virus type 1 Vpu proteins are able to antagonize macaque BST-2 in vitro and in vivo: Vpu-negative simian-human immunodeficiency viruses are attenuated in vivo. J Virol 85: 9708–9715.

29. Gao F, Bailes E, Robertson DL, Chen Y, Rodenburg CM, et al. (1999) Origin of HIV-1 in the chimpanzee Pan troglodytes troglodytes. Nature 397: 436–441.

30. Tokarev A, Guatelli J (2011) Misdirection of membrane trafficking by HIV-1 Vpu and Nef: Keys to viral virulence and persistence. Cell Logist 1: 90–102.

31. Sauter D (2014) Counteraction of the multifunctional restriction factor tetherin. Front Microbiol 5: 163.

32. Mitchell RS, Katsura C, Skasko MA, Fitzpatrick K, Lau D, et al. (2009) Vpu antagonizes BST-2-mediated restriction of HIV-1 release via beta-TrCP and endo-lysosomal trafficking. PLoS Pathog 5: e1000450.

33. Iwabu Y, Fujita H, Kinomoto M, Kaneko K, Ishizaka Y, et al. (2009) HIV-1 accessory protein Vpu internalizes cell-surface BST-2/tetherin through transmembrane interactions leading to lysosomes. J Biol Chem 284: 35060–35072.

34. Douglas JL, Viswanathan K, McCarroll MN, Gustin JK, Fruh K, et al. (2009) Vpu directs the degradation of the human immunodeficiency virus restriction factor BST-2/Tetherin via a {beta}TrCP-dependent mechanism. J Virol 83: 7931–7947.

35. Gupta RK, Hue S, Schaller T, Verschoor E, Pillay D, et al. (2009) Mutation of a single residue renders human tetherin resistant to HIV-1 Vpu-mediated depletion. PLoS Pathog 5: e1000443.

36. Mangeat B, Gers-Huber G, Lehmann M, Zufferey M, Luban J, et al. (2009) HIV-1 Vpu neutralizes the antiviral factor Tetherin/BST-2 by binding it and directing its beta-TrCP2-dependent degradation. PLoS Pathog 5: e1000574.

37. Habermann A, Krijnse-Locker J, Oberwinkler H, Eckhardt M, Homann S, et al. (2010) CD317/tetherin is enriched in the HIV-1 envelope and downregulated from the plasma membrane upon virus infection. J Virol 84: 4646–4658.

38. Goffinet C, Homann S, Ambiel I, Tibroni N, Rupp D, et al. (2010) Antagonism of CD317 restriction of human immunodeficiency virus type 1 (HIV-1) particle release and depletion of CD317 are separable activities of HIV-1 Vpu. J Virol 84: 4089–4094.

39. Miyagi E, Andrew AJ, Kao S, Strebel K (2009) Vpu enhances HIV-1 virus release in the absence of Bst-2 cell surface down-modulation and intracellular depletion. Proc Natl Acad Sci U S A 106: 2868–2873.

40. Dube M, Roy BB, Guiot-Guillain P, Binette J, Mercier J, et al. (2010) Antagonism of tetherin restriction of HIV-1 release by Vpu involves binding and sequestration of the restriction factor in a perinuclear compartment. PLoS Pathog 6: e1000856.

41. Schmidt S, Fritz JV, Bitzegeio J, Fackler OT, Keppler OT (2011) HIV-1 Vpu blocks recycling and biosynthetic transport of the intrinsic immunity factor CD317/tetherin to overcome the virion release restriction. MBio 2: e00036-00011.

42. Andrew AJ, Miyagi E, Strebel K (2011) Differential effects of human immunodeficiency virus type 1 Vpu on the stability of BST-2/tetherin. J Virol 85: 2611–2619.

43. Rollason R, Dunstan K, Billcliff PG, Bishop P, Gleeson P, et al. (2013) Expression of HIV-1 Vpu leads to loss of the viral restriction factor CD317/Tetherin from lipid rafts and its enhanced lysosomal degradation. PLoS One 8: e75680.

44. Jolly C, Booth NJ, Neil SJ (2010) Cell-cell spread of human immunodeficiency virus type 1 overcomes tetherin/BST-2-mediated restriction in T cells. J Virol 84: 12185–12199.

45. Casartelli N, Sourisseau M, Feldmann J, Guivel-Benhassine F, Mallet A, et al. (2010) Tetherin restricts productive HIV-1 cell-to-cell transmission. PLoS Pathog 6: e1000955.

46. Kuhl BD, Sloan RD, Donahue DA, Bar-Magen T, Liang C, et al. (2010) Tetherin restricts direct cell-to-cell infection of HIV-1. Retrovirology 7: 115.

47. Giese S, Marsh M (2014) Tetherin can restrict cell-free and cell-cell transmission of HIV from primary macrophages to T cells. PLoS Pathog 10: e1004189.

48. Zhang J, Liang C (2010) BST-2 diminishes HIV-1 infectivity. J Virol 84: 12336–12343.

49. Arias JF, Heyer LN, von Bredow B, Weisgrau KL, Moldt B, et al. (2014) Tetherin antagonism by Vpu protects HIV-infected cells from antibody

dependent cell-mediated cytotoxicity. Proc Natl Acad Sci U S A 111: 6425–6430.

50. Alvarez RA, Hamlin RE, Monroe A, Moldt B, Hotta MT, et al. (2014) HIV-1 Vpu Antagonism of Tetherin Inhibits Antibody-Dependent Cellular Cytotoxic Responses by Natural Killer Cells. J Virol 88: 6031–6046.

51. Pham TN, Lukhele S, Hajjar F, Routy JP, Cohen EA (2014) HIV Nef and Vpu protect HIV-infected CD4+ T cells from antibody-mediated cell lysis through down-modulation of CD4 and BST2. Retrovirology 11: 15.

52. Pardieu C, Vigan R, Wilson SJ, Calvi A, Zang T, et al. (2010) The RING-CH ligase K5 antagonizes restriction of KSHV and HIV-1 particle release by mediating ubiquitin-dependent endosomal degradation of tetherin. PLoS Pathog 6: e1000843.

53. Adachi A, Gendelman HE, Koenig S, Folks T, Willey R, et al. (1986) Production of acquired immunodeficiency syndrome-associated retrovirus in human and nonhuman cells transfected with an infectious molecular clone. J Virol 59: 284–291.

54. Klimkait T, Strebel K, Hoggan MD, Martin MA, Orenstein JM (1990) The human immunodeficiency virus type 1-specific protein vpu is required for efficient virus maturation and release. J Virol 64: 621–629.

55. Strebel K, Klimkait T, Martin MA (1988) A novel gene of HIV-1, vpu, and its 16-kilodalton product. Science 241: 1221–1223.

56. Nguyen KL, llano M, Akari H, Miyagi E, Poeschla EM, et al. (2004) Codon optimization of the HIV-1 vpu and vif genes stabilizes their mRNA and allows for highly efficient Rev-independent expression. Virology 319: 163–175.

57. Hubner W, Chen P, Del Portillo A, Liu Y, Gordon RE, et al. (2007) Sequence of human immunodeficiency virus type 1 (HIV-1) Gag localization and oligomerization monitored with live confocal imaging of a replication-competent, fluorescently tagged HIV-1. J Virol 81: 12596–12607.

58. Betzig E, Patterson GH, Sougrat R, Lindwasser OW, Olenych S, et al. (2006) Imaging intracellular fluorescent proteins at nanometer resolution. Science 313: 1642–1645.

59. Maldarelli F, Chen MY, Willey RL, Strebel K (1993) Human immunodeficiency virus type 1 Vpu protein is an oligomeric type I integral membrane protein. J Virol 67: 5056–5061.

60. Freed EO, Martin MA (1994) Evidence for a functional interaction between the V1/V2 and C4 domains of human immunodeficiency virus type 1 envelope glycoprotein gp120. J Virol 68: 2503–2512.

61. Hancock JF (1992) COS Cell Expression. Methods Mol Biol 8: 153–158.

62. Martinez NW, Xue X, Berro RG, Kreitzer G, Resh MD (2008) Kinesin KIF4 regulates intracellular trafficking and stability of the human immunodeficiency virus type 1 Gag polyprotein. J Virol 82: 9937–9950.

63. Rai T, Mosoian A, Resh MD (2010) Annexin 2 is not required for human immunodeficiency virus type 1 particle production but plays a cell type-dependent role in regulating infectivity. J Virol 84: 9783–9792.

64. Reed JC, Molter B, Geary CD, McNevin J, McElrath J, et al. (2012) HIV-1 Gag co-opts a cellular complex containing DDX6, a helicase that facilitates capsid assembly. J Cell Biol 198: 439–456.

65. Wen X, Ding L, Wang JJ, Qi M, Hammonds J, et al. (2014) ROCK1 and LIM Kinase Modulate Retrovirus Particle Release and Cell-Cell Transmission Events. J Virol.

66. Varthakavi V, Smith RM, Bour SP, Strebel K, Spearman P (2003) Viral protein U counteracts a human host cell restriction that inhibits HIV-1 particle production. Proc Natl Acad Sci U S A 100: 15154–15159.

67. Krementsov DN, Rassam P, Margeat E, Roy NH, Schneider-Schaulies J, et al. (2010) HIV-1 assembly differentially alters dynamics and partitioning of tetraspanins and raft components. Traffic 11: 1401–1414.

68. Neil SJ, Sandrin V, Sundquist WI, Bieniasz PD (2007) An interferon-alpha-induced tethering mechanism inhibits HIV-1 and Ebola virus particle release but is counteracted by the HIV-1 Vpu protein. Cell Host Microbe 2: 193–203.

69. Douglas JL, Gustin JK, Viswanathan K, Mansouri M, Moses AV, et al. (2010) The great escape: viral strategies to counter BST-2/tetherin. PLoS Pathog 6: e1000913.

70. Kobayashi T, Ode H, Yoshida T, Sato K, Gee P, et al. (2011) Identification of amino acids in the human tetherin transmembrane domain responsible for HIV-1 Vpu interaction and susceptibility. J Virol 85: 932–945.

71. Dube M, Roy BB, Guiot-Guillain P, Mercier J, Binette J, et al. (2009) Suppression of Tetherin-restricting activity upon human immunodeficiency virus type 1 particle release correlates with localization of Vpu in the trans-Golgi network. J Virol 83: 4574–4590.

5

Molecular and Genetic Determinants of the NMDA Receptor for Superior Learning and Memory Functions

Stephanie Jacobs[1,9], **Zhenzhong Cui**[1,9], **Ruiben Feng**[1], **Huimin Wang**[2], **Deheng Wang**[3], **Joe Z. Tsien**[1*]

1 Brain and Behavior Discovery Institute and Department of Neurology, Medical College of Georgia at Georgia Regents University, Augusta, Georgia, United States of America, **2** Shanghai Institute of Functional Genomics, East China Normal University, Shanghai, China, **3** Banna Biomedical Research Institute, Xi-Shuang-Ban-Na Prefecture, Yunnan Province, China

Abstract

The opening-duration of the NMDA receptors implements Hebb's synaptic coincidence-detection and is long thought to be the rate-limiting factor underlying superior memory. Here, we investigate the molecular and genetic determinants of the NMDA receptors by testing the "synaptic coincidence-detection time-duration" hypothesis vs. "GluN2B intracellular signaling domain" hypothesis. Accordingly, we generated a series of GluN2A, GluN2B, and GluN2D chimeric subunit transgenic mice in which C-terminal intracellular domains were systematically swapped and overexpressed in the forebrain excitatory neurons. The data presented in the present study supports the second hypothesis, the "GluN2B intracellular signaling domain" hypothesis. Surprisingly, we found that the voltage-gated channel opening-durations through either GluN2A or GluN2B are sufficient and their temporal differences are marginal. In contrast, the C-terminal intracellular domain of the GluN2B subunit is necessary and sufficient for superior performances in long-term novel object recognition and cued fear memories and superior flexibility in fear extinction. Intriguingly, memory enhancement correlates with enhanced long-term potentiation in the 10–100 Hz range while requiring intact long-term depression capacity at the 1–5 Hz range.

Editor: Ya-Ping Tang, Louisiana State University Health Sciences Center, United States of America

Funding: This work was supported by funds from the National Institute of Mental Health (MH060236), National Institute on Aging (AG024022, AG034663 & AG025918), USAMRA00002, and Georgia Research Alliance (all to JZT). The funders had no role in study design, data collection and analysis, decision to publish, or preparation of the manuscript.

* Email: jtsien@gru.edu

⑨ These authors contributed equally to this work.

Introduction

N-methyl-D-aspartate (NMDA) receptors are known to be the key modulators of synaptic plasticity in the forebrain regions [1–4] and act as the molecular gating switch for learning and memory [5,6]. It is widely accepted that their unique coincidence detection property allows them to impart Hebb's rule on synapses, by requiring the simultaneous pre-synaptic release of glutamate and the depolarization of the postsynaptic membrane to remove the extracellular Mg^{2+} block [7]. NMDA receptors are composed of two GluN1 subunits, as well as two GluN2 subunits [8]. In the adult forebrain regions, GluN2A and GluN2B subunits are the main subunits available in excitatory synapses for receptor complex formation [8,9], and are ideal for coincidence detection due to their strong Mg^{2+} dependency [10–12].

During postnatal brain development, the GluN2B subunits are the predominate subunits expressed specifically in excitatory neurons of the forebrain regions, such as the cortex and hippocampus [9,13]. As the animal develops into adulthood, GluN2B expression decreases while GluN2A expression increases, resulting in an overall decrease in synaptic plasticity. Previously, we have shown that an overexpression of the GluN2B subunit in the forebrain excitatory neurons enhances many forms of learning

and memory in both transgenic mice and rats [14–18]. The prevalent view in the field is that memory enhancement by GluN2B up-regulation is due to its longer opening duration in comparison to that of the GluN2A subunit. This enables the GluN2B-containing NMDA receptors to be better coincidence detectors.

However, the different structural motifs of GluN2 subunits are known to regulate the NMDA receptors' Mg^{2+} dependency, channel opening duration, magnitude of Ca^{2+} influx, as well as, intracellular signaling cascades [10]. For example, the GluN2A and GluN2B subunits have high Mg^{2+} dependency, whereas the GluN2C and GluN2D subunits have much less Mg^{2+} dependency [13]. Thus, the extracellular Mg^{2+} blockade of the GluN2A or GluN2B-containing NMDA receptors suppresses NMDA-mediated Ca^{2+} influx at voltages close to the resting membrane potential allowing the cell to differentiate between correlated synaptic input and uncorrelated activity [7,19,20]. Recent studies have further demonstrated that the C-terminals of GluN2A and GluN2B bind to different downstream signaling molecules. This has led to a greater appreciation for their contribution to synaptic plasticity and behavior [21,22]. To examine the effects of GluN2A on learning and memory, we recently generated CaMKII promoter-driven GluN2A transgenic mice and found profound long-term memory deficits in these mice, while their short-term memories remain unaffected [23]. Therefore, overexpression of GluN2A or GluN2B in the mouse forebrain leads to impaired or enhanced memory function, respectively. These observations have raised several key questions, as to whether enhanced memories in GluN2B transgenic mice or impaired memories in GluN2A transgenic mice were due to their differences in NMDA receptor channel-opening durations or their distinct intracellular signaling processes. Answers to this crucial question can be highly valuable for developing therapeutic strategies for preventing memory loss in patients.

Currently, two hypotheses have been postulated to explain the observed memory enhancement in the GluN2B transgenic animals or memory impairment in GluN2A transgenic mice [4]. One hypothesis, known as the "coincidence-detection" hypothesis, posits that because the GluN2B subunit makes the channel opening duration longer than that of the GluN2A subunit, the GluN2B overexpression allows a greater coincidence detection window, thereby leading to superior memory functions. The shorter coincidence-detection, such as in the GluN2A transgenic mice, underlies impaired long-term memories. The second hypothesis is that the distinct intracellular domain of the GluN2B subunit is responsible for the enhancements observed in the GluN2B transgenic mice [4]. Several recent key observations support this "intracellular domain hypothesis". Biochemical studies have shown that the intracellular C-terminal domains of the GluN2A and GluN2B subunits preferentially interact with different downstream molecules and play distinct roles in synaptic functions [24]. Conversely, studies using genetically truncated GluN2A or GluN2B subunits demonstrate that the C-terminal connections are essential for NMDA receptor function. The truncated subunits often act as functional knockouts of the whole subunit [21]. Although several truncated C-terminal studies have focused on the mechanisms by which the GluN2 subunits mediate the NMDA receptor functions, the structural motifs crucial for learning and memory enhancement remain undefined. It is completely unknown as to whether and what degree the C-terminal domain of the GluN2B would contribute to memory enhancement.

In the present study, we set out to examine the above two hypotheses aimed at determining how and whether the molecular motifs underlying coincidence-detection time duration, or intracellular signaling cascades play a role in enhancing learning and memory. Our strategy is to swap or replace the C-terminal cytoplasmic domain of the GluN2B subunit with the C-terminal domain of the GluN2A subunit, or vice versa. Additionally, we have replaced the C-terminal domain of the GluN2D subunit with the C-terminal domain of the GluN2B subunit effectively reducing the Mg^{2+} dependency of the receptor. We have produced and analyzed five different GluN2 transgenic mouse lines, namely, GluN2A transgenic mice (Tg-GluN2A), Tg-GluN2B$^{2A(CT)}$ transgenic mice, GluN2B transgenic mice (Tg-GluN2B), Tg-GluN2A$^{2B(CT)}$ transgenic mice, and Tg-GluN2D$^{2B(CT)}$ transgenic mice. Our experiments suggest that the C-terminal domain of the GluN2B subunit is necessary and sufficient to produce memory enhancement, as long as it is coupled to Mg^{2+} dependent forms of GluN2 subunits such as GluN2A or GluN2B, while coupling of the GluN2A's C-terminal domain to GluN2B N-terminal and transmembrane domains lead to profound memory impairment. Moreover, coupling of the C-terminal domain of the GluN2B subunit to GluN2D subunit's N-terminal and transmembrane domains lead to memory deficits.

Results

Generation of transgenic mice expressing chimeric GluN2A$^{2B(CTR)}$, GluN2B$^{2A(CTR)}$, or GluN2D$^{2B(CTR)}$ subunits in the forebrain principal neurons

To investigate the potentially distinct roles of the C-terminal domains vs. the N-terminal and membrane domains in GluN2 subunits in mediating memory enhancement, we created constructs encoding chimeric receptors based on GluN2B and GluN2A but with their respective CTDs replaced (denoted as CTR) with each other's (GluN2B$^{2A(CTR)}$ and GluN2A$^{2B(CTR)}$, respectively. We have created three new chimeric GluN2 transgenic mouse lines. We used the same αCaMKII promoter for driving transgene expression in forebrain excitatory neurons as we did for producing the GluN2B [16] and GluN2A transgenic mice [23]. In the first transgenic line, termed Tg-GluN2B$^{2A(CT)}$ chimeric transgenic mice, the C-terminal domain of the GluN2B subunit has been swapped for the counterpart C-terminal domain of the GluN2A and overexpressed in the forebrain excitatory neurons (Figure 1A). This effectively pairs the opening duration of the GluN2B subunit with the signaling domain of the GluN2A subunit. In the second transgenic line, termed Tg-GluN2A$^{2B(CT)}$, the C-terminal domain of the GluN2A subunit has been swapped for the counterpart C-terminal domain of the GluN2B (Figure 1A). This chimeric subunit possesses the GluN2A opening duration but with the signaling domain from the GluN2B subunit. Additionally, to investigate the requirement of the Mg^{2+} dependent synaptic coincidence-detection function for producing GluN2B-mediated intracellular signaling, we created a third transgenic mouse line, namely, Tg-GluN2D$^{2B(CT)}$ mice, in which the C-terminal domain of the GluN2B subunit has been fused to the N-terminal and membrane domain of the GluN2D subunit (which is less Mg^{2+}-dependent) denoted as GluN2D$^{2B(CTR)}$ (Figure 1A).

We confirmed the transgene integration into the genome of their off-spring by Southern Blot analysis using Poly(A) probes (Figure S1A) and Western Blot (Figure S1B). Next, we performed a series of in situ hybridization experiments to determine the expression pattern of the transgenes in the mouse brains. As shown, the transgenes are highly enriched in the cortex, striatum, and hippocampus, but not in hindbrain regions such as the cerebellum (Figure 1C). The high expression transgenic mice were crossed with C57BL/6J wild-type mice for at least 8 generations. These chimeric GluN2 transgenic offspring were found to grow and breed normally, having similar adult weights to their wild-type littermates (Figure 1D) (Wt: n = 11, 29.18±0.985 g; Tg-GluN2A$^{2B(CT)}$: n = 7, 30.09±0.990 g; Tg-GluN2B$^{2A(CT)}$: n = 10, 30.34±0.648 g; Tg-GluN2D$^{2B(CT)}$: n = 9, 29.41±0.940 g), and being visually indistinguishable among them. Additionally, we also produced transgenic GluN2A overexpression mice (Tg-GluN2A mice) [23] and transgenic GluN2B overexpression mice (Tg-GluN2B mice) [14–17] for comparisons on learning and memory tests. These two transgenic mouse lines were also maintained on the same genetic background.

The Tg-GluN2A$^{2B(CT)}$, Tg-GluN2B$^{2A(CT)}$, and Tg-GluN2D$^{2B(CT)}$ mice showed no differences in the open field behavioral paradigm, either in time spent in the center verses the periphery (Figure 1E) (center: Wt: n = 7, 117.91±17.006 s; Tg-GluN2A$^{2B(CT)}$: n = 6, 95.95±12.893 s; Tg-GluN2B$^{2A(CT)}$: n = 5, 78.74±15.154 s; Tg-GluN2D$^{2B(CT)}$: n = 6, 79.15±13.483 s; periphery: Wt: 481.86±16.994 s; Tg-GluN2A$^{2B(CT)}$: 503.98±12.901 s Tg-GluN2B$^{2A(CT)}$: 520.96±15.098 s; Tg-GluN2D$^{2B(CT)}$: 520.65±13.527 s), or in locomotor activity (Figure 1F) (center: Wt: 1184.87±218.32 cm; Tg-GluN2A$^{2B(CT)}$:

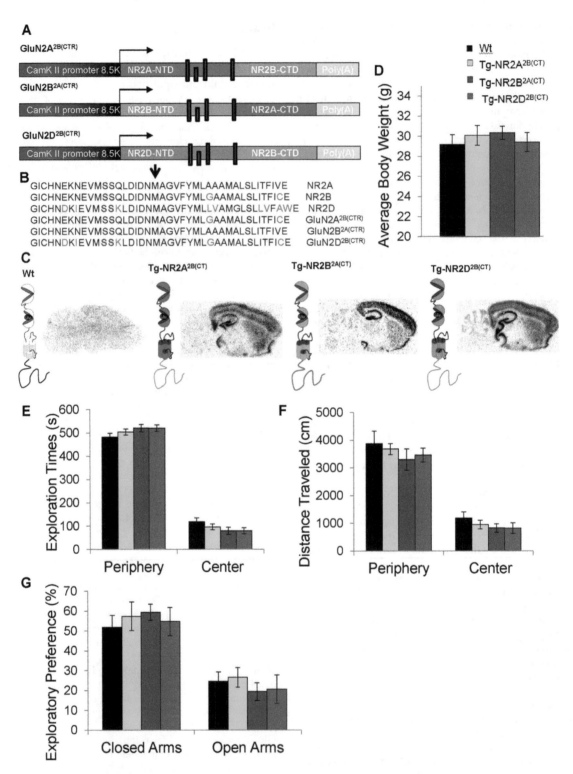

Figure 1. Constructs and basic behavioral assays of the GluN2 chimeric mice. (**A**) Illustration of the constructs used to create the GluN2A[2B(CTR)], GluN2B[2A(CTR)], and GluN2D[2B(CTR)] chimeric subunits. (**B**) A point mutation was made on the cloning vector to induce an Aat II cutting sites to link the N-terminal and membrane domain to the C-terminal domain. After successfully joining the domains, the point mutation was restored to the original sequence. The arrow indicates the fusion position located in trans-membrane domain. (**C**) *In situ* hybridization of the transgene expression in the wild-type mice (Wt), the Tg-GluN2A[2B(CT)] mice, the Tg-GluN2B[2A(CT)] mice and the Tg-GluN2D[2B(CT)] mice using SV-40 probes with a schematic of the receptor subunit expressed in the excitatory neurons. (**D**) No differences were found in the average adult body weight of the wild-type mice, the Tg-GluN2A[2B(CT)], Tg-GluN2B[2A(CT)], and Tg-GluN2D[2B(CT)]mice. (**E**) The chimeric transgenic mice spent similar amounts of time as the wild-type mice in the center verses the periphery of the open field arena. (**F**) The chimeric transgenic mice and the wild-type mice showed similar locomotion in the open field. (**G**) The Tg-GluN2A[2B(CT)], Tg-GluN2B[2A(CT)] and Tg-GluN2D[2B(CT)] mice spent similar amounts of time in the closed arms and the open arms of the elevated plus maze as the wild-type mice.

945.79±158.233 cm; Tg-GluN2B$^{2A(CT)}$: 824.15±147.123 cm; Tg-GluN2D$^{2B(CT)}$: 819.96±195.070 cm; periphery: Wt: 3864.83±460.03 cm; Tg-GluN2A$^{2B(CT)}$: 3675.25±196.950 cm; Tg-GluN2B$^{2A(CT)}$: 3294.38±378.288 cm; Tg-GluN2D$^{2B(CT)}$: 3458.34±249.227 cm). This suggests that these transgenic mice were normal in locomotor activity and anxiety. Additionally, no differences were found in the elevated plus maze paradigm, which also measured for anxiety-like behavior (Figure 1G). Therefore, the chimeric GluN2 mice were indistinguishable from their wild-type littermates in growth, body weights, and these basic behaviors.

Enhancement of long-term object recognition memory in Tg-GluN2A$^{2B(CT)}$ mice but impairments in long-term memory of the Tg-GluN2B$^{2A(CT)}$ and Tg-GluN2D$^{2B(CT)}$ mice

To investigate recognition memory functions in the transgenic mice, we tested the mice in a novel object recognition task for both short-term and long-term memory domains. During training, all transgenic mouse groups showed comparable exploratory behavior and motivation for the task, exploring each object to a similar degree (Figure 2A) (Wt: n = 10, 51.55±3.65%; Tg-GluN2A: n = 10, 50.15±0.932%; Tg-GluN2B: n = 9, 49.17±1.611%; Tg-

GluN2A$^{2B(CT)}$: n = 10, 51.85±3.192%; Tg-GluN2B$^{2A(CT)}$: n = 10, 49.85±0.932%; Tg-GluN2D$^{2B(CT)}$: n = 20, 52.52±2.097%).

At the one hour retention session, the Tg-GluN2A, Tg-GluN2B and Tg-GluN2A$^{2B(CT)}$ mice showed similar interest in the novel object as compared to the wild-type mice (Wt: n = 10, 60.90±4.913%; Tg-GluN2A: n = 11, 59.80±4.270%; Tg-GluN2B: n = 7, 57.38±2.76%; Tg-GluN2A$^{2B(CT)}$: n = 14, 59.15±5.294%) demonstrating no changes in short-term recognition memory. Whereas the Tg-GluN2B$^{2A(CT)}$ and Tg-GluN2D$^{2B(CT)}$ mice show no preference for the novel object (Tg-GluN2B$^{2A(CT)}$: n = 12, 51.15±4.808%; Tg-GluN2D$^{2B(CT)}$: n = 22, 53.85±2.535%), suggesting memory impairment in this test.

At the 24 hour retention session, the Tg-GluN2A mice showed no preference for the novel object and significantly less interest in it than the wild-type mice (Figure 2A) (GluN2A: n = 10, 50.03±3.860%; $F(2, 24) = 7.45$, p = 0.003) as noted previously [23], demonstrating their inability to form a long-term recognition memory. As expected, the Tg-GluN2D$^{2B(CT)}$ mice also showed no preference for the novel object at the 24 hour retention session spending significantly less time exploring the novel object than the wild-type mice (Wt: n = 10, 66.56±3.610%; Tg-GluN2D$^{2B(CT)}$: n = 20, 43.80±2.566%; $F(5, 63) = 6.36$, p = $7.7×10^{-5}$). The Tg-GluN2B$^{2A(CT)}$ mice spent only slightly more time with the novel object than the familiar object (n = 12, 55.10±4.821%). However,

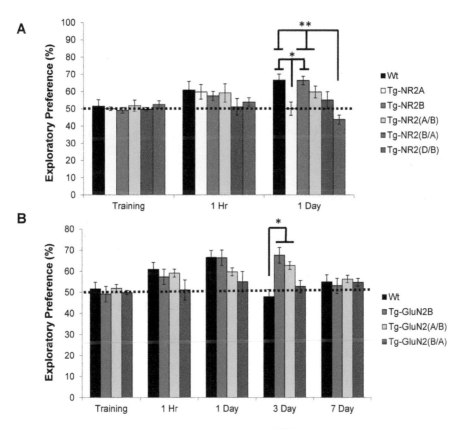

Figure 2. Enhanced long-term recognition memory of the Tg-GluN2A$^{2B(CT)}$ mice and impaired long-term memory on the Tg-GluN2D$^{2B(CT)}$ mice. (A) All groups of mice tested showed similar exploratory behavior in the training session. At the one hour retention session, the Tg-GluN2A, Tg-GluN2B and Tg-GluN2A$^{2B(CT)}$ mice showed similar interest in the novel object as the wild-type mice. Whereas the Tg-GluN2B$^{2A(CT)}$ and Tg-GluN2D$^{2B(CT)}$ mice show almost no preference for the novel object. At the 24 hour retention test, as expected the Tg-GluN2A and Tg-GluN2D$^{2B(CT)}$ mice showed no preference for the novel object. The Tg-GluN2B, The Tg-GluN2A$^{2B(CT)}$ and Tg-GluN2B$^{2A(CT)}$ mice all spent similar amounts of time with the novel object. *p = 0.003, **p = $7.7×10^{-5}$. (B) In addition to the enhancement seen in the Tg-GluN2A$^{2B(CT)}$ mice at the 24 hour recall session, these mice also showed enhanced recognition memory even at 3 days post-training over the wild-type mice. *p = 0.003. Whereas the GLUN2A$^{2B(CT)}$ mice show no preference for the novel object at 3 day or 7 days.

the Tg-GluN2B and Tg-GluN2A$^{2B(CT)}$ mice showed similar memory of the novel object to the wild-type mice (Tg-GluN2B: n = 7, 66.44±2.417%; Tg-GluN2A$^{2B(CT)}$: n = 14, 59.77±3.418%). This demonstrates impaired long-term recognition memory in the Tg-GluN2D$^{2B(CT)}$ mice.

To determine the extent of the enhancement in the Tg-GluN2A$^{2B(CT)}$ mice, separate cohorts of mice were further used to test in their ability to retain the memory of the object over three-day and seven-day periods. Remarkably, at the three day retention tests, the Tg-GluN2A$^{2B(CT)}$ mice, like the Tg-GluN2B mice, spent significantly more time investigating the novel object (Figure 2B) (Tg-GluN2B: n = 7, 67.64±4.337%; Tg-GluN2A$^{2B(CT)}$: n = 14, 62.76±2.968%) than the wild-type mice (Wt: n = 10, 47.93±4.045%, $F(2, 28)$ = 7.05, p = 0.003). However, the Tg-GluN2B$^{2A(CT)}$ mice showed no interest in the novel object, spending approximately equal time with both objects (Tg-GluN2B$^{2A(CT)}$: n = 8, 52.74±2.924%). At the seven day retention session, the Wt, GluN2B, Tg-GluN2B$^{2A(CT)}$, and Tg-GluN2A$^{2B(CT)}$ mice spent similar amounts of time investigating the novel object (Wt: n = 10, 55.05±4.096%; Tg-GluN2B: n = 6, 53.12±3.373%; Tg-GluN2A$^{2B(CT)}$: n = 14, 56.36±3.344%; Tg-GluN2B$^{2A(CT)}$: n = 8, 54.75±1.928%; $F(2,35)$ = 0.21, p = 0.93). This demonstrates the significant enhancement in long-term recognition memory in the Tg-GluN2A$^{2B(CT)}$ mice, to a similar degree as the Tg-GluN2B mice did.

Normal contextual fear memory in the chimeric transgenic mice

To investigate the emotional memory in the chimeric transgenic mice, we tested the mice in a contextual fear conditioning task. This type of fear conditioning is hippocampal-dependent and is often used to test short-term (one-hour) and long-term (one-day) time points. In the training session, all of the mice displayed similar freezing responses immediately after the shock was delivered (Figure 3A) (Wt: n = 13, 26.92±6.126%; Tg-GluN2A: n = 15, 25.75±3.146%; Tg-GluN2B: n = 30.00±3.637%; Tg-GluN2A$^{2B(CT)}$: n = 13, 34.61±5.155%; Tg-GluN2B$^{2A(CT)}$: n = 12, 25.00±7.812%; Tg-GluN2D$^{2B(CT)}$: n = 13, 23.18±7.045%). At the one-hour retention session, the wild-type mice, the Tg-GluN2A, Tg-GluN2B$^{2A(CT)}$ and Tg-GluN2D$^{2B(CT)}$ mice all displayed similar freezing responses when they were returned to the shock chamber in the absence of footshock (Wt: n = 13, 26.28±4.444%; Tg-GluN2A: n = 11, 32.22±4.789%; Tg-GluN2B$^{2A(CT)}$: n = 10, 25.28±6.286%; Tg-GluN2D$^{2B(CT)}$: n = 13, 37.51±5.415%). Interestingly, both the Tg-GluN2B and the Tg-GluN2A$^{2B(CT)}$ mice spent significantly more time freezing than the wild-type mice (Tg-GluN2B: n = 7, 45.06±2.823%; Tg-GluN2A$^{2B(CTR)}$: n = 13, 52.56±3.672%; $F(6, 61)$ = 4.98, p = 0.0007). This suggests that Tg-GluN2A$^{2B(CT)}$ mice, similar to Tg-GluN2B, exhibited enhanced 1-hr contextual fear memory.

At the one-day retention session, the Tg-GluN2B mice still displayed significantly more freezing than the wild-type mice as previously reported (Figure 3A) (n = 8, 55.94±4.911%, p = 0.039), suggesting greater long-term contextual fear memory in these mice. However, Tg-GluN2A$^{2B(CT)}$, Tg-GluN2B$^{2A(CT)}$ and Tg-GluN2D$^{2B(CT)}$ mice displayed similar freezing responses as those of the wild-type mice, indicating that all of these chimeric transgenic mice have the normal 1-day hippocampal-dependent contextual fear memories. (Figure 3A) (Tg-GluN2A$^{2B(CT)}$: n = 13, 44.87±4.779%; Tg-GluN2B$^{2A(CT)}$: n = 12, 41.91±7.003%; Tg-GluN2D$^{2B(CT)}$: n = 12, 47.92±7.050%). On the contrary, the Tg-GluN2A mice demonstrated reduced freezing responses during the contextual recall (Wt: n = 14, 43.65±4.382%; Tg-GluN2A: n = 13, 23.54±3.811%, $F(5, 66)$ = 3.65, p = 0.005), suggesting

the deficit in converting short-term contextual fear memory into long-term contextual fear memory due to expression of GluN2A.

Tg-GluN2B$^{2A(CT)}$ and Tg-GluN2D$^{2B(CT)}$ mice showed significant impairments in long-term cued fear memory, whereas Tg-GluN2A$^{2B(CT)}$ showed enhanced memory

To assess whether and how hippocampal-independent forms of memories are affected by the N-terminal and C-terminal domain properties, we used a new cohort of the mice and tested them in cued fear conditioning task which required the mouse to associate an unconditioned stimulus (a shock) with a conditioned stimulus (a tone). In the cued fear conditioning paradigm, all of the mice exhibited little pre-tone freezing responses during the retention tests as they entered a novel chamber (Figure 3B) (Wt: n = 10, 2.22±0.997%; Tg-GluN2A: n = 15, 6.17±1.734%; Tg-GluN2B: n = 7, 6.17±1.251%; Tg-GluN2A$^{2B(CT)}$: n = 11, 2.02±0.758%; Tg-GluN2B$^{2A(CT)}$: n = 11, 1.26±0.576%; Tg-GluN2D$^{2B(CT)}$: n = 12, 3.47±1.373%). Upon the recall tone, all five types of transgenic mice exhibited significant amounts of freezing responses at the one hour retention session, comparable to that of wild-type mice (Wt: n = 10, 71.67±4.843%; Tg-GluN2A: n = 18, 58.08±4.710%; Tg-GluN2B: n = 7, 62.08±6.801%; Tg-GluN2A$^{2B(CT)}$: n = 10, 67.22±4.83%; Tg-GluN2B$^{2A(CT)}$: n = 10, 60.00±4.833%; Tg-GluN2D$^{2B(CT)}$: n = 9, 61.11±4.856%). This shows that the transgenic mouse lines have normal short-term hippocampal-independent emotional memory and all are able to form an association between the tone (CS) and the shock (US).

For one-day cued fear memory retention tests, a second cohort of similarly trained mice was placed into a novel enclosure and an identical tone to the training tone was presented. Interestingly, the Tg-GluN2A mice, the Tg-GluN2B$^{2A(CT)}$ mice and the Tg-GluN2D$^{2B(CT)}$ mice demonstrated significantly less freezing than the wild-type mice (Figure 3B) (Wt: n = 9, 55.02±3.030%; Tg-GluN2A: n = 10,14.75±5.189%; Tg-GluN2B$^{2A(CT)}$: n = 10, 14.72±8.477%; Tg-GluN2D$^{2B(CT)}$: n = 12, 16.67±5.772%; $F(5, 56)$ = 27.03, p<1.0×10^{-6}). The Tg-GluN2A$^{2B(CT)}$ mice froze significantly more than the wild-type mice (n = 13, 71.58±2.727%). Consistent with the previous studies (Tang et al, 1999), the Tg-GluN2B mice also showed enhanced cued memory over the wild-type mice (Tg-GluN2B: n = 8, 65.75±5.918%, p< 0.05). These data demonstrate that the Tg-GluN2A, Tg-GluN2B$^{2A(CT)}$ and Tg-GluN2D$^{2B(CT)}$ mice have impaired long-term hippocampal-independent fear memory, whereas the Tg-GluN2A$^{2B(CT)}$ and Tg-GluN2B mice exhibited similarly enhanced long-term cued fear memories.

Enhanced cued fear extinction in Tg-GluN2A$^{2B(CT)}$ mice over the wild-type mice

Fear extinction has been widely used as a test for assessing flexible learning behaviors. The extinction of learned fear requires the formation of new flexible relations, instead of forgetting or erasing the established fear memories [25]. Because the Tg-GluN2A, Tg-GluN2B$^{2A(CT)}$, and Tg-GluN2D$^{2B(CT)}$ mice were impaired in the one-day retention session, they were not used for the fear extinction experiment. Instead, we focused our investigation of this form of learning on the Tg-GluN2A$^{2B(CT)}$ mice, the Tg-GluN2B mice, as well as the wild-type mice.

In this fear extinction task, we used a five trial extinction paradigm in which the animals were repeatedly exposed to the training chamber (contextual extinction) or the tone in a novel context (cued extinction) without the delivery of the shock in either context. We first tested the mice in the contextual fear extinction paradigm. We found that the Tg-GluN2B mice initially showed

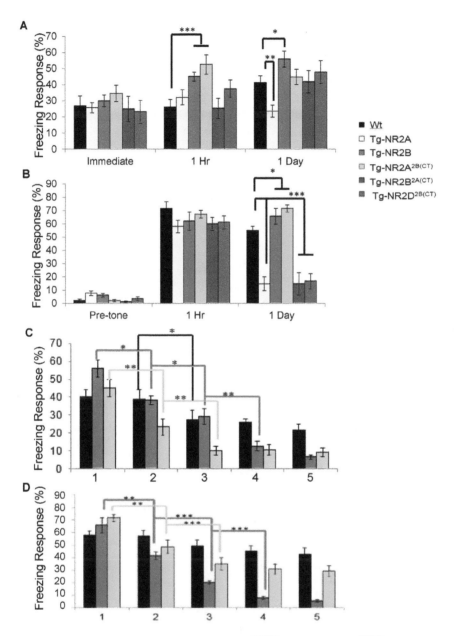

Figure 3. Selectively impaired emotional memory in the Tg-GluN2B$^{2A(CT)}$ and Tg-GluN2D$^{2B(CT)}$ mice. (A) The mice showed similar freezing responses immediately following the US. At the one hour retention session, the wild-type mice, the Tg-GluN2A, Tg-GluN2B$^{2A(CT)}$ and Tg-GluN2D$^{2B(CT)}$ mice all displayed similar freezing responses. Interestingly, both the Tg-GluN2B and the Tg-GluN2A$^{2B(CT)}$ mice spent significantly more time freezing. At the 24 hour recall session only the Tg-GluN2A mice demonstrated a diminished freezing response to the context in which the shock was delivered. *p = 0.005, **p = 0.0007. **(B)** The mice tested also showed similar pre-tone freezing responses and similar freezing at the one hour contextual recall. At the 24 hour recall session the Tg-GluN2A mice, the Tg-GluN2B$^{2A(CT)}$ mice and the Tg-GluN2D$^{2B(CT)}$ mice demonstrated significantly less freezing than the wild-type mice, whereas the Tg-GluN2A$^{2B(CT)}$ mice and Tg-GluN2B froze significantly more than the wild-type mice. *p<0.05, **p<1.0×10^{-6}, *** p = 0.0007. **(C)** The Tg-GluN2A$^{2B(CT)}$ mice showed quicker fear extinction than the wild-type mice in the contextual fear extinction paradigm *p<0.05, **p<0.01. **(D)** The Tg-GluN2A$^{2B(CT)}$ mice showed quicker fear extinction to the CS than the wild-type mice in the contextual fear extinction paradigm *p<0.05, **p<0.01, ***p<0.001.

significantly more freezing than the wild-type mice and the Tg-GluN2A$^{2B(CT)}$ mice 24 hours after the training session (Figure 3C) (1: Wt: n = 15, 40.19±3.852%; Tg-GluN2B: n = 8, 55.94±4.911%; Tg-GluN2A$^{2B(CT)}$: n = 13, 44.87±4.779%). Interestingly, over the extinction trials both the Tg-GluN2B and the Tg-GluN2A$^{2B(CT)}$ mice significantly decreased their freezing responses as early as the second session 2 hours later in comparison to the first trial (2: Tg-GluN2B: 38.29±5.161%; p = 0.014; Tg-GluN2A$^{2B(CT)}$: 23.29±4.409%; p = 0.002), whereas

the wild-type mice did not significantly decrease their freezing response from the first to the second exposure session. It is noted that the Tg-GluN2A$^{2B(CT)}$ mice exhibited significantly less freezing, as determined by ANOVA analysis, than both the Tg-GluN2B mice and their wild-type littermates (Wt: 38.70±5.161%; $F_{(2, 33)}$ = 3.56, p = 0.04). All groups of mice spent significantly less time freezing in the third exposure than the second exposure (3: Wt: 27.41±5.034%, p = 0.01; Tg-GluN2B: 29.12±4.294%, p = 0.03; Tg-GluN2A$^{2B(CT)}$: 10.26±2.632%, p = 0.007). The mice continued to decrease their freezing responses in the fourth (4: Wt: 25.74±2.125%; Tg-GluN2B: 12.69±2.576%; Tg-GluN2A$^{2B(CT)}$: 10.47±2.816%, p = 0.002) and fifth exposures (5: Wt: 21.48±3.325%; Tg-GluN2B: 6.57±1.300%; Tg-GluN2A$^{2B(CT)}$: 9.19±2.276%). ANOVA analysis indicated that in both the fourth and fifth exposures, the wild-type mice had significantly higher freezing responses than both the Tg-GluN2B and Tg-Glu-N2A$^{2B(CT)}$ mice (4: $F_{(2, 33)}$ = 11.91, p = 0.0001; 5: $F_{(2, 33)}$ = 8.08, p = 0.001). These data demonstrate that the Tg-GluN2B mice and the Tg-GluN2A$^{2B(CT)}$ mice had better hippocampal-dependent fear flexibility learning ability than the wild-type mice.

Next, we exposed the mice to the tone in a novel environment for cued fear extinction learning. The Tg-GluN2A$^{2B(CT)}$ mice, similar to the Tg-GluN2B mice, showed significantly faster cued fear extinction than the wild-type mice (Figure 3D). In the first exposure to the tone 24 hours after training, the Tg-GluN2B and Tg-GluN2A$^{2B(CT)}$ mice showed significantly higher freezing responses than the wild-type mice (1: Wt: n = 12, 58.10±3.323%; Tg-GluN2B: n = 8, 70.87±5.017%; Tg-Glu-N2A$^{2B(CT)}$: n = 13, 71.58±2.727%, $F_{(2, 30)}$ = 4.97, p = 0.01). At the second exposure, two hours after the first exposure, the Tg-GluN2B mice and the Tg-GluN2A$^{2B(CT)}$ mice significantly reduced their freezing responses to the presentation of the tone, whereas the wild-type mice did not (2: Wt: 57.41±4.436%; Tg-GluN2B: 41.46±2.836%, p = 0.002; Tg-GluN2A$^{2B(CT)}$: 48.71±5.34%, p = 0.003), again suggesting the faster fear extinction in these transgenic mice. Remarkably, both the Tg-GluN2B and Tg-GluN2A$^{2B(CT)}$ mice further decreased freezing from the second to the third exposures as well (3: Wt: 49.54±4.323%; Tg-GluN2B: 20.14±1.278%, p = 0.0004; Tg-GluN2A$^{2B(CT)}$: 35.04±4.958%, p = 0.0005). The wild-type mice spent significantly more time freezing in the third exposure than the Tg-GluN2B and Tg-GluN2A$^{2B(CT)}$ mice ($F_{(2, 30)}$ = 9.87, p = 0.0005). The Tg-GluN2B mice and the Tg-GluN2A$^{2B(CT)}$ mice continued to decrease their freezing responses in the fourth (4: Wt: 45.14±4.340%; Tg-GluN2B: 7.75±1.175%; Tg-Glu-N2A$^{2B(CT)}$: 30.77±4.118%) and fifth exposures (5: Wt: 42.82±4.972%; Tg-GluN2B: 5.34±0.939%; Tg-GluN2A$^{2B(CT)}$: 29.27±4.342%). It is worth noting that the Tg-GluN2B had the faster extinction learning. ANOVA analysis revealed that the Tg-GluN2B mice demonstrated significantly less freezing than the Tg-GluN2A$^{2B(CT)}$ mice and the wild-type mice in both the fourth ($F_{(2, 30)}$ = 19.32, p = 4.0×10^{-6}) and fifth exposures ($F_{(2, 30)}$ = 16.22, p = 1.7×10^{-5}).

Basic electrophysiological properties in the chimeric GluN2 mice

GluN2A and GluN2B subunits' contribution to synaptic plasticity has been intensely investigated in the CA1 region using both pharmacological and genetic methods [26–29]. We took advantage of the existing knowledge in the literature and investigated and compared how various chimeric transgenic overexpressions would affect the bidirectional control of synaptic plasticity in the CA1 region. To investigate the basic electrophysiological properties in the hippocampus of the Tg-GluN2A$^{2B(CT)}$, Tg-GluN2B$^{2A(CT)}$, and Tg-GluN2D$^{2B(CT)}$ mice, we recorded from the CA1 Schaffer collaterals of the mouse hippocampus. We found the input-output properties (Figure S2A), as well as the paired pulse facilitation (Figure S2B) from each genotype were similar to those of the wild-type controls, thereby demonstrating normal presynaptic function and basal transmissions in these transgenic mice. We then systematically measured the long-term potentiation (LTP) and long-term depression (LTD) in the CA1 slices from each mouse line.

Enhanced 10 Hz induced LTP observed in the Tg-GluN2A$^{2B(CT)}$ CA1 region

We first performed LTP and LTD studies on the Tg-GluN2A$^{2B(CT)}$ mice. In the Tg-GluN2A$^{2B(CT)}$ mice, LTP can be readily induced by 100 Hz stimulation (Figure 4A) (Wt: n = 6/3 (# of slices/# of animals), 135.2±7.6%; Tg-GluN2A$^{2B(CT)}$: n = 7/4, 146.5±8.7%). Interestingly, a significant increase in LTP was observed in the transgenic mice, compared to that of wild-type slices, in response to the 10 Hz frequency stimulation (Figure 4B) (Wt: n = 7/4, 103.2±13.0%; Tg-GluN2A$^{2B(CT)}$: n = 5/3, 150.1±17.0%). Additionally, a significant difference was further observed at 5 Hz stimulation (Figure 4C) (Wt: n = 4/3, 94.4±1.8%; Tg-GluN2A$^{2B(CT)}$: n = 6/4, 115.5±3.9%). There is no statistical difference at the 3 Hz stimulation (Figure 4D) (Wt: n = 7/3, 70.3±11.3%; Tg-GluN2A$^{2B(CT)}$: n = 5/3, 79.2±13.3%) or 1 Hz stimulation (Figure 4E) (Wt: n = 7/5, 82.4±1.6%; Tg-GluN2A$^{2B(CT)}$: n = 6/3, 95.4±15.6%). Overall, we found that the Tg-GluN2A$^{2B(CT)}$ mice show little difference in the LTD, except at 5 Hz, but show significantly enhanced LTP around at 10 Hz frequency (Figure 4F). These data indicate that the GluN2A$^{2B(CT)}$ overexpression produced synaptic changes that were more similar to that of GluN2B overexpression in the transgenic mice and rats [14,16,18].

Enhanced 10 Hz LTP and diminished 1–3 Hz LTD in the Tg-GluN2B$^{2A(CT)}$ CA1 region

We then measured synaptic plasticity in the Tg-GluN2B$^{2A(CT)}$ CA1 slices. Overexpression of GluN2B$^{2A(CT)}$ significantly increased LTP versus their wild-type littermates at both 100 Hz (Figure 5B) (n = 13/6, 176.6±16.2%; Figure 5A) and 10 Hz (n = 5/3, 180.7±33.0%) frequencies. Interestingly, while 10 Hz response did not differ, LTD was also significantly impaired as compared to the wild-type hippocampal slices at 5 Hz (n = 4/2, 121.6±2.2%; Figure 5C), 3 Hz (n = 6/3, 103.6±13.8%; Figure 5D) and 1 Hz (n = 21/13, 114.2±6.6%; Figure 5E). This shows that although the Tg-GluN2B$^{2A(CT)}$ mice have significantly increased LTP, they also have significantly blocked 1 Hz and 3 Hz induced LTD (summarized in Figure 5F). This decrease in LTD in Tg-GluN2B$^{2A(CT)}$ slices was more similar to that seen in the Tg-GluN2A mice [23].

Impaired 5 Hz responses in the Tg-GluN2D$^{2B(CT)}$ CA1 region

Finally, we examined the effects of GluN2D$^{2B(CT)}$ overexpression on CA1 plasticity. Since GluN2D has very weak magnesium dependency but much greater opening duration, it would lead to significantly more Ca^{2+} influx into the postsynaptic sites. We performed LTP and LTD measurements on the Tg-GluN2D$^{2B(CT)}$ hippocampal slices. Interestingly, we found no differences at either the 100 Hz (Figure 6A) (n = 8/6, 128±8.6%) or the 10 Hz frequency between the transgenic and control littermates (Figure 6B) (n = 6/3, 125.5±12.6%). However, at the 5 Hz frequency, a small, but significant LTP was observed in the transgenic slices, in comparison to the LTD induced in the slices from the control littermates (Figure 6C) (n = 5/4, 123.5±3.6%). However, there were no significant differences observed in LTD at 1 Hz stimulation (Figure 6E) (n = 8/4, 88.6±7.6%), 3 Hz (Figure 6D) (n = 7/5, 87.6±12.4%). The summary graphs show the overall similarities between the Tg-GluN2D$^{2B(CT)}$ mice and their wild-type counterparts, except in its 5 Hz frequency response (Figure 6F).

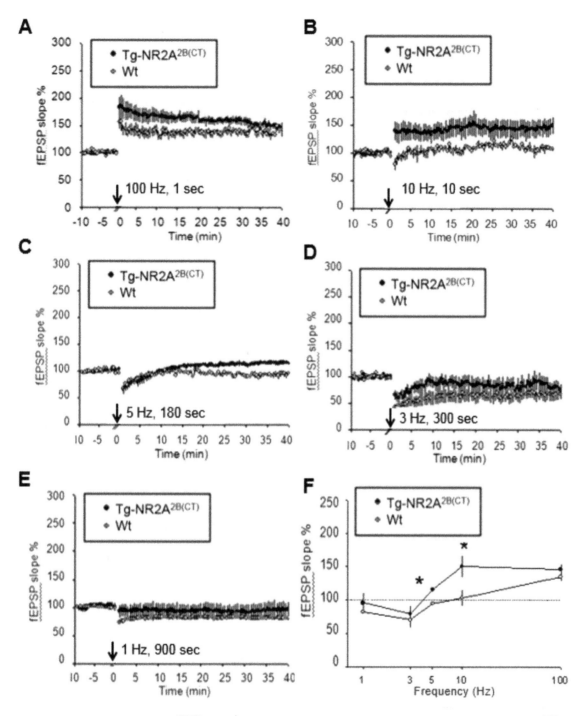

Figure 4. Enhanced LTP in the Tg-GluN2A$^{2B(CT)}$ mouse hippocampal slices. A. Slightly enhanced LTP seen in the Tg-GluN2A$^{2B(CT)}$ mice with a 1 s 100 Hz stimulation. B. Significantly enhanced LTP was seen in the Tg-GluN2A$^{2B(CT)}$ mice when a 10 Hz stimulation was applied from 10 s. C–E. No changes in LTD were seen in the 5 Hz, 3 Hz, or 1 Hz stimulation protocols. F. A summary plot of the % change in fEPSP slope versus the frequencies.

Discussion

The NMDA receptor is widely known as the key coincidence-detector at central synapses to implement Hebb's learning rule. The channel opening-duration and the level of membrane depolarization, determines the amount of Ca^{2+} that influxes into the cell [30,31]. In this study, we have identified the critical molecular motifs of the GluN2 subunits essential for achieving learning and memory enhancement in the adult mouse brain. By systematically analyzing three chimeric GluN2 transgenic mice together with Tg-GluN2A and Tg-GluN2B mice, we have tested two major hypotheses, namely, synaptic coincidence-detection/calcium influx hypothesis vs. GluN2B C-terminal intracellular signaling hypothesis in gating memory enhancement. Our experiments have revealed several novel insights into the relationships between GluN2 subunit motifs, synaptic plasticity, and memory enhancement.

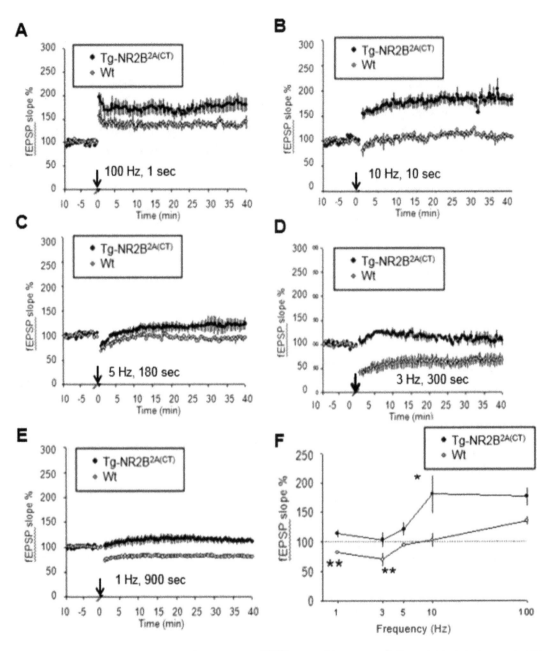

Figure 5. Enhanced LTP and diminished LTD in the Tg-GluN2B$^{2A(CT)}$ mouse hippocampal slices. (A) A slight increase in LTP was seen in the 100 Hz stimulation protocol in the Tg-GluN2B$^{2A(CT)}$ mice. **(B)** When a 10 Hz stimulation was applied for 10 s a significant increase in LTP was seen in the Tg-GluN2B$^{2A(CT)}$ mice over their wild-type littermates. **(C)** LTD was diminished at 5 Hz stimulation in the Tg-GluN2B$^{2A(CT)}$ mice. **(D)** LTD was significantly diminished at the 3 Hz stimulation protocol. **(E)** At the 1 Hz stimulation, the Tg-GluN2B$^{2A(CT)}$ mice show significantly diminished LTD. **(F)** A summary plot of the % change in fEPSP slope versus the frequencies.

The "synaptic coincidence-detection" hypothesis reflects the predominant view in the field as the rate-limiting factor in determining learning and memory capability. It posits that because the GluN2B subunit makes the channel opening duration longer than that of the GluN2A subunit, GluN2B overexpression allows a greater coincidence detection window, thereby leading to superior memory functions [4]. Because the N-terminal and transmembrane domains of the GluN2 subunits are known to be crucial for controlling voltage-gating and ion (Ca^{2+}) influx duration, we replaced the GluN2B N-terminal domains with either GluN2A or GluN2D while retaining its wild-type C-terminal intracellular domain. As such, these two chimeric GluN2

subunits possessed the GluN2B intracellular signaling capability but with the other key properties such as the shorter opening duration and the voltage-dependency from the GluN2A and GluN2D, respectively.

As predicted, because GluN2D has greatly reduced Mg^{2+} dependency, which renders synaptic coincidence-detection ineffective, GluN2D$^{2B(CT)}$ transgenic mice indeed exhibited memory deficits in novel object recognition (in both the short-term and long-term form) and long-term cued fear conditioning memory (although the contextual fear memory seemed to be normal). These observations have provided evidence that synaptic coincidence-detection is necessary for producing memory enhancement

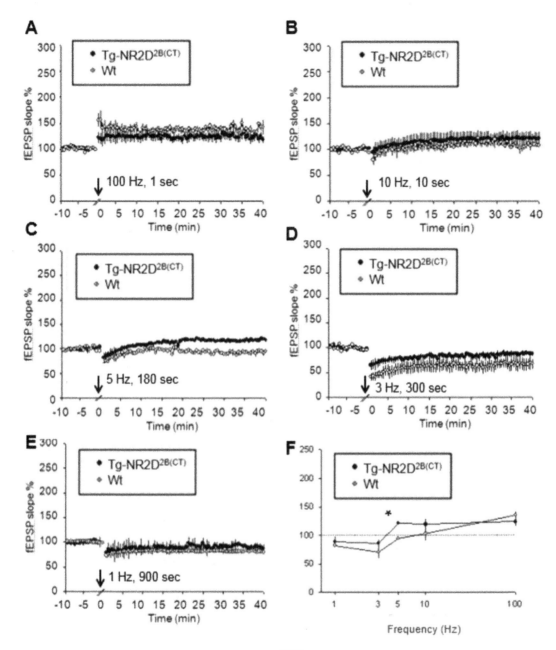

Figure 6. Diminished LTD at the 5 Hz range in the Tg-GluN2D^2B(CT) mouse hippocampal slices. (A) No changes from the wild-type hippocampal slices were seen in the LTP of the Tg-GluN2D^2B(CT) mouse hippocampal slices when a 100 Hz stimulation was applied. **(B)** When a 10 Hz stimulation was applied for 10 s there, again was no significant change in LTP observed in the Tg-GluN2D^2B(CT) mice over their wild-type littermates. **(C)** LTD was diminished at 5 Hz stimulation in the Tg-GluN2D^2B(CT) mice. **(D)** LTD was not significantly diminished at the 3 Hz stimulation protocol. **(E)** At the 1 Hz stimulation, the Tg-GluN2D^2B(CT) mice showed no significant differences in LTD. **(F)** A summary plot of the % change in fEPSP slope versus the frequencies.

via the GluN2B intracellular signaling cascades. Without proper magnesium dependent voltage gating, the presence of the overexpressed GluN2B domain from the chimeric GluN2D^2B(CTD) subunit still could not produce optimal synaptic changes for memory enhancement.

On the other hand, we were quite surprised that the Tg-GluN2A^2B(CT) transgenic mice exhibited a very similar memory enhancement phenotype to those of the Tg-GluN2B mice. The different channel opening-durations derived from GluN2A and GluN2B subunits' N-terminal and transmembrane domains are not the most critical factor in determining the memory

enhancement, as long as the GluN2B C-terminal domain is transducing the signaling. It is important to note here that Punnakkal et al. found little differences in the whole cell currents of similar chimeric constructs, with only a slight decrease in the peak amplitude of a similar GluN2AB construct from that of the GluN2A wildtype subunit [32]. No changes in the peak amplitude of a similar GluN2BA construct over that of the GluN2B wildtype subunit. Additionally, deactivation times remained unchanged between the wildtype and chimeric receptors. Importantly, they also concluded that the peak opening probability appeared to be determined by the GluN2

N-terminal domain [32]. Therefore, these genetic experiments have shown that Mg^{2+}-dependent coincidence-detection function, but not necessarily the opening-duration difference between GluN2A and GluN2B, is prerequisite for achieving learning and memory enhancement in the adult brain.

Interestingly our present study has provided clear evidence supporting the second hypothesis that is known as the "GluN2B intracellular domain" hypothesis [4,33]. Two separate pieces of evidence came from our behavioral analyses of the Tg-GluN2-B$^{2A(CT)}$ and Tg-GluN2A$^{2B(CT)}$ transgenic mice. First, we found that the Tg-GluN2A$^{2B(CT)}$ mice had enhanced object recognition memory and emotional memory. These phenotypes are very similar to those of the Tg-GluN2B mice (and also Tg-GluN2B rats) (Figure 7). On the contrary, when the C-terminal domain of the GluN2B subunit was replaced by that of the GluN2A subunit, as we did in Tg-GluN2B$^{2A(CT)}$ mice, this swap led to profound memory deficits in novel object recognition test and long-term cued fear memories. These memory deficits mirrored those of Tg-GluN2A mice [23]. These subunits-swap experiments, by extending to learning and memory enhancement, are consistent with other reports that the intracellular domains of the GluN2 subunits play critical roles in mediating different functions, such as synaptic localization, clustering, signal transduction, and behaviors [22,33–39]. Therefore, our studies suggest that both "synaptic coincidence-detection" hypotheses and "GluN2B intracellular signaling" hypothesis are mutually complementary in term of explaining the molecular determinants for memory enhancement.

Two additional conceptual insights have also been obtained on how different molecular motifs of the overexpressed GluN2 subunits regulate the levels and degrees of LTP or LTD over a wide range of stimulation frequencies. Despite multiple pharmacological and knockout approaches to analyzing GluN2A and GluN2B on regulating LTP and LTD [40], few were done under the context of examining its relationship with cognitive enhancement [16,18,41]. Here, we consistently found that GluN2B and GluN2A$^{2B(CT)}$ overexpression enhanced LTP in the range of 10 Hz and/or 100 Hz range without significantly affecting 1 Hz or 3 Hz LTD. It is noteworthy to point out that that Tg-GluN2A$^{2B(CT)}$ seemed to produce larger 100 Hz LTP in the initial 20~30 minutes range than that of wild-type slices, but become indistinguishable by the 40-minutes time points. Interestingly, the Tg-GluN2B overexpression tended to produce much larger 100 Hz induced LTP in comparison to the wild-type mice well

beyond 60 minutes [16]. This indicates the longer time duration of channel opening (thereby more calcium influx) via the GluN2B N-terminal and pore does make larger and more stable LTP in response to 100 Hz stimulation [42,43]. However, at the 10 Hz stimulation range, the amount of calcium influx via the GluN2A N-terminal and pore domains (coupled to GluN2B C-terminal region) can produce the similarly larger LTP in the Tg-GluN2A$^{2B(CT)}$ mice as that of the Tg-GluN2B mice in comparison to that of the wild-type controls. This 10 Hz stimulation frequency can be particularly interesting because we have observed that fear conditioning-induced firing increase in CA1 pyramidal cells is mostly in the range of 5~30 Hz [6,44]. This behaviorally relevant frequency range deserves special investigation for memory enhancement in future experiments both in the hippocampus and other brain regions such as the prefrontal cortex and amygdala. In addition, contrary to LTD produced by 5 Hz stimulation in the wild-type slices, Tg-GluN2D$^{2B(CT)}$ slices exhibited a significant switch to LTP. Taken together, these findings have provided additional support for the notion that the GluN2B C-terminal domain plays a key role in regulating LTP [26,45–48], and more importantly, our study has further defined, for the first time, its essential link to memory enhancement.

While it is evident that the C-terminal of the GluN2B subunit plays a crucial role in producing synaptic potentiation, we found that Tg-GluN2B$^{2A(CT)}$ mice had larger, more robust, LTP not only at 10 or 100 Hz. Intriguingly, such a swap also promoted an overall shift toward potentiation even in response to lower frequencies. As a result, the ability to produce LTD at 1–3 Hz frequency range is greatly impaired in Tg-GluN2B$^{2A(CT)}$ slices. These findings show that longer opening duration achieved by the overexpressed GluN2B N-terminal and pore domains, but coupled to the GluN2A intracellular signaling cascade, brings a greater potentiation but at the cost of losing synaptic depression capacity. This is in stark contrast with the normal LTD in Tg-GluN2B or Tg-GluN2A$^{2B(CT)}$ in response to 1 or 3 Hz stimulation. This strongly suggests that increased calcium influx (via the GluN2B N-terminal and core domains) is useful to produce bigger LTP, but its effect on 1 Hz LTD critically depends on whether the downstream signaling cascade is mediated by the GluN2A C-terminal tail or GluN2B C-terminal tail. In other words, under such circumstance, the presence of chimeric GluN2A C-terminal domain, but not chimeric GluN2B C-terminal domain, can override LTD. This

	LTP	LTD	Novel Object Recognition	Contextual Fear Conditioning	Cued Fear Conditioning
Tg-NR2A$^{2B(CT)}$	Enhanced at 10 Hz	Impaired at 5Hz	Enhanced at 3 days	Enhanced	Normal
Tg-NR2B$^{2A(CT)}$	Enhanced	Significantly Impaired	Impaired at 1 hour Normal at 24 hour	Normal	Impaired at 24 hours
Tg-NR2D$^{2B(CT)}$	Normal	Decreased at 5 Hz	Impaired at 24 hours	Normal	Impaired at 24 hours

Figure 7. Summary of LTP, LTD and behavioral tasks results. The Tg-NRA$^{2B(CT)}$ mice had enhanced LTP, as well as enhanced long-term recognition memory and contextual fear conditioning. The Tg-GluN2B$^{2A(CT)}$ mice have significantly impaired LTD resulting in impaired short-term recognition memory and impaired long-term cued fear conditioning. The Tg-GluN2D$^{2B(CT)}$ mice have decreased LTD at 5 Hz and impaired long-term recognition memory and long-term cued fear memory.

novel insight adds to the notion that GluN2A may have a general ability to drive toward LTP [49–54].

In addition, by taking advantage of the correlational analysis between synaptic changes and memory performances, our present study has uncovered two detailed insights into the memory enhancement strategy: first, bigger LTP would lead to better learning and memory, however, only if the LTD ability remains intact. This is supported by the observation that bigger CA1 LTP is associated with better memory in Tg-GluN2B and Tg-GluN2A$^{2B(CT)}$ mice while their LTD was not altered. Second, if LTP enhancement results in overriding or diminishing LTD capacity, such as those observed in Tg-GluN2B$^{2A(CT)}$, it would also lead to memory deficits. The Tg-GluN2B$^{2A(CT)}$ phenotypes are more similar to the knockout of PSD-95 which also leads to larger LTP, lack of 1 Hz LTD, and memory deficits (i.e. [55,56]). Our recent characterization of Tg-GluN2A mice showed that overexpression of GluN2A results in no change in 100 Hz LTP or 1 Hz, but greatly impaired 3 or 5 Hz LTD. These mice also exhibited long-term memory deficits, while short-term memories remained mostly normal. Our Tg-GluN2D$^{2B(CT)}$ mice, which also showed 5 Hz LTP responses instead of either no change or LTD, as in the wild-type mice, were also profoundly impaired in long-term memory. These observations support the "LTD-memory trace sculpting" hypothesis, that the weakening of uncorrelated synaptic connections would reduce the background "noise" while enabling the stabilization (or crystallization) of the learning-related synaptic patterns [23]. However, it is important to note that our current electrophysiological recordings were limited to the CA1 region. Given the fact that we used the CaMKII promoter to drive the transgenes, electrophysiological analyses should be extended in future experiments into other brain regions such as the amygdala and prefrontal cortex from which cued fear learning and fear extinction are processed. Clearly, simple correlation between CA1 synaptic plasticity and memory are likely not sufficient for counting memory enhancement and thus, any extrapolation should be only taken with great caution. In addition, little is known about how any of the artificial stimulation paradigms for producing LTP or LTD can be translated into real-time memory patterns. Recent successful decoding of real-time fear memory traces in the hippocampal CA1 from the wild-type mice and the forebrain excitatory neuron-specific NMDA receptor inducible knockout mice, have revealed many fundamental insights how the NMDA receptors regulate real-time memory code and memory engrams [6,44]. It would be of great interest to use such brain decoding technologies to investigate the various transgenic mice described here.

In summary, our above experiments have identified the key molecular and genetic determinants that would be necessary and sufficient for achieving superior learning and memory ability in the adult brain. Although transgenetic methods are unlikely to be used for human clinical settings, the C-terminal region of the GluN2B subunit contains many important sites for various molecular interaction including with CaMKII, cdk5, and Kinesin superfamily protein 17 (KIF17) [57–60]. Indeed, manipulations of cdk5 and KIF17 which result in upregulation of GluN2B also resulted in memory enhancement [61,62]. More recently, researchers have taken a novel, dietary approach to up-regulate GluN2B expression in the brain via elevating brain magnesium [63]. They showed that the compound, magnesium threonate, can cross the blood brain barrier efficiently and boost GluN2B expression in the neurons, and subsequent memory improvement in both aging and wild-type mice [64,65]. This compound is currently under clinical trials [66]. Therefore, it is conceivable that knowledge gained from the present study will be valuable to the current efforts in developing and optimizing memory enhancement strategies.

Methods

Production of Transgenic Mice

We have produced three chimeric GluN2 subunit constructs for the present study. In the first two constructs, the N-terminal and transmembrane domains of GluN2A or GluN2D subunit were fused with the C-terminal domain from the GluN2B subunit, termed GluN2A$^{2B(CTR)}$ and GluN2D$^{2A(CTR)}$, respectively. In the third construct, we also fused the N-terminal and transmembrane domain of GluN2B subunit with the C-terminal domain from the GluN2A subunit, termed GluN2B$^{2A(CTR)}$ (see Figure 1A). The fusion site was located near the end of the fourth transmembrane domain, just before the C-terminal domains begin. For making the constructs, we first introduced a point mutation to create a unique Aat II cutting site for the fusing of the given C-terminal domain (Figure 1B). Upon successful ligation, the point mutation was mutated back to its original sequence. These chimeric transgene constructs were driven by the forebrain-specific αCaMKII promoter for targeting their expression to the excitatory neurons in the forebrain regions such as the cortex and hippocampus.

The chimeric constructs were created by first introducing a point-mutation at the site to create a unique Aat II cutting site for swapping the NT and CT domains. Upon successful ligation the point mutation was swapped back to the original sequence. The modified subunit was targeted for forebrain expression by the CaM-kinase II (CaMKII) promoter as previously described [16,67]. The founding line of transgenic animals was produced by pronuclear injection of a linearized chimeric transgene vector into C57BL/6J zygotes similar to previously described [16,68]. A total of seven independent mouse founder lines (three lines for GluN2A$^{2B(CTR)}$ termed "Tg-GluN2A$^{2B(CT)}$" mice, two lines for GluN2B $^{2A(CTR)}$ termed "Tg-GluN2B$^{2A(CT)}$" mice, and two lines for the GluN2D$^{2B(CTR)}$ termed "Tg-GluN2D$^{2B(CT)}$" mice). All these lines gave successful germline transmissions. The genotypes of the transgenic mice were determined by PCR analysis of a tail biopsy. The transgene was detected using the SV40 poly(A) sequence, as previously described [16,18,41]. Southern blotting was used to confirm the transgene integration in to the transgenic mouse line. Western blotting of the forebrain regions (cortex and hippocampus) was visualized with either a polyclonal GluN2A C-terminal antibody (Upstate/Millipore) or a polyclonal GluN2B C-terminal antibody (Millipore). For the present electrophysiological and behavioral experiments, we used a high expression line chosen from the Tg-GluN2A$^{2B(CT)}$ mice, the Tg-GluN2B$^{2A(CT)}$ mice, and the Tg-GluN2D$^{2B(CT)}$ mice that have been crossed with C57BL/6J wildtype mice for at least 8 generations. For in situ hybridizations, brains from the transgenic mice and wild-type littermates were isolated and 20 μm sections were prepared using a cryostat. The slices were hybridized to the [α^{35}S] oligonucleotide probe which hybridized to the untranslated artificial intron region in the transgene similarly to previously described [16,23].

Behavioral Experiments

Mice were maintained in a temperature and humidity controlled vivarium with a 12:12 light-dark cycle. All testing was done during the light phase with 3–5 month old animals. Mice were allowed free access to food and water, except during experimental procedures. Mice were extensively handled prior to any testing paradigm. Separate cohorts were used for each study and each recall time point unless otherwise stated. All testing procedures were conducted in sound dampened, dimly lit behavioral rooms. Experimenters were blind to the genotype of the animals. This study was carried out in strict accordance with the recommendations in the Guide for the Care and Use of Laboratory Animals of the National Institutes of Health. The protocols were approved by the Institutional Animal Care and Use Committee of the Georgia Regents University.

Open Field

One cohort of Tg-GluN2A, Tg-GluN2B, Tg-GluN2A$^{2B(CT)}$, Tg-GluN2B$^{2A(CT)}$, and Tg-GluN2D$^{2B(CT)}$ mice and their wild-type littermates were individually placed into a 50 cm L×50 cm W×25 cm H white Plexiglas open field arena. The mouse was allowed to explore for ten minutes. The time that the mouse spent in the center and periphery was determined. The periphery of the open field was considered to be the first four inches along the wall, while the center of the open field was the square inside this area [69]. Additionally, the distance traveled by the mouse was determined using Biobserve Viewer II software.

Elevated Plus Maze

The elevated plus maze consisted of a black Plexiglas "plus" maze approximately 60 cm above the floor, with each arm measuring 30 cm in length and 10 cm wide. Two opposite arms were left open, with the other two arms being enclosed on three sides. The ambient room lighting was 75 lux. The amount of time the mice spent within the enclosed arms was recorded, as well as the amount of time the animal spent in the open arms [69]. The times were used to determine a preference index.

Novel Object Recognition

The behavioral paradigm was the same as previously described [16,70]. The mice were individually habituated to a 50 cm L×50 cm W×25 cm H open field apparatus for 10 minutes a day for three days. On the first testing day, the mice were placed into the open field with two identical objects for 5 minutes. The time they spent exploring each object was recorded. At the described retention time the mice were placed back into the open field arena with one of the familiar objects used in training, and one novel object, and allowed to explore for 5 minutes. The time they spent with each object was recorded and used to determine a preference index. Different groups of mice were used for the each retention session.

Fear Conditioning

An operant chamber (25 cm L×25 cm W×38 cm H) equipped with activity monitors and camera was used. The flooring was a 24 bar shock grid with a speaker, shock generator, and photo-beam scanner (MedAssociates). The chamber was located in a sound damping isolation box. The apparatus was thoroughly cleaned with 70% ethanol between mice to avoid any olfactory cues. Freezing was monitored by the software and confirmed by the experimenter.

Testing procedures were similar to those previously described [16,71,72]. Animals were habituated to the testing environment for 5 minutes one day before testing. On the day of training the mice were placed into the chamber and allowed to explore for 5 minutes. Then the mice were exposed to a conditioned stimulus (CS, 85 dB tone at 2800 Hz), with the unconditioned stimulus (US, a scrambled foot shock at 0.75 mA) occurring the last 2 seconds of the CS. The mice were allowed to stay in the chamber for 30 seconds after the CS/US pairing to monitor immediate freezing.

To test the contextual freezing exhibited by the animal, at the described time (1 hour or 24 hours) the trained mice were placed back into the shock chamber for 5 minutes while their freezing response was monitored. The mice were then placed into a novel chamber and monitored for their freezing response (pre-tone) for 3 minutes before the onset of the CS tone for 3 minutes. During the tone the animal's freezing response was monitored to test the cued fear retention.

Freezing was judged as the complete immobility of the animal, except for movement necessary for respiration. The mice were then returned to their home cage for either 1 hour or 24 hours. At the described time the mice were returned to the chamber for measurement of the contextual freezing. The mice were then placed in a novel chamber and the tone was delivered for 3 min, during which their cued freezing response was monitored. To test the fear extinction of the animals the same recall testing paradigm was repeated at 2 hour intervals for four additional trials.

Statistical analysis of behavioral data

All behavioral data are presented as mean ± SEM. Significance was determined by ANOVA analysis with Tukey-Kramer, or a Student's t-test. P values of <0.05 were considered significant.

Hippocampal Slice Recordings

Transverse slices of the hippocampus were rapidly prepared from wild-type and Tg-GluN2A$^{2B(CT)}$, Tg-GluN2B$^{2A(CT)}$, and Tg-GluN2D$^{2B(CT)}$ mouse lines (3~6 months old) and maintained in an interface chamber at 28°C and were subfused with artificial cerebral spinal fluid (ACSF, 124 mM NaCl, 4.4 mM KCl, 2.0 mM CaCl$_2$, 1.0 mM MgSO$_4$, 25 mM NaCHO$_3$, 1.0 mM Na$_2$HPO$_4$ and 10 mM glucose) and bubbled with 95% O$_2$ and 5% CO$_2$. Slices were kept in the recording chamber for at least two hours. A bipolar tungsten stimulating electrode was placed in the stratum radiatum in the CA1 region. A glass microelectrode (3–12 MΩ) filled with ACSF was used to measure the extracellular field potentials in the stratum radiatum. Test response elicited at 0.02 Hz. Current intensity (0.5–1.2 mA) which produced 30% of maximal response was used for studies of PPF and synaptic plasticity at different frequencies. Various interpulse intervals (20–400 msec) were used for measuring PPF. Low-frequency stimulation of (5 Hz for 3 min, 3 Hz for 300 s, or 1 Hz for 900s) was then used to produce depotentiation [73]. Long term potentiation was induced by tetanic stimulation (100 Hz for 1 s and 10 Hz stimulation for 10 s). Data are expressed as mean ± SEM. One-way ANOVA (with Duncan's multiple range test for *post hoc* comparison) and Student's t-test were used for statistical analysis. The detailed procedures were the same as described (Tang et al. 1999; Shimizu, et al. 2000; Wang et al. 2003; Wang, et al. 2008).

Supporting Information

Figure S1 Conformation of the integration of the transgene. (**A**) Southern Blot analysis of Tg-GluN2(A/B), Tg-GluN2(B/A), and Tg-GluN2(D/B) mice. The numbers indicate the positive control and the copy number. (**B**) Western Blot analysis of the chimeric animals showing enhanced expression of the GluN2A C-terminal domain in the Tg-GluN2A, Tg-GluN2(B/A) mice and no enhancement of the GluN2A C-terminus in the Tg-GluN2(A/B) and Tg-GluN2(D/B) mice. (**C**) Western Blot analysis of the chimeric animals showing enhanced expression of the GluN2B C-terminal tail in the Tg-GluN2(A/B), and Tg-GluN2(D/B) mice of the expression in the wild-type mice.

Figure S2 Electrophysiology of hippocampal slices. (A) There were no significant differences in the basal synaptic transmission as seen in the CA3-CA1 input-output curve between the wildtype mice and the transgenic mice. **(B)**. The paired-pulse facilitation was unchanged between the wildtype and the chimeric transgenic mice indicating that the presynaptic function is unchanged.

Acknowledgments

The authors would like to thank Philip Wang and Shuqin Zhang for his valuable assistances with behavioral experiments, as well as Fengying Huang for animal colony maintenance.

Author Contributions

Conceived and designed the experiments: ZC JZT. Performed the experiments: SJ ZC RF HW DW. Analyzed the data: SJ ZC RF HW DW. Contributed reagents/materials/analysis tools: JZT. Wrote the paper: SJ JZT.

References

1. Bliss TV, Collingridge GL (1993) A synaptic model of memory: long-term potentiation in the hippocampus. Nature 361: 31–39.
2. Stevens CF, Sullivan J (1998) Synaptic plasticity. Curr Biol 8: R151–153.
3. Bear MF, Malenka RC (1994) Synaptic plasticity: LTP and LTD. Curr Opin Neurobiol 4: 389–399.
4. Tsien JZ (2000) Linking Hebb's coincidence-detection to memory formation. Curr Opin Neurobiol 10: 266–273.
5. Wang H, Hu Y, Tsien JZ (2006) Molecular and systems mechanisms of memory consolidation and storage. Prog Neurobiol 79: 123–135.
6. Zhang H, Chen G, Kuang H, Tsien JZ (2013) Mapping and Deciphering Neural Codes of NMDA Receptor-Dependent Fear Memory Engrams in the Hippocampus. PLoS One 8: e79454.
7. Mayer ML, Westbrook GL, Guthrie PB (1984) Voltage-dependent block by Mg2+ of NMDA responses in spinal cord neurones. Nature 309: 261–263.
8. Monyer H, Sprengel R, Schoepfer R, Herb A, Higuchi M, et al. (1992) Heteromeric NMDA receptors: molecular and functional distinction of subtypes. Science 256: 1217–1221.
9. Sheng M, Cummings J, Roldan LA, Jan YN, Jan LY (1994) Changing subunit composition of heteromeric NMDA receptors during development of rat cortex. Nature 368: 144–147.
10. Cull-Candy SG, Leszkiewicz DN (2004) Role of distinct NMDA receptor subtypes at central synapses. Sci STKE 2004: re16.
11. Erreger K, Dravid SM, Banke TG, Wyllie DJ, Traynelis SF (2005) Subunit-specific gating controls rat NR1/NR2A and NR1/NR2B NMDA channel kinetics and synaptic signalling profiles. J Physiol 563: 345–358.
12. Dingledine R, Borges K, Bowie D, Traynelis SF (1999) The glutamate receptor ion channels. Pharmacol Rev 51: 7–61.
13. Monyer H, Burnashev N, Laurie DJ, Sakmann B, Seeburg PH (1994) Developmental and regional expression in the rat brain and functional properties of four NMDA receptors. Neuron 12: 529–540.
14. Cui Y, Jin J, Zhang X, Xu H, Yang L, et al. (2011) Forebrain NR2B overexpression facilitating the prefrontal cortex long-term potentiation and enhancing working memory function in mice. PLoS One 6: e20312.
15. Jacobs SA, Tsien JZ (2012) Genetic overexpression of NR2B subunit enhances social recognition memory for different strains and species. PLoS One 7: e36387.
16. Tang YP, Shimizu E, Dube GR, Rampon C, Kerchner GA, et al. (1999) Genetic enhancement of learning and memory in mice. Nature 401: 63–69.
17. Tang YP, Wang H, Feng R, Kyin M, Tsien JZ (2001) Differential effects of enrichment on learning and memory function in NR2B transgenic mice. Neuropharmacology 41: 779–790.
18. Wang D, Cui Z, Zeng Q, Kuang H, Wang LP, et al. (2009) Genetic enhancement of memory and long-term potentiation but not CA1 long-term depression in NR2B transgenic rats. PLoS One 4: e7486.
19. Nowak L, Bregestovski P, Ascher P, Herbet A, Prochiantz A (1984) Magnesium gates glutamate-activated channels in mouse central neurones. Nature 307: 462–465.
20. Gielen M, Siegler Retchless B, Mony L, Johnson JW, Paoletti P (2009) Mechanism of differential control of NMDA receptor activity by NR2 subunits. Nature 459: 703–707.
21. Sprengel R, Suchanek B, Amico C, Brusa R, Burnashev N, et al. (1998) Importance of the intracellular domain of NR2 subunits for NMDA receptor function in vivo. Cell 92: 279–289.
22. Ryan TJ, Kopanitsa MV, Indersmitten T, Nithianantharajah J, Afinowi NO, et al. (2013) Evolution of GluN2A/B cytoplasmic domains diversified vertebrate synaptic plasticity and behavior. Nat Neurosci 16: 25–32.
23. Cui Z, Feng R, Jacobs S, Duan Y, Wang H, et al. (2013) Increased NR2A: NR2B ratio compresses long-term depression range and constrains long-term memory. Sci Rep 3: 1036.
24. Kennedy MB, Beale HC, Carlisle HJ, Washburn LR (2005) Integration of biochemical signalling in spines. Nat Rev Neurosci 6: 423–434.
25. Falls WA, Miserendino MJ, Davis M (1992) Extinction of fear-potentiated startle: blockade by infusion of an NMDA antagonist into the amygdala. J Neurosci 12: 854–863.
26. Bartlett TE, Bannister NJ, Collett VJ, Dargan SL, Massey PV, et al. (2007) Differential roles of NR2A and NR2B-containing NMDA receptors in LTP and LTD in the CA1 region of two-week old rat hippocampus. Neuropharmacology 52: 60–70.

27. Hrabetova S, Serrano P, Blace N, Tse HW, Skifter DA, et al. (2000) Distinct NMDA receptor subpopulations contribute to long-term potentiation and long-term depression induction. J Neurosci 20: RC81.
28. Xu Z, Chen RQ, Gu QH, Yan JZ, Wang SH, et al. (2009) Metaplastic regulation of long-term potentiation/long-term depression threshold by activity-dependent changes of NR2A/NR2B ratio. J Neurosci 29: 8764–8773.
29. Peng Y, Zhao J, Gu QH, Chen RQ, Xu Z, et al. (2010) Distinct trafficking and expression mechanisms underlie LTP and LTD of NMDA receptor-mediated synaptic responses. Hippocampus 20: 646–658.
30. Wigstrom H, Gustafsson B (1986) Postsynaptic control of hippocampal long-term potentiation. J Physiol (Paris) 81: 228–236.
31. Lynch MA, Littleton JM (1983) Possible association of alcohol tolerance with increased synaptic Ca2+ sensitivity. Nature 303: 175–176.
32. Punnakkal P, Jendritza P, Kohr G (2012) Influence of the intracellular GluN2 C-terminal domain on NMDA receptor function. Neuropharmacology 62: 1985–1992.
33. Halt AR, Dallapiazza RF, Zhou Y, Stein IS, Qian H, et al. (2012) CaMKII binding to GluN2B is critical during memory consolidation. EMBO J 31: 1203–1216.
34. Sheng M (1996) PDZs and receptor/channel clustering: rounding up the latest suspects. Neuron 17: 575–578.
35. Kennedy MB (1997) The postsynaptic density at glutamatergic synapses. Trends Neurosci 20: 264–268.
36. Kennedy PR, Bakay RA (1997) Activity of single action potentials in monkey motor cortex during long-term task learning. Brain Res 760: 251–254.
37. Kornau HC, Seeburg PH, Kennedy MB (1997) Interaction of ion channels and receptors with PDZ domain proteins. Curr Opin Neurobiol 7: 368–373.
38. Steigerwald F, Schulz TW, Schenker LT, Kennedy MB, Seeburg PH, et al. (2000) C-Terminal truncation of NR2A subunits impairs synaptic but not extrasynaptic localization of NMDA receptors. J Neurosci 20: 4573–4581.
39. Martel MA, Ryan TJ, Bell KF, Fowler JH, McMahon A, et al. (2012) The subtype of GluN2 C-terminal domain determines the response to excitotoxic insults. Neuron 74: 543–556.
40. Shipton OA, Paulsen O (2014) GluN2A and GluN2B subunit-containing NMDA receptors in hippocampal plasticity. Philos Trans R Soc Lond B Biol Sci 369: 20130163.
41. Cao X, Cui Z, Feng R, Tang YP, Qin Z, et al. (2007) Maintenance of superior learning and memory function in NR2B transgenic mice during ageing. Eur J Neurosci 25: 1815–1822.
42. Zhang XL, Sullivan JA, Moskal JR, Stanton PK (2008) A NMDA receptor glycine site partial agonist, GLYX-13, simultaneously enhances LTP and reduces LTD at Schaffer collateral-CA1 synapses in hippocampus. Neuropharmacology 55: 1238–1250.
43. Berberich S, Jensen V, Hvalby O, Seeburg PH, Kohr G (2007) The role of NMDAR subtypes and charge transfer during hippocampal LTP induction. Neuropharmacology 52: 77–86.
44. Chen G, Wang LP, Tsien JZ. (2009). Neural population-level memory traces in the mouse hippocampus. PLoS One 4(12): e8256.
45. Clayton DA, Mesches MH, Alvarez E, Bickford PC, Browning MD (2002) A hippocampal NR2B deficit can mimic age-related changes in long-term potentiation and spatial learning in the Fischer 344 rat. J Neurosci 22: 3628–3637.
46. Kohr G, Jensen V, Koester HJ, Mihaljevic AL, Utvik JK, et al. (2003) Intracellular domains of NMDA receptor subtypes are determinants for long-term potentiation induction. J Neurosci 23: 10791–10799.
47. Foster KA, McLaughlin N, Edbauer D, Phillips M, Bolton A, et al. (2010) Distinct roles of NR2A and NR2B cytoplasmic tails in long-term potentiation. J Neurosci 30: 2676–2685.
48. Gardoni F, Mauceri D, Malinverno M, Polli F, Costa C, et al. (2009) Decreased NR2B subunit synaptic levels cause impaired long-term potentiation but not long-term depression. J Neurosci 29: 669–677.
49. Liu HN, Kurotani T, Ren M, Yamada K, Yoshimura Y, et al. (2004) Presynaptic activity and Ca2+ entry are required for the maintenance of NMDA receptor-independent LTP at visual cortical excitatory synapses. J Neurophysiol 92: 1077–1087.
50. Erreger K, Chen PE, Wyllie DJ, Traynelis SF (2004) Glutamate receptor gating. Crit Rev Neurobiol 16: 187–224.

51. Berberich S, Punnakkal P, Jensen V, Pawlak V, Seeburg PH, et al. (2005) Lack of NMDA receptor subtype selectivity for hippocampal long-term potentiation. J Neurosci 25: 6907–6910.
52. Li YH, Han TZ (2008) Glycine modulates synaptic NR2A- and NR2B-containing NMDA receptor-mediated responses in the rat visual cortex. Brain Res 1190: 49–55.
53. Santucci DM, Raghavachari S (2008) The effects of NR2 subunit-dependent NMDA receptor kinetics on synaptic transmission and CaMKII activation. PLoS Comput Biol 4: e1000208.
54. Gerkin RC, Lau PM, Nauen DW, Wang YT, Bi GQ (2007) Modular competition driven by NMDA receptor subtypes in spike-timing-dependent plasticity. J Neurophysiol 97: 2851–2862.
55. Migaud M, Charlesworth P, Dempster M, Webster LC, Watabe AM, et al. (1998) Enhanced long-term potentiation and impaired learning in mice with mutant postsynaptic density-95 protein. Nature 396: 433–439.
56. Carlisle HJ, Fink AE, Grant SG, O'Dell TJ (2008) Opposing effects of PSD-93 and PSD-95 on long-term potentiation and spike timing-dependent plasticity. J Physiol 586: 5885–5900.
57. Gho M, McDonald K, Ganetzky B, Saxton WM (1992) Effects of kinesin mutations on neuronal functions. Science 258: 313–316.
58. Setou M, Nakagawa T, Seog DH, Hirokawa N (2000) Kinesin superfamily motor protein KIF17 and mLin-10 in NMDA receptor-containing vesicle transport. Science 288: 1796–1802.
59. Hawasli AH, Bibb JA (2007) Alternative roles for Cdk5 in learning and synaptic plasticity. Biotechnol J 2: 941–948.
60. Yin X, Takei Y, Kido MA, Hirokawa N (2011) Molecular motor KIF17 is fundamental for memory and learning via differential support of synaptic NR2A/2B levels. Neuron 70: 310–325.
61. Hawasli AH, Benavides DR, Nguyen C, Kansy JW, Hayashi K, et al. (2007) Cyclin-dependent kinase 5 governs learning and synaptic plasticity via control of NMDAR degradation. Nat Neurosci 10: 880–886.
62. Wong RW, Setou M, Teng J, Takei Y, Hirokawa N (2002) Overexpression of motor protein KIF17 enhances spatial and working memory in transgenic mice. Proc Natl Acad Sci U S A 99: 14500–14505.
63. Slutsky I, Abumaria N, Wu LJ, Huang C, Zhang L, et al. (2010) Enhancement of learning and memory by elevating brain magnesium. Neuron 65: 165–177.
64. Abumaria N, Yin B, Zhang L, Li XY, Chen T, et al. (2011) Effects of elevation of brain magnesium on fear conditioning, fear extinction, and synaptic plasticity in the infralimbic prefrontal cortex and lateral amygdala. J Neurosci 31: 14871–14881.
65. Abumaria N, Luo L, Ahn M, Liu G (2013) Magnesium supplement enhances spatial-context pattern separation and prevents fear overgeneralization. Behav Pharmacol 24: 255–263.
66. Cyranoski D (2012) Testing magnesium's brain-boosting effects. Nature News: Nature Publishing Group.
67. Tsien JZ, Chen DF, Gerber D, Tom C, Mercer EH, et al. (1996) Subregion- and cell type-restricted gene knockout in mouse brain. Cell 87: 1317–1326.
68. Tsien JZ, Huerta PT, Tonegawa S (1996) The essential role of hippocampal CA1 NMDA receptor-dependent synaptic plasticity in spatial memory. Cell 87: 1327–1338.
69. Wang LP, Li F, Wang D, Xie K, Shen X, et al. (2011) NMDA receptors in dopaminergic neurons are crucial for habit learning. Neuron 72: 1055–1066.
70. Wang H, Feng R, Phillip Wang L, Li F, Cao X, et al. (2008) CaMKII activation state underlies synaptic labile phase of LTP and short-term memory formation. Curr Biol 18: 1546–1554.
71. Cui Z, Wang H, Tan Y, Zaia KA, Zhang S, et al. (2004) Inducible and reversible NR1 knockout reveals crucial role of the NMDA receptor in preserving remote memories in the brain. Neuron 41: 781–793.
72. Rampon C, Jiang CH, Dong H, Tang YP, Lockhart DJ, et al. (2000) Effects of environmental enrichment on gene expression in the brain. Proc Natl Acad Sci U S A 97: 12880–12884.
73. Brandon EP, Zhuo M, Huang YY, Qi M, Gerhold KA, et al. (1995) Hippocampal long-term depression and depotentiation are defective in mice carrying a targeted disruption of the gene encoding the RI beta subunit of cAMP-dependent protein kinase. Proc Natl Acad Sci U S A 92: 8851–8855.

6

Expression of Cry1Ab and Cry2Ab by a Polycistronic Transgene with a Self-Cleavage Peptide in Rice

Qichao Zhao[1♀], Minghong Liu[1♀], Miaomiao Tan[1], Jianhua Gao[2], Zhicheng Shen[1]*

1 State Key Laboratory of Rice Biology, Institute of Insect Sciences, Zhejiang University, Hangzhou, China, 2 College of Life Science, Shanxi Agricultural University, Taigu, China

Abstract

Insect resistance to *Bacillus thuringiensis* (Bt) crystal protein is a major threat to the long-term use of transgenic Bt crops. Gene stacking is a readily deployable strategy to delay the development of insect resistance while it may also broaden insecticidal spectrum. Here, we report the creation of transgenic rice expressing discrete Cry1Ab and Cry2Ab simultaneously from a single expression cassette using 2A self-cleaving peptides, which are autonomous elements from virus guiding the polycistronic viral gene expression in eukaryotes. The synthetic coding sequences of Cry1Ab and Cry2Ab, linked by the coding sequence of a 2A peptide from either foot and mouth disease virus or porcine teschovirus-1, regardless of order, were all expressed as discrete Cry1Ab and Cry2Ab at high levels in the transgenic rice. Insect bioassays demonstrated that the transgenic plants were highly resistant to lepidopteran pests. This study suggested that 2A peptide can be utilized to express multiple Bt genes at high levels in transgenic crops.

Editor: Mario Soberón, Instituto de Biotecnología, Universidad Nacional Autónoma de México, Mexico

Funding: This research was supported by The National Natural Science Foundation of China (31021003, http://www.nsfc.gov.cn/) received by Zc Shen. The funders had no role in study design, data collection and analysis, decision to publish, or preparation of the manuscript.

Competing Interests: The authors have declared that no competing interests exist.

* Email: zcshen@zju.edu.cn

♀ These authors contributed equally to this work.

Introduction

Since first demonstrated in transgenic tobacco in 1987, Bt crystal toxin genes have been widely utilized in transgenic crops for pest management [1,2]. However, due to the widespread application of Bt toxins, several major insect pests, including fall armyworm (*Spodoptera frugiperda*), dimondback moth (*Plutella xylostella*), maize stalk borer (*Busseola fusca*), cotton bollworm (*Helicoverpa armigera*), western corn rootworm (*Diabrotica virgifera*) and cabbage looper (*Trichoplusia ni*) have developed resistance to Bt toxins in field or greenhouse, which threatens the long-term utilization of Bt crops in the future [3–5].

Several strategies have been proposed and/or deployed to cope with the development of insect resistance to Bt toxins, including high-dose/refuge strategy, discovery of novel insecticidal genes with novel modes of actions, modification of used Bt genes and Bt gene stacking [6–8]. Bt gene stacking strategy introduces more than one Bt genes into plant and has been demonstrated to be an effective way to delay the development of insect resistance to Bt toxins [9–11]. Expression of multiple genes using conventional approaches has several potential limitations, most notably imbalanced expression among different genes and a large T-DNA size required to include multiple genes [12]. Efficient and stoichiometric expression of discrete proteins may be achieved by a polycistronic system involving self-cleaving peptides such as the 2A peptide from foot and mouth disease virus (F2A) or porcine teschovirus-1 (P2A) [12,13]. When expressed in eukaryotic system, the

nascent 2A peptide can interact with the ribosome exit tunnel to dictate a stop-codon-independent termination at the final proline codon of 2A peptide [14]. Subsequently translation is reinitiated on the same proline codon. When linked by a 2A peptide coding sequence, different genes are co-expressed from a single open reading frame.

2A self-cleaving peptides have been extensively studied previously [15]. In this study, the DNA encoding F2A or P2A was used to link two potent insecticidal Bt genes, the truncated *Cry1Ab* encoding the N-terminal 648 amino acids of active Cry1Ab endotoxin (Genbank:AAG16877.1) and the full-length *Cry2Ab* encoding Cry2Ab endoxin (Genbank:AAA22342.1) with 634 amino acid residues, to generate a polycistronic gene for co-expression in transgenic rice. Analysis of the obtained transgenic rice lines revealed that discrete Cry1Ab and Cry2Ab were indeed co-expressed at a level approximately comparable to transgenic plants expressing traditional monocistronic Bt genes. Insect bioassays demonstrated that the transgenic rice generated was highly resistant to its target insects.

Materials and Methods

Rice cultivar

Elite rice (*Oryza sativa spp. Japonica*) cultivar Xiushui 134 originated from the Jiaxing Academy of Agricultural Science in Zhejiang Province was used for *Agrobacterium*-mediated transformation. The homozygous transgenic rice line KMD1, containing a synthetic truncated *Cry1Ab* gene under control of maize ubiquitin promoter [16,17], was kindly provided by Dr. Gongyin Ye from Institute of Insect Sciences, ZheJiang University.

Construction of binary vector for rice transformation

pCambia1300 (Cambia, Canberra, Australia) was used for construction of binary vectors for plant transformation. This vector was first modified by substituting the hygromycin-resistant gene expression cassette with a glyphosate-tolerant *5-enolpyruvylshikimate-3-phosphate synthase* (*EPSPS*) gene expression cassette. The modified vector was named as pCambia1300–GLY, and was further used to clone an expression cassette for a polycistronic gene encoding Cry1Ab and Cry2Ab linked by a 2A peptide (Fig. 1).

Four polycistronic genes of *Cry1Ab* and *Cry2Ab* linked by the coding sequence of F2A or P2A were generated. They were named as *Cry1Ab-F2A-Cry2Ab* (*Cry1-F-Cry2*), *Cry1Ab-P2A-Cry2Ab* (*Cry1-P-Cry2*), *Cry2Ab-F2A-Cry1Ab* (*Cry2-F-Cry1*), and *Cry2Ab-P2A-Cry1Ab* (*Cry2-P-Cry1*), respectively (Fig. 1). *Cry1-F-Cry2* is a fusion gene, in the order from 5' to 3', of coding sequences of Cry1Ab, F2A, and Cry2Ab (GenBank: KJ716232); *Cry1-P-Cry2* is a fusion gene of Cry1Ab, P2A, and Cry2Ab (GenBank: KJ716233); *Cry2-F-Cry1* is a fusion gene of Cry2Ab, F2A, and Cry1Ab (GenBank: KJ716234); and *Cry2-P-Cry1* is a fusion gene of Cry2Ab, P2A, and Cry1Ab (GenBank: KJ716235).

Maize ubiquitin-1 promoter (pUbi) was obtained by PCR from maize genome with primers pUbi-F (5' attaagcttagcttgcatgcctacagtg 3', with *Hind*III restriction site underlined) and pUbi-R (5' taaggatccctctagagtcgacctgca 3', with *Bam*HI restriction site underlined). The pUbi fragment digested with *Hind*III and *Bam*HI, the polycistronic gene fragment digested with *Bam*HI and *Kpn*I and the vector 1300-GLY predigested with *Hind*III and *Kpn*I were ligated to generate plasmid p1300-Cry1-F-Cry2, p1300-Cry1-P-Cry2,p1300-Cry2-F-Cry1 and p1300-Cry2-P-Cry1, respectively (Fig. 1).

Agrobacterium-mediated transformation

The T-DNA plasmids were separately transformed into *Agrobacterium tumefaciens* LBA4404 by electroporation. The *Agrobacterium*-mediated transformation of rice was carried out according to Hiei Y. *et al.*, except that 2 mM glyphosate (Sigma-Aldrich, St. Louis, MO, USA) was used for selection [18].

RT-PCR analysis of transgenic rice

Total RNA was extracted from rice leaf with Trizol reagent (Invitrogen, Carlsbad, CA,USA). Concentration of RNA was determined by a spectrophotometer (Thermo Scientific, DE, USA). One μg RNA was treated with Dnase I and used as template to synthesize the first strand of cDNA using cDNA synthesis Kit (Thermo Scientific, DE, USA). Finally the synthesized cDNA was used as PCR template. The primers P1-2-F (5' CAGCGGCAACGAGGTGTACA 3') and P1-2-R (5' TAGGCGTCGCAGATGGTGGT 3') were used to amplify the joint parts in Cry1-F-Cry2 and Cry1-P-Cry2 (Fig. 2a). The PCR products were expected to be 321 bp. The primers P2-1-F (5' GCCGCTCGACATCAACGTGA 3') and P2-1-R (5' TCAGGCTGATGTCGATGGGG 3') were used to amplify the joint parts in Cry2-F-Cry1 and Cry2-P-Cry1 (Fig. 2a). The PCR products were expected to be 314 bp. The procedure for both PCRs was pre-denaturation at 95°C for 3 min; then 30 cycles of denaturation at 95°C for 40 s, annealing at 50°C for 30 s, and extension at 72°C for 30 s; finally followed with extension at 72°C for 5 min. The PCR products were analyzed by electrophoresis in 1% (w/v) agarose gel.

Western blot analysis

Leaf sample of 0.01 g collected from 1-month-old rice was frozen in liquid nitrogen for 45 s and disrupted with TissueLyserII (Qiagen, Hilden, Germany) at 40 Hz for 45 s. The sample was then homogenized in 200 μL 1×PBS Buffer by shaking vigorously for 30 s and centrifuged at 12000 rpm for 15 min at 4°C. The supernatants were collected as protein samples for western blot analysis. These samples were separated by electrophoresis on 8% polyacrylamide gel and transferred to PVDF membrane (Pall, Ann Arbor, MI, USA). The blotted membrane was blocked in 5% (w/v) skim milk in Tris Buffer Saline with 0.1% Tween-20 (TBST) for 1 h at room temperature. The membrane was then incubated with primary antibody against either Cry1Ab or Cry2Ab (diluted 1:1000 in 1% skim milk/TBST) for 1 h at room temperature with gentle shaking. The polyclonal antisera against Cry1Ab and Cry2Ab were prepared from New Zealand white rabbits immunized with purified recombinant Cry1Ab and Cry2Ab from *Escherichia coli*, respectively. After three times of 5 minute wash in TBST, the membrane was incubated with HRP-conjugated secondary antibody (Promega, Madison, WI, USA) diluted at

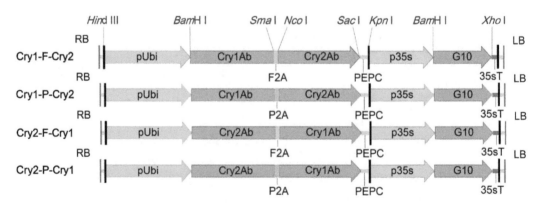

Figure 1. Schematic diagram of T-DNAs for transgenic expression of 2A linked polycistronic genes. Each T-DNA contained two expression cassettes, one for insecticidal polycistronic gene under control of maize ubiquitin-1 promoter (pUbi), the other for glyphosate-tolerant *5-enolpyruvylshikimate-3-phosphate synthase* (*EPSPS*) gene (G10) under control of Cauliflower mosaic virus 35s promoter (p35s). Cry1Ab and Cry2Ab,synthetic Bt inseciticidal gene; F2A and P2A, synthetic DNAs encoding Foot and Mouth Disease Virus 2A and Porcine teschovirus-1 2A, respectively; RB and LB, right border and left border of T-DNA; PEPC, maize *phosphoenolpyruvate carboxylase* gene terminator; 35sT, Cauliflower mosaic virus 35S gene terminator.

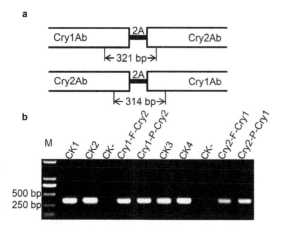

Figure 2. RT-PCR detection of mRNA integrity of the polycistronic genes. a, schematic diagram of the fragments amplified by RT-PCR; **b**, electrophoresis analysis of RT-PCR products. CK1, plasmid containing *Cry1-F-Cry2*; CK2, plasmid containing *Cry1-P-Cry2*; CK3, plasmid containing *Cry2-F-Cry1*; CK4, plasmid containing *Cry2-P-Cry1*; CK-, cDNA prepared from non-transgenic rice; M, DNA marker.

1:5000 with 1% skim milk/TBST for 1 h at room temperature with gentle shaking. After another three times of 5 minute wash in TBST, signals were visualized with DAB substrate (Sigma-Aldrich).

Protein quantification of Cry1Ab and Cry2Ab

Concentrations of Cry1Ab and Cry2Ab in the leaf of transgenic rice at tillering stage were determined by enzyme-linked immunosorbent assay (ELISA) using Envirologix kit AP003 and AP005 (Envirologix, Portland, ME, USA), respectively, according to the manufacturer's recommendation. For each construct, 3 transgenic lines were selected, and the samples prepared with the same method from non-transgenic rice were used to eliminate the basal absorption at 450 nm. After quantification of Cry1Ab and Cry2Ab, the molar ratio of the upstream protein to downstream protein of each transgenic line selected was calculated and analyzed by One-Way ANOVA [19] in SPSS.

Insect bioassay

Insect bioassays were conducted with cotton bollworm (Bt-susceptible and Cry1Ac-resistant), beet armyworm (*Spodoptera exigua*), striped stem borer (*Chilo suppressalis*) and rice leaf roller (*Cnaphalocrocis medinalis*). For beet armyworm and striped stem borer, detached leaf bioassay was carried out [20]. Two leaf blades from the same line were collected at 2-3 cm length at seedling stage and placed in a 70-mm-diameter petri dish lined with a pre-moistened filter paper. The leaf samples in the petri dish were infested with 10 newly hatched neonates. The petri dishes were then sealed with parafilm membrane and placed in dark at 28°C. The result was recorded after 3 days incubation. For cotton bollworm, detached leaf bioassay was carried out except that one leaf blade was placed in a petri dish and infested with one neonate to avoid cannibalism of cotton bollworm. Each line was infested with a total of 10 neonates. Eggs of Bt-susceptible cotton bollworm (CB-S), beet armyworm and striped stem borer were obtained from Genralpest Biotech (Genralpest Biotech, Bejing, China). Eggs of Cry1Ac-resistant cotton bollworm (CB-RR) were kindly provided by Dr. Kongming Wu from Institute of Plant Protection, Chinese Academy of Agricultural Sciences. For rice leaf roller, whole plant assay was carried out. Tillering-stage rice grown in

greenhouse was infested with 10 newly hatched neonates. The result was recorded 2 weeks after infestation. Eggs of rice leaf roller used in the assays were obtained from caged moths collected from rice field. Each assay was repeated for 3 times.

Results

Transgene transcribed as a long intact mRNA

About 30 independent transgenic lines were generated via *Agrobacterium*-mediated transformation for each of the four constructs. To investigate whether the transcript of each polycistronic transgene is a long intact mRNA with sequences encoding both Cry1Ab and Cry2Ab, RT-PCR was used to detect the sequences in the area connecting the two Bt genes, which include sequences from both *Cry1Ab* and *Cry2Ab*, and the 2A peptide coding sequence in between (Fig. 2**a**). The RT-PCR products with expected sizes were detected clearly in the transgenic lines from all of the four constructs (Fig. 2**b**), suggesting that each of the polycistronic genes was transcribed into a long intact mRNA as expected.

Co-expression of discrete Cry1Ab and Cry2Ab from polycistronic transgenes

Transgenic lines from each construct were analyzed by western blot analysis using antisera against Cry1Ab and Cry2Ab,respectively. When detected with antiserum against Cry1Ab, a band of approximately 72 kD was detected among different transgenic lines, indicating that the Cry1Ab protein was expressed as a discrete protein rather than a fusion protein (Fig. 3 Upper panel). When detected by antiserum against Cry2Ab, a band of about 68~70 kD was detected among all the transgenic lines (Fig. 3 Lower panel), indicating a discrete Cry2Ab protein was expressed. A possible fusion protein of "Cry-2A-Cry" would be at a size of approximately 147 kD. However, no significant signal of proteins at such size was detected by antiserum against either Cry1Ab or Cry2Ab, suggesting that there was no significant amount of fusion protein expressed in the transgenic rice. The upstream Cry2Ab detected in the Cry2-F-Cry1 and the Cry2-P-Cry1 transgenic lines was slightly larger than the downstream Cry2Ab in the Cry1-F-Cry2 and the Cry1-P-Cry2. This increased size was likely due to the 2A peptide residues attached to the upstream protein during translation,which has been demonstrated in other 2A polycistronic transgenes [21].

Determination of gene expression level and 2A cleavage efficiency

To determine the expression levels of Bt genes with 2A polycistronic transgene and evaluate the cleavage efficiency of 2A peptide in each construct, the amounts of Cry1Ab and Cry2Ab expressed in the transgenic rice leaves were determined by ELISA. Three transgenic lines at tillering stage from each construct were selected. The expression levels of Cry1Ab and Cry2Ab were different among different lines as expected. The concentrations of the soluble Cry1Ab and Cry2Ab in the leaf were in the range of 0.67 to 1.82 µg/g and 0.69 to 2.31 µg/g of leaf fresh weight (LFW), respectively (Fig. 4**a**).

The molar ratio (R) of the upstream protein to the downstream protein of each construct was calculated to estimate the cleavage efficiency. The R value for polycistronic gene using F2A as 2A cleavage peptide was lower than that using P2A (Fig. 4**b**), indicating that the cleavage efficiency of F2A peptide was higher than P2A. The R value of polycistronic gene with Cry1Ab at the upstream was lower than that with Cry2Ab at the upstream (Fig. 4**b**), suggesting that the Cry1Ab at the upstream was likely

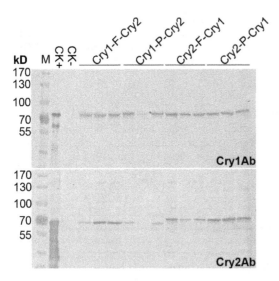

Figure 3. Western bolt analysis of Cry1Ab and Cry2Ab in transgenic rice. Three transgenic lines from each construct were selected for the analysis. Each sample was detected with antiserum against Cry1Ab (upper) and Cry2Ab (bottom) respectively. Cry1Ab or Cry2Ab protein expressed by *E.coli* was used as positive control (CK+). Sample prepared from non-transgenic rice was used as negative control (CK-). M, prestained protein ladder.

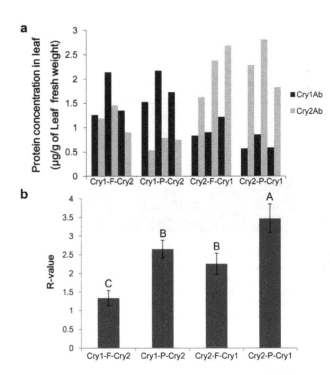

Figure 4. Cry1Ab and Cry2Ab concentrations in different transgenic rice lines and comparison of 2A cleavage efficiency. a, concentrations of Cry1Ab and Cry2Ab in leaves of 3 selected transgenic lines from each construct; **b**, R values of different 2A constructs. R value is the molar ratio of the upstream protein to the downstream protein. Different capital letters on each bar indicate extreme significant differences ($p<0.01$, Fisher's least-significant difference and Duncan's multiple range test).

more efficient in cleavage than the Cry2Ab at the upstream, regardless of whether F2A or P2A was the cleavage peptide in the synthetic construct. The transgenic lines from the construct of Cry1-F-Cry2 had the R value close to 1, indicating that the two Bt proteins were expressed almost equally (Fig. 4**a**).

Insect-resistant activity of transgenic rice

The transgenic rice lines were assayed for their insecticidal activity using neonates of CB-S, striped stem borer and rice leaf roller. Three transgenic lines from each construct were selected and age-matched non-transgenic rice plants were used as control. Mortalities of all the 3 insect species feeding on transgenic rice were 100% while those on the non-transgenic rice were much lower (Fig. 5). In the assays with CB-S and striped stem borer, the transgenic leaves were only slightly bitten by the insects while non-transgenic controls were severely consumed (Fig. 6**a**). In the assays with rice leaf roller, no rolled leaf was observed in the transgenic plant while several leaves were rolled in the non-transgenic plant (Fig. 6**b**). The bioassays demonstrated that transgenic rice plants generated from the 2A constructs were highly insect-resistant.

The transgenic rice lines were also assayed for their activity against beet armyworm and CB-RR. The KMD1 transgenic rice at the same growing stage was used as control. The mortalities of beet armyworm feeding on the 2A transgenic rice lines were obvious higher than that feeding on KMD1 ($p<0.01$) (Figure 7). The KMD1 showed a rather low activity to CB-RR, while the 2A transgenic rice showed much higher activity ($p<0.01$) (Figure 7).

Discussion

In this study, Cry1Ab and Cry2Ab, two potent Bt crystal proteins with different receptors in the insect midgut [22], were expressed by the polycistronic transgene using 2A peptide. To our best knowledge, this is the first application of 2A peptide for expressing two Bt genes in a transgenic crop. Utilization of two or more Bt genes simultaneously is desirable for management of

insect resistance to Bt toxins as well as for broadening insecticidal spectrum of Bt crops. Self-cleaving 2A peptides and 2A-like sequences have also been found in insect virus, and some of them have high self-cleavage activity [23]. While there is no scientific base for any safety concern for using 2A peptide from mammalian virus, it may be better for public perception to use 2A peptides from insect virus for the development of transgenic rice.

Introduction of multiple genes into plant permits complex and sophisticated manipulation of traits in transgenic crops [2]. The number of genes being introduced into crops for genetic engineering is increasing steadily. For instance, the recently released commercial transgenic corn "Genuity SmartStax" (Monsanto, St Louis, MS, USA) has a total of 8 transgenes for insect-resistance and herbicide-tolerance. With more traits under research and development for transgenic improvement, more genes are expected to be utilized for future transgenic crops. Clearly polycistronic strategy using 2A peptide can simplify the process of gene stacking significantly, as it enables us to introduce multiple genes into plants using a single T-DNA with a single promoter [2]. Moreover, polycistronic strategy alleviates the concern of gene silencing induced by the insertion of homologous promoters or multiple T-DNA sequences into recipient genome [2,24]. Additionally, equal expressions of different genes could be achieved by the 2A polycistronic strategy [12,24].

The expression levels of Cry1Ab and Cry2Ab by 2A polycistronic transgene in this study were comparable to traditional monocistronic transgene. The Cry1Ab concentration of transgenic rice KMD1, which is highly resistant to striped stem borer and rice leaf roller [16], was 2.95 μg/g of LFW [25]. The Cry1C concentration in homozygous rice line T1c-19 harboring a

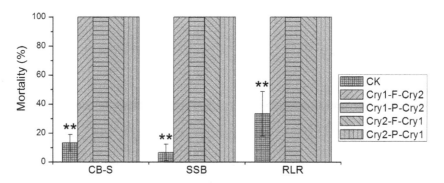

Figure 5. Mortality of Bt-susceptible cotton bollworm, striped stem borer and rice leaf roller feeding on transgenic rice plants. Non-transgenic rice at the same growing stage was used as control (CK). ** on the bar indicate extreme significant differences ($p < 0.01$, Kruskal Wallis test). CB-S, Bt-susceptible cotton bollworm; SSB, striped stem borer; RLR, rice leaf roller.

synthetic Cry1C gene under control of maize ubiquitin promoter was 1.38 µg/g of LFW at the heading stage [26]. The Cry2A protein concentrations of the homozygous transgenic rice lines harboring a synthetic Cry2 gene driven by maize ubiquitin promoter was reported at the range from 9.65 to 12.11 µg/g of LFW [27]. In the transgenic *indica* rice line harboring both *Cry1Ac* and *Cry2Ab* gene, the Cry2Ab concentration was reported to be around 1 µg/g LFW [28]. Thus, the expression levels of the Cry1Ab and Cry2Ab in the transgenic rice lines obtained by this study were well within the range of the expression levels of the monocistronic transgene reported previously.

Although cotton bollworm was not a pest of rice, it has been long a good target insect for evaluating Bt toxicity. Moreover, cotton bollworm is a good model insect for the study of insect resistance to Bt toxins because of the availabilities of both sensitive and resistant lines to Cry1Ac. The 2A transgenic lines obtained in this study showed high insect-resistant activity against cotton bollworm, either sensitive or resistant to Cry1Ac, while the KMD1 transgenic rice had little activity toward Cry1Ac-resistant cotton bollworm. This suggested that little or no cross resistance existed between Cry1Ac and Cry2Ab and the 2A peptide based transgenic Bt rice will be useful for the management of Bt resistance developed by insect pests.

The 2A transgenic lines obtained in this study conferred significantly higher activity toward beet armyworm than the KMD1 transgenic rice containing Cry1Ab only [16], suggesting that the Cry2Ab expressed in the 2A transgenic lines indeed contributed to the enhancement of insecticidal activity. A similar phenomenon was also observed in transgenic cotton BollgardII. While Bollgard cotton only had moderate toxic activity to beet armyworm, BollgardII cotton showed greatly improved activity due to the addition of Cry2Ab [29].

The upstream proteins expressed with 2A polycistronic strategy are usually attached with the 2A peptide residues at the C-terminus [30]. The western blot analysis in this study appeared to agree with it. However, due to the limitation of the western blot analysis in determination of protein size, C-terminal amino acid sequencing and mass spectrum are required to confirm. While no adverse impact was observed for the utilization of 2A peptide in gene stacking [12], the Bt toxins attached with a 2A peptide will be considered as different proteins according to the regulation of transgenic crops, and thus additional safety studies will be required for commercialization of transgenic crops using 2A cleaving peptide.

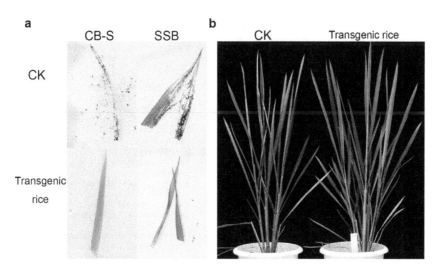

Figure 6. Insect bioassay for transgenic rice with a polycistronic Bt transgene. a, bioassay with Bt-susceptible cotton bollworm (CB-S) and striped stem borer (SSB). **b,** bioassay with rice leaf roller. For each assay, age-matched non-transgenic rice was used as control (CK).

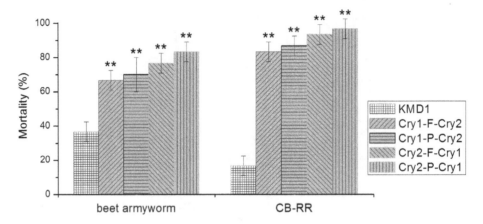

Figure 7. Mortality of beet armyworm and Cry1Ac-resistant contton bollworm feeding on KMD1 and various 2A transgenic rice lines. The transgenic rice KMD1 was used as control. ** on the bar indicate extreme significant difference ($p < 0.01$, Fisher's least-significant difference and Duncan's multiple range test). CB-RR, Cry1Ac-resistant cotton bollworm.

Acknowledgments

We thank Kongming Wu and Gemei Liang for the supply of Cry1Ac-resistant cotton bollworm. We thank Gongyin Ye for the supply of KMD1 transgenic rice.

Author Contributions

Conceived and designed the experiments: QcZ MhL JHG ZcS. Performed the experiments: QcZ MhL MmT. Analyzed the data: QcZ MhL MmT. Contributed reagents/materials/analysis tools: QcZ MhL JHG. Wrote the paper: QcZ MhL JHG ZcS.

References

1. Romeis J, Meissle M, Bigler F (2006) Transgenic crops expressing Bacillus thuringiensis toxins and biological control. Nature Biotechnology 24: 63–71.
2. Halpin C (2005) Gene stacking in transgenic plants – the challenge for 21st century plant biotechnology. Plant Biotechnology Journal 3: 141–155.
3. Gassmann AJ, Petzold-Maxwell JL, Keweshan RS, Dunbar MW (2011) Field-Evolved Resistance to Bt Maize by Western Corn Rootworm. PLoS ONE 6.
4. Tabashnik BE, Van Rensburg JBJ, Carriere Y (2009) Field-Evolved Insect Resistance to Bt Crops: Definition, Theory, and Data. Journal of Economic Entomology 102: 2011–2025.
5. Wang P, Zhao J-Z, Rodrigo-Simon A, Kain W, Janmaat AF, et al. (2006) Mechanism of resistance to Bacillus thuringiensis toxin Cry1Ac in a greenhouse population of cabbage looper, Trichoplusia ni. Applied and Environmental Microbiology: 01834–01806.
6. Bates SL, Zhao JZ, Roush RT, Shelton AM (2005) Insect resistance management in GM crops: past, present and future. Nature Biotechnology 23: 57–62.
7. Shelton AM, Tang JD, Roush RT, Metz TD, Earle ED (2000) Field tests on managing resistance to Bt-engineered plants. Nature Biotechnology 18: 339–342.
8. Tabashnik BE, Carriere Y, Dennehy TJ, Morin S, Sisterson MS, et al. (2003) Insect resistance to transgenic Bt crops: Lessons from the laboratory and field. Journal of Economic Entomology 96: 1031–1038.
9. Cao J, Zhao JZ, Tang JD, Shelton AM, Earle ED (2002) Broccoli plants with pyramided cry1Ac and cry1C Bt genes control diamondback moths resistant to Cry1A and Cry1C proteins. Theoretical and Applied Genetics 105: 258–264.
10. Yang Z, Chen H, Tang W, Hua HX, Lin YJ (2011) Development and characterisation of transgenic rice expressing two Bacillus thuringiensis genes. Pest Management Science 67: 414–422.
11. Zhao JZ, Cao J, Li YX, Collins HL, Roush RT, et al. (2003) Transgenic plants expressing two Bacillus thuringiensis toxins delay insect resistance evolution. Nature Biotechnology 21: 1493–1497.
12. Szymczak-Workman AL, Vignali KM, Vignali DAA (2012) Design and Construction of 2A Peptide-Linked Multicistronic Vectors. Cold Spring Harbor Protocols 2012: pdb.ip067876.
13. Ha S-H, Liang YS, Jung H, Ahn M-J, Suh S-C, et al. (2010) Application of two bicistronic systems involving 2A and IRES sequences to the biosynthesis of carotenoids in rice endosperm. Plant Biotechnology Journal 8: 928–938.
14. Brown JD, Ryan MD (2010) Ribosome "Skipping": "Stop-Carry On" or "StopGo" Translation Recoding: Expansion of Decoding Rules Enriches Gene Expression. InAtkinsJFGestelandRFeditors: New York:Springerpp101–121.
15. Luke G, Escuin H, De Felipe P, Ryan M (2010) 2A to the fore - research, technology and applications. Biotechnology and Genetic Engineering Reviews 26: 223–260.

16. Shu QY, Ye GY, Cui HR, Cheng XY, Xiang YB, et al. (2000) Transgenic rice plants with a synthetic cry1Ab gene from Bacillus thuringiensis were highly resistant to eight lepidopteran rice pest species. Molecular Breeding 6: 433–439.
17. Ye GY, Shu QY, Yao HW, Cui HR, Cheng XY, et al. (2001) Field evaluation of resistance of transgenic rice containing a synthetic cry1Ab gene from Bacillus thuringiensis Berliner to two stem borers. Journal of Economic Entomology 94: 271–276.
18. Hiei Y, Komari T, Kubo T (1997) Transformation of rice mediated by Agrobacterium tumefaciens. Plant Molecular Biology 35: 205–218.
19. Heiberger R, Neuwirth E (2009) One-Way ANOVA: R Through Excel: New YorkSpringerpp165–191.
20. Chen Y, Tian J-C, Shen Z-C, Peng Y-F, Hu C, et al. (2010) Transgenic Rice Plants Expressing a Fused Protein of Cry1Ab/Vip3H Has Resistance to Rice Stem Borers Under Laboratory and Field Conditions. Journal of Economic Entomology 103: 1444–1453.
21. Halpin C, Cooke SE, Barakate A, Amrani AE, Ryan MD (1999) Self-processing 2A-polyproteins – a system for co-ordinate expression of multiple proteins in transgenic plants. The Plant Journal 17: 453–459.
22. Hernandez-Rodriguez CS, Hernandez-Martinez P, Van Rie J, Escriche B, Ferre J (2013) Shared Midgut Binding Sites for Cry1A. 105, Cry1Aa, Cry1Ab, Cry1Ac and Cry1Fa Proteins from Bacillus thuringiensis in Two Important Corn Pests, Ostrinia nubilalis and Spodoptera frugiperda. Plos One 8.
23. Ryan M (2014) 2A and 2A like sequences. Available: http://www.st-andrews.ac.uk/ryanlab/2A_2Alike.pdf/. Accessed 4 September 2014.
24. Sainsbury F, Benchabane M, Goulet M-C, Michaud D (2012) Multimodal Protein Constructs for Herbivore Insect Control. Toxins 4: 455–475.
25. Tian J-C, Chen Y, Li Z-L, Li K, Chen M, et al. (2012) Transgenic Cry1Ab Rice Does Not Impact Ecological Fitness and Predation of a Generalist Spider. PLoS ONE 7: e35164.
26. Tang W, Chen H, Xu C, Li X, Lin Y, et al. (2006) Development of insect-resistant transgenic indica rice with a synthetic cry1C* gene. Molecular Breeding 18: 1–10.
27. Chen H, Tang W, Xu C, Li X, Lin Y, et al. (2005) Transgenic indica rice plants harboring a synthetic cry2A* gene of Bacillus thuringiensis exhibit enhanced resistance against lepidopteran rice pests. Theoretical and Applied Genetics 111: 1330–1337.
28. Bashir K, Husnain T, Fatima T, Latif Z, Mehdi SA, et al. (2004) Field evaluation and risk assessment of transgenic indica basmati rice. Molecular Breeding 13: 301–312.
29. Perlak FJ, Oppenhuizen M, Gustafson K, Voth R, Sivasupramaniam S, et al. (2001) Development and commercial use of Bollgard cotton in the USA – early promises versus today's reality. The Plant Journal 27: 489–501.
30. Szymczak-Workman AL, Vignali KM, Vignali DAA (2012) Verification of 2A Peptide Cleavage. Cold Spring Harbor Protocols 2012: pdb.prot067892.

7

Bt Rice Expressing Cry2Aa Does Not Harm *Cyrtorhinus lividipennis*, a Main Predator of the Nontarget Herbivore *Nilapavarta lugens*

Yu Han[1], Jiarong Meng[1], Jie Chen[1], Wanlun Cai[1], Yu Wang[1], Jing Zhao[1], Yueping He[1], Yanni Feng[2], Hongxia Hua[1]*

1 Hubei Insect Resources Utilization and Sustainable Pest Management Key Laboratory, College of Plant Science and Technology, Huazhong Agricultural University, Wuhan, P.R. China, 2 College of Life Science and Technology, Huazhong Agricultural University, P.R. China

Abstract

T2A-1 is a newly developed transgenic rice that expresses a synthesized *cry2Aa* gene driven by the maize *ubiquitin* promoter. T2A-1 exhibits high resistance against lepidopteran pests of rice. The brown planthopper, *Nilapavarta lugens* (Stål), is a main nontarget sap-sucking insect pest of rice, and *Cyrtorhinus lividipennis* (Reuter) is the major predator of the eggs and young nymphs of planthoppers. As *C. lividipennis* may expose to the Cry2Aa protein via *N. lugens*, it is therefore essential to assess the potential effects of transgenic *cry2Aa* rice on this predator. In the present study, three experiments were conducted to evaluate the ecological risk of transgenic *cry2Aa* rice to *C. lividipennis*: (1) a direct feeding experiment in which *C. lividipennis* was fed an artificial diet containing Cry2Aa at the dose of 10-time higher than that it may encounter in the realistic field condition; (2) a tritrophic experiment in which the Cry2Aa protein was delivered to *C. lividipennis* indirectly through prey eggs or nymphs; (3) a realistic field experiment in which the population dynamics of *C. lividipennis* were investigated using vacuum-suction. Both direct exposure to elevated doses of the Cry2Aa protein and prey-mediated exposure to realistic doses of the protein did not result in significant detrimental effects on the development, survival, female ratio and body weight of *C. lividipennis*. No significant differences in population density and population dynamics were observed between *C. lividipennis* in transgenic *cry2Aa* and nontransgenic rice fields. It may be concluded that transgenic *cry2Aa* rice had no detrimental effects on *C. lividipennis*. This study represents the first report of an assessment continuum for the effects of transgenic *cry2Aa* rice on *C. lividipennis*.

Editor: Guy Smagghe, Ghent University, Belgium

Funding: This research was supported by the National Genetically Modified Organisms Breeding Major Project: Technology of Environmental Risk Assessment on Transgenic Rice (2014ZX08011-001). HH received the funding. The funders had no role in study design, data collection and analysis, decision to publish, or preparation of the manuscript.

Competing Interests: The authors have declared that no competing interests exist.

* Email: huahongxia@mail.hzau.edu.cn

Introduction

Rice, *Oryza sativa* L., is the staple food of more than three billion people of Asia [1]. More than 200 species of insect pests infest rice during its growing season [2], and this causes 15% to 25% yield losses of rice [3,4]. Among these insects, lepidopteran species such as stem borers and leaffolders are particularly serious chronic pests and cause large annual yield losses [5,6]. Traditional management of lepidopteran pests is mainly dependent on the spraying of pesticides. However, excessive or continual applications of pesticide not only cause environmental contamination and the resurgence of herbivores but also reduce populations of the natural enemies of these herbivores [7,8]. Researchers have therefore been encouraged to seek more effective and environmentally friendly methods to control lepidopteran pests.

The *Bacillus thuringiensis* (*Bt*) insecticidal δ-endotoxin has been used as a biological insecticide for more than 50 years, and it is now possible to introduce different *Bt* genes into rice. Transgenic rice plants expressing Bt proteins have been shown to be effective against many lepidopteran insect pests [9] and have led to great reductions in the use of insecticides [10,11]. A series of rice lines that express various *Bt* genes (e.g., *cry1Ab*, *cry1Ab/1Ac*, *cry1C* and *cry2A*) have been developed and can effectively suppress the infestation of target lepidopteran insect pests [12–14]. Despite these successes, concerns have been raised regarding the potential impacts of transgenic *Bt* rice on nontarget herbivores and their natural enemies through tritrophic transmission. It is therefore necessary to conduct an environmental risk assessment of any novel transgenic *Bt* rice prior to its commercialization, and the environmental risks of each rice line must be evaluated on a case-by-case basis.

The brown planthopper, *Nilapavarta lugens* (Stål) (Hemiptera: Delphacidae), is the main nontarget sap-sucking insect pest of rice and causes significant annual reductions in the rice yield [15,16]. The evaluation of nontarget effects of transgenic rice on the natural enemies of *N. lugens* is an important part of the environmental risk assessment, and should be conducted before the commercialization of any novel *Bt* rice. Several previous reports have examined the effects of transgenic *Bt* rice on the predators of *N. lugens*. Although the Cry1Ab protein can be transferred from transgenic rice plants to predators via *N. lugens*, no adverse effects have been found on any of the fitness parameters (survival, developmental time, weight and fecundity) of numerous predators [*Cyrtorhinus lividipennis* (Reuter) (Hemiptera: Miridae), *Pardosa pseudoannulata* (Bösenberg et Strand) (Araneae: Lycosidae), *Pirata subpiraticus* (Bösenberg et Strand) (Araneae: Lycosidae), *Propylea japonica* (Thunberg) (Coleoptera: Coccinellidae) and *Ummeliata insecticeps* (Bösenberg et Strand) (Araneae: Linyphiidae)] that preyed on *N. lugens* reared on *Bt*-transformed rice lines in the laboratory [17–22]. Additionally, the results from field experiments have also shown that the population densities and population dynamics of *U. insecticeps*, *C. lividipennis* and *P. pseudoannulata* are not significantly different between transgenic *cry1Ab* rice and non-*Bt* rice fields [21–23]. All of these results have indicated that the predators of *N. lugens* are not affected by the Cry1Ab protein. T2A-1 is a relatively newly developed transgenic rice line expressing *cry2Aa* driven by the maize *ubiquitin* promoter, and it exhibits high resistance against lepidopteran pests of rice [12]. However, the potential effects of this rice line on the predators of *N. lugens* have received little attention. T2A-1 did not cause direct detrimental effects on the larvae of *Chrysoperla sinica* (Tjeder) (Neuroptera: Chrysopidae), a general predator of *N. lugens* [24]; that study represents the only existing report about the effects of Cry2Aa on the predators of *N. lugens*.

Cyrtorhinus lividipennis (Reuter) (Hemiptera: Miridae) is a major predator of the eggs and young nymphs of planthoppers, and it is a primary factor regulating the population density of *N. lugens* in rice fields [7,25,26]. Thus, *C. lividipennis* may expose to the Bt protein via *N. lugens*. The potential effects of transgenic *Bt* rice on *C. lividipennis* should be evaluated. Transgenic *cry1Ab* rice had no negative effects on the life-table parameters, population density and population dynamics of *C. lividipennis* [18,23]. However, no such study of transgenic *cry2Aa* rice has yet been conducted with *C. lividipennis*. In the current study, we conducted comprehensive experiments in the laboratory and in rice fields to examine the effects of T2A-1 on *C. lividipennis*. The effects of T2A-1 via prey on the life-table parameters and the functional response of *C. lividipennis* to *N. lugens*, and the direct toxicity of the Cry2Aa protein to *C. lividipennis* were evaluated in the laboratory. The effects of T2A-1 on population density and population dynamics were investigated through a 3-year experiment in rice fields. ELISA was used to determine whether the Cry2Aa protein could be transferred to *C. lividipennis* via *N. lugens*.

Materials and Methods

Ethics Statement

All necessary permits were obtained for the described field studies. Permission of small-field test of the transgenic line (T2A-1) at the suburbs of Xiaogan City and the suburbs of Suizhou City during the 2011-2013 was issued by Ministry of Agriculture of the People's Republic of China.

Plant materials

The transgenic *Bt* rice line (T2A-1) and the nontransgenic parental *indica* rice line Minghui 63 were selected for the experiments. T2A-1 expresses a synthesized *cry2Aa* gene driven by the maize *ubiquitin* promoter. T2A-1 is homozygous and

exhibits high resistance against lepidopteran pests of rice [12]. Minghui 63 is an elite *indica* restorer strain, and served a nontransgenic control. Both rice lines were gifted by National Key Laboratory of Crop Genetic Improvement, Wuhan, China.

The two rice lines used for the laboratory experiments were cultured in different plastic tanks (25 cm length×20 cm width×3 cm height) in Yoshida culture solution [27]. Fifteen day-old rice seedlings (approximately 15 cm in height) were used in the experiments. All plants were maintained at 26°C±2°C, and the relative humidity was approximately 80%.

Insects for the laboratory experiments

The original adults of *N. lugens* were collected from paddy fields in Wuhan, Hubei Province, China. Prior to the tritrophic bioassay, independent colonies of *N. lugens* were established on T2A-1 and Minghui 63 and maintained for more than ten generations. The *C. lividipennis* individuals were collected from paddy fields in Xiaogan and reared with eggs of *N. lugens* infested on Minghui 63. The colony was maintained for more than six generations before its use in the present study. The insects were cultured at 28±1°C, RH 70±5% and with a light-dark cycle of 14 h:10 h. Both susceptible strains of *Plodia interpunctella* (Hubner) (Lepidoptera: Pyralidae) and *Cnaphalocrocis medinalis* (Guenee) (Lepidoptera: Pyralidae) were used to confirm the bioactivity of the Cry protein.

Prey egg–mediated effects of transgenic *cry2Aa* rice on the life-table parameters of *C. lividipennis*

Two reproductive females of *N. lugens* were placed into a glass tube (3 cm diameter×25 cm length) containing six or seven 15-day-old rice seedlings (Minghui 63 and T2A-1) and allowed to lay eggs for 2 days. The *N. lugens* adults were then removed, and the eggs of *N. lugens* on the rice seedlings were used as the prey of *C. lividipennis*. Newly hatched nymphs of *C. lividipennis* (<24 h) were placed individually in glass tubes. All tubes were sealed with nylon mesh. During the period from the first to the third instar of *C. lividipennis*, the rice seedlings were refreshed every 2 days, and during the period from the fourth instar to adulthood, the rice seedlings were refreshed every day. Yoshida culture solution [27] was used to keep the rice seedlings fresh. The survival and molting of the *C. lividipennis* nymphs were recorded every day. After the *C. lividipennis* adults emerged, the sex and body weight of these adults were recorded. Thirty-two nymphs of *C. lividipennis* were tested for each rice line.

Prey nymph–mediated effects of transgenic *cry2Aa* rice on the life-table parameters of *C. lividipennis*

Newly molted second-instar nymphs (<24 h) of *C. lividipennis* were placed individually in glass tubes (2 cm diameter×12 cm length) covered with tampons. The bottom of each tube was filled with a piece of wetted sponge to maintain humidity. Newly hatched nymphs of *N. lugens* (24–48 h after hatching), reared on either Minghui 63 or T2A-1 rice plants, were employed as the prey of *C. lividipennis*. During the period from the second to the third instar of *C. lividipennis*, 10 nymphs of *N. lugens* were provided daily, and during the period from the fourth instar to adulthood, 20 nymphs were provided daily. The survival and molting of the *C. lividipennis* nymphs were monitored on a daily basis. After the adults of *C. lividipennis* emerged, the sex and body weight of these adults were recorded. Eighty nymphs of *C. lividipennis* were tested for each rice line.

Cry2Aa contents in rice plants, N. lugens and C. lividipennis

The sheaths of 15-day-old rice seedlings, neonates of *N. lugens* fed on T2A-1 or Minghui 63 for 2 days, *N. lugens* eggs laid by *N. lugens* adults fed on T2A-1 or Minghui 63, and third- or fourth-instar nymphs of *C. lividipennis* that preyed on eggs or nymphs of *N. lugens* fed on T2A-1 or Minghui 63 were collected. Four or five samples were collected for each treatment. Before the assay, the insect samples were washed four times with PBST (PBS/0.55% Tween-20) to remove any Cry protein from their outer surfaces. The Cry protein contents were determined using AP005 ENVIRONLOGIX kits (ENVIRONLOGIX, USA). The kits were used according to the manufacturer's instructions.

Exposure of C. lividipennis to Cry2Aa at high dose

Lyophilized Cry2Aa protein was purchased from the Biochemistry Department Laboratory, School of Medicine, Case Western Reserve University, USA. The Cry2Aa protoxin was expressed by *Escherichia coli*, then the protoxin inclusion bodies were solubilized and trypsinized, subsequently the toxins were purified and lyophilized. The purity of toxins is about 95–98%, and the molecular size of activated toxin is 65 kDa. Potassium arsenate (PA, KH_2AsO_4), which has been previously reported as toxic to insects [24,28], was used as a toxic model compound and was purchased from Sigma-Aldrich (St. Louis, MO).

The results of the preliminary experiments showed that the chemically defined diet for *N. lugens* as described by Fu et al. [29] was sufficient for sustaining the growth and development of *C. lividipennis* from the second instar to adulthood (The survival of *C. lividipennis* from the first instar to adulthood was low). The artificial diet for *N. lugens* was therefore used as the medium to deliver the Cry2Aa protein to the gut of *C. lividipennis*. Second-instar nymphs of *C. lividipennis* were reared on one of three different diets: i) an artificial diet (negative control); ii) an artificial diet containing 300 µg/ml of Cry2Aa (a level over ten times higher than that to which *C. lividipennis* would realistically be exposed in rice fields); iii) an artificial diet containing 40 µg/ml of PA (positive control). The diets were refreshed daily. The molting and survival of *C. lividipennis* were observed daily. When the adults emerged, their genders and body weights were recorded. Thirty-six nymphs of *C. lividipennis* were evaluated for each treatment.

To ensure the stability of the Cry2Aa protein in the artificial diets before and after 24 h of feeding exposure, the Cry2Aa proteins were extracted from the artificial diets, and the concentrations of Cry2Aa protein in the diet were determined using AP005 ENVIRONLOGIX kits.

To examine the bioactivity of this batch of Cry2Aa on lepidopteran insects, Cry2Aa was mixed with an artificial diet for *P. interpunctella*. Each bioassay included five concentrations for Cry2Aa (2, 10, 20, 30, 40 µg/g) plus a control. Forty newly hatched *P. interpunctella* larvae javascript:void(0); were introduced into the artificial diets of each concentration and five replicates were tested. Mortality of the larvae was determined after 1 week. The LC_{50} (concentration resulting in 50% *P. interpunctella* larval mortality) of this batch of Cry2A protein was measured.

To examine the bioactivity of Cry2Aa protein in the artificial diets before and after 24 h of feeding exposure, the artificial diets containing Cry2Aa were appropriately diluted and sprinkled on the leaves of corn. After 2 h of air-drying, 15 *Bt*-susceptible second-instar larvae of *C. medinalis* were distributed onto the leaves for each treatment. Four replicates were tested in each treatment. Mortality of the insects were recorded 48 h later.

Effects of transgenic cry2Aa rice on the functional response of female C. lividipennis

C. lividipennis individuals were reared on T2A-1 or Minghui 63 rice plants infested with *N. lugens* eggs for one generation. Female adults of *C. lividipennis* (2 days after eclosion) were starved for 24 h, and each female was then transferred to one glass tube (2 cm diameter×12 cm length) containing three or four 15-day-old rice seedlings infested with first-instar nymphs of *N. lugens*. The experimental densities of *N. lugens* were 10, 20, 30, 40 and 50 nymphs per tube. The number of *N. lugens* consumed by *C. lividipennis* was recorded after 24 h. The experiment was repeated five times for each density.

Field experiment design

The experiments were conducted during the growing seasons of 2011–2013 at the two different sites, where field trials of *Bt* rice were permitted, in Hubei Province, China. The first site located in the suburbs of Xiaogan City, and the second site located in the suburbs of Suizhou City. The layout of the plots in the field followed a completely randomized block design with four replications for each rice line. Each experimental plot was 150 m^2 (10 m×15 m) and was surrounded by a 1 m wide unplanted border. The entire experimental field was bordered by five rows of the nontransgenic control plants. The *Bt* and non-*Bt* control rice lines were sown in early May and transplanted 1 month after sowing. The seedlings were manually transplanted with one seedling per pot, 13.3 cm between plants within a row, and 29.9 cm between rows. The agronomic practices used for growing the rice, including fertilization and irrigation, were consistent with those followed by local farmers, except that no insecticides were applied during the whole growing season.

Field sampling with a vacuum-suction machine

Sampling of *C. lividipennis* was conducted as described by Xu et al. [30]. Arthropods at both field sites were collected using a vacuum-suction machine, constructed basing on a description by Carino et al. [31] and supplemented by a square sampling box (50 cm length×50 cm width×120 cm height) with a metal frame enclosed by Mylar film. Samples were collected every 10–15 days, starting 1 month after transplantation and continuing until the rice was ripe (as measured by grain maturity and harvest). On each sampling date, a square sampling box was placed at random along the diagonal line of each test plot at each site, with five subsamples per plot. The sample locations in each plot were marked with bamboo stakes to avoid resampling at the same location. Arthropods inside the frame enclosure were collected using the vacuum-suction machine for 5 min at each sampling location and were preserved in 75% ethanol. All samples were taken back to the laboratory and identified to the species level.

Data analysis

ELISA data and body weights were compared using the Student's *t*-tests. The Chi-square test was used for the parameters of preimaginal survival and female ratio. Nymphal developmental time was analyzed using Mann–Whitney *U*-tests, as the data did not fulfill the assumptions required for parametric analyses (normal distribution of residues and homogeneity of error variances). Survival response to the artificial diets containing Cry2Aa was analyzed using the Kaplan-Meier procedure, and the log-rank test was used in the purified toxin experiment.

In the field experiment, the population density and population dynamics determined by vacuum-suction were used to evaluate the impacts of the transgenic *Bt* rice on *C. lividipennis* populations in the fields. The population density of *C. lividipennis* was represented by seasonal means as captured by vacuum-suction. The population dynamics of the predators were measured by the means at each sampling date. Population density and population dynamics were analyzed using the Student's *t*-test.

The data from the functional response experiment were fitted to Holling's "Type II" disc equation, which estimates *Na* as follows: $Na = aTN/(1 + aThN)$, where *Na* is the number of prey attacked, *N* is prey density, and *T* is the duration of the experiment ($T = 1$ day in the present study). The parameters *a* (instantaneous search rate) and *Th* (time required to handle a prey item) were calculated via least-squares nonlinear regression based on the Gauss-Newton method.

The percentage data were arcsine–square root transformed, and all count data were square root $(x+1)$ or $\log_{10}(x+1)$ transformed before being subjected to data analysis. The untransformed means are presented in the results. All statistical analyses were performed using the software package SPSS (version 16.0 for Windows, 2007).

Results

Prey-mediated effects of transgenic *cry2Aa* rice on the life-table parameters of *C. lividipennis*

No significant differences were observed between the developmental time, preimaginal survival, female ratio and fresh body weight of *C. lividipennis* adults reared with eggs or nymphs of *N. lugens* fed on *Bt* and non-*Bt* rice ($P>0.05$) (Tables 1, 2).

The Cry2Aa content in T2A-1 rice sheaths was 13.9±1.2 μg/g. The content of Cry2Aa in nymphs of *N. lugens* fed on T2A-1 was 5.3±0.8 ng/g, only 0.04% of that in the rice sheaths. The content of Cry2Aa in the eggs of *N. lugens* fed on T2A-1 was undetectable (Table 3).

When *N. lugens* nymphs and T2A-1 seedlings were provided simultaneously to *C. lividipennis*, Cry2Aa was detectable, and the concentration of Cry2Aa in *C. lividipennis* was 48.5±13.0 ng/g; this value was much lower than that in the rice sheaths (13.9±1.2 μg/g) but higher than that in the nymphs of *N. lugens* (5.3±0.8 ng/g). However, when the *N. lugens* nymphs fed on T2A-1 were removed from T2A-1 and the nymphs alone were

provided to *C. lividipennis*, no Cry2Aa was detected in the predator. When *N. lugens* eggs and T2A-1 seedlings were provided to *C. lividipennis* simultaneously, Cry2Aa was detectable, and the concentration of Cry2Aa in *C. lividipennis* was 200.2±33.2 ng/g. However, when the *N. lugens* eggs were removed from the T2A-1 seedlings and the eggs alone were provided to *C. lividipennis*, Cry2Aa was not detectable in the predator. As was expected, no Cry2Aa protein was detected in the Minghui 63 rice plants (Table 3).

Purified Cry2Aa protein bioassay

Before and after exposure to *C. lividipennis* for 24 h, the concentrations of Cry2Aa in the artificial diets were 98.4±3.2 μg/g and 91.1±3.7 μg/g, respectively. After 24 h of exposure, the concentration of Cry2Aa in the diets decreased by 7.42%, and no statistical differences were detected between the values before and after exposure (Student's *t*-test, $P = 0.185$). Therefore, it could be concluded that the Cry2Aa protein in the artificial diets was stable.

The LC$_{50}$ of this batch of Cry2Aa for *P. interpunctella* larvae was 14.9 μg/g fresh weight. Before and after exposure to *C. lividipennis* for 24 h, the artificial diets containing 300 μg/ml Cry2Aa resulted in 90% and 80% mortality of *C. medinalis* larvae, respectively; these mortalities were significantly higher than those that occurred in larvae fed with the pure artificial diet (13%). These results indicated that the Cry2Aa protein in the artificial diets for *C. lividipennis* was bioactive.

Survival of the *C. lividipennis* fed with the artificial diet containing 40 μg/ml PA was significantly decreased compared to that of the *C. lividipennis* fed with the pure artificial diet ($P< 0.001$) (Fig. 1). This result indicated that the test system employed in the present study could detect the dietary effects of insecticidal compounds. High-dosage exposure to the Cry2Aa protein had no adverse effects on the survival response of *C. lividipennis* in comparison to the pure artificial diet (negative control) (Fig. 1). Similarly, the developmental time from the second instar to adulthood, preimaginal survival and body weight of *C. lividipennis* were unaffected by the Cry2Aa protein in comparison to the negative control (Table 4).

Table 1. Prey-mediated effects of Cry2Aa on life-table parameters of *Cyrtorhinus lividipennis* preying eggs of *Nilapavarta lugens* reared with T2A-1 or Minghui 63 rice plants.

Parameters	T2A-1	Minghui 63	Statistics
1st instar developmental time (days ± SE)[a]	2.3±0.09 (32)	2.5±0.11 (32)	$U=415$, $P=0.117$
2nd instar developmental time (days ± SE)[a]	1.7±0.09 (30)	1.6±0.10 (30)	$U=400.5$, $P=0.387$
3rd instar developmental time (days ± SE)[a]	1.5±0.09 (29)	1.5±0.10 (30)	$U=435$, $P=1.000$
4th instar developmental time (days ± SE)[a]	1.8±0.09 (29)	1.8±0.07 (30)	$U=400$, $P=0.473$
5th instar developmental time (days ± SE)[a]	2.7±0.09 (29)	2.9±0.14 (29)	$U=364$, $P=0.399$
Whole nymphal stage developmental time (days ± SE)[a]	10.0±0.13 (28)	10.3±0.13 (29)	$U=313$, $P=0.080$
Preimaginal survival (%)[b]	87.5	90.6	$\chi^2=0.002$, $P=0.967$
Female ratio (%)[b]	42.9	55.2	$\chi^2=0.864$, $P=0.352$
Male weight (mg ± SE)[c]	0.55±0.03	0.49±0.03	$t=-1.492$, $P=0.179$
Female weight (mg ± SE)[c]	0.93±0.02	0.92±0.04	$t=-0.193$, $P=0.851$

The experiment started with 32 nymphs per treatment. (n), number of individuals at each development stage.
[a]Mann–Whitney *U*-test.
[b]Chi-square test.
[c]Student's *t*-test.

Table 2. Prey-mediated effects of Cry2Aa on life-table parameters of *Cyrtorhinus lividipennis* preying nymphs of *Nilapavarta lugens* reared with T2A-1 or Minghui 63 rice plants.

Parameters	T2A-1	Minghui 63	Statistics
2nd instar developmental time (days ± SE)[a]	1.9±0.08 (76)	1.8±0.06 (77)	$U = 2.538E3^*$, $P = 0.102$
3rd instar developmental time (days ± SE)[a]	1.8±0.07 (63)	2.0±0.08 (70)	$U = 1.875E3^*$, $P = 0.087$
4th instar developmental time (days ± SE)[a]	2.4±0.09 (50)	2.4±0.13 (48)	$U = 1.180E3^*$, $P = 0.873$
5th instar developmental time (days ± SE)[a]	2.8±0.17 (29)	2.7±0.15 (27)	$U = 367.5$, $P = 0.654$
2nd instar-adult developmental time (days ± SE)[a]	8.8±0.21 (28)	9.2±0.20 (26)	$U = 292$, $P = 0.190$
Preimaginal survival (%)[b]	35.0	32.5	$\chi^2 = 0.112$, $P = 0.738$
Female ratio (%)[b]	42.9	38.5	$\chi^2 = 0.108$, $P = 0.743$
Male weight (mg ± SE)[c]	0.33±0.05	0.30±0.01	$t = -0.420$, $P = 0.696$
Female weight (mg ± SE)[c]	0.46±0.02	0.43±0.09	$t = -0.322$, $P = 0.764$

The experiment started with 80 nymphs per treatment. (n), number of individuals at each development stage.
*E3 = 10^{-3}.
[a]Mann–Whitney U-test.
[b]Chi-square test.
[c]Student's t-test.

Effects of Cry2Aa on the functional response of *C. lividipennis*

The results indicated that the functional response of *C. lividipennis* to *N. lugens* on both rice lines was typically Type II as described by Holling [32,33] (Fig. 2). The instantaneous search rate and handling time were not significantly affected by rice line ($P > 0.05$) (Table 5).

Effects of transgenic *cry2Aa* rice on the population density and dynamics of *C. lividipennis*

The population density of *C. lividipennis* is shown in Table 6. Compared with that on the nontransgenic Minghui 63, the population density of *C. lividipennis* was not significantly influenced by transgenic *cry2Aa* rice at any site or in any year (Student's t-test, $P > 0.05$) (Table 6). The population dynamics (means of each sampling date) of *C. lividipennis* are shown in Fig. 3. No significant differences in the population dynamics of *C. lividipennis* were observed between Minghui 63 and transgenic *cry2Aa* rice fields at any sampling date, at any site or in any year (Student's t-test, all $P > 0.05$). Repeated measures ANOVA analysis showed that the population dynamics were unaffected by rice line ($P > 0.05$).

Discussion

The ecological risk assessment of an insect-resistant transgenic crop for a nontarget arthropod (NTA) should be conducted within a tiered scheme: (i) effects on the NTA at elevated doses in a replicated controlled system; (ii) effects on the NTA at realistic doses in a replicated controlled system; (iii) effects on the population of the NTA at realistic doses in a realistic agricultural system [34]. According to these criteria, three experiments were conducted to evaluate the ecological risk of transgenic *cry2Aa* rice to *C. lividipennis*, a primary predator of *N. lugens* that is the main NTA of transgenic *Bt* rice in the present study: (1) a direct feeding experiment, in which *C. lividipennis* was fed an artificial diet containing Cry2Aa at the dose of 10-time higher than that it may encounter in the realistic field condition; (2) a tritrophic experiment, in which the Cry2Aa protein was delivered to *C. lividipennis* indirectly through prey eggs or nymphs; (3) a realistic field experiment, in which the population dynamics of *C. lividipennis* were investigated by vacuum-suction. Direct exposure to elevated doses of the Cry2Aa protein and prey-mediated exposure to realistic doses of the Cry2Aa protein did not result in significant detrimental effects on the developmental time, preimaginal survival, female ratio and body weight of *C. lividipennis*.

Table 3. Contents of Cry2Aa protein in rice sheath tissue, *Nilapavarta lugens* and *Cyrtorhinus lividipennis*.

Treatments	T2A-1	Minghui 63
Sheath of rice plants	13.9±1.2 µg/g	Not detectable
Eggs of *N.lugens*	Not detectable	Not detectable
Nymphs of *N. lugens*	5.3±0.8 ng/g	Not detectable
C. lividipennis provided with *N. lugens* eggs with rice plants	200.2±33.2 ng/g	Not detectable
C. lividipennis provided with *N. lugens* nymphs with rice plants	48.5±13.0 ng/g	Not detectable
C. lividipennis provided with *N. lugens* eggs without rice plants	Not detectable	Not detectable
C. lividipennis provided with *N. lugens* nymphs without rice plants	Not detectable	Not detectable
C. lividipennis provided with rice plants	17.1±8.6 ng/g	Not detectable

Data are represented as mean ± SE.

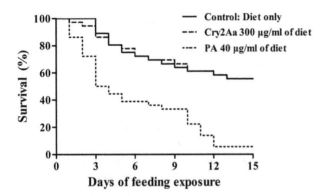

Figure 1. Survival of *Cyrtorhinus lividipennis* fed pure artificial diet or diet containing different insecticidal compounds. 300 μg Cry2Aa and 40 μg PA per ml were incorporated into artificial diets. Pure diet served as a negative control (N = 36).

No significant differences in population density and population dynamics were observed between *C. lividipennis* populations in transgenic *cry2Aa* and nontransgenic rice fields. It could be concluded that transgenic *cry2Aa* rice had no adverse effects on *C. lividipennis*. This study represents the first report of an assessment continuum for the effects of transgenic *cry2Aa* rice on *C. lividipennis*.

Planthoppers are the main nontarget herbivores in transgenic *Bt* rice fields. Determining whether the Bt protein may be transmitted to predators via planthoppers is important for the ecological risk assessment of transgenic *Bt* rice. Several reports have examined Bt protein transmission to predators via predation on planthoppers infesting transgenic *Bt* rice, but their results were inconclusive. Cry1Ab could be transferred to *P. subpiraticus* by predation on *N. lugens* fed on transgenic *cry1Ab* rice, and the content of Cry1Ab in *P. subpiraticus* was significantly higher than that in *N. lugens* fed on transgenic *cry1Ab* rice [19]. Cry1Ab could also be transferred to *U. insecticeps* via predation on *N. lugens* fed on transgenic *cry1Ab* rice, and the concentration of Cry1Ab in *U. insecticeps* was significantly lower than that in *N. lugens* [21]. However, no Cry2Aa was detected in a study of the larvae of a general predator

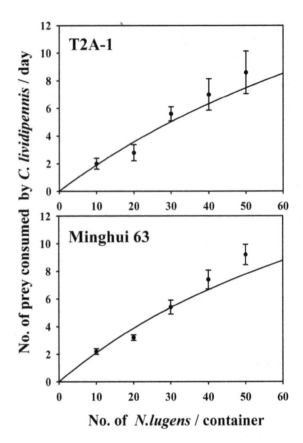

Figure 2. Functional response of *Cyrtorhinus lividipennis* collected from T2A-1 and Minghui 63.

(*C. sinica*) of planthoppers, in which *C. sinica* was provided with *Laodelphax striatellus* fed on transgenic *cry2Aa* rice [24].

In the present study, when *N. lugens* nymphs and T2A-1 seedlings were provided simultaneously to *C. lividipennis*, Cry2Aa was detected in the predator. However, when the *N. lugens* nymphs fed on T2A-1 were removed from the rice and provided alone to *C. lividipennis*, no Cry2Aa was detected in the predator.

Table 4. Effects of purified Cry2Aa incorporating into artificial diet on life-table parameters of *Cyrtorhinus lividipennis*.

Parameters	Treatments		
	Control	300 μg/ml Cry2Aa	40 μg/ml PA
2nd instar developmental time (days ± SE)[a]	2.1±0.15 (35)	2.3±0.13 (35)	2.4±0.21 (23)
3rd instar developmental time (days ± SE)[a]	2.5±0.10 (28)	2.6±0.13(29)	4.2±0.36 (13)**
4th instar developmental time (days ± SE)[a]	3.1±0.17 (21)	3.1±0.24 (22)	7.0±0.58 (3)**
5th instar developmental time (days ± SE)[a]	4.2±0.18 (20)	4.1±0.20 (20)	—
2nd instar-adult developmental time (days ± SE)[a]	11.8±0.40 (20)	11.7±0.40 (20)	—
Preimaginal survival (%)[b]	55.6	55.6	0.0
Male weight (mg ± SE)[c]	0.41±0.02	0.40±0.02	—
Female weight (mg ± SE)[c]	0.58±0.02	0.54±0.02	—

Nymphs of *C. lividipennis* were fed with an artificial diet containing 300 μg/ml Cry2Aa or 40 μg/ml PA (positive control). Pure diet served as a negative control (N = 36). The experiment lasted until adult eclosed. Statistical comparisons were made separately for each of the insecticidal compounds comparing with the control. Asterisks denote significant differences: $P < 0.01$.
[a]Mann–Whitney U-test with Bonferroni correction (adjusted $\alpha = 0.025$).
[b]Chi-square test with Bonferroni correction (adjusted $\alpha = 0.025$).
[c]Student's t-test.

Table 5. Parameters of Type II functional response of *Cyrtorhinus lividipennis* to *Nilapavarta lugens* nymph fed on *Bt* or non-*Bt* rice.

Rice materials	*a*	*Th*	R^2 (%)
T2A-1	0.251±0.028	0.036±0.008	95.83
Minghui 63	0.231±0.025	0.037±0.007	95.20

a: instantaneous search rate (day^{-1}). *Th*: time required to handle a prey (day). Data are represented as mean ± SE.
There was no significant difference between T2A-1 and Minghui 63, based on Student's *t*-test (*P*<0.05).

Similarly, when *N. lugens* eggs and T2A-1 seedlings were provided to *C. lividipennis* simultaneously, Cry2Aa was detected in the predator. However, when the *N. lugens* eggs were removed from the T2A-1 seedlings and provided alone to *C. lividipennis*, no Cry2Aa was detected in the predator. Therefore, it may be inferred that Cry2Aa was not transmitted to *C. lividipennis* via predation on the eggs and nymphs of *N. lugens*; the protein may instead be transferred by the piercing-sucking foraging behavior of *C. lividipennis* on rice. This hypothesis was verified through the results of our supplementary experiments: when *C. lividipennis* nymphs were provided with T2A-1 seedlings alone for 1 day, Cry2Aa was detected in the predator, and the concentration of Cry2Aa in *C. lividipennis* was 17.1±8.6 ng/g (Table 3). Therefore, *C. lividipennis* may serve as a good indicator species in ecological risk assessments of transgenic *Bt* tice.

In the ecological risk assessment of an arthropod-resistant genetically engineered crop, researchers normally use "Tier-1 assays" as the initial step to determine the toxicity of the insecticidal compounds expressed by the transgenic crop on NTAs. In Tier-1 tests, insecticidal compounds are added to artificial diets for the tested NTAs, and the tested organisms are directly exposed to doses of the insecticidal compounds several times higher than those realistically present in the field. Tier-1 tests increase the likelihood that a hazard will be detected if the hazard exists, and therefore provide confidence that minimal risk is present if no adverse effect is detected [34,35]. Three important factors must be considered in Tier-1 assays: (i) the methods for the delivery of the insecticidal proteins to the test organisms; (ii) the need for and selection of the compounds used as positive controls; and (iii) the methods for monitoring the concentration, stability and bioactivity of the insecticidal proteins during the assay [35]. In

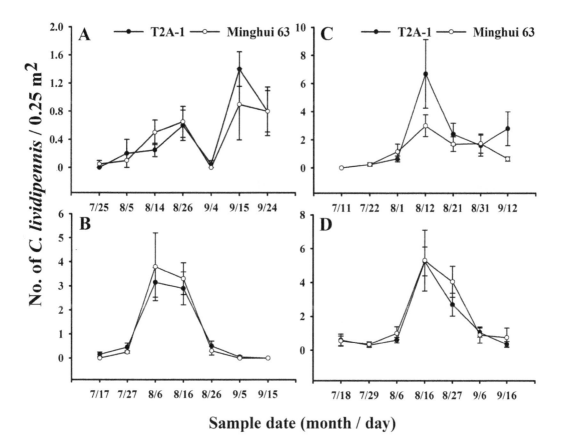

Figure 3. Population dynamics of *Cyrtorhinus lividipennis* collected by vacuum-suction. Data are represented as mean ± SE. (A). Xiaogan, 2011; (B). Xiaogan, 2012; (C). Xiaogan, 2013; (D) Suizhou, 2012. There was no significant difference between *Bt* rice and control plots at the same sampling time, based on Student's *t*-test (N=4). Repeated measures ANOVA: (A). $F_{1,6}=0.059$, $P=0.816$; (B). $F_{1,6}=0.058$, $P=0.777$; (C). $F_{1,6}=1.366$, $P=0.287$; (D). $F_{1,6}=0.964$, $P=0.364$.

Table 6. Population densities (no. of per 0.25 m^2) of *Cyrtorhinus lividipennis* collected by vacuum-suction.

Rice materials	Xiaogan			Suizhou
	2011	**2012**	**2013**	**2012**
T2A-1	0.47±0.10	1.03±0.09	2.06±0.68	1.55±0.06
Minghui 63	0.43±0.14	1.09±0.20	1.21±0.23	1.84±0.29

N = 4 at both sites in 2011, 2012, and 2013. Data are represented as mean ± SE. There was no significant difference between T2A-1 and Minghui 63 field, based on Student's *t*-test.

the present study, a dietary exposure experiment was conducted in which purified Cry2Aa protein was directly fed to *C. lividipennis* nymphs through its incorporation into a previously described artificial diet for *N. lugens* [29]. This artificial diet was sufficient for sustaining the growth and development of *C. lividipennis* from the second instar to adulthood. Before and after exposure to *C. lividipennis* for 24 h, the Cry2Aa protein in the artificial diets was stable and bioactive. The oral poison PA was used as a positive control to validate our dietary exposure assay. The survival of the *C. lividipennis* fed with the artificial diet containing 40 μg/ml PA significantly decreased compared to that of the *C. lividipennis* fed with the pure artificial diet ($P<0.001$) (Fig. 1). This result confirmed that the test system employed in the current study was able to detect the dietary effects of insecticidal compounds. No detrimental effects on the life-table parameters of *C. lividipennis* were observed when the insects were provided with an artificial diet containing Cry2Aa at a concentration that was nearly ten times higher than that measured in the rice sheaths. This study represents the first report of the use of a Tier-1 system to evaluate the potential effects of Cry2Aa on *C. lividipennis*.

The survival rate of *C. lividipennis* preying the nymphs of *N. lugens* was 32–35%, which was much lower than that of *C. lividipennis* preying eggs of *N. lugens* (87–91%) in the tritrophic experiment. So did body weight of adults of *C. lividipennis*. Similar results have been reported by Chua and Mikil and Chen et al., and they concluded that *N. lugens* eggs was an essential food type for *C. lividipennis*, and *N. lugens* nymphs was not an ideal food for the predator [36,37]. The different nutrient composition of prey may cause biological parameters difference of predator. *Harmonia axyridis* (Pallas) (Coleoptera: Coccinellidae) is a polyphagous species, the complete development of this predator can be accomplished using the aphid *Acyrthosiphon pisum* (Harris) (Homoptera: Aphididae) or *Ephestia kuehniella* (Zeller) (Lepidop-

tera: Pyralidae) eggs as substitution prey. Biochemical analyses indicated that amino acids and lipids of *E. kuehniella* eggs were richer than *A. pisum* adults, but, on the contrary, the glycogen of aphids was richer than *E. kuehniella* eggs. Some biological parameters such as larval mortality, adult weight, and fecundity, were modified according to the food eaten [38]. Whether the different survival of *C. lividipennis* in the present study is caused by different nutrient composition of prey needs to be further explored.

Conclusions

In summary, this comprehensive study, involving a Tier-1 examination system, a tritrophic bioassay, functional response experiments in the laboratory and population dynamics determinations in the field, provides the most complete information to date on the impacts of *Bt* rice expressing Cry2Aa on *C. lividipennis*, a major predator in rice ecosystems. These results indicate that *C. lividipennis* is not sensitive to the Cry2Aa protein and that *Bt* rice (T2A-1) poses a negligible risk to this nontarget organism.

Acknowledgments

We thank Prof. Yongjun Lin (National Key Laboratory of Crop Genetic Improvement at Huazhong Agricultural University) for providing the transgenic rice seeds.

Author Contributions

Conceived and designed the experiments: HH Y Han. Performed the experiments: Y Han JM JC YW. Analyzed the data: Y Han WC. Contributed reagents/materials/analysis tools: JZ Y He YF. Wrote the paper: HH Y Han.

References

1. Zeigler RS, Barclay A (2008) The relevance of rice. Rice 1: 3–10.
2. Cheng JA, He J (1996) Rice insect pests. China Agricultural Press, Beijing.
3. Kiritani K (1979) Pest management in rice. Annu Rev Entomol 24: 279–312.
4. Oerke EC, Dehne HW, Schönbeck F, Weber A (1994) Crop production and crop protection: estimated losses in major food and cash crops. Amsterdam: Elsevier.
5. Pathak MD, Khan ZR (1994) Insect Pests of Rice. Los Baños, Laguna, Philippines: International Rice Research Institute.
6. Sheng CF, Wang HT, Gao LD, Xuan JW (2003) The occurrence status, damage cost estimate and control strategies of stem borers in China. Plant Prot 29: 37–39. (in Chinese with English summary).
7. Lou YG, Zhang GR, Zhang WQ, Hu Y, Zhang J (2013) Biological control of rice insect pests in China. Biol Control 67: 8–20.
8. Matteson PC (2000) Insect pest management in tropical Asian irrigated rice. Annu Rev Entomol 45: 549–574.
9. Tu J, Zhang G, Datta K, Xu C, He Y, et al. (2000) Field performance of transgenic elite commercial hybrid rice expressing *Bacillus thuringiensis*-endotoxin. Nat Biotechnol 18: 1101–1104.
10. Huang JK, Hu RF, Rozelle S, Pray C (2005) Insect-resistant GM rice in farmers' fields: assessing productivity and health effects in China. Science 308: 688–690.
11. Shelton AM, Zhao JZ, Roush RT (2002) Economic, ecological, food safety, and social consequences of the deployment of *Bt* transgenic plants. Ann Rev Entomol 47: 845–881.
12. Chen H, Tang W, Xu CG, Li XH, Lin YJ, et al. (2005) Transgenic indica rice plants harboring a synthetic *cry2A** gene of *Bacillus thuringiensis* exhibit enhanced resistance against lepidopteran rice pests. Theor Appl Genet 111: 1330–1337.
13. Cheng X, Sardana R, Kaplan H, Altosaar I (1998) Agrobacterium-transformed rice plants expressing synthetic *cry1A(b)* and *cry1A(c)* genes are highly toxic to yellow stem borer and striped stem borer. Proc Natl Acad Sci U S A 95: 2767–2772.
14. Tang W, Chen H, Xu CG, Li XH, Lin YJ, et al. (2006) Development of insect-resistant transgenic indica rice with a synthetic *cry1C** gene. Mol Breeding 18: 1–10.
15. Heinrichs EA (1979) Control of leafhopper and planthopper vectors of rice viruses. In: Moramorosch K, Arris KF, Editors. Leafhopper vectors and planthopper disease agents. New York: Academic Press. pp. 529–558.
16. Sogawa K, Liu GJ, Shen JH (2003) A review on the hyper susceptibility of Chinese hybrid rice to insect pests. Chinese J Rice Sci 17, 23–30. (in Chinese with English summary).

17. Bai YY, Jiang MX, Cheng JA (2005) Effects of transgenic *cry1Ab* rice pollen on fitness of *Propylea japonica* (Thunberg). J Pest Sci 78: 123–128.

18. Bernal CC, Aguda RM, Cohen MB (2002) Effect of rice lines transformed with *Bacillus thuringiensis* toxin genes on the brown planthopper and its predator *Cyrtorhinus lividipennis*. Entomol Exp Appl 102: 21–28.

19. Chen M, Ye GY, Lu XM, Hu C, Peng YF, et al. (2005) Biotransfer and bioaccumulation of Cry1Ab insecticidal protein in rice plant-brown planthopper-wolf spider food chain. Acta Entomol Sinica 48: 208–213. (in Chinese with English summary).

20. Jiang YH, Fu Q, Cheng JA, Zhu ZR, Jiang MX, et al. (2004) Dynamics of Cry1Ab protein from transgenic *Bt* rice in herbivores and their predators. Acta Entomol Sinica 47: 454–460. (in Chinese with English summary).

21. Tian JC, Liu ZC, Chen M, Chen Y, Chen XX, et al. (2010) Laboratory and field assessments of prey-mediated effects of transgenic Bt rice on *Ummeliata insecticeps* (Araneida: Linyphiidae). Environ Entomol 39: 1369–1377.

22. Tian JC, Chen Y, Li ZL, Li K, Chen M, et al. (2012) Transgenic Cry1Ab rice does not impact ecological fitness and predation of a generalist spider. PLoS ONE. 7: e35164.

23. Chen M, Liu ZC, Ye GY, Shen ZC, Hu C, et al. (2007) Impacts of transgenic *cry1Ab* rice on non-target planthoppers and their main predator *Cyrtorhinus lividipennis* (Hemiptera: Miridae)-A case study of the compatibility of Bt rice with biological control. Biol Control 42: 242–250.

24. Li YH, Wang YY, Romeis J, Liu QS, Lin KJ, et al. (2013) *Bt* rice expressing Cry2Aa does not cause direct detrimental effects on larvae of *Chrysoperla sinica*. Ecotoxicology 22: 1413–1421.

25. Chen JM, Cheng JA, He JH (1992) Review on *Cyrtorrhinus livdipennis* Reuter. Entomol Knowledg 25: 370–373.

26. Sigsgaard L (2007) Early season natural control of the brown planthopper, *Nilaparvata lugens*: the contribution and interaction of two spider species and a predatory bug. Bull Entomol Res 97: 533–544.

27. Yoshida S, Forno DA, Cock JH, Gomez KA (1976) Laboratory manual for physical studies of rice, third editon. Los Baños, Laguna, Philippines: International Rice Research Institute. pp. 61–65.

28. Duan JJ, Head G, McKee MJ, Nickson TE, Martin JW, et al. (2002) Evaluation of dietary effects of transgenic *cry1Ab* rice pollen on a nontarget ladybird beetle, *Coleomegilla maculata*. Entomol Exp Appl 104: 271–280.

29. Fu Q, Zhang ZT, Hu C, Lai FX, Sun ZX (2001) A chemically defined diet enables continuous rearing of the brown planthopper, *Nilaparvata lugens* (Stål) (Homoptera: Delphacidae). Appl Entomol Zool 36: 111–116.

30. Xu XL, Han Y, Wu G, Cai WL, Yuan BQ, et al. (2011) Field evaluation of effects of transgenic *cry1Ab/cry1Ac*, *cry1C* and *cry2A* rice on *Cnaphalocrocis medinalis* and its arthropod predators. Sci China Life Sci 54: 1019–1028.

31. Carino FO, Kenmore PE, Dyck VA (1979) The farmcop suction sampler for hoppers and predators in flooded rice fields. Int Rice Res News 4: 21–22.

32. Holling CS (1959) Some characteristics of simple types of predation and parasitism. Can Entomol 91: 385–398.

33. Holling CS (1961) Principles of insect predation. Annu Rev Entomol 6: 163–182.

34. Romeis J, Bartsch D, Bigler F, Candolfi MP, Gielkens MMC, et al. (2008) Assessment of risk of insect-resistant transgenic crops to nontarget arthropods. Nat Biotechnol 26: 203–208.

35. Li YH, Romeis J, Wu KM, Peng YF (2013) Tier-1 assays for assessing the toxicity of insecticidal proteins produced by genetically engineered plants to nontarget arthropods. Insect Sci DOI: 10.1111/1744-7917.12044. Available: http://onlinelibrary.wiley.com/doi/10.1111/1744-7917.12044/abstract. Accessed 2013 Dec 6.

36. Chen JM, Cheng JA, He JH (1994) Effects of temperature and food on the development, survival and reproduction of *Cyrtorrhinus livdipennis* (Reuter). Acta Entomol Sinica 37: 63–70. (in Chinese with English summary).

37. Chua TH, Mikil E (1989) Effects of prey number and stage on the biology of *Cyrtorhinas lividipennis* (Hemiptera: Miridae): a predator of *Nilaparvata lagens* (Homoptera: Delphacidae). Environ Entomol 18: 251–255.

38. Specty O, Febvay G, Grenier S, Delobel B, Piotte C, et al. (2003) Nutritional plasticity of the predatory ladybeetle *Harmonia axyridis* (Coleoptera: Coccinellidae): comparison between natural and substitution prey. Arch Insect Biochem 52: 81–91.

8

Transgene Detection by Digital Droplet PCR

Dirk A. Moser[1,3], **Luca Braga**[2], **Andrea Raso**[2], **Serena Zacchigna**[2], **Mauro Giacca**[2], **Perikles Simon**[3*]

1 Faculty of Psychology, Genetic Psychology, Ruhr-University-Bochum, Bochum, Germany, 2 International Centre for Genetic Engineering and Biotechnology (ICGEB), Molecular Medicine, Trieste, Italy, 3 Department of Sports Medicine, Disease Prevention and Rehabilitation, Johannes Gutenberg-University Mainz, Mainz, Germany

Abstract

Somatic gene therapy is a promising tool for the treatment of severe diseases. Because of its abuse potential for performance enhancement in sports, the World Anti-Doping Agency (WADA) included the term 'gene doping' in the official list of banned substances and methods in 2004. Several nested PCR or qPCR-based strategies have been proposed that aim at detecting long-term presence of transgene in blood, but these strategies are hampered by technical limitations. We developed a digital droplet PCR (ddPCR) protocol for Insulin-Like Growth Factor 1 (*IGF1*) detection and demonstrated its applicability monitoring 6 mice injected into skeletal muscle with AAV9-*IGF1* elements and 2 controls over a 33-day period. A duplex ddPCR protocol for simultaneous detection of Insulin-Like Growth Factor 1 (*IGF1*) and Erythropoietin (*EPO*) transgenic elements was created. A new DNA extraction procedure with target-orientated usage of restriction enzymes including on-column DNA-digestion was established. *In vivo* data revealed that *IGF1* transgenic elements could be reliably detected for a 33-day period in DNA extracted from whole blood. *In vitro* data indicated feasibility of *IGF1* and *EPO* detection by duplex ddPCR with high reliability and sensitivity. On-column DNA-digestion allowed for significantly improved target detection in downstream PCR-based approaches. As ddPCR provides absolute quantification, it ensures excellent day-to-day reproducibility. Therefore, we expect this technique to be used in diagnosing and monitoring of viral and bacterial infection, in detecting mutated DNA sequences as well as profiling for the presence of foreign genetic material in elite athletes in the future.

Editor: Domenico Coppola, H. Lee Moffitt Cancer Center & Research Institute, United States of America

Funding: Work of PS and MG is carried out with the support of the World Anti-Doping Agency (WADA). The funders had no role in study design, data collection and analysis, decision to publish, or preparation of the manuscript.

Competing Interests: The authors have declared that no competing interests exist.

* Email: simonpe@uni-mainz.de

Introduction

Somatic gene therapy represents a promising tool to treat inherited or acquired diseases by transferring genetic material in order to compensate for defective genes, to produce a therapeutic substance, or to specifically trigger the immune system [1]. Despite its potential to treat life-threatening diseases, this technique might also be abused to improve physical performance [2]. Animal studies demonstrated successful viral transfer of potential performance-enhancing genes such as Insulin-Like Growth Factor 1 (*IGF1*; [3,4]) and Erythropoietin (*EPO*; [5]) but severe adverse events were also reported [6–9]. This has raised concerns about the illicit use of gene transfer technologies in elite sports (i.e. gene doping), and demonstrated the necessity to prohibit the use of gene transfer aimed at enhancing performance (WADA 2004), with the consequent need to monitor for the presence of transgenic DNA in routine gene doping tests.

The main obstacle for gene-doping detection is that the athlete's body would be enabled to produce doping substances that, in most cases, would be indistinguishable from endogenous proteins [10,11]. Therefore, currently proposed direct detection methods rely on specific sequence characteristics in transgenic DNA constructs that allow unbiased discriminability between genomic and exogenous DNA. Since skeletal muscle (the most likely target tissue for gene doping applications) would be difficult to harvest for

routine doping testing, several studies explored the possibility to detect minute amounts of transgenic DNA in the blood after somatic (intramuscular) gene transfer.

In these studies different PCR-based approaches, such as nested PCR [12–14] and TaqMan qPCR [8,15–17], were applied in order to detect minute amounts of transgenic DNA in blood. These methods selectively detect and discriminate the intron-less transgene from genomic DNA by PCR. Thus, detection of transgenic DNA in whole blood samples could provide a suitable means to support conviction of unscrupulous athletes. However, current PCR-based detection approaches are either highly sensitive but require a nested PCR procedure with a laborious workflow [12–14], or they display weaknesses to sensitively detect minute amounts of the transgene in a single round qPCR, [8,15,17,18]. In addition, qPCR relies on external standard curves, which further complicates inter-lab comparability of results. It is also important to point out that DNA amplification efficiency is highly dependent on template structure in the way that circular DNA amplifies poorly compared to linear DNA [19,20]. To date, all aforementioned methods use undigested DNA as a template which can result in poor amplification of circular DNA structures and consequently in failure to detect the specific transgenic sequence in a huge background of endogenous DNA. As recombinant DNA is integrated into the host genome or persists as episomal circular supercoiled DNA, predigestion of

DNA should improve PCR sensitivity and increase the likelihood of transgene detection.

Digital droplet PCR (ddPCR) is a new method that enables the absolute measure of target DNA. Its principle is based on the portioning of PCR mixture into thousands of droplets per reaction [21]. As ddPCR can also handle huge amounts of background DNA in the reaction, it represents a convenient method to find the transgenic "needle in a haystack". Additional benefit of ddPCR over qPCR also includes absolute quantification, which does not rely on external standard curve, leading to excellent day-to-day reproducibility. This is why ddPCR is likely to become a favourite tool for accurate routine analysis, especially when performed at multi-site laboratories.

Here we describe a new digital droplet PCR assay for transgene *IGF1* detection after intramuscular AAV9 gene transfer. DNA was purified from whole blood of 6 intramuscularly AAV9-*IGF1* transduced mice and 2 uninjected control animals, which were monitored for 33 days for the detectability of transgene elements by ddPCR.

We also developed a duplex ddPCR protocol for *IGF1* and *EPO* transgene elements to simultaneously detect two candidate genes for gene doping in a single assay. We further established a method to test for ddPCR efficiency by the addition of an internal control standard (ICS) at a defined copy number to the reactions, a method already described for qPCR elsewhere [17]. This ICS differs only from the transgene in its probe binding site and can be detected in parallel to the transgene, which makes monitoring of ddPCR efficiency feasible in each reaction.

To minimize sample handling and to optimize for transgene detectability, we also developed a new DNA extraction protocol, which includes on-column DNA restriction enzyme digestion and allows the elution of fragmented high quality DNA as optimal target for ddPCR. This procedure leads to further improved sensitivity of the assay and might also find application for assays where preferential amplification of target sequences is warranted.

Thus we aim at presenting a straightforward and highly sensitive approach for DNA detection, which could lead to the next generation of transgene detection.

Methods

DNA extraction, digestion and purification

Human genomic DNA (hgDNA) for spike-in experiments was extracted at large scale using the salting-out procedure as described by Miller et al. [22]. PCR standards of the *IGF1* and *EPO* coding-sequence were generated using the primers and resulting amplicon lengths as indicated in Table 1. To generate circular standards, purified PCR products were cloned into the PCRII-TOPO vector according to the manufacturer's recommendations. All standards were Sanger-sequenced to check for correctness of the sequences. DNA was spiked with defined copy numbers of freshly prepared ddPCR quantified *IGF1*- or *EPO* standard, and digested using restriction enzymes DdeI and RsaI (2 units/μg hgDNA) in NEB buffer 2 for 1 h at 37°C followed by 20 minutes heat inactivation at 65°C.

Production, purification, and characterization of rAAV vectors

The human hepatic *IGF1-IA* (ref. seq. NM_000618.3) was PCR amplified and cloned into the pZac recombinant AAV expression vector, generating the pAAV-*IGF1* vector. Viral particles were produced by the AAV Vector Unit at ICGEB Trieste (http://www.icgeb.org/avu-core-facility.html). Methods for production and purification were previously described [23].

AAV9 titers were in the range of 1×10^{13} genome copies per milliliter.

AAV9-*IGF1* injection into the skeletal muscle

Animal care and treatments were conducted in conformity with institutional guidelines in compliance with national and international laws and policies, upon approval by the ICGEB Ethical Committee and by the Italian Minister of Health (EEC Council Directive 86/609, OJL 358, December 12, 1987). Animals were provided with housing in an enriched environment, with at least some freedom of movement, food, water and daily care and cleaning. Experiments were performed under general or local anesthesia, and with constant use of analgesics. At the end of any experiment, competent authorized persons decided the proper time and most appropriate humane method for animal sacrifice. All experiments were performed in male CD1 mice, 4–6 weeks of age. As a model of gene doping, tibialis anterior and gastrocnemius muscles were injected with 50 μl of either PBS or a viral suspension containing 10^{11} viral particles of AAV9-*IGF1*, and harvested after the indicated periods of time.

DNA extraction and digestion from mouse blood

DNA was extracted from 100 μl mouse blood using the Qiagen DNA microkit, which enables the elution of DNA in variable volumes between 20–100 μl. DNA was eluted in a volume of 25 μl and digested with 5 U DdeI and RsaI in a final volume of 30 μl at 37°C for 1 h followed by heat inactivation at 65°C for 20 minutes.

Restriction enzymes DdeI and RsaI were chosen according to the following criteria (see also Table 2):

- they do not cut the coding sequence between the primers
- they cut the *IGF1* and *EPO* coding sequence 5′ and/or 3′ of the PCR-amplicon which linearizes potential circular viral or plasmid constructs irrespective of any knowledge of the vector sequence
- they frequently cut the *IGF1* and *EPO* intronic region to prevent background amplification of their genomic locus during PCR
- in order to exclude the effects of endogenous CpG methylation on restriction enzyme activity, no CpG sites are allowed to be present in the restriction enzyme recognition sites

qPCR

IGF1-specific primers and probe were chosen to target all *IGF1* mRNA isoforms including mechano-growth factor (*MGF*) using UCSC Genome Browser (http://genome.ucsc.edu/) and Primer 3 [24]. Then, *IGF1* qPCR was optimized and tested for the limit of detection/limit of quantification (LOD/LOQ) as described by Burns et al. [25] using standard calibrators in a background of 500 ng hgDNA. The LOD was estimated by 2-fold serial dilutions between ~2000 and 1 calibrator copies per reaction (See Figure S1). We defined the LOD as the lowest copy number that gives a detectable PCR amplification product at least 95% of the time. The LOQ was defined as the lowest concentration that could be quantified with >80% accuracy, and LOD was defined as the minimum copy number for which all replicates of the same dilution could be successfully detected. qPCR mixture contained 10 μl SsoFast probes supermix (Bio-Rad) and primers and probes as indicated in Table 1. Two-step PCR protocol using CFX384 (Bio-Rad) started with 2 min at 98°C followed by 45 cycles at 95°C melting and 30 sec annealing/extension at 64°C.

Table 1. Primer and Probe sequences.

Gene	Primer and Probe sets in 5→3 orientation [concentration]	Amplicon size
Erythropoietin (*EPO*); NM_000799	Fw: TGAATGAGAATATCACTGTCCCAGAC [900 nM] Rev: CTTCCGACAGCAGGGCC (900 nM); P: [Hex]AAG[+A]GG[+A]TG[+G]AG[+G]TCGG[BHQ1] [250 nM]; Sigma	114 bp
	Coding sequence primers: Fw: ATGggggtgcacgaatgt; Rv: TCAtctgtcccctgtcctg	582 bp
Insulin-like growth factor 1; (IGF1) NM_000618.3	Fw: GCTGGTGGATGCTCTTCAGTT [900 nM] Rev: TCCGACTGCTGGAGCCATAC [900 nM] P:[FAM]CTT[+T]TA[+T]TT[+C]AA[+C]AA[+G]CC[+C]AC[BHQ1] [250 nM]; Sigma	83 bp
	Coding sequence primers: Fw: ATGggaaaaatcagcagtcttc; Rv: CTAcatcctgtagttcttgtttcctg	462 bp
Internal Control Standard (ICS)	Fw: GCTGGTGGATGCTCTTCAGTT [900 nM] Rev: TCCGACTGCTGGAGCCATAC [900 nM] P: [VIC]TGCTCCAGAGAAGAAACCAC[MGB-NFQ] [250 nM]; Life Technologies	82 bp

Forward (Fw), reverse (Rev) primer-, and probe (P) sequences inclusive corresponding amplicon lengths. Start and stop codons in the coding sequence primers are capitalized.

To test for putative template effects on amplification efficiencies, two different sized plasmids containing the 462 bp *IGF1* coding sequence (pAAV9-*IGF1*-5237 bp; Topo PCR 2.1-*IGF1*-4339 bp) and a PCR-generated *IGF1* standard were subjected to qPCR with or without DdeI and RsaI double-digestion (See Figure S2).

Nested-qPCR

As a positive control, nested qPCR was performed to test for the presence of AAV9-*IGF1* elements in DNA isolated from mouse blood. The first round PCR was done using 4 µl of DNA using primers and conditions as described earlier [13]. PCR-products were 1:50 diluted in water and two microliter subjected to second round amplification (in triplicates) using the primers and conditions as described above. After qPCR, nested-qPCR products were analysed on a 2.5% agarose gel.

ddPCR

Each PCR reaction consisted of a 20 µL solution containing 10 µL ddPCR supermix for probes (Bio-Rad), 900 nM primers, 250 nM probe and 4 µl template DNA. Droplets (~20.000/reaction) were generated on the Bio-Rad QX-100 following the manufacturer's instructions. Samples were transferred on a 96 well-plate and thermal cycled to the endpoint (T100 Thermal Cycler; Bio-Rad) using a standard protocol; initial denaturation at 95°C for 10 min, followed by 40 cycles of melting at 95°C for 30 seconds and annealing/elongation at 61°C for 1 minute, before droplet stabilisation by 10 min incubation at 98°C. After cycling, the 96 well-plate was immediately transferred on a QX100 Droplet Reader (Bio-Rad) where flow cytometric analysis determined the fraction of PCR-positive droplets vs. the number of PCR-negative droplets in the original sample. Data were analysed using Poisson statistics to determine the target DNA template concentration in the original sample. Optimal annealing temper-

Table 2. Number of restriction sites for *Dde*I and *Rsa*I in the human/mouse genome and at the *IGF1* and *EPO* gene locus.

	*Dde*I (CTNAG)	*Rsa*I (GTAC)
Approximate number of cutting sites per Mbase in the human genome	4,844.1*	1,764.1*
average fragment length	206*	567*
Approximate number of cutting sites per Mbase in the mouse genome	5,291.6*	2,122.4*
average fragment length	189*	471*
IGF1		
IGF1 gene locus (human); chr12: 102789645–102874378	432	147
Igf1 gene locus (mouse); chr10: 87859056–87937047	430	129
Human IGF1 gene locus (56035 bp between primers); chr12: 102813436–102869470	293	100
IGF1 mRNA (7321 nucleotides)	28	12
IGF1 cds (462 nucleotides)	3	1
IGF1 amplicon (83 nukleotides)	0	0
EPO		
EPO gene locus (human); chr7: 100318423–100321323	19	4
Human EPO gene locus (729 bp between primers); chr7: 100319610-100320338	5	0
EPO mRNA (1340 nucleotides)	10	2
EPO cds (582 nucleotides)	1	2
EPO amplicon (114 nucleotides)	0	0

* Information taken from http://tools.neb.com.

atures were assayed carrying out gradient PCR for all primers and their specific targets (See Figure S3). Subsequently, we tested for potential inhibitory effects of restriction buffer on ddPCR and observed inhibitory effects only in the cases when more than 5 µl of DNA restriction solution (2.5 mM NaCl; 500 nM Tris-HCl; 500 nM MgCl$_2$; 50 nM DTT) were subjected to 20 µl ddPCR (Figure S4 a). Increasing amounts (up to 1500 ng) of genomic DNA in the background of the reaction were also assayed and did not show any inhibitory effects on ddPCR (Figure S4 b). Consequently, all ddPCRs were performed using a final volume of 4 µl template DNA (47 ng–1500 ng), which avoided potential salt and DNA inhibitory effects.

ddPCR detection of *IGF1* transgene from mouse blood and human spike-in experiments

For transgene detection, human/mouse genomic DNA was extracted as described above and 4 µl subjected to ddPCR using the primers and probes at concentrations as indicated in Table 1.

IGF 1 and *EPO* ddPCR-duplex assay

We also established a duplex protocol for parallel *IGF1* and *EPO* ddPCR transgene detection. *EPO*-specific primers and probe were used as described elsewhere [17]. Serially diluted standards (~5000 copies were 1:5, 1:2, 1:5, 1:2, 1:5 diluted) were assayed using ddPCR chemistry containing 900 nM *IGF1* and *EPO* forward and reverse primers, supplemented with differentially labelled *IGF1* and *EPO*-specific LNA TaqMan-probes at 250 nM final concentration (see Table 1). *EPO*-specific primers and probe were used as described elsewhere [17]. Samples were assayed in a background of 500 ng DdeI and RsaI fragmented human genomic DNA containing serial dilutions of either *IGF1* or *EPO* standards only; both standards at decreasing levels and also containing *IGF1* at decreasing and *EPO* at increasing amounts. Digital droplet PCR was performed under standard conditions as described above.

ICS (internal control standard)

To check for uniform PCR efficiency all reactions were spiked with an internal standard (similar to the internal threshold control (ITC) as described for qPCR [17]). This artificial standard (purchased and synthesized by MWG Eurofins) was designed to have the identical 5′ and 3′ sequence compared to the *IGF1* amplicon, but with the probe binding site replaced by a sequence taken from the ancestral organism Cyanobacterium stanieri (NC_019778.1). This 20 nucleotide sequence was blasted against the human genome to confirm that there was no presence of either identical or similar sequence, which could lead to false positive detection. A VIC-labelled TaqMan probe (Life Technologies) was designed to target this sequence in a duplex PCR when *IGF1* transgenic elements are also detected.

Subsequently, *IGF1* standard was assayed using ddPCR chemistry supplemented with an internal-control standard at defined concentration (100 copies/reaction).

DNA extraction by on-column digestion

Duplicates of whole blood samples of 5 human donors (100 µl) were spiked with the same amount of circular standard DNA. Using the Qiagen DNA microkit, DNA was extracted and was finally processed using 3 different procedures as follows:

a) DNA was eluted conventionally with 30 µl H$_2$O.

b) Water-eluted DNA was DdeI and RsaI digested (5 U each) for 1 h at 37°C.

c) DNA was on-column digested for 1 h at 37°C with 30 µl of a solution containing 10 U DdeI and RsaI in 1 x buffer 2.

All samples were adjusted to the same salt concentration and dilution, heat inactivated for 20 min at 65°C, Nano-dropped, agarose-gel visualized and subsequently ddPCR quantified under the conditions as described above.

Results

As a preliminary step towards ddPCR, qPCR experiments were performed to identify those primers and probes, which led to best PCR efficiencies, and which did not produce artefacts such as excessive by-products or false positive signals. Accordingly, primers and probes as indicated in Table 1 were used.

qPCR

We designed a new assay for *IGF1* transgene detection with primers that resulted in an 83 bp amplicon, in which the exon2/3 boundary was targeted by a 6-FAM-labelled LNA-probe. As illustrated in Figure S1, *IGF1* qPCR efficiency was 96.7% with a linearity of $r^2 = 0.98$-. The LOQ was defined as the lowest concentration that could be quantified with >80% accuracy, and set to 16 copies per reaction (See Figure S1); LOD was found to be 4 copies. We then tested for differential PCR efficiency dependent on DNA structure. We assayed serial dilutions (ranging from 10^6– 10 copies) from 2 vectors carrying the *IGF1* coding sequence (pAAV9-*IGF1*-5237 bp and TOPO-*IGF1*-4397 bp) compared to linear PCR product. Improved amplification for the linear PCR standard (Ct difference of more than 3), compared to circular, supercoiled vectors was observed. Subsequently, we tested digested plasmid vs. undigested plasmid compared to linear PCR-standard and were able to amplify the digested plasmid with efficiencies close to the linear PCR-standard (See Figure S2).

ddPCR optimization

To determine best target-specific annealing temperatures, all primers used in ddPCR were initially tested by gradient ddPCR. As indicated in Figure S3, all primers worked well between 66°C– 60°C with highest amplitude and best sensitivity at ~61°C. Subsequently, all ddPCR assays were conducted at 61°C annealing/extension temperature. Furthermore, we verified that genomic DNA, up to 1500 ng, in the ddPCR reaction did not affect results (See Figure S4 a), and that 1 x restriction solution, less than 5 µl, added to ddPCR did not inhibit the reaction (See Figure S4 b). Consequently, 4 µl of restriction solution were routinely used for transgene detection.

ddPCR for AAV9-*IGF1* transgene detection in mice

Our aim was to detect minute amounts of *IGF1* transgenic DNA using ddPCR, a method so far never described for gene-doping detection. From 100 µl whole blood, DNA was extracted with 25 µl H$_2$O and digested in NEB buffer 2 using the restriction enzymes DdeI and RsaI, in a final volume of 30 µl. Four µl of digested DNA were subjected to PCR in 3 independent trials, and *IGF1* was detected as illustrated in Figure 1. Results provided specific indication of *IGF1*-transgene detection in all transduced animals over a 33 day-period. As displayed in Table S1, copy numbers were in the range of 3200–164800 copies/reaction (which indicates 240–12263 viral elements/µl of whole blood) also indicating excellent ddPCR reproducibility with a tendency to lower values, due to additional thawing and freezing cycles. Control animals remained negative with one single false positive event (Fig 1-day30 and Figure S5). However, this false positive

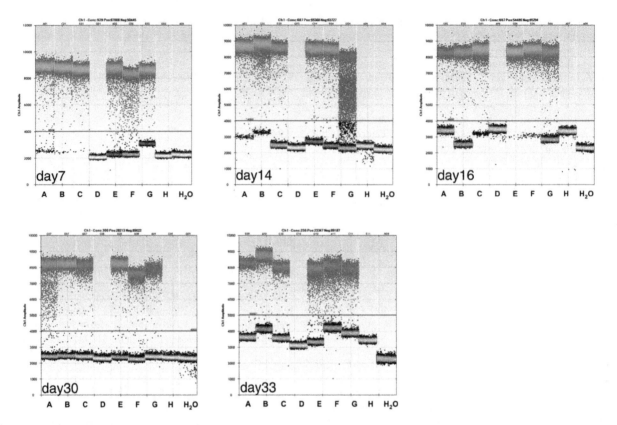

Figure 1. *IGF1* transgene detection by ddPCR. Eight mice, 6 AAV9-IGF1 muscle transduced (A, B, C, E, F, G) and two controls (D and H) were screened for 33 days by ddPCR for *IGF1* transgene detectability.

event could be clearly discriminated from true positives by manual re-adjustment of the threshold to a value that defined the lower limit of the positive control, a best practice for ddPCR, as also discussed earlier elsewhere [26].

Digital droplet PCR results were confirmed by nested-qPCR and by gel-electrophoresis of the corresponding PCR-products. PCR products showed by-products of similar size as the expected *IGF1* PCR-products, however, these by products were not detected by the *IGF1*-specific probe during qPCR (data not shown).

Spike-in duplex assay for simultaneous *IGF1* and *EPO* detection at low copy numbers

To ameliorate the use of the limited amount of DNA extracted from blood and to optimize the flow capacity of a potential routine transgene detection setting, we also developed a new duplex ddPCR assay for simultaneous *IGF1* and *EPO* detection.

Serially diluted *IGF1* and *EPO* standards were spiked into 500 ng human genomic DNA and analysed by ddPCR after DdeI and RsaI digestion. Digital droplet PCR was performed using a mixture containing *IGF1* and *EPO* primers and probes. This test should reveal sensitivity and linearity of the assay for single transgene detection and also in a duplex approach at low copy numbers (range from ~5000–10 copies/reaction). As indicated in Figure 2 (A–E), both transgenes were detected showing linearity values from 0.9997 to 1 for *IGF1* and 0.9995 to 0.9998 for *EPO* under the conditions tested. Sensitivity and linearity of the duplex assay did not differ from those obtained by the single gene assay.

IGF1 standard was also assayed in the presence of an internal control standard at ~100 copies/reaction. As indicated in

Figure 2E, ICS was clearly detected in all reactions, and could be used as a reporter for PCR efficiency in each run and for all reactions. This minimizes probability of false negatives due to potential PCR inhibitors being present in some samples.

DNA extraction by on-column digestion

We also optimized a protocol of on-column digestion prior to DNA fragmentation. This procedure led to fragmented high quality DNA, with final yields that were on average 3 times higher compared to conventionally eluted DNA as revealed by Nano-Drop 1000, agarose-gel analysis (See Figure S6).

In addition, testing DNA extracted from whole blood (a-conventionally extracted; b- conventionally extracted followed by DNA digestion; c- on-column digested) that was spiked with the same amount of plasmid standard by ddPCR revealed more sensitive detection for digested DNA, and most sensitive detection for samples subjected to on-column digestion. As indicated in Figure 3, we could achieve 1.9–2.8 fold increased ddPCR sensitivity comparing digested DNA to undigested DNA. When ddPCR was performed from the same blood after on-column digestion, ddPCR performance was further increased 2.9–19 fold compared to DNA eluted using water (Figure 3).

Discussion

Digital droplet PCR represents a new technical approach for applications where conventional PCR meets its technical limits, such as rare event detection in the presence of high amounts of genomic DNA. Compared to conventional qPCR, quantification by ddPCR does not rely on external standard curves and is more tolerant to inhibitors and variation in amplification efficiencies

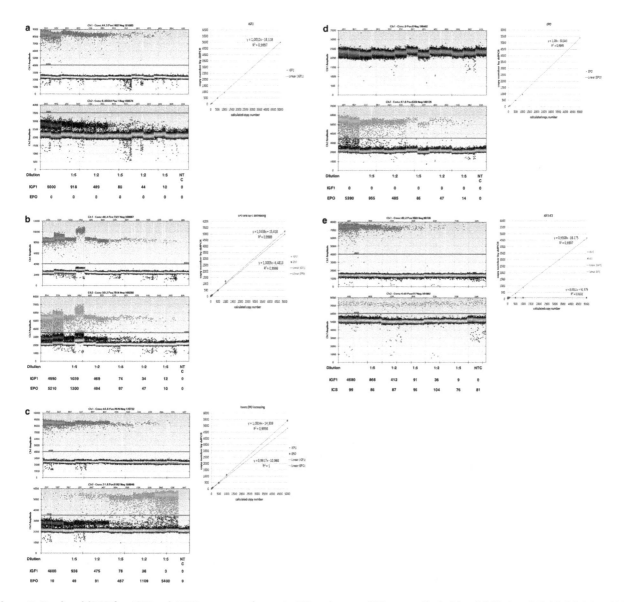

Figure 2. Duplex ddPCR for *IGF1* and *EPO* transgene elements. 500 ng human gDNA were spiked with serial dilutions (1:5; 1:2; 1:5; 1:2 and 1:5) of *IGF1*-standard (A), decreasing amounts of *IGF1* and *EPO*-standard (B), inverse amounts of *IGF1* (decreasing) and *EPO* (increasing) (C), *EPO* only (D) and *IGF1* multiplexed with ICS at ~100 copies/reaction (E). Displayed are amplitudes, copy numbers per 20 µl reaction and linearity of the signals.

Thus, ddPCR is well suited for the identification of transgenic elements with the objective of gene-doping detection.

Here we present a new ddPCR method for *IGF1* and *EPO* transgene detection with high sensitivity and reliability, in combination with an effective technique to isolate high amounts of fragmented DNA with minimal effort. The aim of this study was to test and optimize detection of transgenic elements by the use of ddPCR. Accordingly, we initially tested *IGF1*-specific primers and probes by qPCR for optimal primer annealing, amplicon generation and specific probe detection. Quantitative PCR-experiments revealed a LOQ of 16 and a LOD of 4, which indicated highly efficient PCR conditions (See Figure S1). This qPCR-assay was subsequently transferred to ddPCR and re-optimized for best annealing temperature (See Figure S3) and controlled for either salt or DNA inhibition using (See Figure S4) ddPCR specific chemistry.

Eight mice, of which 6 were transduced with AAV9-*IGF1* and 2 controls, were then monitored over a 33-day period for the detectability of transgenic elements extracted from whole blood. Strong signals could be detected for all AAV9-*IGF1* transduced mice at all points in time, with a tendency of signal reduction over time (Figure 1 and Table S1). We also created a ddPCR duplex assay aimed at the simultaneous detection of *IGF1* and *EPO* detection. As indicated in Figure 2 (A–E), both transgenes were sensitively detected showing linearity from 0.9997 to 1 for *IGF1* and 0.9995 to 0.9998 for *EPO* with copy numbers from 5400 to 3. These data are in accordance with published data describing linearity close to 1 with a limit of detection of 5 when performing duplex ddPCR assays for other genes [16].

The integration of an artificially generated internal control standard into ddPCR enabled us to control for PCR efficiency in each well. As indicated in Figure 2E, the documentation of ddPCR efficiency for each reaction represents a valuable tool for

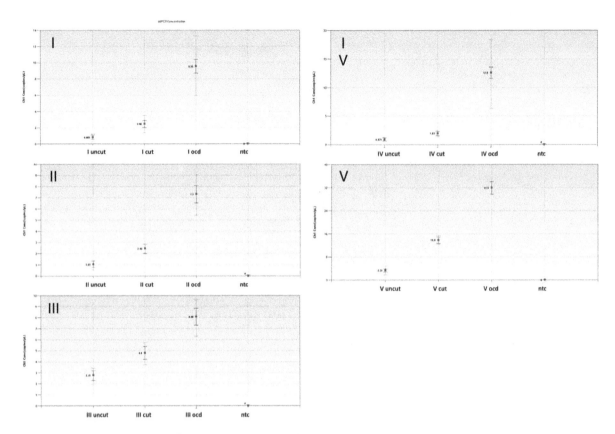

Figure 3. DdPCR results following 3 different DNA extraction procedures. Blood of 5 donors was spiked with the same amount of circular plasmid containing the *IGF1* coding sequence. DNA was extracted using 3 different procedures: uncut, DdeI and RsaI cut, and on-column DdeI and RsaI digested (ocd). Results are displayed as copies detected per µl for 5 subjects (I–V for uncut; conventionally digested (cut), on-column digested DNA (ocd), and no-template control (ntc). Results derive from two independent experiments performed in duplicates.

routine analyses as it allows for identification of false negatives, potentially originating from PCR inhibitors being present in some reactions.

On the genomic level, *IGF1* primers target exonic elements separated by a ~56 kb sequence (56035 bp), containing 293 DdeI and 138 RsaI cutting sites. Primers and probe for EPO detection were used as described by others [17] and could generate a genomic fragment of 726 bp which is cut 5 times by DdeI during digestion. DdeI and RsaI restriction enzyme digestion lead to fragmentation of the respective gene loci and inhibition of primer extension after digestion. This, and the design of *IGF1* and *EPO* specific probes which specifically target exon/exon boundaries, makes detection of false positive signals generated from *IGF1* and *EPO* genomic loci unlikely. Additionally, our data demonstrate that amplification of linearized DNA has superior PCR efficiency over circular DNA elements (Figure 3 and Figure S2), which further supports the use of restriction enzyme digestion of DNA prior to PCR. Our optimized DNA-extraction procedure includes an on column-digest at the final step of DNA extraction. This involves minimal handling of samples and results in much higher DNA yield as compared to conventional DNA extraction procedure (Figure 3 and Figure S6). This may be explained by improved elution of fragmented DNA from the column as compared to undigested high molecular DNA stretches. Comparing eluates of blood samples that were spiked with the same amount of circular *IGF1* standard after conventional DNA purification with, and without restriction to on-column digested DNA by ddPCR, we found the latter to be the most efficient

substrate for transgene detection. Accordingly, we recommend use of on-column digestion for all applications where minute amounts of DNA need to be detected, as it increases DNA yield and, thus, the probability of target elution. As displayed in Figure 3, linearization of circular target DNA improved (dd-) PCR detection efficiency, which was further enhanced by on-column digestion due to significantly improved DNA elution from the column.

In analogy to trials which aim to isolate low levels of a mutant sequence in a high background of DNA, such as ARMS-PCR (Amplification Refractory Mutation System, or allele-specific PCR; ASP [27]), we believe on-column digestion using carefully chosen enzymes may help to increase detection of specific DNA elements by PCR in the future.

We did not, however, observe a consistent Limit of Blank (LOB) for *IGF1* and *EPO* analysis, as recently described for the detection of mutated *KRAS* elements [28]. This difference might be attributed to different PCR systems, or bigger differences between wild type and transgene detection than present in mutational detection. Occasional false positives could be explained by DNA carry-over due to sample handling as previously described for other high sensitive detection methods [13].

If we compare our ddPCR assay to other transgene detection approaches, such as qPCR or agarose-gel based nested PCR, we observe similar sensitivities, as all methods can identify less than five copies per PCR. However, ddPCR offers to advantages to be independent of an artificial standard curve, with all its confounding variables, and to minimise DNA carry-over by strict reduction of DNA handling steps.

Testing for practicability and cost value of ddPCR, Morisset and co-workers [29] describe similar or improved throughput and cost-effectiveness for ddPCR when compared to qPCR. Additionally, increased usage of ddPCR can be expected to lead to further cost reduction in the future. Guidelines for standardization purposes and creation of Good Laboratory Practice (GLP) procedures for ddPCR have already been described [26] and were applied in this study.

Implementation of ddPCR for transgene detection is further encouraged by data recently presented by Strain and co-workers [30], who showed 5–20 fold improved precision comparing ddPCR to qPCR for the detection of total DNA originating from the human immunodeficiency virus (HIV) and its episomal 2-LTR (long terminal repeat) circles.

Hence, we present a new approach for transgene detection at high and minute copy amounts, including a new DNA extraction and digestion method, which delivers high quality, concentrated and fragmented DNA as an appropriate target for ddPCR.

Digital PCR as presented here, including appropriate restriction enzyme digestion of genomic DNA, represents a promising approach for the detection of minimal DNA fractions. Digital PCR will therefore most probably find implementation to monitor gene-therapy trials, and the possible abuse as gene doping, but also to screen for and monitor viral and bacterial infection, and to analyse food and feed products for xenogeneic components.

Supporting Information

Figure S1 Standard curve, amplification plot and calculation for experimental estimation of IGF1 LOD and LOQ.

Figure S2 PCR efficiency for undigested vs. digested IGF1 standards - Linearization of circular DNA leads to improved PCR efficiency and better detection at low copy numbers.

Figure S3 Temperature gradient for ddPCR detection of IGF1, EPO and ICS.

Figure S4 ddPCR efficiency under various conditions.

Figure S5 Qualitative assessment of ddPCR results.

Figure S6 Concentration, purity and integrity of DNA comparing three different DNA extraction procedures.

Table S1 IGF1 transgene copy numbers as detected by ddPCR for 6 AAV9-IGF1 transduced mice and 2 controls at 5 different days.

Acknowledgments

The authors are grateful to Dr. Pia Scheu (Bio-Rad) for her expert introduction into ddPCR technology. Furthermore, the authors thank Olga Moser and Marco Zahn for their excellent technical assistance and Thomas Beiter and Mike Bramwell for critical reading of the manuscript.

Author Contributions

Conceived and designed the experiments: DAM LB AR SZ MG PS. Performed the experiments: DAM LB AR SZ MG PS. Analyzed the data: DAM LB AR SZ MG PS. Contributed reagents/materials/analysis tools: DAM LB AR SZ MG PS. Wrote the paper: DAM LB AR SZ MG PS.

References

1. Kay MA (2011) State-of-the-art gene-based therapies: the road ahead. Nat Rev Genet 12: 316–328.
2. Schneider AJ, Friedmann T (2006) Gene doping in sports: the science and ethics of genetically modified athletes. Adv Genet 51: 1–110.
3. Barton-Davis ER, Shoturma DI, Musaro A, Rosenthal N, Sweeney HL (1998) Viral mediated expression of insulin-like growth factor I blocks the aging-related loss of skeletal muscle function. Proc Natl Acad Sci U S A 95: 15603–15607.
4. Macedo A, Moriggi M, Vasso M, De Palma S, Sturnega M, et al. (2012) Enhanced athletic performance on multisite AAV-IGF1 gene transfer coincides with massive modification of the muscle proteome. Hum Gene Ther 23: 146–157.
5. Rivera VM, Gao GP, Grant RL, Schnell MA, Zoltick PW, et al. (2005) Long-term pharmacologically regulated expression of erythropoietin in primates following AAV-mediated gene transfer. Blood 105: 1424–1430.
6. Chenuaud P, Larcher T, Rabinowitz JE, Provost N, Cherel Y, et al. (2004) Autoimmune anemia in macaques following erythropoietin gene therapy. Blood 103: 3303–3304.
7. Gao G, Lebherz C, Weiner DJ, Grant R, Calcedo R, et al. (2004) Erythropoietin gene therapy leads to autoimmune anemia in macaques. Blood 103: 3300–3302.
8. Ni W, Le Guiner C, Gernoux G, Penaud-Budloo M, Moullier P, et al. (2011) Longevity of rAAV vector and plasmid DNA in blood after intramuscular injection in nonhuman primates: implications for gene doping. Gene Ther 18: 709–718.
9. Zhou S, Murphy JE, Escobedo JA, Dwarki VJ (1998) Adeno-associated virus-mediated delivery of erythropoietin leads to sustained elevation of hematocrit in nonhuman primates. Gene Ther 5: 665–670.
10. Baoutina A, Alexander IE, Rasko JE, Emslie KR (2007) Potential use of gene transfer in athletic performance enhancement. Mol Ther 15: 1751–1766.
11. Lippi G, Guidi G (2003) New scenarios in antidoping research. Clin Chem 49: 2106–2107.
12. Beiter T, Zimmermann M, Fragasso A, Armeanu S, Lauer UM, et al. (2008) Establishing a novel single-copy primer-internal intron-spanning PCR (spiPCR) procedure for the direct detection of gene doping. Exerc Immunol Rev 14: 73–85.
13. Beiter T, Zimmermann M, Fragasso A, Hudemann J, Niess AM, et al. (2011) Direct and long-term detection of gene doping in conventional blood samples. Gene Ther 18: 225–231.
14. Moser DA, Neuberger EW, Simon P (2012) A quick one-tube nested PCR-protocol for EPO transgene detection. Drug Test Anal 4: 870–875.
15. Baoutina A, Coldham T, Bains GS, Emslie KR (2010) Gene doping detection: evaluation of approach for direct detection of gene transfer using erythropoietin as a model system. Gene Ther 17: 1022–1032.
16. Baoutina A, Coldham T, Fuller B, Emslie KR (2013) Improved Detection of Transgene and Nonviral Vectors in Blood. Hum Gene Ther Methods.
17. Ni W, Le Guiner C, Moullier P, Snyder RO (2012) Development and utility of an internal threshold control (ITC) real-time PCR assay for exogenous DNA detection. PLoS One 7: e36461.
18. Baoutina A, Coldham T, Fuller B, Emslie KR (2013) Improved detection of transgene and nonviral vectors in blood. Hum Gene Ther Methods 24: 345–354.
19. Hou Y, Zhang H, Miranda L, Lin S (2010) Serious overestimation in quantitative PCR by circular (supercoiled) plasmid standard: microalgal pcna as the model gene. PLoS One 5: e9545.
20. Lin CH, Chen YC, Pan TM (2011) Quantification bias caused by plasmid DNA conformation in quantitative real-time PCR assay. PLoS One 6: e29101.
21. Hindson BJ, Ness KD, Masquelier DA, Belgrader P, Heredia NJ, et al. (2011) High-throughput droplet digital PCR system for absolute quantitation of DNA copy number. Anal Chem 83: 8604–8610.
22. Miller SA, Dykes DD, Polesky HF (1988) A simple salting out procedure for extracting DNA from human nucleated cells. Nucleic Acids Res 16: 1215.
23. Arsic N, Zentilin L, Zacchigna S, Santoro D, Stanta G, et al. (2003) Induction of functional neovascularization by combined VEGF and angiopoietin-1 gene transfer using AAV vectors. Mol Ther 7: 450–459.
24. Untergasser A, Cutcutache I, Koressaar T, Ye J, Faircloth BC, et al. (2012) Primer3–new capabilities and interfaces. Nucleic Acids Res 40: e115.
25. Burns M, Valdivia H (2008) Modelling the limit of detection in real-time quantitative PCR. European Food Research and Technology 226: 1513–1524.
26. Huggett JF, Foy CA, Benes V, Emslie K, Garson JA, et al. (2013) The digital MIQE guidelines: Minimum Information for Publication of Quantitative Digital PCR Experiments. Clin Chem 59: 892–902.

27. Little S (2001) Amplification-refractory mutation system (ARMS) analysis of point mutations. Curr Protoc Hum Genet Chapter 9: Unit 9 8.

28. Taly V, Pekin D, Benhaim L, Kotsopoulos SK, Le Corre D, et al. (2013) Multiplex Picodroplet Digital PCR to Detect KRAS Mutations in Circulating DNA from the Plasma of Colorectal Cancer Patients. Clin Chem.

29. Morisset D, Stebih D, Milavec M, Gruden K, Zel J (2013) Quantitative analysis of food and feed samples with droplet digital PCR. PLoS One 8: e62583.

30. Strain MC, Lada SM, Luong T, Rought SE, Gianella S, et al. (2013) Highly precise measurement of HIV DNA by droplet digital PCR. PLoS One 8: e55943.

9

Reduction of T Cell Receptor Diversity in NOD Mice Prevents Development of Type 1 Diabetes but Not Sjögren's Syndrome

Joanna Kern, Robert Drutel¤, Silvia Leanhart, Marek Bogacz, Rafal Pacholczyk*

Center for Biotechnology and Genomic Medicine, Georgia Regents University, Augusta, Georgia, United States of America

Abstract

Non-obese diabetic (NOD) mice are well-established models of independently developing spontaneous autoimmune diseases, Sjögren's syndrome (SS) and type 1 diabetes (T1D). The key determining factor for T1D is the strong association with particular MHCII molecule and recognition by diabetogenic T cell receptor (TCR) of an insulin peptide presented in the context of I-A^{g7} molecule. For SS the association with MHCII polymorphism is weaker and TCR diversity involved in the onset of the autoimmune phase of SS remains poorly understood. To compare the impact of TCR diversity reduction on the development of both diseases we generated two lines of TCR transgenic NOD mice. One line expresses transgenic TCRβ chain originated from a pathogenically irrelevant TCR, and the second line additionally expresses transgenic TCRα^{mini} locus. Analysis of TCR sequences on NOD background reveals lower TCR diversity on Treg cells not only in the thymus, but also in the periphery. This reduction in diversity does not affect conventional CD4$^+$ T cells, as compared to the TCR^{mini} repertoire on B6 background. Interestingly, neither transgenic TCRβ nor TCR^{mini} mice develop diabetes, which we show is due to lack of insulin B:9–23 specific T cells in the periphery. Conversely SS develops in both lines, with full glandular infiltration, production of autoantibodies and hyposalivation. It shows that SS development is not as sensitive to limited availability of TCR specificities as T1D, which suggests wider range of possible TCR/peptide/MHC interactions driving autoimmunity in SS.

Editor: John A. Chiorini, National Institute of Dental and Craniofacial Research, United States of America

Funding: This research was supported by grants from National Institute of Allergy and Infectious Diseases of the National Institutes of Health (R01AI081798) and Juvenile Diabetes Research Foundation (RP). The funders had no role in study design, data collection and analysis, decision to publish, or preparation of the manuscript.

Competing Interests: The authors have declared that no competing interests exist.

* Email: rpacholczyk@gru du

¤ Current address: Medical University of South Carolina, College of Medicine, Charleston, South Carolina, United States of America

Introduction

NOD mice serve as well-established models of independently developing autoimmune diseases, Type 1 Diabetes (T1D) and Sjögren's syndrome (SS) [1,2]. T1D is characterized by autoimmune attacks against the pancreatic beta-cells with T cells playing an essential role in the initiation and progression of the disease, leading to hyperglycemia and vascular complications [3,4]. SS is an autoimmune disease with local and systemic manifestations, characterized by mononuclear infiltrates into salivary and lacrimal glands leading to clinical symptoms of dry mouth and dry eyes [5,6]. Glandular infiltrates consist mostly of CD4$^+$ T cells with lesser amounts of CD8$^+$ T cells and B cells. Although factors like viral or bacterial infections, aberrant glandular development or cytokine production are important in the initial phase of the pathogenesis of SS, CD4$^+$ T cells are important players in the onset of autoimmunity and disease progression.

Autoimmunity in NOD mice is attributed to several different events occurring in the thymus and in the periphery. Studies in these mice showed a defect in negative selection [7], perturbed αβ/γδ lineage decision leading to a shift in selection niches [8], reduced relative diversity of thymic Treg cells [9], peripheral

hyper-responsiveness of effector CD4$^+$ T cells [10], multiple binding registers of insulin B:9–23 peptide resulting in poor negative selection in the thymus [11,12], or peripheral post-translational modification of self-peptides/neo-antigens [13]. Despite genetic predispositions, the key component in the development of autoimmune diseases is the recognition of a particular antigen in the context of MHC Class II molecule by CD4$^+$ T cells. The development of diabetes in NOD mice is associated with the key I-A^{g7} molecule (HLA-DQ8 in humans) in the absence of a functional I-E molecule [14,15]. Co-expression of other MHC molecules with I-A^{g7} can prevent development of diabetes in a dominant fashion [14,15]. Replacement of I-A^{g7} with other MHC molecules, like I-A^b, I-A^p or I-A^q, does not promote the development of diabetes yet mice continue to develop autoimmune exocrinopathy and the severity of the SS and the profile of antibodies' specificities vary between congenic mice [16]. In large-scale association study of SS in humans, HLA was found to have the strongest linkage to the disease [17].

The strict dependence of T1D on the particular MHC allele correlates with its primary antigen requirement where insulin B:9–23 peptide has been identified as the epitope necessary for onset of the disease in NOD mice [18]. In SS, no key epitope(s) are identified, although several proteins have been implicated as a source of antigens: Ro/SSA 52 kDa, αFodrin, Muscarinic Acetylcholine 3 Receptor (M3R), α-amylase, islet cell autoantigen-69, kallikrein-13 [19–24]. Recently it has been shown that the transfer of T cells from M3R-immunized $M3R^{-/-}$ mice into $Rag^{-/-}$ mice leads to development of sialadenitis, showing pathogenic potential of M3R specific T cells [25].

Despite the strict requirement of the presence of the insulin B:9–23/I-A^{g7} combination, the development of T1D in NOD mice proceeds even when total TCR diversity and precursor frequency of diabetogenic TCRs is limited. The reduction of TCR diversity by use of TCRβ transgenic mice [26], or great reduction of precursor frequency relying on allelic exclusion escapees on NOD background does not prevent development of T1D [27], although not all endogenous TCRβ chains are permissive for the development of insulin B:9–23 specific TCRs [28]. In SS it is not clear as to what role diversity of interactions between TCRs and different peptide/MHCII complexes play in the onset and development of the disease. Previous studies in patients with SS found that TCR repertoire of infiltrating T cells is to some extent restricted with different dominant clonotypes of Vβ families [29–33]. Despite the lack of dominant Vα/Vβ families or dominant specificity in different patients these studies show clonal expansion of infiltrating T cells, which suggests that the number of epitopes participating in the autoimmunity of the disease is limited [34,35]. However, the weaker dependence of SS on MHC polymorphism suggests broad diversity of possible TCR/peptide/MHCII interactions participating in the pathogenesis of the disease. As the diversity of antigenic specificities and TCR repertoire on T cells involved in SS development is not well understood, we wanted to compare sensitivity of development of SS versus T1D to the diminishing diversities of TCRs in the presence of the same I-A^{g7} molecule in NOD mice. To reduce the diversity of TCRs in NOD mice we generated two types of transgenic strains, in which all T cells express either one transgenic TCRβ chain or additionally co-express TCRα chains from TCRα^{mini} locus [36]. Interestingly, these mice do not develop T1D but still develop SS with glandular infiltrations, autoantibody presence and hyposalivation. We investigated the reasons for the lack of T1D development and the role of reduced diversity on generated repertoires of TCRs on conventional and regulatory T cells in the thymus and periphery of NOD mice.

Materials and Methods

Ethics statement

All mice used in this study were housed in the animal care facility at the Georgia Regents University (GRU). All work involving animals was conducted under protocols approved by the Animal Care and Use Committee at the GRU (#2008−0231). All efforts were made to minimize suffering. Mice were euthanized by CO2 followed by cervical dislocation.

Mice

Production of TCRβ^{Tg} and TCRα^{mini} constructs and generation of transgenic mice on C57BL/6 (B6) background was described previously (Pacholczyk 2006). A similar strategy was used to microinject one or both DNA constructs into zygotes of NOD mice (Transgenic Mice Core Facility, GRU). To eliminate expression of endogenous TCRα chains, NOD.TCRβ^{Tg}.TCRα^{mini} transgenic mice were crossed with NOD.TCRα$^{-/-}$ (NOD.129P2(C)-Tcratm1Mjo/DoiJ) mice purchased from The Jackson Laboratory (Bar Harbor, ME). To facilitate identification

of Treg cells both transgenic lines then were crossed with NOD.FoxP3GFP/cre mice (NOD/ShiLt-Tg(Foxp3-EGFP/cre)1Jbs/J) purchased from The Jackson Laboratory. The B6.Aec1Aec2 (B6.DC) mice were kindly provided by Dr. Ammon Peck [37].

Histology

Organs were removed from each mouse at the time of euthanasia, placed in 10% phosphate-buffered formalin for at least 24 h and then embedded in paraffin. Sections were taken at 5 μm of thickness 200 μm apart. The tissue sections were stained with hematoxylin and eosin (H&E) at the Histology Core Laboratory, GRU. One infiltrate was defined as a cluster of at least 50 nucleated cells, scoring described in figures.

Measurement of saliva flow

Mice were given an i.p. injection of 100 μl of pilocarpine (0.05 mg/ml in PBS) per 20 g of body weight. Saliva was collected for 10 min., starting 1 min. after injection of pilocarpine. The volume of saliva was measured and normalized to the mouse body weight.

Detection of auto-antibodies

Auto-antibodies and total IgG1 were measured using mouse serum with the following ELISA kits: αFodrin (American Research Products), ANA, ssDNA, dsDNA (Immuno-Biological Laboratories) and IgG1 (Immunology Consultants Laboratory). Assays were performed according to manufacturer's protocols with serum dilution 1:100 and 1:50,000 for IgG1. OD_{450} values of negative samples were subtracted from the OD_{450} of experimental samples.

Cell preparation, flow cytometry and cell sorting

Cells were isolated from peripheral lymph nodes (axillary, brachial and inguinal) and thymii by mechanical disruption through nylon mesh. Salivary mandibular and extraorbital lacrimal glands were first cut and digested using collagenase (1 ug/ml) for 30 min in 37°C. Cells were washed and counted (Countess, Invitrogen) and used for staining with monoclonal antibodies: CD4 (clone RM4-5), CD8α (53-6.7), B220 (RA3-6B2), TCRVα2 (B20.1), TCRVβ14 (14-2), TCRβ (H57-597), CD25 (PC61), CD45RB (16A), CD62L (MEL-14), all from BD Biosciences. Stained cells were either analyzed using FACS Canto (BD Biosciences) or sorted on MoFlo Sorter (Cytomation). Dead cells were excluded using forward vs side scatter dot plots and doublets discrimination was accomplished using forward scatter height vs width dot plots. Purities of all sorted populations were above 98%.

Immunization and generation of T cell hybridomas

Mice were immunized at the base of the tail with 50 μg of insulin B:9–23 peptide emulsified in Complete Freund's Adjuvant (Thermo Scientific). One week later lymphocytes were isolated from draining lymph nodes and cultured in vitro for 3–4 days with insulin B:9–23 peptide (50 μg/ml), followed by 3–4 days of expansion with murine recombinant IL-2 (20 U/ml, Peprotech). For generation of allo-specific T cell hybridomas, CD4$^+$ T cells sorted from un-immunized experimental mice were stimulated in vitro by co-culture with splenocytes from B6.TCRα$^{-/-}$ mice, followed by expansion with IL-2. Activated T cells were fused to BW5147 TCRα-β- NFAT-EGFP cells as previously described [38]. BW NFAT-EGFP fusion partner expresses GFP protein under the minimal human IL-2 promoter, which contains NFAT-binding sites [39]. T cell hybridomas were selected using HAT

(Cellgro) selection media by limiting dilution method. For stimulation, cloned T cell hybridomas (10^5 cells) were co-cultured overnight with 5×10^5 splenocytes (from NOD.TCRα−/− or B6.TCRα−/− mice) with or without insulin B:9–23 peptide (50 µg/ml) or anti-CD3 antibody (1 µg/ml). Specific activation of the hybridomas was measured by detection of GFP-positive cells or by detection of IL-2 in culture supernatant using CBA Mouse IL-2 Flex Set (BD Biosciences).

MTT assay

The proliferation of the cells was measured using 3-(4,5-dimethythiazol-2-yl)-2,5-diphenyltetrazolium bromide (MTT; Sigma) assay. On the third day of *in vitro* stimulation 10 µl of MTT (5 mg/ml) was added to each well and incubated for 4 h. After discarding supernatant, the remaining formazan precipitates were dissolved in 150 ul of 70% isopropanol solution (70% isopropanol, 30% water, 0.02N hydrochloric acid) overnight. Absorbance was measured at 570 nm using Microplate Reader (Biotek Synergy HT).

Sequencing

Single cell sorting and sequencing was done as previously described [36]. High-throughput sequencing was done using Ion Torrent platform (Life Technologies) by Genomic Core Facility at GRU. Libraries of the TCRs were prepared according to protocol. Shortly, CDR3α regions were amplified using primers specific for Vα2 and Cα segments with integrated adapters and barcodes, provided by manufacturer. Before sequencing, consistency of samples was checked by 2D-F-SSCP analysis of PCR products from three aliquots of cDNA per each sample [36]. Only samples with three similar/identical profiles were considered without PCR bias and were used for further purification using Agencourt AMPure XP reagent (Beckman Coulter) and used for sequencing. FASTQ files with sequences were processed and analyzed using custom-written program in Pearl (ActivePearl, ActiveState Software Inc.). Sequences with quality score of CDR3α region above 27 were used for analysis. The length of CDR3α region was defined by counting from the third amino acid after the invariant C residue in all Vα regions (Y-L/F-C-A-X-first) to the amino acid immediately preceding common Jα motif (last-F-G-X-F-G-T).

Statistical analysis

The similarity, diversity and richness estimators were calculated using programs designed to measure biodiversity: EstimateS8.2 (Colwell, R. EstimateS: Biodiversity Estimation Software. Program and User's Guide at http://viceroy.eeb.uconn.edu/estimates) and SPADE (Chao, A. and Shen, T.-J. (2010) Program SPADE (Species Prediction And Diversity Estimation). Program and User's Guide published at http://chao.stat.nthu.edu.tw).

Results

Development of T cells in NOD^{βTg} and NOD^{mini} mice

We generated TCRβ transgenic mice, in which all T cells use the same TCRβ^{Tg} chain (Vβ14Dβ2Jβ2.6) and one of the endogenous TCRα chains. The TCRβ transgenic chain originated from I-A^b restricted TCR specific to Ep63 K peptide an analog of Eα 52–68 peptide in which the residue at position 63(I) was substituted with lysine [40]. To further reduce diversity of TCRs we used TCRα^{mini} transgenic construct, used previously to create TCR^{mini} mouse on B6 background (B6^{mini}) [36]. The TCRα^{mini} transgene allows a single Vα2.9 segment to rearrange to one of the two Jα (Jα26 and Jα2) segments. The generated transgenic mice were further crossed with TCRα$^{−/−}$ mice to ensure that all

developing T cells use transgenic TCRα^{mini} locus. To track CD4$^+$ Foxp3$^+$ regulatory T (Treg) cells we crossed transgenic mice with NOD.Foxp3.EGFP/cre mice as described in methods. Final characteristics of mice used in this paper were NOD^{mini} (NOD.TCRα^{mini}.TCRβ^{Tg}.TCRα$^{−/−}$.Foxp3EGFP/cre), NOD^{βTg} (TCRβ^{Tg}.TCRα$^{+/−}$.Foxp3EGFP/cre) and NOD (NOD.Foxp3EGFP/cre).

In NOD^{mini} and NOD^{βTg} transgenic mice thymic development of T cells proceeds normally, and selection of single positive (SP) thymocytes is very efficient with bias toward CD4$^+$ T cells in the thymus and in the periphery, similarly to a previously reported bias on B6 background (Fig. 1A, 1B) [36]. TCRβ transgenic mice are characterized by allelic exclusion, resulting in all T cells to express only one transgenic TCRβ chain. It has been proposed that allelic exclusion in NOD mice is less efficient [27]. In our mice all SP thymocytes and peripheral T cells have exclusive expression of transgenic TCRβ^{Tg} chain (TCRVβ14) in both types of transgenic mice and exclusive expression of TCRα^{mini} transgene (TCRVα2) in NOD^{mini} mice (Fig. 1A, 1B). Also we did not observe emergence of the cells bearing TCRs with Vβ segments other than Vβ14 even in older mice. Furthermore, we observed lower percentages of Treg cells in thymii and periphery of transgenic mice as compared to NOD mice and the percentages correlated with diversities of TCRs in analyzed mice (Fig. 1C, 1E). Nevertheless, the total numbers of peripheral Treg cells were similar in all of analyzed mice and reduced percentages of Treg cells were due to more efficient selection of CD4$^+$ T cells, rather than diversity of TCRs, as we observed this correlation on B6 background and in other types of mice that differ by efficiency of CD4$^+$ T cells selection (Fig. 1C, 1E) [36,41]. Finally, to check whether this particular TCRVβ14 transgenic chain can impose an unusual restriction on the TCRα repertoire, we evaluated the frequency of individual Vα families. We found no bias in TCR repertoire as the frequency of Vα families usage by CD4$^+$Vβ14$^+$ T cells was similar between NOD and NOD^{βTg} mice (Fig. 1D).

Lack of T cells specific to insulin B:9–23 in NOD transgenic mice

To test development of diabetes we measured blood glucose levels in experimental mice. Surprisingly, neither NOD^{βTg} nor NOD^{mini} transgenic mice develop diabetes (Fig. 2A). Evaluation of H&E stained pancreatic sections showed lack of lymphocytic infiltrates in 25 week old transgenic mice, with only insignificant infiltrates found in a few sections of NOD^{βTg} mice (Fig. 2B). As a control we used non-transgenic littermates of NOD^{βTg}, NOD.T-CRα$^{+/−}$ mice. These mice developed diabetes and by 30 weeks of age more than 75% of females had high levels of blood glucose (Fig. 2A). One of the possibilities was that the lack of development of diabetes in transgenic mice is due to perturbed proportions of regulatory to effector T cell ratios, which could result in more efficient suppression of potentially diabetogenic T cells. We took advantage of cyclophosphamide treatment, which selectively affects numbers and function of Treg cells and accelerates development of diabetes in NOD mice [42,43]. This treatment however did not cause insulitis nor diabetes in transgenic mice indicating inability of effector T cells to initiate autoimmunity (unpublished data).

The onset of spontaneous diabetes in NOD mice is dependent upon presence of effector T cells specific to insulin B:9–23 antigen and lack of such specificity results in lack of early infiltrates into pancreatic islets [18]. As our mice did not develop islet-infiltrates, we tested their ability to respond to stimulation by insulin B:9–23 peptide (Fig. 3). CD4$^+$ T cells isolated from NOD^{mini}, NOD^{βTg} and NOD mice responded to stimulation by allogeneic spleno-

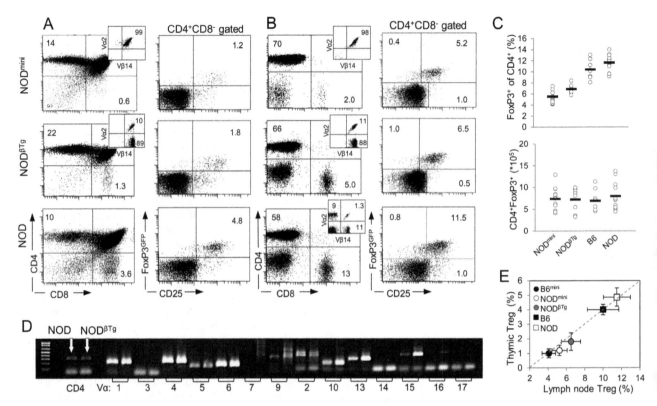

Figure 1. Efficient selection of CD4⁺ T lymphocytes in NOD^mini and NOD^βTg mice. Lymphocytes isolated from thymii (A) and lymph nodes (B) of indicated mice were stained with monoclonal antibodies and analyzed by flow cytometry. Numbers in quadrants are representative percentages of at least six mice (6 week old) per group. The numbers of thymocytes and lymphocytes ± SD recovered from NOD^mini, NOD^βTg and NOD mice were: from thymii 78.5±16.7×10⁶, 72.05±13.8×10⁶ and 70.1±14.0×10⁶, and from lymph nodes (axillary, brachial and inguinal) 17.2±5.8×10⁶, 14.1±2.5×10⁶, and 13.2±3.0×10⁶, respectively. (C) Percentages (top) and total numbers (bottom) of CD4⁺Foxp3⁺ T cells in peripheral lymph nodes of 6 week old mice; each circle represents individual mouse. (D) Expression of mRNA of TCRVα genes in sorted CD4⁺ T cells isolated from NOD and NOD^βTg mice. Analysis was done by RT-PCR using primers specific to indicated Vα segments and Cα region. (E) Comparison of CD4⁺Foxp3⁺ T (Treg) cells from thymii and peripheral lymph nodes of transgenic and wild type B6 and NOD mice. Mean percentage and SD of six young mice per group are shown.

cytes, however only CD4⁺ T cells from immunized NOD mice were able to respond to B:9–23 peptide in the presence of syngeneic splenocytes (Fig. 3). As of note, B:9–23 peptide does not bind to I-A^b molecule and in the presence of B6 splenocytes does not lead to CD4 T cell response [44]. We were also unable to generate B:9–23 specific T cell hybridomas from immunized NOD^mini and NOD^βTg mice, however we had no problem generating allo-specific T cell hybridomas from transgenic mice. These results showed that changes introduced by transgenic chains to the TCR repertoire in NOD mice resulted in lack of peripheral specificity to the key antigen required for onset of T1D.

Development of SS in NOD^βTg and NOD^mini mice

The lack of T1D in transgenic mice prompted us to test whether reduced TCR diversity will affect development of Sjögren's syndrome – secondary autoimmune disease in NOD mice. It is characterized by lymphocytic infiltrates into salivary and lacrimal glands, production of auto-antibodies and the loss of saliva and tear production by 20 weeks of age [5,45]. Histological evaluation of tissue sections of submandibular salivary and extraorbital lacrimal glands showed focal lymphocytic infiltrates, in both transgenic NOD^mini and NOD^βTg and parental NOD mice (Fig. 4). Similarly to the NOD mice, males of both transgenic mice had milder infiltration of salivary glands than females, while the infiltration of lacrimal glands was more prominent in male

NOD^βTg and NOD^mini mice. This infiltration is organ specific, as we did not detect infiltrates in lungs, kidney, liver, as well as sublingual and parotid salivary glands and harderian lacrimal glands of 20 week old transgenic mice, similarly to NOD mice (Fig. 4B). Of note, in some of 30 week old NOD^mini females, but not NOD^βTg mice, we noticed development of lymphoprolifera-tion, manifested by enlarged lymph nodes and spleen and infiltration into multiple organs (unpublished data). FACS analysis of lymphocytic infiltrates in affected exocrine glands showed that all of the infiltrating CD4⁺ T cells expressed transgenic Vβ14 chain in NOD^βTg mice and Vβ14/Vα2 transgenic chains in NOD^mini mice (Fig. 4C). This also showed the aforementioned stability of expression of transgenic TCRs in experimental mice.

Development of SS in humans is often correlated with presence of anti-nuclear antibodies (ANA), anti-Ro/SSA, anti-La/SSB, anti-dsDNA, anti-αFodrin and anti-M3R [5,22,46]. In the NOD mouse model of SS, anti-SSA and anti-SSB auto-antibodies are rarely present and they are found at the very low levels [47]. To determine the presence of auto-antibodies in experimental mice, we used ELISA-based assay. For negative control, we used sera from NOD.TCRα⁻/⁻ mice, that do not have T cells. As shown in Fig. 5, transgenic and parental NOD mice from 14–17 week old group had elevated levels of antibodies against αFodrin, ssDNA, dsDNA and ANA. Additionally, we determined the staining pattern of ANA auto-antibodies using immunofluorescent staining of HEp-2 cells (Fig. 5B). The majority (>90%) of NOD^mini and

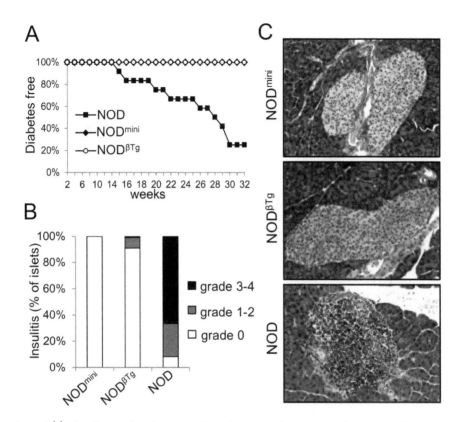

Figure 2. NOD$^{\beta Tg}$ and NOD^{mini} mice do not develop T1D. (A) Incidence of diabetes in mice shown as Kaplan–Meier survival curve. Mice with three consecutive measurements of blood glucose level above 250 mg/dL were considered diabetic. At least twelve mice per group were analyzed; p<0.05. (B) H&E staining of pancreatic tissue sections were analyzed for lymphocytic infiltrates and percentages of islets with grade of insulitis were counted. Insulitis was graded based on following criteria: no infiltrates – grade 0; peri-insulitis - grade 1; insulitis <25% - grade 2; insulitis <50% - grade 3; insulitis >50% - grade 4. At least six mice at 20 weeks of age were analyzed with the total number of 140 (NOD^{mini}), 155 (NOD$^{\beta Tg}$) and 160 (NOD) islets. (C) Example of H&E staining of pancreatic tissue sections of indicated mice at 20 weeks of age.

NOD$^{\beta Tg}$ mice tested had a speckled nuclear staining pattern, characteristic for SS development and found in parental NOD mice.

The onset of salivary gland dysfunction and presence of autoantibodies relies on B cells involvement and is dependent on IL-4 mediated IgM to IgG1 class switching [48–50]. Analysis of IgG1 concentration in sera of experimental mice revealed increased levels in older mice with the highest level in NOD^{mini} mice (Fig. 5C). This increase in IgG1 titer correlates with detection of auto-antibodies. Finally, to evaluate glandular dysfunction, we quantified secretion of saliva in 20 wks old mice. Both NOD$^{\beta Tg}$ and NOD^{mini} mice had reduced saliva flow as compared to healthy B6 mice and at the level of the reference parental NOD mice (Fig. 5D). Taken together, glandular infiltration, autoantibody production, and defective salivary secretion is diagnostic of Sjögren's syndrome in transgenic NOD$^{\beta Tg}$ and NOD^{mini} mice, similarly to the parental strain of NOD mice. Interestingly, the timing of infiltrates and levels of autoantibodies and total IgG1 production vary between analyzed mice, which differ only by TCR repertoire diversity.

Diversity of TCR repertoire on CD4$^+$ T cells in NOD^{mini} mice

Lack of certain specificities in NOD^{mini} TCR repertoire and the previously reported lower TCR diversity of thymic Treg cells on NOD background prompted us to take a closer look at the similarity and diversity of TCRα^{mini} repertoires. To determine the influence of TCR diversity reduction on the selection efficiency of TCRs in NOD mice we started with a single cell analysis to compare to a previously analyzed similar model of TCR^{mini} transgenic mice on "healthy" C57BL/6 background (B6^{mini}) [36]. We compared the similarity between TCR sequences from single cell sorted T$_N$ (CD4$^+$Foxp3$^-$CD45RB$^+$CD62L$^+$) and Treg (CD4$^+$Foxp3$^+$) cells from thymii and peripheral lymph nodes of NOD^{mini} mice. As expected, based on the Morisita-Horn index, the highest similarity was observed between thymic and peripheral subsets of T$_N$ or Treg cells, whereas comparison between populations of T$_N$ and Treg cells showed mostly non-overlapping repertoires with values similar to those observed on B6 background (Fig. 6A and [36]). Previously we've shown that based on abundance coverage estimator (ACE), estimated richness (total unique CDR3α clonotypes in the population) of Treg cells in B6^{mini} mice significantly exceeded estimated richness of T$_N$ cells [36]. Although ACE underestimates true richness at low sample size, it accounts for "unseen sequences" based on low abundance data and is suitable for comparative analyses. We combined thymic and peripheral sequences for each population and calculated the ACE index, based on 578 DNA sequences per subset. The ACE values for T$_N$ and Treg cells were respectively 1187 and 994 for NOD^{mini} mice, and 1184 and 1815 for B6^{mini} mice. Interestingly, estimated ACE value for TCRs on T$_N$ cells from NOD^{mini} mice was comparable to the ACE value for TCRs on T$_N$ cells from B6^{mini} mice. However when we compared the ratio of ACE values for Treg cells between NOD^{mini} and B6^{mini}

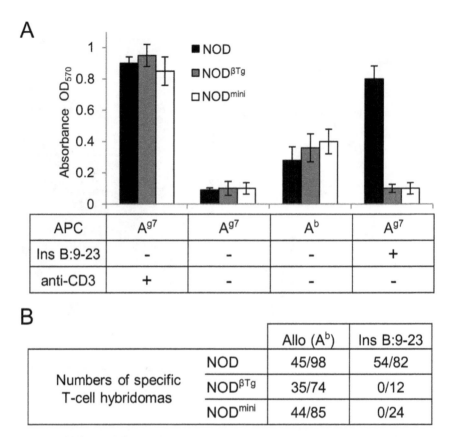

Figure 3. Lack of response to insulin B:9–23 peptide in transgenic NOD mice. (A) 5×10^4 of CD4$^+$ T cells sorted from lymph nodes of NOD, NOD$^{\beta Tg}$ and NOD^{mini} mice were cultured in the presence of 5×10^5 splenocytes from NOD.TCRα$^{-/-}$ (A^{g7}) or B6.TCRα$^{-/-}$ (A^b) mice and soluble anti-CD3 (1 µg/ml) or insulin B:9–23 peptide (50 µg/ml), as indicated. Proliferation of cells was measured after 3 days by MTT assay [38]. Experiments were done twice with 3 mice per group. (B) T-cell hybridomas specific to allo-antigens or insulin B:9–23 peptide were generated from indicated mice. For generation of B:9–23 specific hybridomas, mice were immunized with the peptide 7 days prior to isolation of lymph nodes for *in vitro* blasts generation [38]. Generated hybridomas were tested for their ability to respond to syngeneic (NOD.TCRα$^{-/-}$) or allogeneic (B6.TCRα$^{-/-}$) splenocytes with or without B:9–23 peptide or anti-CD3. Table shows numbers of identified hybridomas specific to indicated antigens and numbers of hybridomas responding to anti-CD3 stimulation. Table is representative of 3 independent experiments with two mice per group.

mice the number of possible unique TCRs on NOD background was reduced by almost 50% (Fig. 6B).

Previously it has been reported that based on analysis of two selected VJ (TCRα) or one VDJ (TVRβ) rearrangements in wild type mice, TCR repertoire of Treg cells in NOD mice was less diverse as compared to conventional T cells in the thymus, but also less diverse in the thymus of B6 mice [9]. These differences were observed based on calculation of Shannon entropy and normalization to the logarithm of unique sequences [9]. Such transformation is a measure of distribution of frequency of individual species and is a good measure of relative evenness of assemblage [51,52]. Together with richness, evenness is a descriptive measure of diversity not the diversity *per se* [51]. Therefore, as explained by Jost, to put the estimates in perspective, we converted Shannon entropy (diversity index) to "true diversity" by calculating "numbers equivalent" also called "effective number of species" (ENS), to preserve linear scale of comparison [51,53]. The ENS measure represents diversity of a particular sample, and the numeric value represents the theoretical number of equally common unique sequences in the assemblage. Comparison of "true diversity" between B6^{mini} and NOD^{mini} mice showed almost reversal of ratios of diversities between TCRs on T$_N$ and Treg cells, with the differences more profound in the thymus than in the periphery (Fig. 6C). These differences in NOD^{mini} mice were confirmed by high throughput sequencing, and were consistent

regardless of total numbers of sequences analyzed (Fig. 6C, D). Lower diversity of Treg TCR repertoire was visualized empirically by plotting accumulation curves of observed sequences from peripheral T$_N$ and Treg cells (Fig. 6E). These curves show that accumulation of unique CDR3 regions from the first 120 thousands of sequences for each population gives twice as many unique DNA clonotypes in T$_N$ (9757) as compared to Treg cells (4968). This ratio is reversed in comparison to accumulation curves observed in B6^{mini} mice [36]. Collectively our data show that although TCR diversity on Treg cells in NOD^{mini} mice is reduced, conventional T cells retain diverse TCR repertoire at least at the levels found on B6 background.

Discussion

In this study we investigated the impact of the reduction of TCR diversity on the development of two autoimmune diseases in NOD mice; T1D and SS. Previously it has been shown that T1D can develop despite use of transgenic TCRβ chains or reduction of precursor frequency of potentially diabetogenic T cell clones. In our model overall diversity of TCRs was reduced by allelic exclusion caused by use of transgenic TCRβ chain that was not only pathogenically irrelevant, but also was originally selected in B6 mice on I-A^b molecule. Despite normal distribution of Vα families in NOD$^{\beta Tg}$ mice, neither insulitis nor diabetes developed.

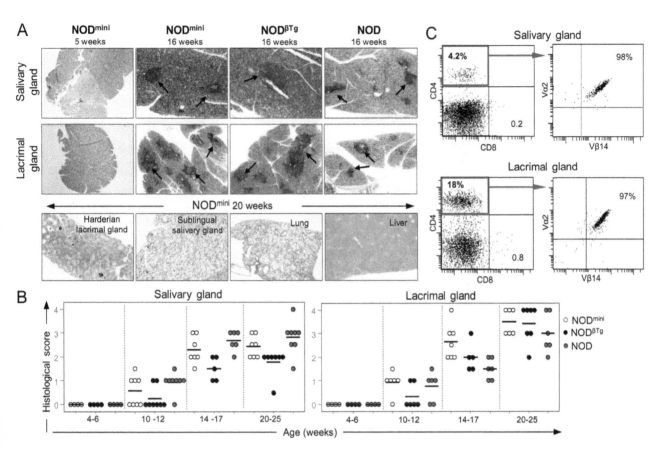

Figure 4. Lymphocytic infiltrates in mandibular salivary and extraorbital lacrimal glands. (A) H&E staining of tissue sections from indicated organs of analyzed mice, showing lymphocytic infiltrates indicated by arrows. (B) Histological score of infiltrated glands in indicated age groups. Scoring criteria: score 0, no infiltrates; score 1–1.5, 1–2 foci per section; 2–2.5, 3–5 foci per section; score 3, 6–10 foci per section; score 4, more than 10 foci per section. Infiltrate is considered as focus when number of infiltrating cells in continuous space is greater than 50. Three sections at different anatomical locations per organ were analyzed with at least 5 mice per age group. (C) FACS analysis of CD4 T cells infiltrating into salivary and lacrimal glands in 16 week old NOD^{mini} mice. Dot plots on the right show expression of transgenic TCR on CD4$^+$ gated cells.

Conversely, these mice developed infiltrates in salivary and lacrimal glands, leading to autoantibody production and exocrine gland dysfunction. Further reduction of TCR diversity by generation of transgenic mice with TCR^{mini} repertoire, where one Vα segment is allowed to rearrange to only two Jα segments, did not prevent development of SS. Our results indicate that the difference between T1D and SS regarding the dependence on MHC polymorphism is directly correlated to the magnitude of possible TCR/peptide/MHCII interactions participating in the autoimmune phase of the disease.

We show that the lack of development of diabetes or even insulitis in our NOD$^{\beta Tg}$ or NOD^{mini} transgenic mice is due to lack of specificity to the key immunodominant insulin B:9–23 peptide, which is known to be instrumental for the onset of the T1D in NOD mice. The use of transgenic TCRVβ14 chain in our mice did not dramatically influence the ability of its binding to different TCRα chains, as T cells from NOD$^{\beta Tg}$ mice use all Vα families with frequencies found in NOD mice (Fig. 1). This includes efficient amplification of the Vα13 family which contains TRAV5D-4 chain (Vα13s1) that was shown to be sufficient to elicit anti-insulin autoimmunity without bias toward particular Vβ family of TCRβ chain partners [28,54]. Moreover, previous studies show that T cells using Vβ14 family were found on T cells specific to insulin antigen, T cells expanding in pancreatic lymph nodes or in T cells infiltrating pancreatic islets, showing that the Vβ14 family is not negatively influencing the development of

diabetogenic TCRs [55–58]. One cannot exclude the possibility that this particular transgenic TCRVβ14 chain may be unable to pair with appropriate TCRα chain, preventing the ability of the expressed $\alpha\beta$TCR to recognize the B:9–23 peptide. This selective requirement for a TCRβ chain would reinforce our observation that development of SS is less dependent than development of T1D on overall TCR diversity and a particular peptide/MHC combination.

Development of T1D relies on different insulin B:9–23 register recognition, allowing escape of specific T cells due to register shifting [11–13]. The lack of peripheral recognition of insulin B:9–23 in our transgenic mice can also be due to the impact of the limited TCR repertoire. Reduction of overall TCR diversity can influence (reduce) the precursor frequency of DP thymocytes bearing potentially autoreactive TCRs, resulting in more efficient negative selection in the thymus of NOD mice. It has been shown that early expression of transgenic $\alpha\beta$TCR, due to ERK1/2 defect on NOD background results in greater commitment of the DN thymocytes to $\alpha\beta$ lineage "overcrowding" DP compartment [8]. Our TCRα^{mini} transgene has natural timing of expression, similar to B6^{mini} and polyclonal NOD mice, where pre-TCR signaling is not perturbed by early expression of the transgene [36]. As suggested by Mingueneau et. al., in the polyclonal repertoire on the NOD background, the ERK1/2 defect increases the affinity threshold of positive selection, shifting the selection window of thymocytes toward self-reactivity, however not impacting the

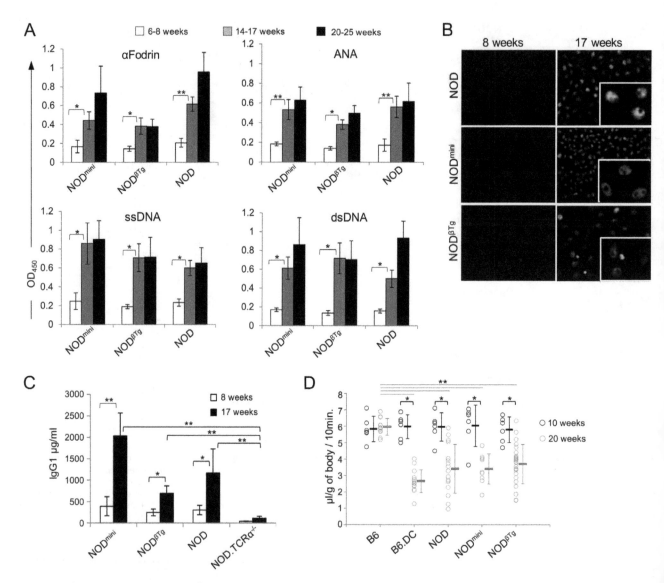

Figure 5. Detection of autoantibodies and hyposalivation in NOD^mini and NOD^βTg mice. (A) ELISA assay was performed using mouse serum (1:100) from indicated age groups. Each graph represents mean value of OD_450 and standard deviation for indicated antigens. At least six mice were used per each group. *p<0.05, **p<0.005. (B) Detection of ANAs pattern using Hep-2 cell line. Sera from mice were diluted 1:40, incubated with HEp-2-fixed slides and evaluated under fluorescent microscope at x20 magnifications. Representative images are shown. (C) Quantitative ELISA analysis of IgG1 levels in sera (1:50,000) from indicated mice. Each bar represents mean value and standard deviation from six mice per experimental groups. (D) Salivary flow rates after pilocarpine injection in indicated mice at 10 and 20 wks of age. Double congenic B6.NODIdd3.NODIdd5 (B6.DC) mice, that develop SS on B6 genetic background, were used as a control. Saliva volume was measured and calculated in mg per mouse body mass. Each circle represents one mouse and horizontal lines indicate mean values of the experimental groups. T-test was used to calculate differences between groups. *p<0.002, **p<0.0002.

efficiency of negative selection. This results in higher overall self-reactivity of peripheral effector T cells and possibly explains lower diversity of thymic Treg cells on NOD background [8,9]. Considering partial overlap of specificities between Treg and autoreactive T cells, one could suggest similar effect of lower diversity on autoreactive population. However weak and unstable peptide-binding property of I-A^g7 molecule does not favor the elimination or inactivation of autoreactive T cells [59]. This instability may have additional influence on "a leak" of autoreactive T cells, but also on inefficient generation of Treg population, which may require longer or stronger interactions with self MHC/peptide complexes [60,61]. Therefore the shift in selection window will not impact autoreactive T cells as much as it

will impact Treg cells, after all, Treg development relies on recognition of self-peptide/MHCII complexes in the thymus, whereas thymic escape of autoreactive T cells relies on avoidance of such complexes during negative selection. It is possible that in our model, we reached the threshold of diversity required to generate autoreactive TCR repertoire without "holes" in specificities. Therefore despite reduced TCR diversity of Treg cells mice do not develop diabetes and we were unable to detect insulin B:9–23 specific T cells in the periphery. Similarly, the comparison of TCR^mini repertoires between B6 and NOD backgrounds shows minimal impact of the NOD genotype on TCR diversity of conventional CD4^+ T cells however it substantially reduces the TCR diversity on Treg cells in NOD mice.

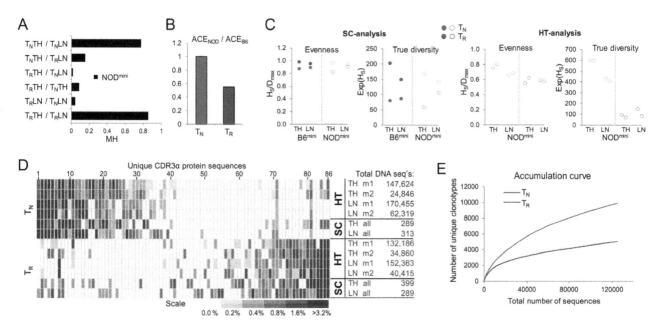

Figure 6. TCR repertoire of naïve and regulatory T cells in NOD^{mini} mice. (A) Similarity between indicated populations (first 289 single cell sequences per each population) was estimated based on Morisita-Horn index (MH). T_N - naïve CD4$^+$Foxp3$^-$CD45RB$^+$CD62L$^+$ T cells, T_R - regulatory CD4$^+$Foxp3$^+$ T cells, TH - thymus, LN - lymph nodes. (B) Ratio of richness of TCRs on T_N and T_R cells between NOD^{mini} and B6^{mini} mice. Abundance coverage estimator (ACE) was calculated based on 578 sequences for each population combined from thymus and peripheral lymph nodes. (C) Evenness and effective number of species (ENS) for analyzed TCR repertoires. Shannon evenness index was calculated as Shannon entropy (H$_s$) divided by maximum diversity D$_{max}$, where D$_{max}$ equals natural logarithm of number of unique sequences in analyzed population. ENS (true diversity) was calculated as exponential of Shannon entropy. (D) Comparison of frequency of 20 most dominant unique protein CDR3 clonotypes found in each population indicated on the right of the heat map. Table indicates analyzed populations from lymph nodes and thymii by single cell analysis (SC) and high throughput sequencing (HT). Experimental mouse 1 and 2 are marked as m1 and m2. Numbers next to each population indicate total numbers of DNA sequences analyzed. All 86 unique CDR3α protein sequences in the heat map are shown in Table S1. (E) Accumulation curve of unique DNA clonotypes observed after accumulation of 124,730 sequences for T_N cells and 124,696 for T_R cells. (A–C) All indices were computed based on DNA sequences for each population using software SPADE and EstimateS8.2.

Development of SS in NOD^{mini} mice is especially interesting, since CD4$^+$ T cells are instrumental in immunopathogenesis and their recognition of self antigens is essential for the onset and progression of the disease [34,62]. It shows that the TCR^{mini} repertoire is diverse enough not only to drive glandular infiltration and activation of the CD4$^+$ T cells but also the repertoire is still diverse enough to support the full development of the disease with production of Th2-dependent IgG1 pathogenic autoantibodies (Fig. 5C) [50]. Moreover, we noticed differences in timing of infiltrates, levels of autoantibodies and total IgG1 production between transgenic mice and parental NOD mice, which indicates different frequencies of certain TCR specificities between mice. In congenic strains of NOD mice models of SS (NOD.B10-H2^b, NOD.H2^p, NOD.H2^q, NOD.H2^{h4}) replacement of I-A^{g7} with other MHC molecules does not prevent salivary and lacrimal gland infiltration and decreased saliva and tear production [16,48,63]. Interestingly, in NOD.H2^{h4}, contrary to the parental NOD strain, there is a high frequency of ANA with a high proportion of SSA/Ro and SSB/La observed [63]. Also, NOD.H2^q mice exhibit increased production of lupus-like types of autoantibodies and develop nephritis, as compared to NOD and NOD.H2^p mice [16]. This weak dependence of SS on a particular MHC haplotype in NOD mice correlates with our data that show development of SS despite limited TCR diversity. In human studies it was suggested that production profiles of certain autoantibodies were associated with HLA-DR haplotypes rather than with clinical manifestations [64], however studies of familial inheritance in patients with SS showed a linkage between particular HLA and disease susceptibility [65,66]. The most

recent comprehensive analysis by Sjögren's Genetics Network showed that HLA has the strongest linkage to the SS, although it is not on the level of T1D [17]. Certain HLA haplotypes will influence binding diversity of self or environmental peptides and the nature of antigen presentation to T cells during thymic development or during immune responses in the periphery. Our results from the mouse model emphasize that SS may be less affected by requirement of a unique key antigen/MHCII combination but rather may be more influenced by a wider range of overall TCR/peptide/MHC interactions involved in the onset/progression of the disease. It can be due to a combination of cross-reactivity of the TCR repertoire on SS-specific T cells, wider range of antigens presented by MHCII molecules, higher peripheral self-reactivity of effector T cells, increased tissue expression of MHCII complexes on salivary epithelial cells and de novo expression or post-translational modification of self-antigens [67–69].

Acknowledgments

Cell sorting was performed by Jeanene Pihkala in the GRU FACS Core, the sequencing was performed by John Nechtman in the GRU Genomic Core, and transgenic mice were generated by Gabriela Pacholczyk in the GRU Transgenic Core.

Author Contributions

Conceived and designed the experiments: JK RP. Performed the experiments: JK RD SL MB. Analyzed the data: JK RD SL MB RP. Contributed to the writing of the manuscript: JK RP.

References

1. Chaparro RJ, Dilorenzo TP (2010) An update on the use of NOD mice to study autoimmune (Type 1) diabetes. Expert Rev Clin Immunol 6: 939–955.
2. Lavoie TN, Lee BH, Nguyen CQ (2011) Current concepts: mouse models of Sjogren's syndrome. J Biomed Biotechnol 2011: 549107.
3. Anderson MS, Bluestone JA (2005) The NOD Mouse: A Model of Immune Dysregulation. Annu Rev Immunol 23: 447–485.
4. Mathis D, Vence L, Benoist C (2001) beta-Cell death during progression to diabetes. Nature 414: 792–798.
5. Nguyen CQ, Peck AB (2009) Unraveling the pathophysiology of Sjogren syndrome-associated dry eye disease. Ocul Surf 7: 11–27.
6. Fox RI (2005) Sjogren's syndrome. Lancet 366: 321–331.
7. Kishimoto H, Sprent J (2001) A defect in central tolerance in NOD mice. Nat Immunol 2: 1025–1031.
8. Mingueneau M, Jiang W, Feuerer M, Mathis D, Benoist C (2012) Thymic negative selection is functional in NOD mice. J Exp Med 209: 623–637.
9. Ferreira C, Singh Y, Furmanski AL, Wong FS, Garden OA, et al. (2009) Non-obese diabetic mice select a low-diversity repertoire of natural regulatory T cells. Proc Natl Acad Sci U S A 106: 8320–8325.
10. D'Alise AM, Auyeung V, Feuerer M, Nishio J, Fontenot J, et al. (2008) The defect in T-cell regulation in NOD mice is an effect on the T-cell effectors. Proceedings of the National Academy of Sciences 105: 19857–19862.
11. Stadinski BD, Zhang L, Crawford F, Marrack P, Eisenbarth GS, et al. (2010) Diabetogenic T cells recognize insulin bound to IAg7 in an unexpected, weakly binding register. Proceedings of the National Academy of Sciences 107: 10978–10983.
12. Mohan JF, Petzold SJ, Unanue ER (2011) Register shifting of an insulin peptide–MHC complex allows diabetogenic T cells to escape thymic deletion. J Exp Med 208: 2375–2383.
13. Marrack P, Kappler JW (2012) Do MHCII-Presented Neoantigens Drive Type 1 Diabetes and Other Autoimmune Diseases? Cold Spring Harbor Perspectives in Medicine 2.
14. Wicker LS, Appel MC, Dotta F, Pressey A, Miller BJ, et al. (1992) Autoimmune syndromes in major histocompatibility complex (MHC) congenic strains of nonobese diabetic (NOD) mice. The NOD MHC is dominant for insulitis and cyclophosphamide-induced diabetes. J Exp Med 176: 67–77.
15. Li X, Golden J, Faustman DL (1993) Faulty major histocompatibility complex class II I-E expression is associated with autoimmunity in diverse strains of mice. Autoantibodies, insulitis, and sialadenitis. Diabetes 42: 1166–1172.
16. Lindqvist AKB, Nakken B, Sundler M, Kjellen P, Jonsson R, et al. (2005) Influence on Spontaneous Tissue Inflammation by the Major Histocompatibility Complex Region in the Nonobese Diabetic Mouse. Scand J Immunol 61: 119–127.
17. Lessard CJ, Li H, Adrianto I, Ice JA, Rasmussen A, et al. (2013) Variants at multiple loci implicated in both innate and adaptive immune responses are associated with Sjogren's syndrome. Nat Genet 45: 1284–1292.
18. Nakayama M, Abiru N, Moriyama H, Babaya N, Liu E, et al. (2005) Prime role for an insulin epitope in the development of type[thinsp]1 diabetes in NOD mice. Nature 435: 220–223.
19. Arakaki R, Ishimaru N, Saito I, Kobayashi M, Yasui N, et al. (2003) Development of autoimmune exocrinopathy resembling Sjögren's syndrome in adoptively transferred mice with autoreactive CD4+ T cells. Arthritis Rheum 48: 3603–3609.
20. Takada K, Takiguchi M, Konno A, Inaba M (2005) Autoimmunity against a tissue kallikrein in IQI/Jic Mice: a model for Sjogren's syndrome. J Biol Chem 280: 3982–3988.
21. Winer S, Astsaturov I, Cheung R, Tsui H, Song A, et al. (2002) Primary Sjögren's syndrome and deficiency of ICA69. Lancet 360: 1063–1069.
22. Haneji N, Nakamura T, Takio K, Yanagi K, Higashiyama H, et al. (1997) Identification of alpha-fodrin as a candidate autoantigen in primary Sjögren's syndrome. Science 276: 604–607.
23. Naito Y, Matsumoto I, Wakamatsu E, Goto D, Ito S, et al. (2006) Altered peptide ligands regulate muscarinic acetylcholine receptor reactive T cells of patients with Sjögren's syndrome. Ann Rheum Dis 65: 269–271.
24. Matsumoto I, Maeda T, Takemoto Y, Hashimoto Y, Kimura F, et al. (1999) Alpha-amylase functions as a salivary gland-specific self T cell epitope in patients with Sjögren's syndrome. Int J Mol Med 3: 485–490.
25. Iizuka M, Wakamatsu E, Tsuboi H, Nakamura Y, Hayashi T, et al. (2010) Pathogenic role of immune response to M3 muscarinic acetylcholine receptor in Sjogren's syndrome-like sialadenitis. J Autoimmun 35: 383–389.
26. Lipes MA, Rosenzweig A, Tan KN, Tanigawa G, Ladd D, et al. (1993) Progression to diabetes in nonobese diabetic (NOD) mice with transgenic T cell receptors. Science 259: 1165–1169.
27. Serreze DV, Johnson EA, Chapman HD, Graser RT, Marron MP, et al. (2001) Autoreactive diabetogenic T-cells in NOD mice can efficiently expand from a greatly reduced precursor pool. Diabetes 50: 1992–2000.
28. Zhang L, Jasinski JM, Kobayashi M, Davenport B, Johnson K, et al. (2009) Analysis of T cell receptor beta chains that combine with dominant conserved TRAV5D-4*04 anti-insulin B: 9–23 alpha chains. J Autoimmun 33: 42–49.
29. Dwyer E, Itescu S, Winchester R (1993) Characterization of the primary structure of T cell receptor beta chains in cells infiltrating the salivary gland in the sicca syndrome of HIV-1 infection. Evidence of antigen-driven clonal selection suggested by restricted combinations of V beta J beta gene segment usage and shared somatically encoded amino acid residues. J Clin Invest 92: 495–502.
30. Matsumoto I, Okada S, Kuroda K, Iwamoto I, Saito Y, et al. (1999) Single cell analysis of T cells infiltrating labial salivary glands from patients with Sjogren's syndrome. Int J Mol Med 4: 519–527.
31. Pivetta B, De Vita S, Ferraccioli G, De RV, Gloghini A, et al. (1999) T cell receptor repertoire in B cell lymphoproliferative lesions in primary Sjogren's syndrome. J Rheumatol 26: 1101–1109.
32. Sumida T, Yonaha F, Maeda T, Tanabe E, Koike T, et al. (1992) T cell receptor repertoire of infiltrating T cells in lips of Sjögren's syndrome patients. J Clin Invest 89: 681–685.
33. Yonaha F, Sumida T, Maeda T, Tomioka H, Koike T, et al. (1992) Restricted junctional usage of T cell receptor V beta 2 and V beta 13 genes, which are overrepresented on infiltrating T cells in the lips of patients with Sjögren's syndrome. Arthritis Rheum 35: 1362–1367.
34. Singh N, Cohen PL (2012) The T cell in Sjögren's syndrome: Force majeure, not spectateur. J Autoimmun 39: 229–233.
35. Sumida T, Tsuboi H, Iizuka M, Hirota T, Asashima H, et al. (2014) The role of M3 muscarinic acetylcholine receptor reactive T cells in Sjögren's syndrome: A critical review. J Autoimmun.
36. Pacholczyk R, Ignatowicz H, Kraj P, Ignatowicz L (2006) Origin and T cell receptor diversity of Foxp3+CD4+CD25+ T cells. Immunity 25: 249–259.
37. Cha S, Nagashima H, Brown VB, Peck AB, Humphreys-Beher MG (2002) Two NOD Idd-associated intervals contribute synergistically to the development of autoimmune exocrinopathy (Sjögren's syndrome) on a healthy murine background. Arthritis Rheum 46: 1390–1398.
38. Pacholczyk R, Kern J, Singh N, Iwashima M, Kraj P, et al. (2007) Nonself-antigens are the cognate specificities of Foxp3+ regulatory T cells. Immunity 27: 493–504.
39. Kisielow J, Tortola L, Weber J, Karjalainen K, Kopf M (2011) Evidence for the divergence of innate and adaptive T-cell precursors before commitment to the alphabeta and gammadelta lineages. Blood 118: 6591–6600.
40. Kraj P, Pacholczyk R, Ignatowicz L (2001) Alpha beta TCRs differ in the degree of their specificity for the positively selecting MHC/peptide ligand. J Immunol 166: 2251–2259.
41. Pacholczyk R, Kraj P, Ignatowicz L (2002) Peptide specificity of thymic selection of CD4+CD25+ T cells. J Immunol 168: 613–620.
42. Lutsiak MEC, Semnani RT, De Pascalis R, Kashmiri SVS, Schlom J, et al. (2005) Inhibition of CD4+25+ T regulatory cell function implicated in enhanced immune response by low-dose cyclophosphamide. Blood 105: 2862–2868.
43. Harada M, Makino S (1984) Promotion of spontaneous diabetes in non-obese diabetes-prone mice by cyclophosphamide. Diabetologia 27: 604–606.
44. Michels AW, Ostrov DA, Zhang L, Nakayama M, Fuse M, et al. (2011) Structure-Based Selection of Small Molecules To Alter Allele-Specific MHC Class II Antigen Presentation. The Journal of Immunology 187: 5921–5930.
45. Cha S, Peck AB, Humphreys-Beher MG (2002) Progress in understanding autoimmune exocrinopathy using the non-obese diabetic mouse: an update. Crit Rev Oral Biol Med 13: 5–16.
46. Atkinson JC, Travis WD, Slocum L, Ebbs WL, Fox PC (1992) Serum anti-SS-B/La and IgA rheumatoid factor are markers of salivary gland disease activity in primary Sjögren's syndrome. Arthritis Rheum 35: 1368–1372.
47. Skarstein K, Wahren M, Zaura E, Hattori M, Jonsson R (1995) Characterization of T cell receptor repertoire and anti-Ro/SSA autoantibodies in relation to sialadenitis of NOD mice. Autoimmunity 22: 9–16.
48. Robinson CP, Yamachika S, Bounous DI, Brayer J, Jonsson R, et al. (1998) A novel NOD-derived murine model of primary Sjögren's syndrome. Arthritis Rheum 41: 150–156.
49. Brayer JB, Cha S, Nagashima H, Yasunari U, Lindberg A, et al. (2001) IL-4-dependent effector phase in autoimmune exocrinopathy as defined by the NOD.IL-4-gene knockout mouse model of Sjögren's syndrome. Scand J Immunol 54: 133–140.
50. Gao J, Killedar S, Cornelius JG, Nguyen C, Cha S, et al. (2006) Sjögren's syndrome in the NOD mouse model is an interleukin-4 time-dependent, antibody isotype-specific autoimmune disease. J Autoimmun 26: 90–103.
51. Jost L (2010) The Relation between Evenness and Diversity. Diversity 2: 207–232.
52. Pielou EC (1966) The measurement of diversity in different types of biological collections. J Theor Biol 13: 131–144.

53. Adelman MA (1969) Comment on the H Concentration Measure as a Numbers-Equivalent. The Review of Economics and Statistics 51: 99–101.

54. Nakayama M, Castoe T, Sosinowski T, He X, Johnson K, et al. (2012) Germline TRAV5D-4 T-Cell Receptor Sequence Targets a Primary Insulin Peptide of NOD Mice. Diabetes 61: 857–865.

55. Simone E, Daniel D, Schloot N, Gottlieb P, Babu S, et al. (1997) T cell receptor restriction of diabetogenic autoimmune NOD T cells. Proceedings of the National Academy of Sciences 94: 2518–2521.

56. Baker FJ, Lee M, Chien Y-h, Davis MM (2002) Restricted islet-cell reactive T cell repertoire of early pancreatic islet infiltrates in NOD mice. Proceedings of the National Academy of Sciences 99: 9374–9379.

57. Marrero I, Hamm DE, Davies JD (2013) High-Throughput Sequencing of Islet-Infiltrating Memory CD4+ T Cells Reveals a Similar Pattern of TCR Vβ Usage in Prediabetic and Diabetic NOD Mice. PLoS ONE 8: e76546.

58. Petrovc Berglund J, Mariotti-Ferrandiz E, Rosmaraki E, Hall H, Cazenave P-A, et al. (2008) TCR repertoire dynamics in the pancreatic lymph nodes of non-obese diabetic (NOD) mice at the time of disease initiation. Mol Immunol 45: 3059–3064.

59. Carrasco-Marin E, Shimizu J, Kanagawa O, Unanue ER (1996) The class II MHC I-Ag7 molecules from non-obese diabetic mice are poor peptide binders. J Immunol 156: 450–458.

60. Aschenbrenner K, D'Cruz LM, Vollmann EH, Hinterberger M, Emmerich J, et al. (2007) Selection of Foxp3+ regulatory T cells specific for self antigen expressed and presented by Aire+ medullary thymic epithelial cells. Nat Immunol 8: 351–358.

61. Fontenot JD, Rasmussen JP, Williams LM, Dooley JL, Farr AG, et al. (2005) Regulatory T cell lineage specification by the forkhead transcription factor foxp3. Immunity 22: 329–341.

62. Sumida T, Tsuboi H, Iizuka M, Nakamura Y, Matsumoto I (2010) Functional role of M3 muscarinic acetylcholine receptor (M3R) reactive T cells and anti-M3R autoantibodies in patients with Sjögren's syndrome. Autoimmunity Reviews 9: 615–617.

63. Burek CL, Talor MV, Sharma RB, Rose NR (2007) The NOD.H2h4 mouse shows characteristics of human Sjogren's Syndrome. J Immunol 178: S232–S223d.

64. Gottenberg JE, Busson M, Loiseau P, Cohen-Solal J, Lepage V, et al. (2003) In primary Sjögren's syndrome, HLA class II is associated exclusively with autoantibody production and spreading of the autoimmune response. Arthritis Rheum 48: 2240–2245.

65. Fox RI, Kang HI (1992) Pathogenesis of Sjogren's syndrome. RheumDisClinNorth Am 18: 517–538.

66. Manoussakis MN, Georgopoulou C, Zintzaras E, Spyropoulou M, Stavropoulou A, et al. (2004) Sjogren's syndrome associated with systemic lupus erythematosus: clinical and laboratory profiles and comparison with primary Sjogren's syndrome. Arthritis Rheum 50: 882–891.

67. Anderton SM (2004) Post-translational modifications of self antigens: implications for autoimmunity. Curr Opin Immunol 16: 753–758.

68. Engelhard VH, Altrich-Vanlith M, Ostankovitch M, Zarling AL (2006) Post-translational modifications of naturally processed MHC-binding epitopes. Curr Opin Immunol 18: 92–97.

69. Moutsopoulos HM, Hooks JJ, Chan CC, Dalavanga YA, Skopouli FN, et al. (1986) HLA-DR expression by labial minor salivary gland tissues in Sjogren's syndrome. Ann Rheum Dis 45: 677–683.

10

Overexpression of a Soybean Ariadne-Like Ubiquitin Ligase Gene *GmARI1* Enhances Aluminum Tolerance in Arabidopsis

Xiaolian Zhang, Ning Wang, Pei Chen, Mengmeng Gao, Juge Liu, Yufeng Wang, Tuanjie Zhao, Yan Li*, Junyi Gai*

National Key Laboratory of Crop Genetics and Germplasm Enhancement, National Center for Soybean Improvement, Key Laboratory for Biology and Genetic Improvement of Soybean (General, Ministry of Agriculture), Nanjing Agricultural University, Nanjing, Jiangsu, China

Abstract

Ariadne (ARI) subfamily of RBR (Ring Between Ring fingers) proteins have been found as a group of putative E3 ubiquitin ligases containing RING (Really Interesting New Gene) finger domains in fruitfly, mouse, human and Arabidopsis. Recent studies showed several RING-type E3 ubiquitin ligases play important roles in plant response to abiotic stresses, but the function of ARI in plants is largely unknown. In this study, an ariadne-like E3 ubiquitin ligase gene was isolated from soybean, *Glycine max* (L.) Merr., and designated as *GmARI1*. It encodes a predicted protein of 586 amino acids with a RBR supra-domain. Subcellular localization studies using Arabidopsis protoplast cells indicated GmARI protein was located in nucleus. The expression of *GmARI1* in soybean roots was induced as early as 2–4 h after simulated stress treatments such as aluminum, which coincided with the fact of aluminum toxicity firstly and mainly acting on plant roots. In vitro ubiquitination assay showed GmARI1 protein has E3 ligase activity. Overexpression of *GmARI1* significantly enhanced the aluminum tolerance of transgenic Arabidopsis. These findings suggest that *GmARI1* encodes a RBR type E3 ligase, which may play important roles in plant tolerance to aluminum stress.

Editor: Ji-Hong Liu, Key Laboratory of Horticultural Plant Biology (MOE), China

Funding: This work was supported by the National Hightech R & D Program of China (2013AA102602), the National Key Basic Research Program of China (2011CB1093), the Program for Changjiang Scholars and Innovative Research Team in University (PCSIRT13073), the Program for New Century Excellent Talents in University (NCET-12-0891), the MOE 111 Project (B08025), the Program for High-level Innovative and Entrepreneurial Talents in Jiangsu Province, and the Jiangsu Higher Education PAPD Program. The funders had no role in study design, data collection and analysis, decision or preparation of the manuscript.

Competing Interests: The authors have declared that no competing interests exist.

* Email: yanli1@njau.edu.cn (YL); sri@njau.edu.cn (JG)

Introduction

Ubiquitination is an enzymatic, protein post-translational modification by which proteins are selectively targeted for a variety of cellular processes including DNA transcription and repair, cell cycle and division, response to stresses and many others [1]. This process is carried out by three types of enzyme, including an ubiquitin-activating enzyme (E1), an ubiquitin-conjugating enzyme (E2), and an ubiquitin protein ligase (E3) [2]. Encoded by a large gene family of widely divergent isoforms [3], E3 ligases play important roles in governing the ubiquitin signaling pathway by transferring ubiquitin from E2 conjugation to specific protein substrates. E3 ligases are generally divided into two families, with either a HECT or RING-finger domain(s) [1], [4]. The RING-type E3 ubiquitin ligases are generally identified by the presence of conserved cysteine- and histidine- rich RING finger motifs that coordinate zinc atoms [5]. Recently several RING-type E3 ubiquitin ligases were found to play important roles in plant responses to abiotic and biotic stresses. The pepper E3 ubiquitin ligase RING1 gene, *CaRING1*, is required for cell death and the salicylic acid (SA)-dependent defense response [6]. *AtAIRP1* and *AtAIRP2* play roles in abscisic acid (ABA)-mediated drought stress

responses in Arabidopsis [7]. In soybean, a RING-finger protein encoded by *GmRFP1* was identified and shown to be involved in ABA signaling and stress responses through the ubiquitin-proteasome pathway [8].

RBR (Ring Between Ring fingers) proteins are characterized by the presence of their RING1 – IBR – RING2 supra-domain, which is composed of two RING finger domains plus an IBR (In Between Rings) domain [9]. Many RBR proteins are known to have E3 ubiquitin ligase activity [10]. ARIADNE (ARI) proteins, a subclass of RBRs, have been identified in fruitfly [11], mouse [12], [13], human [14–16], and Arabidopsis [17], [18]. ARI proteins are characterized by the presence of an N-terminal acid-rich cluster, followed by a C3HC4 RING-finger motif, a central IBR or B-box, a second C3HC4 RING-finger structure, and Leu-rich domain at the C terminus. ARI proteins share their RBR domain with PARKIN, a protein involved in autosomal recessive familial Parkinson's disease [9]. PARKIN functions as E2-dependent ubiquitin-protein ligase [17]. Recent studies suggest that the ARI/PARKIN proteins define a new class of RING-finger E3 ligases [19]. There are only few studies on ARI proteins in plants

Figure 1. The protein structure of GmARI1. Acid: acid-rich cluster; **Leu**: Leu-rich cluster; **RING1**: a C3HC4 RING-finger; **IBR**: In Between RING fingers (IBR); **RING2**: a second C3HC4 RING finger; **Coile~**: Coiled coil.

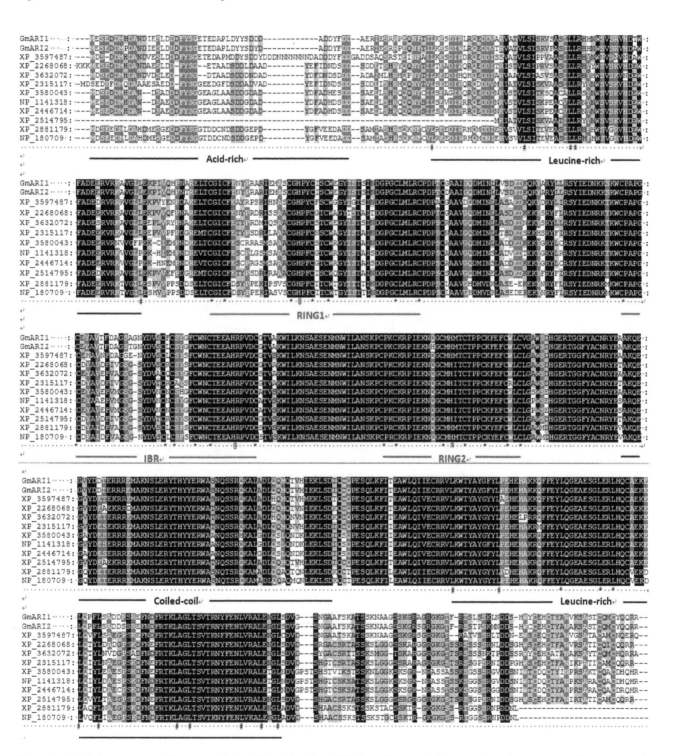

Figure 2. Multiple sequence alignment of GmARI1 with other RBR domain containing proteins. XP_3597487: *Medicago truncatula*; XP_2268068 and XP_3632072: *Vitis vinifera*; XP_2315117: *Populus trichocarp*; XP_3580043: *Brachypodium distachyon*; NP_1141318: *Zea mays*; XP_2446714: *Sorghum bicolor*; XP_2514795: *Ricinus communis*; XP_2881179: *Arabidopsis lyrata*; NP_180709: *Arabidopsis thaliana*; *: Cys; @: His; #: Leu and Ile.

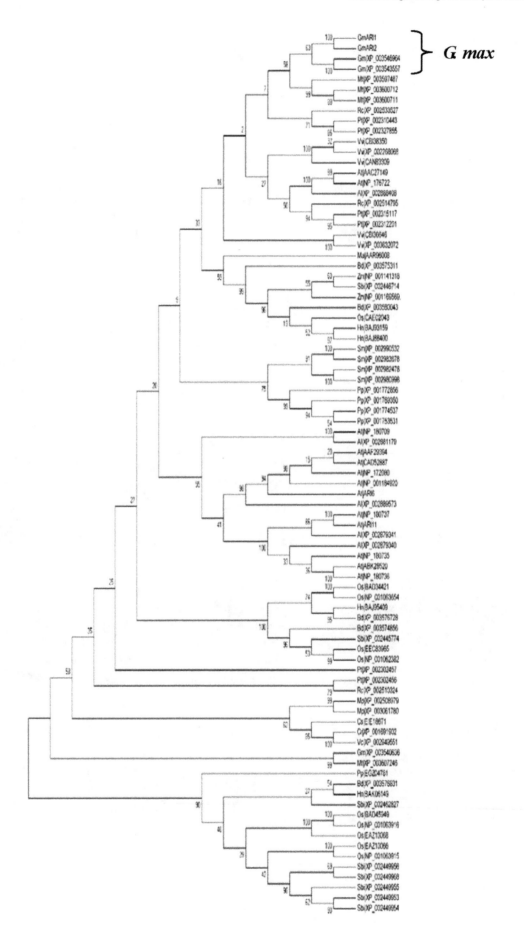

Figure 3. Phylogenetic relationships among soybean ARI proteins and other RBR domain containing proteins. The numbers on the tree indicate percent bootstrap values. The species abbreviations are listed as follows; Al: Arabidopsis lyrata; At: Arabidopsis thaliana; Bd: Brachypodium distachyon; Cr: Chlamydomonas reinhardtii; Cs: Coccomyxa subellipsoidea; Hn: Hordeum vulgare; Ma: Musa acuminata; Mp: Micromonas pusilla; Mt: Medicago truncatula; Nt: Nicotana plumbaginifolia; Os: Oryza sativa subsp. Japonica; Pp: Physcomitrella patens; Pt: Populus trichocarpa; Rc: Ricinus communis; Sbi: Sorghum bicolor; Sm: Selaginella moellendorffii; Vc: Volvox carteri, Vvi: Vitis vinifera.

Recently *AtARI12* in Arabidopsis was identified to be involved in UV-B signaling pathway [20].

Soybean (*Glycine max* [L.] Merr.) is widely grown as a major source of vegetable oil and protein. Soybean quality and yield are affected by various abiotic and biotic stresses. Soybean is also an important crop grown in South China, where acid soils comprise approximately 21% of the total land area [21], [22]. Aluminum (Al) toxicity is a major limiting factor of plant growth and crop production on acidic soils. There is large variation in Al tolerance among soybean varieties, and most of the Al tolerant varieties are from China [23]. Another study showed many Al tolerant varieties are from South China [24]. However, the genes underlying the Al tolerance in soybean remain largely unknown, except recently a soybean malate transporter gene *GmALMT1* which was shown to mediate root malate efflux which underlies soybean Al tolerance [25].

Increasing evidence indicates that RING-type E3 ubiquitin ligases play important roles in plant response to abiotic stresses. However, to date, there are no reports on the functions of soybean ARI proteins. Therefore, in this study, an ariadne-like E3 ubiquitin ligase gene *GmARI1* was cloned from soybean, and its gene expression patterns in different soybean tissues were studied. The transcriptional changes of *GmARI1* in response to various stress such as aluminum (Al) and plant hormone treatments were investigated using real-time quantitative PCR (qRT-PCR). We further characterized the *GmARI1* gene function by its subcellular location, in vitro ubiquitination assay, and performance of the transgenic Arabidopsis overexpressing *GmARI1* under Al stress. The possible mechanisms and signal pathways involved in soybean response to Al are also discussed.

Materials and Methods

Plant materials

Seeds of the soybean (*Glycine max* [L.] Merr.) cultivar Nannong 1138-2, provided by the National Center for Soybean Improvement (Nanjing, China), were germinated in sand under 25°C, 60% relative humidity (RH) and a photoperiod of 16 h/8 h (light/dark) cycle (light intensity was about 110 μmol photons. $m^{-2}s^{-1}$). Nannong 1138-2 is a released cultivar adapted to South China, which has good agronomic traits and moderate Al tolerance. The

soybean plants at VE stage (emergence) were transferred to the 'standard' nutrient solution [26], and grown for another ten days before various stress and hormone treatments. The nutrient solution was renewed every five days.

Isolation of the *GmARI1* gene from soybean

The full-length opening reading frame of the *GmARI1* gene was obtained by RT-PCR using soybean RNA. Total RNA was isolated using Trizol Reagent (Invitrogen, USA) according to the user's manual. 0.2 μg of the purified total RNA was used to synthesize first-strand cDNA by the MMLV-reverse transcriptase (TaKaRa). The primers: 5'-TCCCAATTCTTCTTCTGCCC-TAG-3' and 5'-GCAACCTTTCTTCCAAG CCTTAC -3' were designed to amplify the *GmARI1* gene located on Chromosome 11. The PCR products were cloned into the pGEM-T vector (Promega) and sequenced (Invitrogen). The sequencing results showed that two *ARI* genes, *GmARI1* and *GmARI2*, were isolated using above primers, due to their high similarity of 97%.

Sequence analysis

Protein domains were analyzed by the SMART (Simple Modular Architecture Research Tool) (http://smart.embl-heidelberg.de/) and Pfam (Protein families database of alignments and HMMs). The molecular mass, isoelectric point and secondary structure were predicted using ProtParam and SOPMA on the ExPASY(http://www.expasy.org/tools/). The BLASTP program at GenBank (http://www.ncbi.nlm.nih.gov/blast) was used to search the homologous sequences of GmARI1/GmARI2 from Non-Redundant (NR) database. Alignment was performed with ClustalW2 (http://www.ebi.ac.uk/Tools/msa/clustalw2/) and MUSCLE (http://www.ebi.ac.uk/Tools/msa/muscle/) using the default settings. The phylogenetic tree was constructed by the neighbor-joining algorithm (NJ) using MEGA version 5 with 1000 bootstraps.

Semi-quantitative RT-PCR

To study the tissue expression pattern of *GmARI1*, soybean roots, stems, leaves, and shoot apical meristem (SAM) were collected from 15-day-old plants, flowers and pods were collected from plants at 20 days after flowering (DAF). All tissues were frozen immediately in liquid nitrogen and stored at −80°C. The semi-quantitative RT-PCR assay was performed with 0.1 μg RNA as template for cDNA synthesis. Primers 5'-CTCCATTCTC-CATTCTCCTCCTTTGC-3' and 5'-GTCGTCGTCGCTG-TAGTAGT CC -3' were used for *GmARI1*. Primers:5'-ATCTCATTCCCTTCCCTCGTCTG-3' and 5'-CTGCCT-CTGTGAACTCCATCTCG -3' were used for *Tubulin-3* (GeneBank accession No. U12286) as the internal control. The PCR products were examined by electrophoresis in 2.0% agarose gel.

Transient expression of the GmARI1-GFP fusion protein

GmARI1-GFP was cloned into pMDC83 vector, with the expression driven by the cauliflower mosaic virus 35S promoter. The ORF of *GmARI1* was amplified by PCR using primers:5'-GGGGACAAGTTTGTACAAAAAA GCAGGCTTCCCAAT-TCTTCTTCTGCCCTAG-3' and 5'-GGGGACCACTTTG-

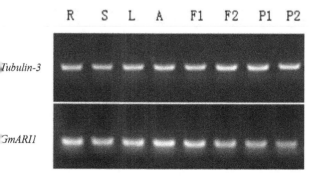

Figure 4. Tissue expression patterns of *GmARI1* in soybean. R: Roots; S: Stems; L: Leaves; A: Apex (SAMs); F1, F2: Flowers; P1, P2: Pods.

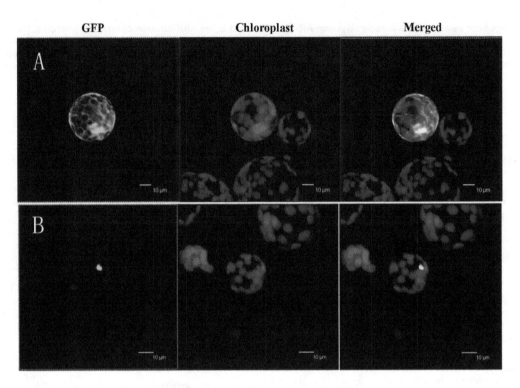

Figure 5. Sub-cellular localizations of GmARI1 protein. A: 35S-GFP; **B**: 35S-GmARI1-GFP.

TACA   AGAAAGCTGGGTCTCGACGTTGTTGATAGCA-CATCTG -3′, without the stop codon. The *35S-GmARI1-GFP* in-frame fusion construct and control vector of *35S-GFP* were introduced into the Arabidopsis protoplasts cells by PEG-mediated protocol [27], [28]. After culturing in dark at 23°C, the localization of GFP was observed with a confocal microscope the next day (Leica TCS SP2).

Stress treatments

The plants were put in the 'standard' nutrient solution [26] with one of the following chemicals for various treatments: Al stress (10 µM Al(NO$_3$)$_3$, pH 4.3), drought (osmotic) stress (400 mM mannitol), salt stress (200 mM NaCl), abscisic acid (ABA,100 µM), indoleacetic acid (IAA, 100 µM), jasmonic acid (JA, 100 µM), and salicylic acid (SA, 150 µM), with 'standard' nutrient solution (Mg^{2+} was withdrew for Al treatment) as control. The leaves and roots were harvested at 0, 0.5, 1, 2, 4, 6, 8, 12, 24, and 48 h after each treatment. Each sample was the mixture of three seedlings and each treatment was repeated three times. All samples were immediately frozen in liquid nitrogen and stored at −80°C for later use.

Real-time quantitative PCR

RNAs of different treatments were extracted using Trizol reagent (Invitrogen) and purified with RNase-free DNase I. The cDNA was synthesized from 0.2 µg RNA in a 10 µl reaction volume using PrimeScript R 1st Strand cDNA Synthesis kit (TaKaRa). Primers for *Tubulin-3* (GeneBank accession No. U12286) were 5′- TCATTCCCTTCCCTCGTCTGC-3′ and 5′-CCTCCTTGGTGCTCATCTTGC-3′. Primers for *GmARI1* were 5′-CGCTGGTTCCTGAATTTCCCTTG-3′ and 5′- GTCGTCGTCGCTGTAGTAGTCC-3′. Quantitative real-time PCR was performed with SYBR Green method on ABI 7500 Fast Real-Time PCP system. The following procedure was

used for qPCR: 95°C for 5 min; 40 cycles of 95°C for 3 sec; 60°C for 30 sec; 72°C for 30 sec. Data was analyzed using the 2$^{-\Delta\Delta CT}$ method as described by Livak and Schmittgen [29].

In vitro E3 ubiquitin ligase activity assay of GmARI1 protein

The full-length cDNA of GmARI1 with SalI/XhoI restriction enzyme sites was amplified by PCR, using primers ARI1-F: 5′-ACGCGTCGACATGGAGTCAGAGGATATGCAC-3′ and ARI1-R: 5′-CCGCTCGAGTCGACGTTGTTGATAGCA-CATCTG -3′. The fragment was cloned into the vector pET28a via the SalI/XhoI restriction sites, with 6×His tag fused to GmARI1 at the N-terminal. The expression construct (pET28a-His$_6$-GmARI1) was transformed into *E.coli* BL21 (DE3) cells to produce a recombinant His$_6$-GmARI1 fusion protein with an expected mass of about 66 kDa. The transformed cells harboring pET28a-His$_6$-GmARI1 were grown at 37°C with vigorous shaking until an OD600 of 0.4–0.6 is reached and induced with 0.1 mM isopropylthio-b-galactoside (IPTG) for 12 h at 16°C. The overexpressed His$_6$-GmARI1 was purified using Ni–NTA resin according to the supplier's instructions (GE life sciences). The protein concentration was determined as described by Bradford [30] using BSA as a standard.

For the autoubiquitination assay, each reaction (30 µl final volume) contained 10 µg of recombinant ubiquitin (Ub, Sigma), 0.1 µg rabbit E1 (Boston Biochemicals), 0.2 µg human E2 (UbcH5b, Boston Biochemicals), 2 mM ATP, 50 mM Tris–HCl (pH 7.5), 5 mM MgCl2, and 2 mM DTT contained 500 ng purified His6-GmARI1 [31]. After incubation at 30°C for 2–3 h, the reactions were stopped with sodium dodecyl sulfate-polyacrylamide gel electrophoresis (SDS-PAGE) loading buffer at 95°C for 5 min. The reaction samples were electrophoretically separated on 12% SDS-PAGE gels and transferred on two PVDF membrane separately. The two membranes were blocked and thereafter

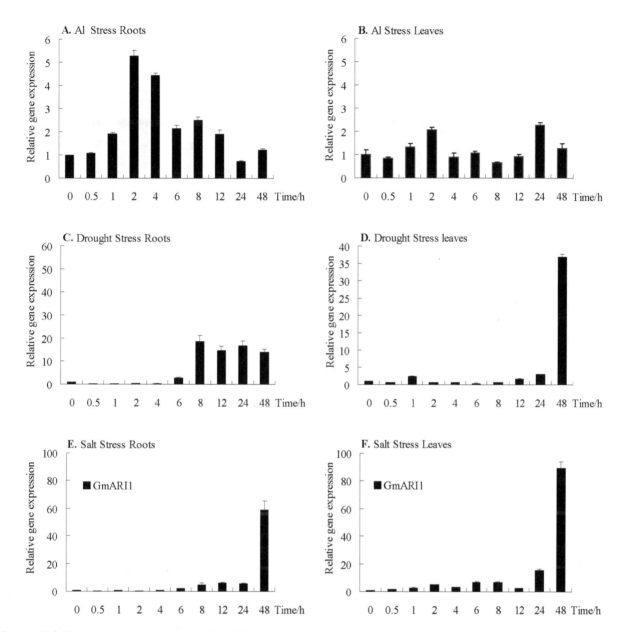

Figure 6. Relative gene expression levels of *GmARI1* in soybean under simulated stress treatments. A. relative gene expression in roots under Al stress (10 µM Al(NO$_3$)$_3$, pH4.3) **B**. relative gene expression in leaves under Al stress (10 µM Al(NO$_3$)$_3$, pH4.3) **C**. relative gene expression in roots under drought stress (400 mM mannitol) **D**. relative gene expression in leaves under drought stress (400 mM mannitol) **E**. relative gene expression in roots under salt stress (200 mM NaCl) **F**. relative gene expression in leaves under salt stress (200 mM NaCl). Error bars represent the standard error of three replicates.

blotted with an anti-ubiquitin monoclonal antibody (Santa Cruz Biotechnology) and an anti-His$_6$ monoclonal antibody (Sigma, USA) for 6 h at a 1:3000 dilution, respectively. After extensive washing, each of the bound primary antibody was detected with a horseradish peroxidase-conjugated goat anti-rabit IgG secondary antibody using the 3,3'-diaminobenzidine (DAB) development kit according to the manufacturer's protocol (Bio Basic Inc, Canada).

Generation of *GmARI1* transgenic Arabidopsis

The *GmARI1* gene was amplified by RT-PCR as described above and cloned into the plant expression vector pMDC83 under the control of CaMV 35S promoter by Gateway technology (Invitrogen), and the recombined plasmid was transferred into *A.*

tumefaciens strain EHA105. Arabidopsis plants (Col-0 ecotype) were transformed using the floral dip method [32]. Twenty transgenic lines of *GmARI1* were obtained. Eight T3 lines of the transgenic *GmARI1* Arabidopsis were examined by RT-PCR to select positive transgenic lines for further analyses.

Al-tolerance of the *GmARI1* transgenic Arabidopsis

Homozygous T$_3$ seeds of the transgenic lines and wild type plants were used for Al-tolerance analysis. Seeds were surfaced-sterilized as described before [33] and germinated on 1/2 MS medium for 7 days. The seedlings were transferred to 1/2 MS without Mg^{2+} but with 0 or 15 µM Al(NO$_3$)$_3$ and 8% Agar, pH 4.3, and then put on the medium with the plates placed

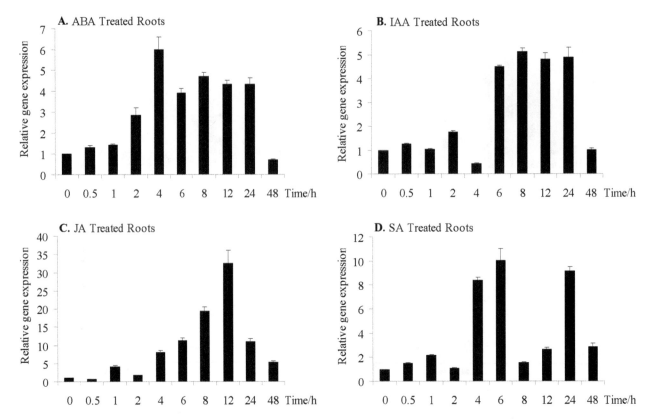

Figure 7. Relative gene expression levels of *GmARI1* in the roots of soybean after exogenous application of plant hormones. A. 100 μM ABA treatment **B**. 100 μM IAA treatment **C**. 100 μM JA treatment **D**. 150 μM SA treatment. Error bars represent the standard error of three replicates.

vertically. After 15 days, the lengths of the roots were measured, and the Relative Root Growth (RRG) in each line was calculated as: RRG (%) = $(RL_{Alt}- RL_{Al0})/(RL_{ct}- RL_{c0})$, in which RL_{Al0} represent the root length before Al treatment, RL_{Alt} represent the root length after 15 days of Al treatment (15 μM $Al(NO_3)_3$, pH 4.3), RL_{c0} represent the root length before growth on the control medium, and RL_{ct} represent the root length after growth on the control medium (0 μM $Al(NO_3)_3$, pH 4.3). The statistical analysis of the experimental data was conducted by t-test with SPSS [34].

Results

Isolation and sequence analysis of *GmARI1* gene

The cDNA of *GmARI1* gene is 2043 bp in length, containing an open reading frame (ORF) of 1761 bp, a 112 bp 5′-untranslated region (UTR), and a 170 bp 3′-UTR. Its homologous gene, *GmARI2*, shares 97% of similarity in the nucleotide sequences of the ORF region. The *GmARI1* and *GmARI2* nucleotide sequence and the predicted amino acid sequence have been deposited in GenBank (accession number JX392390 and JX392391). The genomic sequence of *GmARI1* and *GmARI2* variants in cultivar Williams 82 have 15 exons and 14 introns (http://www.phytozome.net/soybean.php), which located on chromosome 11 and 12, respectively.

The deduced protein of *GmARI1* comprises 586 amino acids with the predicted molecular mass of 66.99 kDa and isoelectric point of 5.37. GmARI1 protein has a RBR domain, which contains an IBR (C6HC) domain flanked by RING1 and RING2 (C3HC4) (Fig. 1). The secondary structure of GmARI1 protein is predicted to be composed of 45.56% alpha helix, 11.09% extended strand, 41.64% random coil, and 1.71% beta turn.

In order to determine the relationship of GmARI1 and other RBR proteins, Blastp on Uniprot was used to search the homologues proteins of GmARI1. The top 87 amino acid sequences with RBR conserved domain were selected from different plant species, including *Arabidopsis thaliana*, *Medicago truncatula*, *Ricinus communis*, and *Zea mays L.* Multiple sequence alignment of these 87 amino acid sequences was performed. In addition to the ARI proteins from soybean which showed high similarity with *GmARI1*, the protein from *M. truncatula* showed 80% similarity with *GmARI1* (Fig. 2) The phylogenetic tree was drawn using MEGA 5.0 program based on Neighbor-Joining (NJ) with 1000 bootstrap replications (Fig. 3), which also showed ARI proteins from *G. max* was closely related to the protein from *M. truncatula*.

Tissue expression pattern of *GmARI1* and the subcellular localization of its protein

GmARI1 gene was expressed in roots, stems, leaves, SAMs, flowers, and pods (Fig. 4). We determined the subcellular localization of the GmARI1 protein by transient expression of *35S-GmARI1-GFP* in Arabidopsis protoplasts. GmARI1 was located in the nucleus, while the GFP control was mainly located in the cytoplasm (Fig. 5).

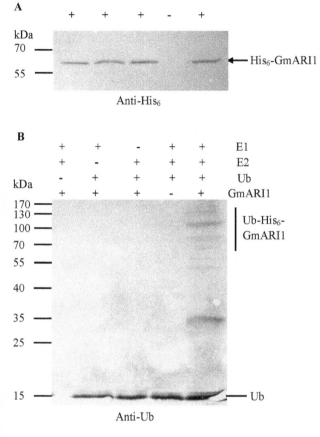

Figure 8. Western blot analysis of GmARI1 protein. A. Western blot using anti-His6 monoclonal antibody in the presence (+) or absence (−) of the purified His6 tag-GmARI1 proteins. **B.** In vitro E3 ubiquitin ligase activity assay of GmARI1 protein. His6 tag-GmARI1 fusion proteins were assayed for E3 activity in the presence of rabbit E1, human E2 (UbcH5b), and ubiquitin (Ub). The left numbers denote the molecular masses of marker proteins in kilodaltons.

Expression pattern of *GmARI1* under stress and plant hormone treatments

Real-time quantitative PCR was carried out to examine the expression pattern of *GmARI1* gene in soybean under different treatments including various abiotic stresses and plant hormones (Fig. 6 and 7). Under 10 µM Al (pH4.3) stress, the transcripts of *GmARI1* in soybean roots increased and peaked during 2 h to 4 h, but the transcripts in leaves did not change significantly (Fig. 6A, B). When treated with 400 mM mannitol (osmotic stress), the *GmARI1* gene expression was induced from 6 h to 48 h in roots, while showed delayed induction in leaves (Fig. 6C, D). Under 200 mM NaCl stress, the *GmARI1* gene expression showed a high induction after 24 h in both roots and leaves (Fig. 6E, F).

The transcriptional changes of *GmARI1* in response to ABA, IAA, JA and SA were also investigated in soybean roots. The transcripts of *GmARI1* were up-regulated during 2-24 h after ABA treatment (Fig. 7A). *GmARI1* was induced after 6 h IAA treatment and its transcript abundance was maintained at a higher level until 24 h (Fig.7B). The expression of *GmARI1* showed continual increase from 4 h and peaked around 12 h after JA treatment (Fig. 7C). Increased transcripts of *GmARI1* were detected at 4 h, 6 h and 24 h after SA treatment (Fig. 7D).

E3 ubiquitin ligase activity of GmARI1 protein

To test if GmARI1 has the E3 ubiquitin ligase activity, a full-length GmARI1 protein with maltose binding protein (6× His tag) was expressed in *E. coli* and subsequently affinity-purified (His6-GmARI1) from the soluble fraction (Fig. S1). The purified recombinant His6-GmARI1 protein was about 66 kDa as expected, and the western blotting with anti-His6 monoclonal antibody also showed the purified target recombinant protein had the right molecular weight of 66 kDa (Fig. 8A). In vitro self-ubiquitination assays were performed in the presence of rabbit E1, human E2 (UbcH5b), and Ub (Fig. 8B). Polyubiquitination was detected only in the presence of E1, E2, Ub and His6-GmARI1. A negative result was observed if either E1, E2, Ub or His6-GmARI1 was omitted in the reaction. These results indicate that GmARI1 has E3 ubiquitin ligase activity.

Performance of transgenic plants over-expressing *GmARI1* under Al stress

The gene expression of *GmARI1* was induced early by Al stress, therefore we further studied its function in transgenic plants. An expression plasmid vector of *pMDC83-GmARI1* was constructed and introduced into Arabidopsis plants using floral dip method. Transgenic T_3 Arabidopsis over-expressing *GmARI1* were generated and the positive transgenic lines were identified by RT-PCR (Fig. S2). Seeds of three T_3 homozygous transgenic lines (*GmARI1-1*, *GmARI1-2*, and *GmARI1-3*) and wild type Col-0 were germinated on 1/2 MS medium. After 10 days, the seedlings were transferred to 1/2 MS (pH4.3) containing 15 µM Al or 0 µM Al as a control. Fifteen days later, the root growth of the wild type Col-0 was severely inhibited by 15 µM Al as compared with the control medium (0 µM), while the transgenic lines were little affected by Al (Fig. 9 A). The relative root growth (RRG) of the transgenic lines was significantly ($p<0.01$) longer than the wild type under Al treatment (Fig. 9B). The relative abundance of *GmARI1* in the transgenic lines of *GmARI1-1* and *GmARI1-2* is higher than *GmARI1-3*, which coincided with RRG result (Fig. S3 and Fig. 9B).

Discussion

RING-type E3 ubiquitin ligases play important roles in plant responses to abiotic stresses [7], [8], [35]. RBR subclass of RING-containing E3 ligases were recently shown as an important group of proteins since the discovery of *parkin*, a mutation causing the familial autosomal-recessive juvenile parkinsonism (AR-JP) [36]. However, there is little research on the plant RBR family. The RBR family was classified into 14 subfamilies, including Ariadne (ARI), ARA54, Dorfin, parkin, PlantI, PlantII, and XAP3 [9], but the function of ARI class is largely unknown. In this study, *GmARI1* gene was isolated and characterized from soybean. Based on the analysis of predicted protein domains, *GmARI1* gene belongs to the Ariadne subfamily, characterized by the presence of an N-terminal acid-rich cluster, followed by a RBR domain and a Coiled coil region at the C-terminus [17]. The origin of Ariadne can be traced back earlier in time, as it was found not only in animal, fungal, and plant, but also in some protist [37]. The blastp result indicated among the RBR domain containing protein from other species, *Medicago truncatula* had the highest similarity (80%) with *GmARI1* and other ARI proteins from soybean.

GmARI1 gene was expressed ubiquitously in roots, stems, SAMs, leaves, flowers, and pods of soybean plants. The GmARI1 protein was located in the nucleus of the cell. Real-time quantitative PCR showed that the expression level of *GmARI1* in soybean root under Al (10 µM, pH 4.3) treatment reached the

A

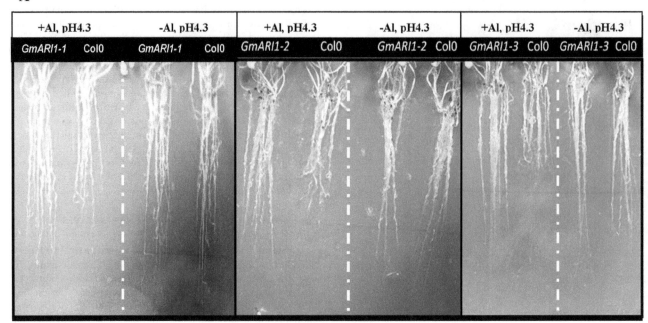

B

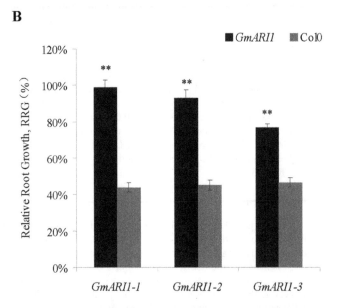

Figure 9. Performance of the transgenic Arabidopsis under 15 μM Al treatment. A. Root phenotypes of the *35S: GmARI1* overexpression lines *GmARI1-3* and the wild type Col-0 grown on 15 μM Al in 1/2 MS (+Al, pH4.3), and 0 μM Al in 1/2 MS (-Al, pH4.3). **B**. Relative root growth (RRG, %) of the transgenic Arabidopsis lines (*GmARI1-1, GmARI1-2, GmARI1-3*) and the wild type (Col-0). RRG was calculated by the root growth length under Al treatment (15 μM Al in 1/2 MS, pH4.3) divided by the root growth length under control (0 μM Al in 1/2 MS, pH4.3). Error bars represent the standard error (SE), ** indicate the significance level of 0.01 by t-tests.

peak during 2 to 4 hours, but no significant change was detected in the leaves. This coincides with the fact that Al toxicity was first and mainly acting on plant roots [38], limiting water and nutrition absorption [39], which further inhibiting the development of whole plant and reducing yield [40], [41]. Therefore in the early stage (less than 4 hours) of Al stress, induced gene expression of *GmARI1* in soybean roots may play important roles to trigger downstream signaling pathways to protect root cells from Al toxicity.

It has been reported that Al induces oxidative stress and DNA damage in plant cells [42–44]. Reactive oxygen species (ROS)

constantly attack DNA, leading to oxidative DNA damage [45]. The cell cycle checkpoint regulators could detect and respond to such damage, leading to inhibition of root growth [46]. Protein ubiquitination is involved in DNA transcription and repair, cell cycle and division [1], and is emerging as a critical regulatory mechanism of DNA damage response [47]. Several RING domain-containing E3 ubiquitin ligases play an essential role in response to DNA damage [48]. In vitro ubiquitination assay showed GmARI1 has E3 ligase activity (Fig. 8), therefore we hypothesize that GmARI might be involved in the oxidative DNA damage repair to confer Al tolerance. We investigated the co expression pattern of *GmARI* (http://bioinformatics.cau.edu.cn/SFGD/), and found Glyma02g15070 was on the top list of coexpression genes with *GmARI*. The homolog gene of Glyma02g15070 in arabidopsis is AT1G49670, which was shown to be involved in oxidative stress tolerance (http://www.arabidopsis.org/). Suppression of oxidative stress might help plants reduce the damage or root growth inhibition [49], [50]. Therefore, the Al tolerance observed in *GmARI* overexpression lines might be due to the improved tolerance to oxidative stress, or/and other signaling cascades.

The activities of cell wall-bound peroxidases in the annual legume, *Cassia tora*, significantly increased with Al concentrations, and were regulated by JA [51]. Another study showed that the SA-signaling and SA-dependent expression of a respiratory burst oxidase homolog gene is involved in Al responsive oxidative burst in Arabidopsis [52]. Here in this study, the expression of *GmARI1* was induced by Al stress, as well as JA and SA treatments in soybean roots. These suggested *GmARI1* might mediate soybean response to Al through oxidative species signals, which may overlap with plant hormone signaling pathways. The T_3 transgenic Arabidopsis plants over-expressing the *GmARI1* gene showed significant improvement in Al tolerance comparing with wild type plants, which further support the important role of *GmARI1* gene in plant response to Al stress.

Supporting Information

Figure S1   Expression and purification of the recombinant GmARI1 proteins. The recombinant His6-GmARI1 proteins were expressed in *E.coli* BL21 (DE3) and analyzed by SDS–PAGE. Lane 1, total proteins from *E. coli* cells before IPTG induction; lane 2, total proteins containing pET28a-GmARI1 from *E. coli* cells after induction by IPTG; lane 3, purified recombinant His6-GmARI1 protein.

Figure S2   RT-PCR confirmation of the transgenic Arabidopsis T3 lines GmARI 1 to 8. (−): Arabidopsis wild ecotype Col-0; (+): plasmid pMDC83-GmARI1.

Figure S3   Expression of the *GmARI1* gene in 2-week-old Arabidopsis plants quantified by qRT-PCR using actin (*AtACT2*) as the reference gene. The Arabidopsis plants were germinated and grown on 1/2 MS medium (pH5.8) for two weeks and then transferred to 1/2 MS medium with 25 μM AlCl₃ (pH4.3). Two hours later, tissues were sampled from the wild type Arabidopsis Col0 and homozygous transgenic lines separately (each sample was the mixture of four plants). Error bars are the standard errors from three replications.

Acknowledgments

We thank the staff at JiangPu Experimental Station for their assistance with the soybean seed production, and PLoS One editors and reviewers for their valuable comments and suggestions.

Author Contributions

Conceived and designed the experiments: JG YL XZ NW PC YW. Performed the experiments: XZ NW MG JL. Analyzed the data: XZ NW MG PC YL JG. Contributed reagents/materials/analysis tools: JG TZ YL. Wrote the paper: XZ NW YL JG.

References

1. Hershko A, Ciechanover A (1998) The ubiquitin system. Annu Rev Biochem 67: 425–479.
2. Pickart CM (2001) Mechanisms underlying ubiquitination. Annu Rev Biochem 70: 503–533.
3. Stone SL, Hauksdottir H, Troy A, Herschleb J, Kraft E, et al. (2005) Functional analysis of the RING-type ubiquitin ligase family Arabidopsis. Plant Physiol 137: 13–30.
4. Moon J, Parry G, Estelle M (2004) The ubiquitin-proteasome pathway and plant development. Plant Cell 16: 3181–3195.
5. Barlow PN, Luisi B, Milner A, Elliott M, Elliott R (1994) Structure of the C3H4 domain by 1H-nuclear magnetic resonance spectroscopy: a new structural class of zinc-finger. J Mol Biol 237: 201–211.
6. Lee DH, Choi HW, Hwang BK (2011) The Pepper E3 ubiquitin ligase RING gene CaRING1 is required for cell death and the Salicylic Acid-dependent defense response. Plant Physiol 156: 2011–2025.
7. Cho SK, Ryu MY, Seo DH, Kang BG, Kim WT (2011) The Arabidopsis RING E3 ubiquitin ligase AtAIRP2 plays combinatory roles with AtAIRP1 in Abscisic Acid-mediated drought stress responses. Plant Physiol 157: 2240–2257.
8. Du QL, Cui WZ, Zhang CH, Yu DY (2010) GmRFP1 encodes a previously unknown RING-type E3 ubiquitin ligase in soybean (Glycine max). Mol Biol Rep 37: 685–693.
9. Marin I, Ferrus A (2002) Comparative genomics of the RBR family, including the Parkinson's disease-related gene parkin and the genes to the Ariadne subfamily. Mol Biol Evol 19: 2039–2050.
10. Eisenhaber B, Chumak N, Eisenhaber F, Hauser MT (2007) The ring between ring fingers (RBR) protein family. Genome Biol 8: 209.
11. Aguilera M, Oliveros M, Matinez-Padron M, Barbas JA, Ferrus A (2000) Ariadne-1: A vital drosophila gene is required in development and defines a new conserved family of RING-finger proteins. Genetics 155: 1231–1244.
12. Itier JM, Ibanez P, Mena MA, Abbas N, Cohen-Salmon C, et al. (2003) Parkin gene inactivation alters behaviour and dopamine neuro-transmission in the mouse. Hum Mol Genet 12: 2277–2291.
13. Bromann PA, Weiner JA, Apel ED, Lewis RM, Sanes JR (2004) A putative ariadne-like E3 ubiquitin ligase (PAUL) that interacts with the muscle-specific kinase(MuSK). Gene Expr Patterns 4: 77–84.
14. Moynihan TP, Ardley HC, Nuber U, Rose SA, Jones PF, et al. (1999) The ubiquitin-conjugating enzymes UbcH7 and UbcH8 interact with RING finger/IBR motif-containing domains of HHARI and H7-AP1. J Biol Chem 274(43): 30963–30968.
15. Ardley HC, Tan NG, Rose SA, Markham AF, Robinson PA (2001) Features of the parkin/ariadne-like ubiquitin ligase, HHARI, that regulate its interaction with the ubiquitin-conjugating enzyme, Ubch7. J Biol Chem 276(22): 19640–19647.
16. Marteijn JA, van Emst L, Erpelinck-Verschueren CA, Nikoloski G, Menke A, et al. (2005) The E3 ubiquitin-protein ligase Triad1 inhibits clonogenic growth of primary myeloid progenitor cells. Blood 106: 4114–4123.
17. Mladek C, Guger K, Hauser MT (2003) Identification and characterization of the ARIADNE gene family in Arabidopsis. A group of puptative E3 ligase. Plant Physiol 131: 27–40.
18. Marin I (2010) Diversification and specialization of plant RBR ubiquitin ligase. PLoS ONE 5(7): e11579.
19. Wenzel DM, Klevit RE (2012) Following Ariadne's thread: a new perspective on RBR ubiquitin ligases. BMC Biol 10: 24.
20. Lang-Mladek C, Xie L, Nigam N, Chumak N, Binkert M, et al. (2012) UV-B signaling pathways and fluence rate dependent transcriptional regulation of ARIADNE12. Physiol Plant 145: 527–539.
21. Huang BQ, Bai JH, Xue XQ (2001) Advances in studies on aluminum toxicity and tolerance in plants. Chinese Bulletin of Botany 18(4): 385–395.
22. Liu Q, Zheng SJ, Lin XY (2004) Plant physiological and molecular biological mechanism in response to aluminium toxicity. Chinese Journal of Applied Ecology 15: 1641–1649.
23. Sapra VT, Mebrahtu T, Mugwira LM (1982) Soybean germplasm and cultivar aluminum tolerance in nutrient solution and bladen clay loam soil. Agron J 74: 687–690.
24. Nian H, Huang H, Yan XL, Lu YG (1998) Studies on adaptability of soybean to acidic and aluminum toxicity I Study on the screening and identification for acidic and aluminum-resistance soybeans. Soybean Science 16(3): 191–196.
25. Liang CY, Pineros M, Tian J, Yao ZF, Sun LL, et al. (2013) Low pH, aluminum and phosphorus coordinately regulate malate exudation through GmALMT1 to improve soybean adaptation to acid soils. Plant Physiol 161: 1347–61.
26. Tocquin P, Corbesier L, Havelange A, Pieltain A, Kurtem E, et al. (2003) A novel high efficiency, low maintenance, hydroponic system for synchronous growth and flowering of Arabidopsis thaliana. BMC Plant Biol 3: 2.
27. Yoo SD, Cho YH, Sheen J (2007) Arabidopsis mesophyll protoplasts: a versatile cell system for transient gene expression analysis. Nat Protoc 2: 1565–1572.
28. Wu FH, Shen SC, Lee LY, Lee SH, Chan MT, et al. (2009) Tape-Arabidopsis Sandwich - a simpler Arabidopsis protoplast isolation method. Plant Methods 5: 16.

29. Livak KJ, Schmittgen TD (2001) Analysis of Relative Gene Expression Data Using Real- Time Quantitative PCR and the $2^{-\Delta\Delta CT}$ method. Methods 25: 402–408.

30. Bradford MM (1976) A rapid and sensitive method for the quantification of microgram quantities of protein utilizing the principal of protein-dye binding. Anal. Biochem 72: 248–254.

31. Zhang Y, Yang C, Li Y, Zheng N, Chen H, et al. (2007) SDIR1 is a RING finger E3 ligase that positively regulates stress-responsive abscisic acid signaling in Arabidopsis. Plant Cell 19(6): 1912–1929.

32. Clough SJ, Bent AF (1998) Floral dip: a simplified method for Agrobacterium-mediated transformation of Arabidopsis thaliana. Plant J 16: 735–743.

33. Ardie SW, Xie L, Takahashi R, Liu S, Tahano T (2009) Cloning of a high-affinity K+ transporter gene PutHKT2; 1 from Puccinellia tenuiflora and its functional comparison with OsHKT2; 1 from rice in yeast and Arabidopsis. J Exp Bot 60: 3491–3502.

34. Norusis MJ (1990) SPSS advanced statistics user's guide: SPSS Chicago.

35. Cheng MC, Hsieh EJ, Chen JH, Chen HY, Lin TP (2012) Arabidopsis RGLG2, Functioning as a RING E3 Ligase, Interacts with AtERF53 and Negatively Regulates the Plant Drought Stress Response. Plant Physiol 158: 363–375.

36. Beasley SA, Hristova VA, Shaw GS (2007) Structure of the Parkin in-between-ring domain provides insights for E3-ligase dysfunction in autosomal recessive Parkinson's disease. Proc Natl Acad Sci USA 104: 3095–3100.

37. Marin I, Lucas JI, Gradilla AC, Ferrus A (2004) Parkin and relatives: the RBR family of ubiquitin ligases. Physiol Genomics 17: 253–263.

38. Taylor GJ (1988) The physiology of aluminum tolerance in higher plants. Commun. Soil Sci. Plant Anal 19: 1179–1194.

39. Barcelo J, Poschenrieder C (2002) Fast root growth responses, root exudates, and internal detoxification as clues to the mechanisms of aluminium toxicity and resistance: A review. Environ Exp Bot 48: 75–92.

40. Foy CD (1996) Tolerance of barley cultivars to an acid, aluminum-toxic subsoil related to mineral element concentrations in their shoots. J Plant Nutr 19: 1361–1380.

41. Ma JF, Furukawa J (2003) Recent progress in the research of external Al detoxification in higher plants: a minireview. J Inorg Biochem 97: 46–51.

42. Achary VM, Jena S, Panda KK, Panda BB (2008) Aluminium induced oxidative stress and DNA damage in root cells of Allium cepa L. Ecotoxicol. Environ. Saf 70: 300–310.

43. Achary VM, Panda BB (2010) Aluminium-induced DNA-damage and adaptive response to genotoxic stress in plant cells are mediated through reactive oxygen intermediates. Mutagenesis 25: 201–209.

44. Yamamoto Y, Kobayashi Y, Devi SR, Rikiishi S, Matsumot H (2002) Aluminum toxicity is associated with mitochondrial dysfunction and the production of reactive oxygen species in plant cells. Plant Physiol 128: 63–72.

45. Markkanen E, Hübscher U, van Loon B (2012) Regulation of oxidative DNA damage repair: the adenine: 8-oxo-guanine problem. Cell Cycle 11: 1070–1075.

46. Nezames CD, Sjogren CA, Barajas JF, Larsen PB (2012) The Arabidopsis cell cycle checkpoint regulators TANMEI/ALT2 and ATR mediate the active process of aluminum-dependent root growth inhibition. Plant Cell 24(2): 608–21.

47. Pinder JB, Attwood KM, Dellaire G (2013) Reading, writing, and repair: the role of ubiquitin and the ubiquitin-like proteins in DNA damage signaling and repair. Front Genet 4: 45.

48. Bartocci C, Denchi EL (2013) Put a RING on it: regulation and inhibition of RNF8 and RNF168 RING finger E3 ligases at DNA damage sites. Front Genet 4: 128.

49. Panda SK, Sahoo L, Katsuhara M, Matsumoto H (2013) Overexpression of alternative oxidase gene confers aluminum tolerance by altering the respiratory capacity and the response to oxidative stress in Tobacco Cells. Mol Biotechnol 54: 551–63.

50. Yin L, Mano J, Wang S, Tsuji W, Tanaka K (2010) The involvement of lipid peroxide-derived aldehydes in aluminum toxicity of tobacco roots. Plant Physiol 152: 1406–17.

51. Xue YJ, Tao L, Yang ZM (2008) Aluminum-induced cell wall peroxidase activity and lignin synthesis are differentially regulated by jasmonate and nitric oxide. J Agric Food Chem 56 (20): 9676–9684.

52. Kunihiro S, Hiramatsu T, Kawano T (2011) Involvement of salicylic acid signal transduction in aluminum-responsive oxidative burst in Arabidopsis thaliana cell suspension culture. Plant Signal Behav 6: 611–616.

11

Production of Phytotoxic Cationic α-Helical Antimicrobial Peptides in Plant Cells Using Inducible Promoters

Nuri Company, Anna Nadal, Cristina Ruiz, Maria Pla*

Institute for Food and Agricultural Technology, University of Girona, Girona, Spain

Abstract

Synthetic linear antimicrobial peptides with cationic α-helical structures, such as BP100, have potent and specific activities against economically important plant pathogenic bacteria. They are also recognized as valuable therapeutics and preservatives. However, highly active BP100 derivatives are often phytotoxic when expressed at high levels as recombinant peptides in plants. Here we demonstrate that production of recombinant phytotoxic peptides in transgenic plants is possible by strictly limiting transgene expression to certain tissues and conditions, and specifically that minimization of this expression during transformation and regeneration of transgenic plants is essential to obtain viable plant biofactories. On the basis of whole-genome transcriptomic data available online, we identified the *Os.hsp82* promoter that fulfilled this requirement and was highly induced in response to heat shock. Using this strategy, we generated transgenic rice lines producing moderate yields of severely phytotoxic BP100 derivatives on exposure to high temperature. In addition, a threshold for gene expression in selected tissues and stages was experimentally established, below which the corresponding promoters should be suitable for driving the expression of recombinant phytotoxic proteins in genetically modified plants. In view of the growing transcriptomics data available, this approach is of interest to assist promoter selection for specific purposes.

Editor: Zhengguang Zhang, Nanjing Agricultural University, China

Funding: This work was financially supported by the Spanish Ministerio de Economía y Competitividad (reference AGL2010-17181/AGR). The research group belongs to 2014SGR697, recognized by the Catalonian Government. N.C. and C.R. received fellowships from Generalitat de Catalunya. The funders had no role in study design, data collection and analysis, decision to publish, or preparation of the manuscript.

Competing Interests: The authors have declared that no competing interests exist.

* Email: maria.pla@udg.edu

Introduction

Antimicrobial peptides (AMPs) are key components of innate immunity in plants and animals, and are also produced by microbes in antibiosis processes. A significant proportion are strongly cationic and have linear structures that adopt an amphipathic α-helical conformation that binds to the phospholipid membranes of target microbes before the hydrophobic face is inserted into the membrane bilayer. This unique mode of action explains the lack of resistance in target pathogens, and makes AMPs valuable novel therapeutic agents against bacteria, fungi, viruses, parasites and tumor cells. Improved or synthetic AMPs have been designed with increased potency against selected pathogens [1–5]. As an example, the synthetic undecapeptide BP100 (KKLFKKILKYL-NH$_2$) is effective against *Xanthomonas vesicatoria* in pepper, *Erwinia amylovora* in apple and *Pseudomonas syringae* in pear [3] with the same efficacy as standard antibiotics, while being biocompatible, as determined by acute oral toxicity tests in mice [6]. Plant expression of recombinant cationic α-helical peptides such as BP100 is preferable for industrial and phytosanitary applications. We have recently showed that active BP100-derived peptides can be expressed as recombinant peptides in plants [7,8], demonstrated by the increased resistance of GM plants to some rice pathogens [7] and by *in vitro* growth inhibition assays [8]. These recombinant BP100-derived peptides include

endoplasmic reticulum (ER) retention motifs to minimize toxicity to the host plant. This did not affect the antimicrobial activity of the products *in vitro*, demonstrated using bacterial growth inhibition tests. Following transient expression in *Nicotiana benthamiana* and stable expression in *Arabidopsis thaliana* seedlings, the peptides accumulated in large ER-derived vesicles, along with typical ER luminal proteins [8]. However, the authors found that the recombinant peptides were often toxic to the host during later developmental stages. Similarly, most transgenic *Oryza sativa* plants constitutively expressing recombinant ER-targeted BP100-derived peptides failed to achieve maturity, with only a few peptides coupling potent antimicrobial and low hemolytic activity accumulating in transgenic rice lines, with a yield of up to 0.5% total soluble protein (TSP) [8]. Not many cationic α-helical peptides can be expressed in transgenic plants following this strategy, although they have potent activities against other types of pathogenic cells making them valuable as novel therapeutics and preservatives.

High temperature stress is one of the most common abiotic stresses among many world crops. Plants have evolved various physiological and molecular mechanisms to resist heat stress. Based on the expression data from different plant species, it has been estimated that high temperatures affect approximately 2% of the plant genome (review in [9]). Exhaustive identification of heat stress-responsive genes has been carried out by means of transcriptomics [10–15]. Two groups of genes have been found, signaling components (e.g. protein kinases and transcription factors) and functional genes such as heat shock proteins (Hsps) [16], although many heat shock genes still have unknown functions. Hsps are functionally linked to molecular chaperones that are essential for maintenance and restoration of protein homeostasis. Protein denaturation occurring during stress triggers high transcription of *hsp* genes by the binding of active heat shock factors (Hsfs) to heat shock elements. Alterations in expression of a high number of genes in response to stress occur in complex systems. Cross-talk seems to exist between the regulatory pathways in response to different abiotic stresses such as temperature, and osmotic or oxidative, and biotic stresses [17]. On fusing several heat-shock promoters to reporter genes, they have been found to be regulated during various stress situations, have organ and developmental stage-specific basal expression levels and are induced by stress [18].

Heat-shock induced genes can be highly up-regulated at the transcriptional level by exposure to temperatures above those of normal plant growth. Therefore, induction can be easily carried out, is inexpensive and does not require the use of hormones or chemicals. Here we explored the use of heat-shock inducible promoters with minimal basal activity during normal plant development, to allow expression of recombinant phytotoxic peptides in transgenic rice. Rice has emerged as a powerful platform for large-scale production of recombinant proteins; it has easy cropping conditions and is a self-pollinating crop [19,20]. It is a most suitable crop for genetic manipulation due to its small genome size and the well-established gene transfer technology. Moreover, the availability of the rice complete genome sequence, the exponential growth of profiling studies reporting gene expression data for rice, and the development of databases and bioinformatics tools, provide a possible way to identify genes related to heat stress responses, and with specific expression patterns.

Materials and Methods

In silico identification of candidate heat shock promoters

Miamexpress-EMBL-EBI (http://www.ebi.ac.uk/miamexpress/) and GEO-NCBI (http://nlmcatalog.nlm.nih.gov/geo/) were used to identify published microarray experiments questioning the late response of rice seedlings to heat shock stress. In the experiment with accession number GSE14275, the authors [10] state that rice Affymetrix microarrays were hybridized with RNA extracted from 14-day-old rice seedlings grown in a growth chamber with a daily photoperiodic cycle of 14 h light and 10 h dark, at between 28–30°C, with or without being subjected to 42 °C for 3 h. The data from this experiment was extracted using the RMA software [21], which includes background adjustment, quantile normalization and summarization, and used to identify probes for subsequent analysis.

Rice gene expression data in different organ and developmental stages, and in response to various stress conditions, was obtained from the CREP database (Collection of Rice Expression Profiles; http://crep.ncpgr.cn/crep-cgi/home.pl) and the microarray hybridization results available at the Miamexpress-EMBL-EBI and GEO-NCBI websites.

The NetAffx-analysis center (www.affymetrix.com) allowed identification of the gene associated to every probe, and the sequence was retrieved from GenBank (http://www.ncbi.nlm.nih.gov/genbank). Likely promoter sequences were identified using the PlantProm DB, an annotated and non-redundant database of proximal promoter sequences [22,23].

Primer design and amplification of rice promoter sequences

Primer blast (http://www.ncbi.nlm.nih.gov/tools/primer-blast/) was used to design primer pairs specifically targeting the identified promoter sequences, blocking the reverse primer at the −1 position (immediately upstream of the ATG translation start codon). Two primer pairs (Table S1) were designed to amplify 553 and 1016 bp fragments of the pHsp18 and pHsp82 promoters. They each included an enzyme restriction site (*Kpn*I and *Spe*I at the distal and proximal promoter end, respectively) to facilitate the subsequent cloning steps.

Oryza sativa L. ssp. *japonica*, v. Senia was grown under controlled conditions at $28 \pm 1°C$ and with a 16 h light / 8 h dark photoperiod with fluorescent Sylvania Cool White lamps. Genomic DNA was extracted from 1 g of leaf samples, obtained from young plants, using the commercial NucleoSpin Plant II kit (Macherey-Nagel, Düren, Germany) according to the manufacturer's instructions.

PCR was in a final volume of 50 μl 1x reaction buffer with the appropriate concentrations of $MgCl_2$ and primers (Fisher Scientific SL, Madrid, Spain) (Table S1), 200 μM dNTPs and 1 unit Expand High Fidelity DNA polymerase (Roche Diagnostics GmbH, Mannheim, Germany). The reaction conditions were as follows: 3 min at 94°C; 10 cycles of 15 s at 94°C, 30 s at the appropriate annealing temperature (Table S1) and 1 min at 72°C; 20 cycles of 15 s at 94°C, 30 s at the same annealing temperature and 1 min, plus an additional 5 s for each successive cycle, at 72°C; and a final extension of 10 min at 72°C.

Construction of plant transformation vectors

The pHsp18 and pHsp82 promoters were separately subcloned into the *Kpn*I and *Spe*I sites of a pBluescriptIIKS + derived vector having a polylinker fragment with the *Kpn*I, *Spe*I and *Bam*HI restriction sites, followed by the *A. tumefaciens* nopaline synthase *nos* terminator sequence. The sequences encoding BP100.2 [7], BP100-DsRed-tag54 and DsRed-tag54 [8] were subcloned into these plasmids using the *Spe*I and *Bam*HI restriction sites.

For the BP100.2 plasmids, the sequence encoding the *N. tabacum* pathogenesis related protein PR1a signal peptide [24] fused to BP100.2 [7] was PCR amplified using pAHC17-bp100.2 [7] as template and the primers PR1a_Spe and BP100KDEL_Bam, with the *Spe*I and *Bam*HI restriction sequences respectively. For DsRed-tag54 control plasmids, the sequence encoding the *Petrosilinum hortense* chalcone synthase 5′ untranslated region plus the codon-optimized leader peptide derived from the heavy chain of the murine mAb24 monoclonal antibody [25] fused to DsRed [26], the epitope tag54 [27] and the KDEL ER retention motif, were PCR amplified using pCdsred-tag54 [8] as template and the primers CHS_Spe and TagKDEL_Bam, that included the required restriction sites. As an additional control, a BP100-DsRed-tag54 plasmid was generated similarly to DsRed-tag54, and including the BP100 antimicrobial sequence placed N-terminal to DsRed-tag54. The coding sequence was obtained by amplification of the pCbp100-dsred-tag54 DNA [8] with the same CHS_Spe and TagKDEL_Bam primers. PCR was in a final volume of 50 μl 1x reaction buffer with the appropriate concentrations of $MgCl_2$ and primers (Fisher Scientific SL, Madrid) (Table S1), 200 μM dNTPs and 1 unit Expand High Fidelity DNA polymerase (Roche Diagnostics Corporation). Reaction conditions were as follows: 3 min at 94°C; 10 cycles of 15 s at 94°C, 30 s at 49°C and 1 min at 72°C; 20 cycles of 15 s at 94°C, 30 s at 58°C and 1 min, plus additional 5 s each successive cycle, at 72°C; and a final extension of 10 min at 72°C.

On sequence verification (Macrogen, Seoul, Korea) using the CLCbio software (Aarhus, Denmark), the complete constructs, *bp100.2*, *bp100-dsred-tag54* and *dsred-tag54* plus promoter and terminator elements, were directionally inserted into the *Kpn*I and *Sbf*I sites of pCAMBIA1300. The resulting binary vectors were transferred into *A. tumefaciens* strain EHA105 by cold shock [28].

Production of transgenic rice plants

Commercial *japonica* rice (*Oryza sativa* L.) var. Senia was transformed by *A. tumefaciens* to obtain transgenic rice lines expressing the chimeric proteins mentioned above, using hygromycin resistance as the selection trait. The control plasmid, pCambia1300 (transferring only the *hpt*II selection gene), was transformed in parallel. Embryonic calluses derived from mature embryos were transformed as previously described [29]. Hygromycin resistant T_0 plants were grown to maturity under standard greenhouse conditions to obtain the T_1 generation. Leaf samples of individual T_0 plants at the 5-leaf vegetative stage were used to extract genomic DNA and assess the presence and copy number of the transgene by real-time PCR (qPCR), targeting the coding region of every transgene as previously described [8]. The *actin* endogenous gene was used to normalize the Ct values. The number of fertile genetically modified (GM) plants containing every plasmid was recorded to calculate the transformation efficiency compared to that of the control pCambia 1300 plasmid (number of fertile GM plants obtained per initial callus. This value was then normalized with that obtained for the control plasmid).

Analysis of gene expression: plant material, RNA isolation and RT-PCR

Plant material: heat shock treatment. Wild type rice seeds (var. Senia) were surface sterilized and germinated *in vitro* in sterile MS medium [30], including vitamins (2.2 g/L MS medium, 8 g/L agar and 30 g/L sucrose), under controlled conditions ($28\pm1°C$ temperature and a 16 h light / 8 h dark photoperiod with fluorescent Sylvania Cool White lamps). After one week (two-leaf vegetative stage, V2), groups of 15 plants were treated at 42°C for 0, 1, 2, 4, 8 and 16 h. For each treatment, three groups of five plants were collected as biological replicates, immediately frozen in liquid nitrogen and stored at −80°C.

For each GM event, 15 T_1 seeds were surface sterilized and *in vitro* cultured, as mentioned above, up to the V2 stage. They were treated at 42°C for 0 and 2 h and their leaves individually frozen in liquid nitrogen. The presence of the transgene was individually assessed in every plant using the Phire Plant Direct PCR Kit (Thermo Scientific, Lithuania) combined with the DsRed_for and Nos.te_rev (constructs with *dsred-tag54* and *bp100-dsred-tag54*) or the P82_for and Nos.te_rev (constructs with *bp100.2*) primer pairs (Table S1), according to the manufacturer's instructions. Samples giving a positive PCR signal (i.e. harboring the transgene) were mixed for subsequent analyses.

Plant material: monitoring of the transformation process. The transformation to obtain transgenic rice was monitored using *A. tumefaciens* strain EHA105, with or without the basic pCambia1300 vector (with *hpt*II). Embryonic calluses derived from rice mature embryos were transformed and, when *hpt*II was transformed, selected with hygromycin [29]. Samples of ten independent calluses or events were taken at eight different stages for each transformation.

For embryo extraction, 120 hygromycin-resistant seeds (harboring *hpt*II) and 80 non-transformed seeds were surface-sterilized and their embryos were immediately extracted and frozen, or germinated in 500 μL water or 500 μL water with 45 mg/L hygromycin B (40 seeds per treatment and embryo type).

Germination was in a culture chamber ($28\pm1°C$ with a 16 h light/8 h dark photoperiod, under fluorescent Sylvania Cool White lamps) for two and five days prior to embryo extraction and immediate freezing in liquid nitrogen. Untransformed seeds were not treated with hygromycin B.

RNA isolation and reverse transcription (RT) coupled to qPCR. Samples were homogenized in liquid nitrogen and 100 mg was used to extract RNA with the Trizol reagent (Invitrogen, Karlsruhe, Germany) based protocol. The DNase I digestion (Ambion, Grand Island, NY) was carried out according to the manufacturer's protocol, and RNA concentration and quality was assessed by UV absorption at 260 and 280 nm using a NanoDrop ND1000 spectrophotometer (Nanodrop technologies, Wilmington, DE). RT-qPCR was carried out as previously described [7]. For each sample, cDNA was synthesized with random primers in duplicate and qPCR reactions targeting *bp100.2* and *dsred-tag54* transgenes, and *Os.hsp18* and *Os.hsp82* rice genes were carried out in triplicate. The qPCRs were in a final volume of 20 μl containing 1X SYBR Green PCR Master Mix (Applied Biosystems, Foster City, CA, USA), the appropriate concentrations of primers (Fisher Scientific SL, Madrid, Spain) (Table S1) and 1 μl cDNA. The reaction conditions were as follows: 10 min at 95°C for initial denaturation; 50 cycles of 15 s at 95°C and 1 min at 60°C; and a final melting curve program of 60 to 95°C with a heating rate of 0.5°C/s. Melting curve analyses produced single peaks, with no primer-dimer peaks or artifacts, indicating the reactions were specific. All reactions had a linearity coefficient exceeding 0.995 and efficiency values above 0.95. The *ef1α* gene was used for normalization, its suitability having been confirmed using the geNORM v3.4 statistical algorithm ([31]; *M* values below 0.5 in our samples). Triplicate biological samples, each containing at least five plants, were analyzed in each case.

Protein extraction and western blot analyses

For each GM event, 15 T_1 seeds were surface sterilized and *in vitro* cultured as mentioned above, up to the V2 stage. Rice seedlings were treated at 42°C for 2 h, then transferred to $28\pm1°C$ with a 16 h light/8 h dark photoperiod under fluorescent Sylvania Cool White lamps for two days, to allow recombinant protein accumulation and DsRed maturation, and then immediately frozen in liquid nitrogen. The Phire Plant Direct PCR Kit (Thermo Scientific, Lithuania) was used in combination with the SYDsRed_for and SYDsRed_rev primers (Table S1), according to the manufacturer's instructions, to individually identify transgenic T_1 plants. Transgenic seedlings were homogenized in liquid nitrogen and TSP extracted as previously described [8]. Insoluble proteins in the samples expressing *bp100-dsred-tag54* were re-extracted by boiling for 10 min in the same buffer supplemented with 8 M urea and 1% SDS. Protein concentration was determined using the Sigma Bradford Reagent and bovine serum albumin as standard. Twenty μg TSP were separated by PAGE, using 18% (w/v) SDS polyacrylamide gels, and electrotransferred to nitrocellulose membranes. Twenty-five to 1 pmol chemically synthesized controltag54 (GQNIRDGIIKAGPAVAVVGQAT-QIAKAGPAKDWEHLKDEL), mixed with 20 μg TSP extracted from untransformed rice seedlings, were included to allow quantification of the recombinant peptides. They were hybridized overnight with the mAb54k monoclonal antibody [27] (1:1,500 dilution) at 4°C, and with horseradish peroxidase-conjugated anti-mouse IgG as the secondary antibody (GE Healthcare Life sciences) (1:10,000 dilution) for 1 h at room temperature. The hybridization signal was detected by ECL chemiluminescence (Luminata Forte HRP Chemiluminescence Detection Reagents, Millipore – Darmstadt, Germany).

Phenotype evaluation

T1 seeds carrying pHsp82::bp100.2, pHsp82::dsred-tag54 (three independent events per construct) or hptII, and conventional Senia seeds, were surface sterilized and *in vitro* cultured as mentioned above, up to the V2 stage (height, 6±1 cm). They were all treated at 42°C for 2 h and allowed to recover for three days under the same conditions. Plant growth was monitored daily by measuring the height of the aerial part. Transgenic plants were identified using the Phire Plant Direct PCR Kit (Thermo Scientific, Lithuania) as mentioned above, and only those harboring the transgene (at least five plants per event) were included in statistical analyses. As an additional control, five untransformed Senia plants were grown in parallel and not subjected to heat shock.

Confocal microscopy

Transgenic rice seedlings, obtained as mentioned above, were treated at 42°C for 2 h and transferred to the standard conditions for two days to allow accumulation of the recombinant protein and DsRed maturation. Radicles were observed under an FV1000 Olympus confocal microscope and red fluorescent images collected with 543 nm excitation using a 550–600 nm emission window. The ImageJ software (http://rsb.info.nih.gov/ij/) was used to calculate the number and size of fluorescent spots and quantify the DsRed fluorescence.

Results

Rationale of the approach and selection of suitable promoters

With the increasingly abundant transcriptomics analytical tools and datasets available, we used an *in silico* approach to initially select a small number of candidate heat shock promoters. The EMBL-EBI Array Express website was searched for datasets on transcription profiling of the *Oryza sativa* in response to heat shock. A single microarray experiment was found (E-GEOD-14275, [10]) that questioned 14-day old seedlings, grown at 28–30°C, and with or without heat stress at 42°C. Expression data was available for a total of 22,905 sequences. Those with the highest expression levels had fluorescence intensities up to 13,800, and 16,300 units in microarrays hybridized with RNA from control and heat-shock treated seedlings, respectively. Only 15 and 17 sequences had values above 10,000 fluorescence units. Up to 19 sequences, corresponding to 16 genes, had fluorescence intensities more than 5,500 units higher in heat-shock treated than in control seedlings (Table S2). This included three sequences (AK071613.1, AK071240.1 and CB621753) with very high expression values under induction conditions (above 10,000 fluorescence units) and sequences with the greatest changes in expression in response to heat stress treatment (up to 615-fold, AK063751.1). These sequences were initially selected to drive the expression of phytotoxic peptides in rice.

The transcription patterns of these 16 genes were subsequently compared with the Collection of Rice Expression Profiles (CREP) database (http://crep.ncpgr.cn/crep-cgi/home.pl), a compilation of quantitative transcriptomes of 39 tissues of various rice genotypes, from the *indica* and *japonica* subspecies, obtained using the commercial Affymetrix Rice GeneChip microarray. We specifically focused on the tissues and developmental stages with special relevance in transformation and regeneration to obtain GM plants. They included callus at different steps of agrotransformation and hygromycin based selection, and imbibed seeds and seedlings at various developmental stages. Multiple gene expression profiles were retrieved using the Multi-genes Chronologer

tool (Figure 1). The Os.519.1.S1_at sequence (AK071240.1, encoding a putative 18 kDa Class II Heat Shock Protein) combined low expression levels in all 14 analyzed tissues considered relevant for transformation and very high expression in heat-shock treated seedlings (0.073 and 1.190 normalized fluorescence units). Although it had some expression in control seedlings (0.052 normalized fluorescence units), this promoter was chosen to drive the expression of phytotoxic peptides in plants. A second selected sequence, Os.11039.1.S1_s_at (AK063751.1, encoding a putative Heat Shock Protein 82), had extremely low expression values in the 14 tissues and control seedlings but only moderate expression in response to high temperature treatment (0.033, 0.001 and 0.650 normalized fluorescence units, respectively). There was no induction of Os.519.1.S1_at or Os.11039.1.S1_s_at upon cold, salt or drought stresses, as assessed *in silico* using profiling datasets with GSE6901 as the reference, reporting transcriptome analyses of 7-day old seedlings of Indica rice variety IR64 subjected to 4±1°C, 200 mM NaCl or insufficient moisture in their roots for 3 h (Figure S1).

Expression patterns of the two selected genes in response to heat shock and during the process of *Agrobacterium tumefaciens* mediated stable transformation of rice using the *hpt*II selection gene

Specific real-time PCR (qPCR) assays were developed to target the coding sequences of *Os.hsp18* and *Os.hsp82* and, coupled to reverse transcription, used to experimentally assess the expression patterns of the selected genes in a range of developmental stages and conditions. Treatment of rice plants with high temperature resulted in increased expression of the two genes (Figure 2). *Os.hsp18* and *Os.hsp82* were highly induced by exposure to 42°C for 2 to 4 h, reaching expression levels of 294 and 516-fold that of *ef1α* in leaves and 326 and 796-fold in roots. There was a decrease in gene expression after these time points.

Additionally, their expression levels were determined during *A. tumefaciens*-mediated transformation and selection using the hygromycin resistance phenotype (Figure 3). Senia wild type seeds were induced to generate primary callus and the standard transformation procedure was followed, using *A. tumefaciens* carrying pCAMBIA1300 (with the *hpt*II selection gene) to produce transgenic plants. *A. tumefaciens* not transformed with the binary plasmid was used as control, without hygromycin as a selection agent. Mature and imbibed embryos from untransformed and transgenic S-hptII plants were also analyzed. *Os.hsp18* was clearly expressed at higher levels than *Os.hsp82* in all tested tissues and stages. Expression was up to 28-fold that of the reference gene in transformed callus incubated for two weeks in hygromycin selection medium and that of mature embryos, and up to 15-fold that of *ef1α* upon *A. tumefaciens* infection. Seed imbibition rapidly resulted in mRNA decline. *Os.hsp82* had a similar expression pattern, although the level was less than that of *ef1α* in virtually all tested samples. As a control, we measured the mRNA levels of the *hpt*II transgene, regulated by the CaMV 35S strong constitutive promoter. They were substantially higher than those of *Os.hsp18*, and especially *Os.hsp82*, at all stages in which there were uniquely transgenic cells (i.e. after hygromycin selection).

Os.hsp18 and *Os.hsp82* promoter fragments regulate the expression of transgenes in response to heat treatment

The sequences corresponding to the promoters of *Os.hsp18* and *Os.hsp82* were retrieved from the Plant Promoter Database, and 553 and 1,016 bp promoter fragments of *Os.hsp18* and *Os.hsp82*,

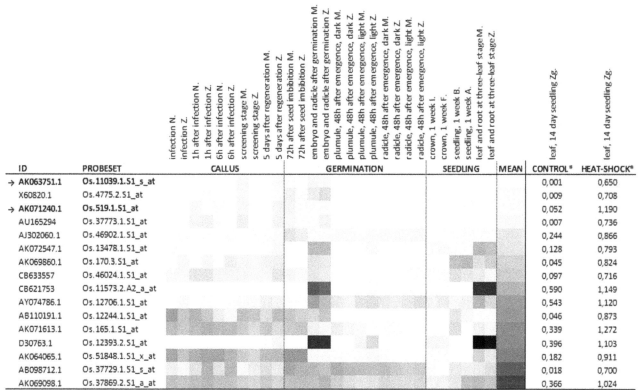

ID	PROBESET	CONTROL*	HEAT-SHOCK*
→ AK063751.1	Os.11039.1.S1_s_at	0,001	0,650
X60820.1	Os.4775.2.S1_at	0,009	0,708
→ AK071240.1	Os.519.1.S1_at	0,052	1,190
AU165294	Os.37773.1.S1_at	0,007	0,736
AJ302060.1	Os.46902.1.S1_at	0,244	0,866
AK072547.1	Os.13478.1.S1_at	0,128	0,793
AK069860.1	Os.170.3.S1_at	0,045	0,824
CB633557	Os.46024.1.S1_at	0,097	0,716
CB621753	Os.11573.2.A2_a_at	0,590	1,149
AY074786.1	Os.12706.1.S1_at	0,543	1,120
AB110191.1	Os.12244.1.S1_at	0,046	0,873
AK071613.1	Os.165.1.S1_at	0,339	1,272
D30763.1	Os.12393.2.S1_at	0,396	1,103
AK064065.1	Os.51848.1.S1_x_at	0,182	0,911
AB098712.1	Os.37729.1.S1_s_at	0,018	0,700
AK069098.1	Os.37869.2.S1_a_at	0,366	1,024

* normalized with the *ef1α* gene

Figure 1. A summary of the expression profiles of a selection of sequences induced by high temperature, in callus and seedlings of eight rice genotypes [Bala (B), FL478 (F), IR29 (I), Minghui 63 (M), Zhenshan 97 (Z), *indica* subspecies, and Azucena (A), Nipponbare (N) and ZhongHua 11 (Zg), *japonica* subspecies], from published *in silico* data. Cluster analysis of 16 genes with the highest expression or induction in response to heat-shock, in a total of 14 tissues considered relevant to the process of obtaining transgenic plants (grey scale). The overall mRNA levels in these tissues and developmental stages were estimated using the mean expression values (green scale). Dark to light color scale represents high to low expression levels, the darkest color corresponding to 90162 (grey scale) and 19788 (green scale) normalized fluorescence units. Arrows indicate the two sequences whose promoters were selected to produce transgenic plants.

respectively, were PCR amplified from rice var. Senia genomic DNA. Their 3′ end was at position −1 relative to the ATG translation start codon. *Os.hsp18* and *Os.hsp82* promoter fragments, pHsp18 and pHsp82, were cloned and sequence verified. The activity of these promoter fragments was experimentally assessed using the *Discosoma* spp. red fluorescent reporter

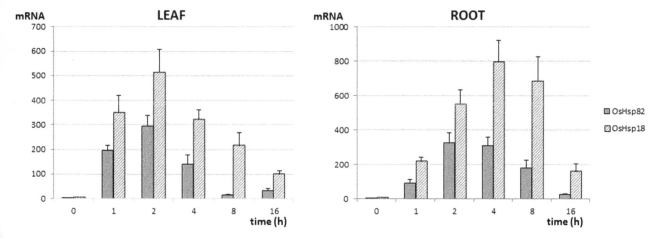

Figure 2. Transcriptional response to high temperature of *Os.hsp18* and *Os.hsp82* in rice var. Senia. Plants at the two leaf vegetative stage were exposed to 42°C for 0, 1, 2, 4, 8 and 16 h, then the mRNA levels of the selected sequences, from leaf and root samples, were analyzed by RT-qPCR. Three biological replicates, each of five plants, with two experimental replicates, were analyzed per treatment. The *ef1α* reference gene (gNORM M value <0.5 in these samples) was used for normalization and the given mRNA values correspond to fold expression vs. the reference gene.

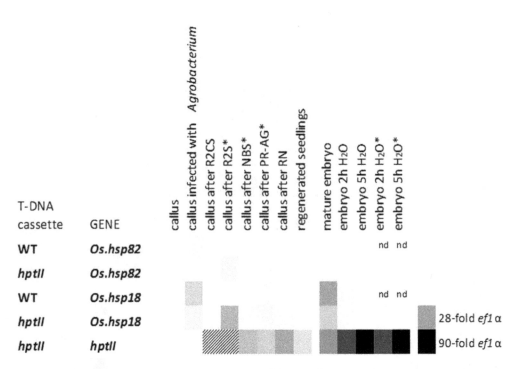

Figure 3. mRNA expression profiles of *Os.hsp18*, *Os.hsp82* and the *hptII* transgene, driven by the CaMV 35S promoter, in rice var. Senia. Different stages during transformation with *A. tumefaciens*, either carrying pCAMBIA1300 (*hptII*) or no T-DNA (WT), and expression in mature and imbibed embryos, are shown. The following stages of callus are shown: 6 weeks after induction of embryogenic callus; immediately after 15 min infection with *A. tumefaciens*; after a 3-day incubation in R2CS medium; after a 2-week incubation in R2S selection medium; 3 weeks after transfer to NBS selection medium; after an 8 to 10-day incubation in PR-AG selection medium; 3–4 weeks after induction of regeneration in RN medium; and two weeks after seedling culture in P medium. Embryos extracted from mature seeds, either before or after imbibition in water for two or five hours, were also analyzed. When the *hptII* DNA cassette was transformed, transgenic cells were selected using hygromycin (asterisk). Three biological replicates, each of five calluses, with two experimental replicates, were analyzed at each stage. The *ef1α* reference gene (M values <0.5 in these samples) was used for normalization. Dark to light color scale represents high to low expression level. All values were in the 90 to 0.01-fold range. nd: not determined. Striped: underestimated values due to callus samples containing a mixture of transgenic and non-transgenic cells.

protein, DsRed [26], and the epitope tag54 sequence to detect recombinant proteins through the specific mAb54 antibody [27]. The sequence encoding the DsRed-tag54-KDEL reporter (hereafter, DsRed-tag54) was placed in-frame with the sequence encoding the signal peptide of the murine monoclonal antibody mAb24 under the control of the *Os.hsp18*, *Os.hsp82* or *Zm.ubi* [32] promoters and *nos* terminator. All constructs were sequence verified, introduced in *A. tumefaciens* and used to transform rice plants, along with the *hptII* marker for hygromycin selection. All constructs yielded hygromycin-resistant plants and those derived from different calluses were considered independent events.

The transgene copy number and expression profile were assessed in three independent events representing each promoter. The ratio of *dsred-tag54* to the *actin* reference gene sequence was close to 0.5 (mean and SD, 0.63±0.27), as determined by qPCR using leaf genomic DNA from T_0 plants, suggesting single-copy insertions. Transgene expression was quantified by RT-qPCR in leaf samples of transgenic T_1 plants at the two-leaf vegetative stage (V2), with or without exposure to 42°C for 2 h (Figure 4). Significant amounts of *dsred-tag54* mRNA were only observed when plants were kept under control temperature conditions, where the transgene was driven by the constitutive pUbi promoter. The expression values, normalized with the *ef1α* constitutive gene, were 0.47 to 2.5, 0.73 to 18.84 and 603 to 904 for pHsp82, pHsp18 and pUbi, respectively. One-way ANOVA, $P = 0.000$, gave two groups in Tukey's b posttest with α<0.05. Upon heat treatment, the pHsp18 promoter reached transgene mRNA levels from 214 to 1,117-fold that of *ef1α*, similar to those reached with

the pUbi promoter. The levels of expression of the transgene was slightly lower with the pHsp82 promoter, 208 to 636-fold that of *ef1α*, but statistically similar (one-way ANOVA $P = 0.279$). As expected, no mRNA encoding DsRed-tag54 was detected in untransformed rice plants. This proved that, after exposure to heat stress, the pHsp82, and especially the pHsp18 promoter fragments, had the capacity to drive the expression of a reporter gene to levels similar to those achieved by a frequently used strong constitutive promoter.

Use of *Os.hsp18* and *Os.hsp82* promoters to drive the expression of phytotoxic BP100 derivatives in rice

The capacity of the pHsp18 and pHsp82 promoter fragments to drive the expression of phytotoxic cationic α-helical antimicrobial peptides in stably transgenic plants was evaluated using BP100.2. The antimicrobial peptide has high phytotoxic and hemolytic activities and its constitutive expression in transgenic rice is incompatible with the survival of GM plantlets [7]. The sequence encoding BP100.2 was placed in-frame with the sequence encoding the signal peptide of the *N. tabacum* pathogenesis related protein PR1a, under the control of the pHsp18 and pHsp82 promoters. After sequence verification, the constructs were introduced in *A. tumefaciens* and used to transform rice plants along with the *hptII* marker. The empty vector, with only the *hptII* selection gene, was transformed in parallel. Joint analysis of the transformation efficiencies achieved for constructs encoding DsRed-tag54 and BP100.2, under the control of the pHsp18 pHsp82 and pUbi promoters, showed that four out of five

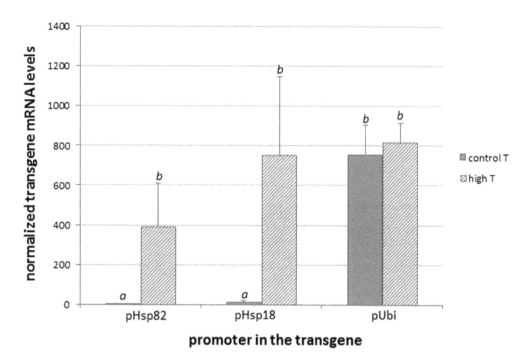

Figure 4. Transgene mRNA expression levels in leaves of GM rice plants in which the reporter *dsred-tag54* was regulated by three different promoters, pHsp82, pHsp18 and pUbi. Three independent events were analyzed carrying every construct. For each event, T_1 seeds were germinated under control conditions and sets of eight plants at the V2 stage were either subjected to 42°C for 2 h, or not, and their leaves were individually sampled. Sets of five plants harboring the transgene (i.e. giving positive transgene signal using the Phire Plant kit) were collectively analyzed by RT-qPCR targeting the sequence encoding DsRed-tag54, and normalized against *ef1α*. Mean and SD values corresponding to each promoter are shown. Letters indicate statistically different transgene mRNA values (One-way ANOVA, Tukey's b posttest α<0.05).

constructs yielded hygromycin-resistant plants with efficiencies similar to that of the empty vector (mean efficiency and SD, 97±18%, Table S3). As expected, this included all vectors encoding the non-phytotoxic DsRed-tag54 reporter, irrespective of the specific promoter regulating its expression (constitutive pUbi or heat-shock inducible pHsp18 and pHsp82).

Remarkably, the vector encoding BP100.2 under the control of the high temperature inducible promoter pHsp82 produced fertile transgenic rice plants with the same efficiency. The presence of the *bp100.2* or *dsred-tag54* transgene was confirmed in all T_0 events using qPCR. Conversely, no GM plants were obtained with the construct encoding the highly phytotoxic BP100.2 antimicrobial peptide under the control of the pHsp18 promoter (transformation efficiency <3%). The efficiency of this transformation was at least 30-fold below that of the control plasmid, which can be considered not workable. This is in agreement with the higher activity of the pHsp18 promoter, compared to pHsp82, during transformation.

Transgene expression profiles were assessed in three independent events carrying sequences encoding BP100.2 under the control of pHsp82 (pHsp82::*bp100.2*), using transgenic plants encoding DsRed-tag54 regulated by the same promoter (pHsp82::*dsred-tag54*) as the control. Hygromycin-resistant plants derived from different calluses were considered independent events. The ratio of *bp100.2* to the reference gene sequence was close to 0.5 in these events (mean and SD, 0.53±0.29) as determined by qPCR using leaf genomic DNA from T_0 plants, suggesting single-copy insertions. Transcript levels of the *bp100.2* transgene were, prior to heat shock, 2.18 to 5.70-fold the level of *ef1α* (Figure 5), i.e. similar to those of *dsred-tag54* (One-way ANOVA P=0.161). Upon heat shock, pHsp82 drove transcription to similar levels for the phytotoxic *bp100.2* (275 to 357-fold the level of *ef1α*) and the control *dsred-tag54* (One-way ANOVA

P=0.592). Thus, the pHsp82 promoter is a suitable tool to achieve transcription of transgenes encoding phytotoxic recombinant peptides in stably transformed GM plants.

Recombinant BP100.2 is difficult to detect due to its low extinction coefficient (low content of aromatic amino acids) and lack of immunogenicity [8]. We used its phytotoxic character to indirectly confirm the synthesis of BP100.2 in GM plants in response to heat treatment. We predicted that, in the presence of the phytotoxic recombinant BP100.2, transgenic plants would show altered phenotype after heat shock as compared to control plants. As shown in Figure 6, untransformed plants not subjected to heat treatment were taller than any temperature-stressed groups. Twenty-four hours after heat shock, plants expressing *bp100.2* were shorter than those either untransformed or expressing *dsred-tag54* or *hptII* (One-way ANOVA P=0.000, three groups after Tukey b posttest with α<0.05) and had more serious symptoms of leaf wilting. This strongly suggested that phytotoxic recombinant BP100.2 was produced in these plants.

Production of recombinant reporter proteins in GM rice plants using high temperature inducible promoters

Phytotoxic recombinant BP100 derivatives in transgenic plants were quantified by fusion of BP100 to a reporter moiety encompassing the fluorescent DsRed and the tag54 epitope. A construct was obtained in which the pHsp82 promoter drove the expression of BP100-DsRed-tag54-KDEL (hereafter, BP100-DsRed-tag54), with the same signal peptide and regulatory elements as the pHsp82::*bp100.2* construct. It was sequence verified and used to transform rice plants using the same *Agrobacterium*-based strategy, achieving similar transformation efficiency as the control (109%). Synthesis of *bp100-dsred-tag54* mRNA was assessed by RT-qPCR in three independent events

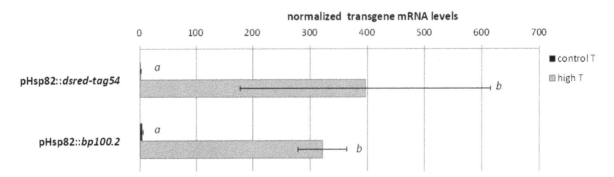

Figure 5. Transgene mRNA expression levels in leaves of GM rice plants harboring pHsp82::bp100.2 or pHsp82::dsred-tag54 constructs. Three independent events of each construct were analyzed. For each event, T₁ seeds were germinated under control conditions and sets of eight plants at the V2 stage were either subjected to 42°C for 2 h (black bars) or not (grey bars), and their leaves individually sampled. Sets of five plants harboring the transgene (i.e. giving positive transgene signal using the Phire Plant kit) were collectively analyzed by RT-qPCR targeting the sequence encoding DsRed-tag54 or BP100.2, and normalized against *ef1α*. Mean and SD values corresponding to every transgene are shown. Different letters indicate statistically different transgene mRNA values (One-way ANOVA, Tukey's b posttest α<0.05).

with single copy insertions (as determined by qPCR of T₀ leaves using *ef1α* as reference, 0.52±0.15). Residual mRNA levels in leaves of V2 plants grown under control temperature were in the range of 0.07 to 0.3-fold actin mRNA levels, but increased to 142 to 275-fold after a 2 h exposure to heat shock. These values were not statistically distinguishable from those of pHsp82-regulated

transgenes expressing *bp100.2* and *dsred-tag54* in the same conditions (One-way ANOVA, $P = 0.524$ and $P = 0.205$ for control and heat treated plants, respectively).

Transgenic plants from three independent events carrying pHsp82::*bp100-dsred-tag54* were grown to the V2 stage under controlled conditions, heat-induced for 2 h at 42°C and allowed to

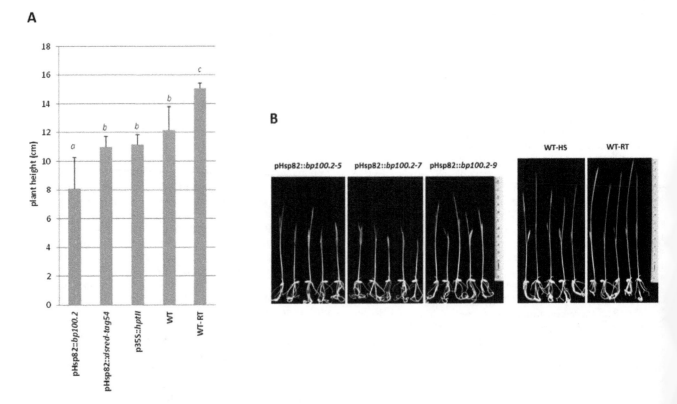

Figure 6. Growth of transgenic rice plants carrying pHsp82::bp100.2 after heat shock. T₁ seeds of rice plants carrying pHsp82::bp100.2, pHsp82::dsred-tag54 (three independent events per construct) and p35S::hptII were germinated for 1 day in the presence of hygromycin, and five plants per event were then *in vitro* cultured, without the selection agent, up to the V2 stage (height, 6±1 cm). Five untransformed Senia plants were grown in parallel (WT). They were all treated at 42°C for 2 h and the length of their aerial parts was measured after 24 h incubation under standard conditions. Five untransformed plants were not subjected to heat stress, as an additional control (WT-RT). **(A)** Means and SD of the height of the different groups are shown. Letters indicate statistically significant differences (One-way ANOVA, Tukey b posttest with α value <0.05). **(B)** The phenotypic effects of heat treatment on transgenic plants carrying pHsp82::bp100.2 (three independent events) and untransformed plants (WT-HS) are shown. WT-RT is shown as control.

accumulate recombinant BP100-DsRed-tag54 for an additional two days. Three events of transgenic plants expressing *dsred-tag54* under the regulation of the pHsp82 and pHsp18 promoters were used as control. Plants with the constitutive pUbi promoter driving the expression of the same recombinant protein were included in the assay. As shown in Figure 7A, TSP extracts from all transgenic plants expressing *dsred-tag54* produced a band of 29.9 kDa in western blot assays, which is similar to the DsRed-tag54 anticipated size, and a band of about 36.8 kDa. There was an additional secondary band of 22.1 kDa that probably corresponds to a cleavage product originated by boiling TSP before SDS-PAGE [33]. TSP extracted from all three samples expressing *bp100-dsred-tag54* under the control of pHsp82 produced a band of an apparent MW of 37.9 kDa (Figure 7B), which is in agreement with the 36.8 kDa band detectable in extracts of plants producing recombinant DsRed-tag54. When higher amounts of protein extracts were analyzed, a faint band of the expected size (31.5 kDa) was also visible (Figure S2). The BP100 portion of the recombinant molecule seemed to strongly enhance the proportion of the slower band, which was probably the result of incomplete denaturation or interaction of different molecules. Subsequent analysis of the insoluble protein fraction showed that a significant proportion of immune-reactive proteins, with apparent MW of 37.9, 48.5 and 70.6 kDa, could only be extracted in an SDS and urea-based, strongly denaturing buffer (Figure 7B). As expected, no immune-reactive bands were observed upon analysis of TSP from untransformed plants, those uniquely expressing the hygro-mycin resistance selection marker, or those transformed with pHsp82::*dsred-tag54* and not subjected to high temperature (Figure 7A). Thus, phytotoxic recombinant BP100 derivatives were produced and accumulated after heat induction in transgenic rice using the pHsp82 heat inducible promoter. Quantification of the immune-reactive bands (Figure 7C) showed that the pUbi promoter led to approximately 0.9% TSP of DsRed-tag54, while promoters derived from *Os.hsp18* and *Os.hsp82* led to recombinant DsRed-tag54 levels up to 0.3 and 0.15% TSP, respectively, after heat shock. Finally, use of the same pHsp82 promoter to drive the expression of the phytotoxic BP100 fusion protein led to similar accumulation levels, up to 0.12% TSP in these conditions.

One week after germination, pHsp82::*bp100-dsred-tag54* and control pHsp82::*dsred-tag54* transgenic plantlets from three different events per construct were induced by heat shock, and the recombinant chimeric protein allowed to accumulate and mature for three days prior to confocal microscopy of the plantlet radicles. All analyzed transgenic plants displayed red fluorescence, further confirming the accumulation of the recombinant proteins. Fluorescence intensities were measured in four fields per radicle. The p82::*bp100-dsred-tag54* and control p82::*dsred-tag54* plants expressed similar levels of recombinant proteins, with mean and SD values of 1258±420 and 981±286 fluorescence units per field, respectively (One-way ANOVA $P = 0.135$). This demonstrated that recombinant phytotoxic BP100-derived peptides could be accumulated to levels similar to those of the DsRed-tag54 reporter in stable GM plants constructed with the pHsp82 temperature inducible promoter. The DsRed-tag54 fluorescence showed a typical ER reticular pattern accompanied by a few spherical structures, attributable to the known effects of high temperatures on the internal organization of the cell, such as fragmentation of the Golgi system and the ER [34]. In contrast, fluorescence in pHsp82::*bp100-dsred-tag54* radicles was essentially in numerous and widely distributed vesicles, obscuring the ER network (Figure S3), in agreement with BP100 inducing ER-derived protein bodies [8].

Discussion

Biotechnological production of cationic α-helical antimicrobial peptides that have potent activities against pathogenic microor-ganisms is of great interest [35], but their constitutive expression is not compatible with the viability and fertility of host plants [7]. This problem has been previously approached by expression of peptides (i) with suboptimal antimicrobial activity linked to lower phytotoxicity [36], and (ii) with rational sequence modification to achieve a targeted reduction in phytotoxicity without affecting antimicrobial potency [8]. Modifications often alter the properties of the peptides not only in terms of antibacterial activity but also the range of target bacterial species [3,5,37]. This makes the design of a peptide with optimal properties difficult [38]. A tool to produce highly active peptides of interest in plants, irrespective of their phytotoxic character, is desirable. Inducible promoters could potentially reduce the negative impact of toxic recombinant peptides on the transgenic plant by forcefully restraining the presence of the peptide during plant development, while allowing high expression upon induction. The probable harmful effects of accumulation of recombinant phytotoxic compounds after induc-tion, to the plant biofactory, are irrelevant since the main objective is recovery of the recombinant product. We predicted that it would only be possible to successfully generate transgenic plants carrying transgenes that encoded phytotoxic peptides by using promoters extremely inactive throughout transformation and regeneration, due to the vulnerability of the plant material during this process. Due to the practicality and low cost of using high temperature as an inducing agent in plants, and the extensive literature on the plant response to heat shock [39–44], we focused on heat shock promoters to assess the feasibility of obtaining recombinant phytotoxic antimicrobial peptides in rice, using as model, highly active and phytotoxic peptides derived from BP100.

Profiling technologies such as microarray hybridization have been widely used to systematically investigate transcriptomic changes during plant development, in different tissues and in response to diverse environmental conditions, including heat shock [45–47]. We selected candidate promoters *in silico* on the unique basis of the massive transcriptome information on publically available web sites (MIAMExpress, CREP), specifically focusing on studies on prolonged heat shock transcriptional response [10]. A similar strategy has recently been used to identify abiotic stress-inducible promoters intended to drive the overexpression of genes encoding nontoxic, stress resistance proteins in transgenic plants, with the aim of reducing the stunted growth and decrease in yield that may be caused by constitutive overexpression of these transgenes [48]. In contrast, here we selected candidate promoters for the expression of phytotoxic recombinant peptides, in plants, to drive high gene expression after heat treatment with minimal gene expression under any other condition, especially focusing on callus during agrotransformation and different tissues in the germination and seedling developmental stages.

The four sequences which best fulfilled these requirements [AK063751.1 (*Os.hsp82*), X60820.1, AU165294 and AK071240 (*Os.hsp18*), Figure 1] encoded heat shock proteins belonging to the Hsp90 (*Os.hsp82*) and the small Hsp (sHsp) families (X60820.1, AU165294 and *Os.hsp18*, UniProt database). Heat shock proteins are found in every organism [49], and play important roles in cell protection against deleterious effects of stress e.g. acting as molecular chaperones [49–55] and in cellular functions related to growth and development [49,56]. Small Hsps are the most abundant, heterogeneous and stress-responsive group of Hsp in higher plants [57–60]. The genes for more than 20 sHsps have been identified in rice, with specific spatial and temporal

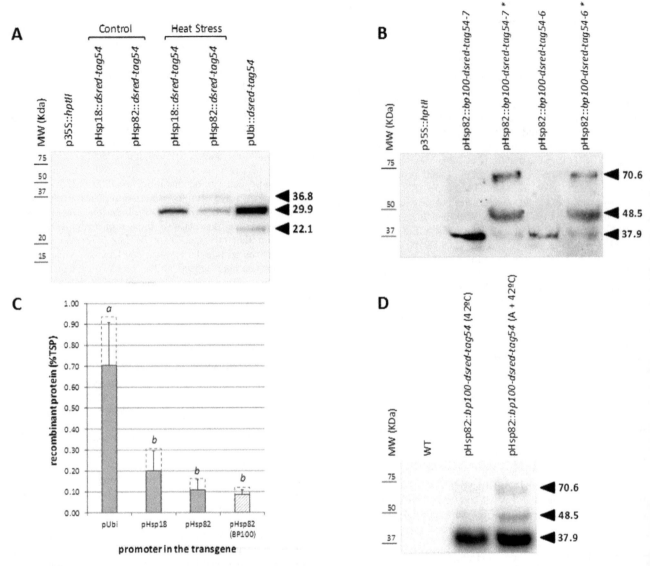

Figure 7. Recombinant DsRed-tag54 and BP100-DsRed-tag54 accumulation in transgenic rice seedlings. Western blot analysis of proteins from GM plants carrying: (**A**) p35S::*hptII*, pUbi::*dsred-tag54*, pHsp18::*dsred-tag54* and pHsp82::*dsred-tag54*, either treated at 42°C (heat shock) or not (control); (**B**) pHsp82::*bp100-dsred-tag54* (two independent events are shown) treated at 42°C. Recombinant protein in TSP was extracted from rice seedlings (five plants per event) and 20 µg of TSP per lane was boiled for five minutes and separated by SDS-PAGE before transfer to nitrocellulose filters. In lanes marked with an asterisk, insoluble pHsp82::*bp100-dsred-tag54* pellets were re-extracted in 8 M urea and 1% SDS buffer. Recombinant proteins were detected using the mAb54k antibody (diluted 1:1,500) and the horseradish peroxidase-labeled anti-mouse IgG secondary antibody (diluted 1:10,000) followed by ECL chemiluminescent detection. Five to 20 pmol of chemically synthesized controltag54 mixed with 20 µg TSP from wild type rice was used as standards for quantification. GM rice carrying p35S::*hptII* was used as control. (**C**) Recombinant protein accumulation values. Means and SD of three independent events per construct are shown in filled boxes; letters indicate statistically different values; values corresponding to the highest producer event are shown in dashed boxes. (**D**) Western blot analysis of proteins from GM plants of a single event carrying pHsp82::*dsred-tag54*, either directly subjected to 42°C or with a 6 h initial step of gradual temperature increase up to 42°C (A+42°C) (five plants per condition). Recombinant proteins were extracted in 8 M urea and 1% SDS buffer. Senia (WT) was used as control.

regulation under stress and in developmental stages [42,61]. Specifically, the *Os.hsp18* gene, encoding the cytosolic class II Hsp18.0, had very high expression upon induction, although our microarray analysis showed that it retained some activity in developmental stages and conditions that are fundamental to the production of GM plants. Hsp90 are, in general, constitutively expressed and among the most abundant proteins in cells, although their expression increases in response to stress [39]. *Os.hsp82* had extremely low expression in control tissues and reached high levels of expression in response to heat, but *Os.hsp18*

and especially *Os.hsp82* were not significantly induced by other stress conditions, such as cold, NaCl and drought, in agreement with previous publications [10,62,63]. Their moderate mRNA levels in seeds, as determined using the CREP platform, would most probably allow fertility of the plants. Experimental assessment of the expression patterns of these two genes in the rice genotype of interest, Senia, allowed ratification of the *Os.hsp18* and *Os.hsp82* candidate sequences [10,62,64]. In a thorough analysis of *A. tumefaciens*-mediated transformation of Senia rice and selection using the hygromycin resistance trait, *Os.hsp18* and

Os.hsp82 mRNA levels were consistently 20 and 60-fold lower, respectively, than in leaves under induction conditions (the highest values were a transient response to *Agrobacterium* infection and hygromycin treatment) and about 10 and 300-fold lower, respectively, than those of the *hptII* transgene regulated by the p35S constitutive promoter. We therefore considered that *Os.hsp82*, and perhaps *Os.hsp18*, promoters would be able to drive the expression of transgenes encoding phytotoxic peptides in rice, possibly the latter with higher yields, and allow survival of GM plants.

Five hundred to 1,000 base pair promoter fragments of *Os.hsp18* and *Os.hsp82* regulated the expression of a reporter sequence in leaves of transgenic rice plants, in a similar manner as the endogenous genes. The low transgene mRNA levels measured prior to heat shock in GM plants carrying pHsp18::*dsred-tag54* and pHsp82::*dsred-tag54* (about 10 and 1-fold those of the reference gene, respectively) increased to approximately 700 and 400-fold that of the reference gene 2 hours after heat treatment. Even though different GM events had different transgene mRNA levels, globally the expression of endogenous *Os.hsp18* and *Os.hsp82* genes was similar to, respectively, that of the *dsred-tag54* transgene driven by the pHsp18 and pHsp82 promoters (One-way ANOVA $P = 0.465$ and $P = 0.463$, respectively). In addition, western blot analyses and confocal microscopy showed accumulation of recombinant DsRed-tag54 specifically upon heat shock in leaves and roots carrying pHsp18::*dsred-tag54* (data not shown) and pHsp82::*dsred-tag54*. The selected pHsp18 and pHsp82 promoter fragments retained the essential elements to drive heat shock induction in these tissues, and they performed as expected in GM plants. This is in agreement with previous publications where regulatory elements including the TATA box were *in silico* predicted in the two 5′ proximal regions [65,66].

Use of the pHsp18 promoter to drive the expression of the phytotoxic peptide BP100.2 did not allow survival of fertile transgenic plants. Remarkably, transgenic plants producing BP100.2 and BP100-DsRed-tag54, also phytotoxic [8], were obtained with pHsp82, with the same efficiency as those expressing a non-toxic recombinant protein such as the reporter DsRed-tag54. After heat treatment, leaves of plants carrying pHsp82::*bp100.2* and pHsp82::*bp100-dsred-tag54* produced levels of transgene mRNA similar to those harboring the *dsred-tag54* reporter under the control of the same pHsp82 promoter, indicating that these transgenes were fully functional. As BP100 is difficult to detect due to lack of antigenicity and a low extinction coefficient [8], a phenotypic approach was indirectly used to prove the synthesis of BP100.2 in these cells. High temperatures are known to cause wilting and a significant decline in relative growth rate [67]. Growth of pHsp82::*bp100.2* plants after a high temperature shock was slower than that of untransformed or transgenic control plants and there was considerably more wilting of their leaves, which supported the production of the phytotoxic recombinant BP100.2 peptide in these cells.

In view of the somewhat higher expression of *Os.hsp18* than *Os.hsp82* during transformation and regeneration (about 5–10 fold), the finding that pHsp82, but not pHsp18, was able to drive the expression of phytotoxic recombinant peptides in plants supports our initial hypothesis and experimental approach. It is essential that promoter sequences to regulate the expression of phytotoxic transgenes in plants are chosen on the basis of their extremely low basal activity in the tissues, the developmental stages and under the conditions for production of GM plants and germination. No viable transgenic plants were obtained that expressed BP100.2 under the control of 1000-bp promoter fragments of the *Os.37773.1.s1_at* (AU165294) and *Os.165.1.s1_at* genes, with mRNA levels above

those of *Os.hsp18* during the transformation process (Figure 1). These results place the threshold of basal expression of a phytotoxic transgene during production of the GM plant at very low levels, similar to those of the *Os.hsp82* gene. We then re-evaluated the initially selected heat stress-responsive genes on the basis of this threshold. Five additional sequences (AB110191.1, AU165294, AK069860.1, X60820.1 and AB098712.1, Table S2) had higher expression than *Os.hsp82* after heat shock, and mRNA levels below *Os.hsp18* in control conditions. Four of them were expressed at levels above the *Os.hsp18* threshold during transformation (Figure 1), so their promoters would not be suitable for producing GM plants by driving the expression of phytotoxic components. As X60820.1 had expression patterns similar to *Os.hsp82*, its promoter most probably would be suitable to drive this expression. However, it is unlikely that yields would be substantially above those of pHsp82, as deduced from its moderate transcriptional response to heat shock.

Recombinant BP100 derivatives were produced in transgenic plants in response to heat treatment through regulation of the pHsp82 promoter. Three days after induction, BP100-DsRed-tag54 represented 0.12% of seedling proteins solubilized in urea and SDS containing buffer, determined by western blot using the tag54 epitope. Using the same promoter to drive the expression of DsRed-tag54 yielded up to 0.15% TSP, harsher denaturing conditions not significantly increasing the yield (data not shown). Although an anti-tag54 reactive band of the anticipated size was visible in TSP from heat-treated pHsp82::*bp100-dsred-tag54* leaves, the major bands, especially those only extracted under strongly denaturing conditions, had slower mobility, which is compatible with remaining BP100-driven molecular interactions. This could be explained by the unusual physicochemical properties of BP100 (highly cationic, pI = 11.5), known to result in interactions with many components of the cell and reduce the efficiency of isolation from complex matrices [8]. In agreement, after the same temperature treatment, BP100-DsRed-tag54 was accumulated at statistically similar levels to the reporter DsRed-tag54, as determined by quantification of the fluorescence signal in confocal micrographs (i.e. when no extraction is needed). We have previously described [8] how the BP100 sequence induces the formation of ER-derived vesicles or protein bodies (PB), functioning similarly to the tandemly repeated VPGXG elastin-like motif [68], hydrophobins [69] and the γ-zein derived Zera polypeptide [70]. Recombinant BP100-DsRed-tag54 accumulates in these vesicles, together with other luminal ER proteins, in transiently transformed *N. benthamiana* leaves [8]. Here we observed that recombinant BP100-DsRed-tag54 had a similar vesicular pattern in radicles of transgenic rice after high temperature induction, further establishing that recombinant BP100 derivatives accumulate in newly formed ER-derived vesicles in different host plant species.

It is therefore clear that phytotoxic compounds such as BP100 can be expressed in stably transformed plants using a strategy based on inducible promoters. Constitutive expression of this type of transgene has only previously been achieved in transient systems, the harmful effects of the recombinant proteins strongly interfering with normal growth and development of the transgenic plants. With the pHsp82 promoter, we estimate a yield of about 0.1% TSP in our initial induction conditions. Larkindale and Vierling [71] have showed that different acclimation schemes resulted in different yields of Hsp in *Arabidopsis*. Notably, a gradual temperature increase led to higher transcript levels than heat shock without acclimation. Here, use of an acclimation treatment (6 h gradual increase prior to 2 h at 42°C) increased the recombinant protein yield 3-fold (Figure 7D). Thus, this approach allowed obtaining up to about 0.3% total extracted phytotoxic recombinant proteins. This is about half the yield of the BP100.gtag peptide, specifically designed to display extremely low toxicity, achieved using the constitutive maize pUbi promoter in the same rice system [8]. Recombinant Cry1B has been produced in rice at similar yields (0.2% TSP) under the control of the maize proteinase inhibitor (Mpi) wound-inducible promoter [72]. Higher yields of recombinant *Acidothermus cellulolyticus* endoglucanase E1, 1.3% TSP, have been achieved in transgenic tobacco using the tomato Rubisco small subunit (RbcS-3C) light-inducible promoter [73]. Note that this type of highly inducible promoter does not allow production of phytotoxic compounds.

In a prospective exercise, our experimental approach was used to identify alternative promoter sequences with increased performance to drive the expression of phytotoxic peptides in plants. Transcriptome-wide datasets are available for rice seedlings under control and a range of abiotic stress conditions, such as salinity, desiccation, suboptimal temperature (4°C), and various minerals, elicitor and hormone treatments, and were analyzed according to the above mentioned criteria. A total of 23 sequences were identified with high expression under induction conditions (above that of *Os.hsp82*) and minimal expression levels in control conditions (below that of *Os.hsp18*) (Figure S4). On assessment of their expression patterns in callus and during the transformation process and production of transgenic plants (taken as the mean of the microarray normalized data in these tissues), there were only four sequences with mRNA levels below *Os.hsp82*, strongly suggesting that their promoters would serve to drive the expression of phytotoxic proteins in transgenic plants. They were induced by drought, NaCl, phosphorus and the heavy metals chromium (VI) and arsenic. A single sequence, Os.49245.1.S1_at, reached very high levels after drought and NaCl induction (similar to *Os.hsp18* after heat shock), making its promoter the best candidate. Five additional sequences had mRNA levels below *Os.hsp18* but above *Os.hsp82* in the analyzed tissues and developmental stages. Their promoters might also be suitable to drive the expression of toxic compounds in GM plants, but their expression levels after induction were below that of *Os.hsp18*. The best use of our promoter-selection strategy would most likely be achieved by using longer-term expression data to select sequences with the longest span of transcriptional response to the stimulus, which we could speculate would result in higher accumulation of the recombinant proteins.

In conclusion, we produced phytotoxic α-helical antimicrobial peptides derived from BP100 in plants on the basis of strict regulation of transgene transcription. A requirement was that transcription was below a rigorous threshold in specific plant tissues and developmental stages, particularly during transformation and regeneration of GM plants. The heat shock induced promoter, pHsp82, fulfilled this condition and was capable of driving production of phytotoxic recombinant peptides in plants, although the yield is rather low for most commercial applications. In addition, we demonstrate that, thanks to the increasing information available, *in silico* analysis of transcriptome profiles is a suitable and inexpensive approach to select promoters with specific activity patterns.

Supporting Information

Figure S1 A summary of the expression profiles of 16 genes selected as having the highest expression or induction in response to heat-shock in 7-day IR64 (*indica*) seedlings subjected to cold, salt and drought stress for 3 h. Data on 14-day ZhongHua 11 (*japonica*)

seedlings subjected to heat shock for 3 h are also shown. Dark to light color scale represents high to low expression levels, black corresponding to 11,309 normalized fluorescence units. Arrows indicate the two sequences whose promoters were selected to produce transgenic plants.

Figure S2 Recombinant BP100-DsRed-tag54 accumulation in transgenic rice seedlings. Western blot analysis of proteins from GM plants carrying pHsp82::*dsred-tag54* and pHsp82::*bp100-dsred-tag54* (two independent events taken as example) treated at 42°C. Recombinant protein in TSP was extracted from rice seedlings (five plants per event) and 30 μg of TSP per lane was boiled for five minutes and separated by SDS-PAGE before transfer to nitrocellulose filters. Recombinant proteins were detected using the mAb54k antibody (diluted 1:1,500) and the horseradish peroxidase-labeled anti-mouse IgG secondary antibody (diluted 1:10,000) followed by ECL chemiluminescent detection.

Figure S3 Confocal micrographs of rice radicles from transgenic plantlets carrying pHsp82::*bp100-dsred-tag54* (A and C) and control pHsp82::*dsred-tag54* (B and D), at the V2 developmental stage, subjected to 42°C for 2 hours and further incubated under control growth conditions for three days in a culture chamber. (A and B), DsRed fluorescence; (C and D), bright field. Scale bars: 0.5 μm.

Figure S4 A summary of the expression profiles of a selection of sequences induced by temperature, drought and NaCl stress, and hormone treatment, in callus and seedlings of rice, based on the following *in silico* published data [74–79]. Genes with mRNA levels below that of *Os.hsp18* in control seedlings (control) and above that of *Os.hsp82* after induction (induced) are shown. Normalized expression levels in 28 tissues, relevant to the process of obtaining transgenic plants, of three rice genotypes: Minghui 63 (M, *indica*), Zhenshan 97 (Z, *indica*) and Nipponbare (N, *japonica*) (grey scale). The mRNA levels in these tissues and developmental stages were estimated using the mean expression values (green scale). Dark to light color scale represents high to low expression levels, the darkest corresponding to 37,405 (grey scale) and 12,480 (green scale) normalized fluorescence units.

Table S1 Primer sequences and PCR conditions used. Additional restriction sites are underlined.

Table S2 Sequences with the greatest expression changes in response to heat stress in rice, extracted from publically available microarray hybridization data. Details on the sequence (representative public ID, Affymetrix code and description) and the mRNA expression in response to treatment at 42°C for 3 h [normalized fluorescence units in rice seedlings under control (control) and heat-shock (heat-shock) conditions; fold change (fold) and difference (HS-C) of normalized fluorescence intensities in the two conditions are also indicated]. Note that the sequences with the same Affymetrix number correspond to the same gene. Dark to light scale of shading represents high to low expression level. The promoters of the two sequences indicated in bold were selected to produce transgenic plants.

Table S3 Transformation efficiencies of the different constructs encoding the DsRed-tag54 reporter or the phytotoxic BP100.2 AMP, under the control of a rice heat-shock (pHsp18 and pHsp82) or a maize constitutive (pUbi) promoter.

Acknowledgments

We thank J. Messeguer, E. Melé (IRTA, Cabrils) and E. Montesinos (UdG) for valuable suggestions; M. Amenós (CRAG, Barcelona) for technical

assistance in confocal microscopy; and the scientific writer S. Burgess for revision of the manuscript.

Author Contributions

Conceived and designed the experiments: MP AN. Performed the experiments: NC CR. Analyzed the data: AN MP NC CR. Contributed reagents/materials/analysis tools: AN NC MP. Wrote the paper: MP AN NC.

References

1. Monroc S, Badosa E, Besalú E, Planas M, Bardají E, et al. (2006) Improvement of cyclic decapeptides against plant pathogenic bacteria using a combinatorial chemistry approach. Peptides 27: 2575–2584.
2. Cavallarin L, Andreu D, San Segundo B (1998) Cecropin A-derived peptides are potent inhibitors of fungal plant pathogens. Mol Plant Microbe Interact 11: 218–227.
3. Badosa E, Ferré R, Planas M, Feliu L, Besalú E, et al. (2007) A library of linear undecapeptides with bactericidal activity against phytopathogenic bacteria. Peptides 28: 2276–2285.
4. Marcos JF, Muñoz A, Pérez-Paya E, Misra S, López-García B (2008) Identification and rational design of novel antimicrobial peptides for plant protection. Annu Rev Phytopathol 46: 273–301.
5. López-García B, Pérez-Paya E, Marcos JF (2002) Identification of novel hexapeptides bioactive against phytopathogenic fungi through screening of a synthetic peptide combinatorial library. Appl Environ Microbiol 68: 2453–2460.
6. Montesinos E, Bardaji E (2008) Synthetic antimicrobial peptides as agricultural pesticides for plant-disease control. Chem Biodivers 5: 1225–1237.
7. Nadal A, Montero M, Company N, Badosa E, Messeguer J, et al. (2012) Constitutive expression of transgenes encoding derivatives of the synthetic antimicrobial peptide BP100: impact on rice host plant fitness. BMC Plant Biol 12: 159.
8. Company N, Nadal A, La Paz J-L, Martínez S, Rasche S, et al. (2014) The production of recombinant cationic α-helical antimicrobial peptides in plant cells induces the formation of protein bodies derived from the endoplasmic reticulum. Plant Biotechnol J 12: 81–92.
9. Qu A-L, Ding Y-F, Jiang Q, Zhu C (2013) Molecular mechanisms of the plant heat stress response. Biochem Biophys Res Commun 432: 203–207.
10. Hu W, Hu G, Han B (2009) Genome-wide survey and expression profiling of heat shock proteins and heat shock factors revealed overlapped and stress specific response under abiotic stresses in rice. Plant Sci 176: 583–590. doi:10.1016/j.plantsci.2009.01.016.
11. Qin D, Wu H, Peng H, Yao Y, Ni Z, et al. (2008) Heat stress-responsive transcriptome analysis in heat susceptible and tolerant wheat (Triticum aestivum L.) by using Wheat Genome Array. BMC Genomics 9: 432. doi:10.1186/1471-2164-9-432.
12. Ginzberg I, Barel G, Ophir R, Tzin E, Tanami Z, et al. (2009) Transcriptomic profiling of heat-stress response in potato periderm. J Exp Bot 60: 4411–4421. doi:10.1093/jxb/erp281.
13. Mittal D, Madhyastha DA, Grover A (2012) Genome-wide transcriptional profiles during temperature and oxidative stress reveal coordinated expression patterns and overlapping regulons in rice. PLoS One 7. doi:10.1371/journal.pone.0040899.
14. Li Y-F, Wang Y, Tang Y, Kakani VG, Mahalingam R (2013) Transcriptome analysis of heat stress response in switchgrass (Panicum virgatum L.). BMC Plant Biol 13: 153.
15. Zhang X, Rerksiri W, Liu A, Zhou X, Xiong H, et al. (2013) Transcriptome profile reveals heat response mechanism at molecular and metabolic levels in rice flag leaf. Gene 530: 185–192. doi:10.1016/j.gene.2013.08.048.
16. Todaka D, Nakashima K, Shinozaki K, Yamaguchi-Shinozaki K (2012) Toward understanding transcriptional regulatory networks in abiotic stress responses and tolerance in rice. Rice 5: 6. doi:10.1186/1939-8433-5-6.
17. Hadiarto T, Tran L-SP (2011) Progress studies of drought-responsive genes in rice. Plant Cell Rep 30: 297–310. doi:10.1007/s00299-010-0956-z.
18. Khurana N, Chauhan H, Khurana P (2013) Wheat Chloroplast Targeted sHSP26 Promoter Confers Heat and Abiotic Stress Inducible Expression in Transgenic Arabidopsis Plants. PLoS One 8: e54418. doi:10.1371/journal.pone.0054418.
19. Ramessar K, Capell T, Christou P (2008) Molecular pharming in cereal crops. Phytochem Rev 7: 579–592. doi:10.1007/s11101-008-9087-3.
20. Wakasa Y, Takaiwa F (2013) The use of rice seeds to produce human pharmaceuticals for oral therapy. Biotechnol J 8: 1133–1143. doi: 10.1002/biot.201300065.
21. Irizarry RA, Hobbs B, Collin F, Beazer-Barclay YD, Antonellis KJ, et al. (2003) Exploration, normalization, and summaries of high density oligonucleotide array probe level data. Biostatistics 4: 249–264.
22. Yamamoto YY, Obokata J (2008) ppdb: a plant promoter database. Nucleic Acids Res 36: D977–D981. doi:10.1093/nar/gkm785.
23. Hieno A, Naznin HA, Hyakumachi M, Sakurai T, Tokizawa M, et al. (2013) Ppdb: Plant Promoter Database Version 3.0. Nucleic Acids Res: 1–5.
24. Cornelissen BJ, Horowitz J, van Kan JA, Goldberg RB, Bol JF (1987) Structure of tobacco genes encoding pathogenesis-related proteins from the PR-1 group. Nucleic Acids Res 15: 6799–6811.
25. Vaquero C, Sack M, Chandler J, Drossard J, Schuster F, et al. (1999) Transient expression of a tumor-specific single-chain fragment and a chimeric antibody in tobacco leaves. Proc Natl Acad Sci USA 96: 11128–11133.
26. Matz MV, Fradkov AF, Labas YA, Savitsky AP, Zaraisky AG, et al. (1999) Fluorescent proteins from nonbioluminescent Anthozoa species. Nat Biotechnol 17: 969–973.
27. Rasche S, Martin A, Holzem A, Fischer R, Schinkel H, et al. (2011) One-step protein purification: Use of a novel epitope tag for highly efficient detection and purification of recombinant proteins. Open Biotechnol J 5: 1–6.
28. Sambrook J, Russell D (2001) Molecular Cloning: a laboratory manual. Cold Spring Harbor, N.Y.: Cold Spring Harbor Laboratory Press.
29. Sallaud C, Meynard D, van Boxtel J, Gay C, Bes M, et al. (2003) Highly efficient production and characterization of T-DNA plants for rice (Oryza sativa L.) functional genomics. Theor Appl Genet 106: 1396–1408. doi:10.1007/s00122-002-1184-x.
30. Murashige T, Skoog F (1962) A revised medium for rapid growth and bioassays with tobacco tissue cultures. Physiol Plant 15: 473–497.
31. Vandesompele J, De Preter K, Pattyn F, Poppe B, Van Roy N, et al. (2002) Accurate normalization of real-time quantitative RT-PCR data by geometric averaging of multiple internal control genes. Genome Biol 3: RESEARCH0034. doi:10.1186/gb-2002-3-7-research0034.
32. Christensen AH, Quail PH (1996) Ubiquitin promoter-based vectors for high-level expression of selectable and/or screenable marker genes in monocotyledonous plants. Transgenic Res 5: 213–218.
33. Gross LA, Baird GS, Hoffman RC, Baldridge KK, Tsien RY (2000) The structure of the chromophore within DsRed, a red fluorescent protein from coral. Proc Natl Acad Sci USA 97: 11990–11995. doi:10.1073/pnas.97.22.11990.
34. Richter K, Haslbeck M, Buchner J (2010) The Heat Shock Response: Life on the Verge of Death. Mol Cell 40: 253–266. doi:10.1016/j.molcel.2010.10.006.
35. Rajasekaran K, Cary J, Jaynes J, Montesinos E (2012) Small wonders: peptides for disease control. Washington, DC: Oxford University Press: American Chemical Society: Oxford University Press.
36. Osusky M, Zhou G, Osuska L, Hancock RE, Kay WW, et al. (2000) Transgenic plants expressing cationic peptide chimeras exhibit broad-spectrum resistance to phytopathogens. Nat Biotechnol 18: 1162–1166.
37. Monroc S, Badosa E, Feliu L, Planas M, Montesinos E, et al. (2006) De novo designed cyclic cationic peptides as inhibitors of plant pathogenic bacteria. Peptides 27: 2567–2574.
38. Badosa E, Moiset G, Montesinos L, Talleda M, Bardají E, et al. (2013) Derivatives of the Antimicrobial Peptide BP100 for Expression in Plant Systems. PLoS One 8: e85515.
39. Efeoğlu B (2009) Heat Shock Proteins and Heat Shock Response in Plants. J Sci 22: 67–75.
40. Kotak S, Larkindale J, Lee U, von Koskull-Döring P, Vierling E, et al. (2007) Complexity of the heat stress response in plants. Curr Opin Plant Biol 10: 310–316. doi:10.1016/j.pbi.2007.04.011.
41. Sarkar NK, Thapar U, Kundnani P, Panwar P, Grover A (2013) Functional relevance of J-protein family of rice (Oryza sativa). Cell Stress Chaperones 18: 321–331.
42. Sarkar NK, Kim Y-K, Grover A (2009) Rice sHsp genes: genomic organization and expression profiling under stress and development. BMC Genomics 10: 393. doi:10.1186/1471-2164-10-393.
43. Sarkar NK, Kundnani P, Grover A (2013) Functional analysis of Hsp70 superfamily proteins of rice (Oryza sativa). Cell Stress Chaperones 18: 427–437.
44. Singh A, Singh U, Mittal D, Grover A (2010) Genome-wide analysis of rice ClpB/HSP100, ClpC and ClpD genes. BMC Genomics 11: 95. doi:10.1186/1471-2164-11-95.
45. Sarkar NK, Kim Y-K, Grover A (2014) Coexpression network analysis associated with call of rice seedlings for encountering heat stress. Plant Mol Biol 84: 125–143.
46. Jung Y-J, Lee S-Y, Moon Y-S, Kang K-K (2012) Enhanced resistance to bacterial and fungal pathogens by overexpression of a human cathelicidin

antimicrobial peptide (hCAP18/LL-37) in Chinese cabbage. Plant Biotechnol Rep 6: 39–46. doi:10.1007/s11816-011-0193-0.

47. Yamakawa H, Hakata M (2010) Atlas of rice grain filling-related metabolism under high temperature: joint analysis of metabolome and transcriptome demonstrated inhibition of starch accumulation and induction of amino acid accumulation. Plant Cell Physiol 51: 795–809. doi:10.1093/pcp/pcq034.

48. Rerksiri W, Zhang X, Xiong H, Chen X (2013) Expression and promoter analysis of six heat stress-inducible genes in rice. ScientificWorldJournal 2013: 397401.

49. Lindquist S, Craig E (1988) The heat-shock proteins. Annu Rev Genet 22: 631–677.

50. Lee GJ, Roseman AM, Saibil HR, Vierling E (1997) A small heat shock protein stably binds heat-denatured model substrates and can maintain a substrate in a folding-competent state. EMBO J 16: 659–671. doi:10.1093/emboj/16.3.659.

51. Blumenthal C, Bekes F, Wrigley C, Barlow E (1990) The Acquisition and Maintenance of Thermotolerance in Australian Wheats. Aust J Plant Physiol 17: 37. doi:10.1071/PP9900037.

52. Feder ME, Hofmann GE (1999) Heat-shock proteins, molecular chaperones, and the stress response: evolutionary and ecological physiology. Annu Rev Physiol 61: 243–282. doi:10.1146/annurev.physiol.61.1.243.

53. Iba K (2002) Acclimative response to temperature stress in higher plants: approaches of gene engineering for temperature tolerance. Annu Rev Plant Biol 53: 225–245. doi:10.1146/annurev.arplant.53.100201.160729.

54. Sorensen JG, Kristensen TN, Loeschcke V (2003) The evolutionary and ecological role of heat shock proteins. Ecol Lett 6: 1025–1037.

55. Young R, Elliott T (1989) Stress proteins, infection, and immune surveillance. Cell 59: 5–8.

56. Vinocur B, Altman A (2005) Recent advances in engineering plant tolerance to abiotic stress: Achievements and limitations. Curr Opin Biotechnol 16: 123–132. doi:10.1016/j.copbio.2005.02.001.

57. Vierling E (1991) The Roles of Heat Shock Proteins in Plants. Annu Rev Plant Physiol Plant Mol Biol 42: 579–620.

58. Siddique M, Gernhard S, von Koskull-Döring P, Vierling E, Scharf K-D (2008) The plant sHSP superfamily: five new members in Arabidopsis thaliana with unexpected properties. Cell Stress Chaperones 13: 183–197. doi:10.1007/s12192-008-0032-6.

59. Waters ER, Lee GJ, Vierling E (1996) Evolution, structure and function of the small heat shock proteins in plants. J Exp Bot 47: 325–338.

60. Sun W, Van Montagu M, Verbruggen N (2002) Small heat shock proteins and stress tolerance in plants. Biochim Biophys Acta 1577: 1–9.

61. Ouyang Y, Chen J, Xie W, Wang L, Zhang Q (2009) Comprehensive sequence and expression profile analysis of Hsp20 gene family in rice. Plant Mol Biol 70: 341–357. doi:10.1007/s11103-009-9477-y.

62. Sarkar NK, Kim YK, Grover A (2009) Rice sHsp genes: genomic organization and expression profiling under stress and development. BMC Genomics 10: 393.

63. Ham D-J, Moon J-C, Hwang S-G, Jang CS (2013) Molecular characterization of two small heat shock protein genes in rice: their expression patterns, localizations, networks, and heterogeneous overexpressions. Mol Biol Rep 1–12.

64. Chang PFL, Jinn TL, Huang WK, Chen Y, Chang HM, et al. (2007) Induction of a cDNA clone from rice encoding a class II small heat shock protein by heat

65. Civán P, Svec M (2009) Genome-wide analysis of rice (Oryza sativa L. subsp. japonica) TATA box and Y Patch promoter elements. Genome 52: 294–297. doi:10.1139/G09-001.

66. Yamamoto YY, Ichida H, Matsui M, Obokata J, Sakurai T, et al. (2007) Identification of plant promoter constituents by analysis of local distribution of short sequences. BMC Genomics: 8:8:67. doi:10.1186/1471-2164-8-67.

67. Wahid A, Gelani S, Ashraf M, Foolad MR (2007) Heat tolerance in plants: An overview. Environ Exp Bot 61: 199–223. doi:10.1016/j.envexpbot.2007.05.011.

68. Conley AJ, Joensuu JJ, Menassa R, Brandle JE (2009) Induction of protein body formation in plant leaves by elastin-like polypeptide fusions. BMC Biol 7: 48.

69. Joensuu JJ, Conley AJ, Lienemann M, Brandle JE, Linder MB, et al. (2010) Hydrophobin fusions for high-level transient protein expression and purification in Nicotiana benthamiana. Plant Physiol 152: 622–633. doi:10.1104/pp.109.149021.

70. Torrent M, Llompart B, Lasserre-Ramassamy S, Llop-Tous I, Bastida M, et al. (2009) Eukaryotic protein production in designed storage organelles. BMC Biol 7: 5.

71. Larkindale J, Vierling E (2008) Core genome responses involved in acclimation to high temperature. Plant Physiol 146: 748–761. doi:10.1104/pp.107.112060.

72. Breitler JC, Vassal JM, del Mar Catala M, Meynard D, Marfà V, et al. (2004) Bt rice harbouring cry genes controlled by a constitutive or wound-inducible promoter: protection and transgene expression under Mediterranean field conditions. Plant Biotechnol J 2: 417–430. doi:10.1111/j.1467-7652.2004.00086.x.

73. Dai Z, Hooker BS, Anderson DB, Thomas SR (2000) Expression of Acidothermus cellulolyticus endoglucanase E1 in transgenic tobacco: biochemical characteristics and physiological effects. Transgenic Res 9: 43–54. doi:10.1023/A:1008922404834.

74. Jain M, Khurana JP (2009) Transcript profiling reveals diverse roles of auxin-responsive genes during reproductive development and abiotic stress in rice. FEBS J 276: 3148–3162. doi:10.1111/j.1742-4658.2009.07033.x.

75. Jain M, Nijhawan A, Arora R, Agarwal P, Ray S, et al. (2007) F-box proteins in rice. Genome-wide analysis, classification, temporal and spatial gene expression during panicle and seed development, and regulation by light and abiotic stress. Plant Physiol 143: 1467–1483. doi:10.1104/pp.106.091900.

76. Sharma R, Mohan Singh RK, Malik G, Deveshwar P, Tyagi AK, et al. (2009) Rice cytosine DNA methyltransferases - gene expression profiling during reproductive development and abiotic stress. FEBS J 276: 6301–6311. doi:10.1111/j.1742-4658.2009.07338.x.

77. Zheng L, Huang F, Narsai R, Wu J, Giraud E, et al. (2009) Physiological and transcriptome analysis of iron and phosphorus interaction in rice seedlings. Plant Physiol 151: 262–274. doi:10.1104/pp.109.141051.

78. Dubey S, Misra P, Dwivedi S, Chatterjee S, Bag SK, et al. (2010) Transcriptomic and metabolomic shifts in rice roots in response to Cr (VI) stress. BMC Genomics 11: 648. doi:10.1186/1471-2164-11-648.

79. Kang K, Park S, Natsagdorj U, Kim YS, Back K (2011) Methanol is an endogenous elicitor molecule for the synthesis of tryptophan and tryptophan-derived secondary metabolites upon senescence of detached rice leaves. Plant J 66: 247–257. doi:10.1111/j.1365-313X.2011.04486.x.

stress, mechanical injury, and salicylic acid. Plant Sci 172: 64–75. doi:10.1016/j.plantsci.2006.07.017.

12

Reduction of the Cytosolic Phosphoglucomutase in Arabidopsis Reveals Impact on Plant Growth, Seed and Root Development, and Carbohydrate Partitioning

Irina Malinova[1,2], Hans-Henning Kunz[3,5], Saleh Alseekh[4], Karoline Herbst[1], Alisdair R. Fernie[4], Markus Gierth[5], Joerg Fettke[1,2]*

1 Plant Physiology, University of Potsdam, Potsdam-Golm, Germany, 2 Biopolymers analytics, University of Potsdam, Potsdam-Golm, Germany, 3 Plant Physiology, Washington State University, Pullman, Washington, United States of America, 4 Max-Planck-Institute of Molecular Plant Physiology, Potsdam-Golm, Germany, 5 Department of Botany II, University of Cologne, Cologne, Germany

Abstract

Phosphoglucomutase (PGM) catalyses the interconversion of glucose 1-phosphate (G1P) and glucose 6-phosphate (G6P) and exists as plastidial (pPGM) and cytosolic (cPGM) isoforms. The plastidial isoform is essential for transitory starch synthesis in chloroplasts of leaves, whereas the cytosolic counterpart is essential for glucose phosphate partitioning and, therefore, for syntheses of sucrose and cell wall components. In Arabidopsis two cytosolic isoforms (PGM2 and PGM3) exist. Both PGM2 and PGM3 are redundant in function as single mutants reveal only small or no alterations compared to wild type with respect to plant primary metabolism. So far, there are no reports of Arabidopsis plants lacking the entire cPGM or total PGM activity, respectively. Therefore, *amiRNA* transgenic plants were generated and used for analyses of various parameters such as growth, development, and starch metabolism. The lack of the entire cPGM activity resulted in a strongly reduced growth revealed by decreased rosette fresh weight, shorter roots, and reduced seed production compared to wild type. By contrast content of starch, sucrose, maltose and cell wall components were significantly increased. The lack of both cPGM and pPGM activities in Arabidopsis resulted in dwarf growth, prematurely die off, and inability to develop a functional inflorescence. The combined results are discussed in comparison to potato, the only described mutant with lack of total PGM activity.

Editor: Miyako Kusano, RIKEN Center for Sustainable Resource Science, Japan

Funding: The authors acknowledge financial support by the Deutsche Forschungsgemeinschaft to J.F. (DFG-Az. FE 1030/1-1 and FE 1030/2-1). H.-H.K. was supported by the Human Frontier Science Program Long-Term fellowship and an Alexander von Humboldt Feodor Lynen fellowship. The funders had no role in study design, data collection and analysis, decision to publish, or preparation of the manuscript.

* Email: fettke@uni-potsdam.de

Introduction

Phosphoglucomutase (PGM) catalyzes the reversible interconversion of glucose 6-phosphate (G6P) and glucose 1-phosphate (G1P). In higher plants PGM activity is verifiable in two compartments, the plastidial stroma and the cytosol. The plastidial isoform is essential for the formation of glucose 1-phosphate a substrate of ADPglucose pyrophosphorylase and, therefore, for starch synthesis. Lack of this isoform results in dramatically diminished starch levels [1,2]. Furthermore, mutants lacking the ability to form starch displayed a higher amount of soluble sugars, like glucose and sucrose [3,4]. The latter carbohydrate is the main transport form in higher plants and supplies non-photosynthetic tissues and organs of the plant with energy and carbon. Sucrose is formed in the light from triose-phosphates exported from chloroplasts. During the formation of sucrose the cytosolic PGM (cPGM) is essential as it converts G6P into G1P, which is the substrate for the UDPglucose pyrophosphorylase.

Also in the dark, when the photosynthetic driven export of carbon from the chloroplast is absent, the formation of sucrose is dependent on cPGM activity [5,6]. Furthermore, this pathway is linked to starch breakdown products. By the action of various enzymes, in most cases hydrolyzing enzymes, the transitory starch is degraded and the major carbohydrates released from the chloroplasts are glucose and maltose [5,7,8]. Starch derived maltose enters the cytosol via maltose exporter 1 (MEX1; [9]) and is further metabolized by disproportionating enzyme 2 (DPE2; [10,11,12]). DPE2 transfers one of the glucosyl residues (the non-reducing) of maltose on cytosolic heteroglycans and releases the second as free glucose. The glucosyl residues of the cytosolic heteroglycans can be released as G1P by the action of the cytosolic phosphorylase (AtPHS2; [13,14]). However, the starch derived glucose is exported from the chloroplast via pGlcT [15,16]. Both the exported glucose and the glucose released by the action of DPE2 are thought to be immediately converted into G6P by the action of hexokinase [5]. The cPGM controls partitioning of both sugar phosphates in the cytosol. G6P is used primarily in respiratory pathways, whereas G1P is linked to sucrose metabolism and in addition to cell wall synthesis. *Arabidopsis thaliana*, tobacco and maize contain one plastidial and two cytosolic isoforms; for potato and spinach only one plastidial and one cytosolic isoform were reported [17,18,19,20,21]. Recently, potato plants with antisense repression of cytosolic phosphoglucomutase were analyzed. These plants displayed a stunted phenotype, diminished root growth and reduced tuber yield [20]. Antisense plants were also characterized by reduced rates of photosynthesis and dramatic reduction in nucleotide level compared to the wild type [22]. Moreover, transgenic lines with altered cPGM activity revealed alterations in starch-related cytosolic heteroglycans. From these results it was concluded that elevated levels of cPGM activity favor the cytosolic phosphorylase-mediated conversion of glucosyl residues from the cytosolic heteroglycans into the cytosolic hexose-phosphate pools during starch degradation [23].

The two genes encoding cytosolic phosphoglucomutase activities in *Arabidopsis thaliana* At1g23190 (PGM 3) and At1g70730 (PGM2) [24,17] reveal high sequence homology as well as possess similar exon/intron structures. Indeed, they encode two isoforms with 91% sequence identity at the amino acid level. Egli *et al.* [24] reported that *pgm2* and *pgm3* mutants deficient in one of the cytosolic isoforms grown under standard 12 h light/12 h dark regime displayed phenotypes similar to that of wild type. The authors suggested that under these conditions the functions of the isoforms were redundant to one another and the loss of one isoform did not affect plant metabolism. Unfortunately, the generation of double mutants was unsuccessful, as formation of homozygous seeds was prevented. Therefore, it was concluded that an absolute lack of cPGM activity compromises gametophyte development [24].

Not so long ago, transgenic potato lines with strongly decreased total PGM activities were identified. Transgenic plants were reduced in growth, tuber yield, and revealed lower levels of starch and sucrose in leaves compared to wild type [25]. Interestingly, rate of starch synthesis was similar to the wild type [26]. A possible explanation for this phenotype is a direct G1P transport over the plastidial membranes, which has been verified for both potato and Arabidopsis [27,1].

However, until now no *A. thaliana* transgenic plants with a strong reduction of both cPGM isoforms or the simultaneous reduction of plastidial and cytosolic phosphoglucomutases have been reported. For this reason, we generated and analyzed Arabidopsis lines with *amiRNA* (artificial micro RNA) repression of both cPGMs. Furthermore, the cPGM *amiRNA* construct was introduced into *pgm1* mutants by Agrobacterium mediated

transformation to explore whether a similar bypass to that observed in potato also occurred in Arabidopsis. In order to test this, the generated plants were assessed at the level of isoform specific activity as well as carbohydrate and metabolite content and phenotypic characterization of vegetative growth and propagative development. Results are discussed in the context of current understanding of the importance of the reactions catalyzed by phosphoglucomutase.

Materials and Methods

Plant material and growth conditions

The *pgm1* mutants were as described in [17]. The *pgm2* [SALK_068481 (AR)] and *pgm3* [SALK_023069 (AZ)] mutants were ordered from NASC. Mutants were identified by PCR amplification using the primers presented in Table S1 in File S1.

pgm2 pgm1 and *pgm3 pgm1* were generated by crosses between individual homozygous mutants and the resulting F1 generations were allowed to self-pollinate. Double mutants were identified in the F2 generation by native PAGE and PGM activity staining (see below). For further analyses plants of F3 or F4 generation were used. Plants were grown either in 14, 10, 8 or 7 h light (110 μmol m^{-2}s^{-1}, 22°C; dark, 18°C, humidity 60%) or in a 12 h diurnal cycle (12 h light [110 μmol m^{-2}s^{-1}], 20°C; 12 h dark, 16°C, humidity 60%). For all Arabidopsis lines used, the genetic background was Col-0.

Generation of *amiRNA* c-pgm plants

The PGM2/3-specific *amiRNA* (tctgttaagataaatgcgcct) was designed and amplified by three consecutive PCR reactions according to guidelines found at http://wmd3.weigelworld.org using the vector pRS300 as template [28] (Table S1 in File S1). The final PCR product including the cPGM-specific *amiRNA* was subcloned into the pENTR/D-TOPO vector and sequence identity was verified. Subsequently, the *amiRNA* was recombined by L/R reaction into pGWB2 [29] to obtain the binary expression plasmid *p35S:amiRNA cPGM*. The binary vector was transformed into *Argobacterium tumefaciens* strain GV3101 and used for plant transformation.

Plant transformation was performed using the floral dip method [30]. Agrobacterium strains were grown in 1 L of LB medium containing antibiotics rifampicin (100 mg/L), kanamycin (50 mg/L), gentamycin (25 mg/L), hygromycin (50 mg/L) at 28°C for 24 h. Cells were collected by centrifugation at 4,000 **g** for 15 min at room temperature (RT) and gently resuspended in 1 L of freshly made 5% [w/v] sucrose solution containing 0.02% [v/v] Silwet L-

Table 1. Carbohydrate content.

carbohydrate	growth photoperiod [h light]	Col-0	pgm3	pgm2
starch [mg/g FW]	7 h	6.1±0.3	5.4±0.6	6.1±0.3
	8 h	9.2±0.4	7.7±0.3	9.4±0.2
	10 h	6.9±0.4	5.0±0.1	6.1±0.2
	14 h	7.0±0.7	6.7±0.6	6.4±0.6
sucrose content [μmol/g FW]	7 h	1.55±0.11	1.45±0.03	1.81±0.03*
	8 h	1.17±0.04	1.04±0.05	1.87±0.33*
	10 h	1.86±0.08	1.96±0.09	2.71±0.05*
	14 h	2.58±0.19	2.46±0.21	2.90±0.03**

Leaves were harvested one hour before beginning of the dark phase. Values are means of four replicates representing a mix of 7–10 plants ± SD. Asterisks denote the significance levels as comparing mutants to Col-0 : * $p \leq 0.01$;** $p \leq 0.05$.

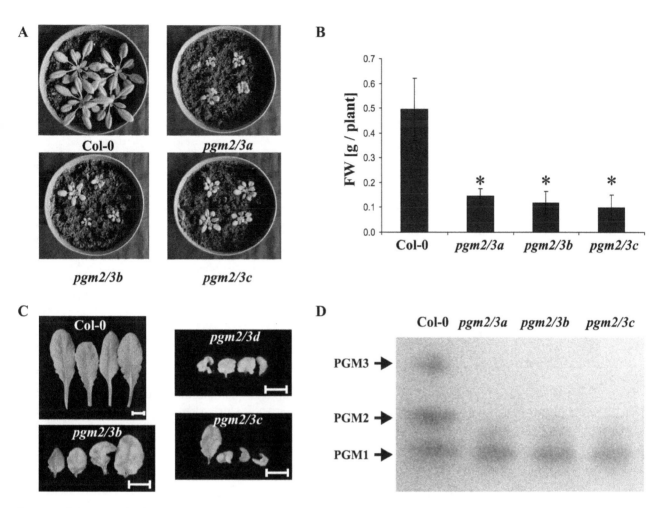

Figure 1. Phenotype of Col-0 and *pgm2/3* plants in 12 h light/12 h dark regime. A, Growth phenotypes. Photographs were taken of six-week-old plants. Bar = 1 cm. B, Fresh weight of plant rosettes. Values are means ± SD (n = 29–30). Plants were harvested after five weeks. Asterisks indicate significant difference from Col-0 (Student Test, P≤0.01). C, Leaf form from Col-0 and transgenic plants. Leaves were harvested from the middle of rosettes from six-week-old plants. Bar = 1 cm. D, Phosphoglucomutase activity in Col-0 and *pgm2/3* plants. Crude extracts were subjected to native PAGE and subsequent PGM activity staining. Separation gel was 7.5% [T] and 25 µg protein was loaded per lane.

77 (Lehle Seeds, USA). Col-0 and *pgm1* plants (approximately four to five weeks after germination) were used for transformation. On reaching the mature stage plants were transferred to a 14 h light/10 h dark regime until mature silique stage.

Screening of *amiRNA* plants

Dry seeds from transformed plants were collected and sterilized. Seeds were immersed in 70% [v/v] ethanol for 5 min, followed by a 20 min soaking in 2.4% [w/v] sodium hypochlorite, 0.02% [v/v] Triton X-100. Seeds were rinsed six times with sterile water and dried under sterile conditions. Seeds were screened on MS-plates with sucrose (4.3 g/L MS salt (Duchefa, Haarlem, Netherlands), 2.5 mM MES, pH 5.7 (NaOH), 1% [w/v] sucrose, 0.8% [w/v] Agar-agar) except where indicated. Selective antibiotics were added: hygromycin (50 mg/L), kanamycin (50 mg/L). Plates were placed in growth chambers and plants were germinated under 12 h light/12 h dark, except otherwise stated. Transformants with well developed leaves (four leaves stage) and roots were planted in soil and grown under standard conditions (12 h light/12 h dark). Seeds of at least four plants were harvested separately and used for generation of four plant lines (*pgm2/3 a* to *d*). Analyses were performed with the F3 to F5 generation of the respective lines.

Phosphoglucomutase assay and PGM activity staining

Buffer-soluble proteins were extracted as described elsewhere [12]. Phosphoglucomutase activity measurement was performed as described [23]. However, in the reaction mixture soluble starch and rabbit muscle phosphorylase were omitted. Measurement was started by addition of 17.5 mM G1P to the reaction mixture. Native PAGE and PGM activity staining were performed according to Fettke *et al.* [23].

Carbohydrate quantification

Starch was extracted and measured as described [1]. Monosaccharides, disaccharides and sugar phosphates were determined according to Stitt *et al.* [31].

Isolation and analysis of cell wall matrix polysaccharides

Leaf material, frozen in liquid nitrogen, was homogenized and resuspended in ice-cold 20% [v/v] ethanol, mixed thoroughly, and centrifuged for 10 min at 20,000 **g** (4°C). Pellets were washed with 20% [v/v] ethanol two times, finally resuspended in 70% [v/v] ethanol and centrifuged (as above). Subsequently, pellets were resuspended in chloroform/methanol (1:1 [v/v]) and incubated for 20 min under continuous stirring followed by centrifugation (as

A

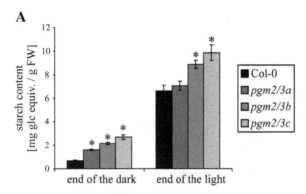

B

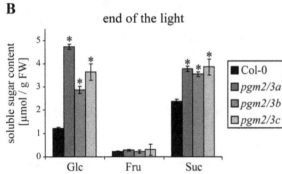

C

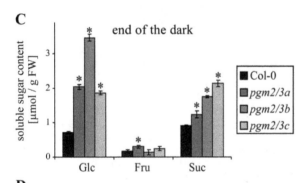

D

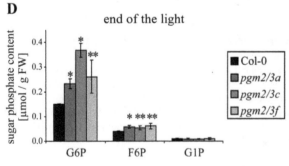

E

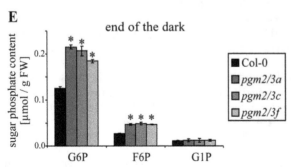

Figure 2. Carbohydrate analysis of Col-0 and *pgm2/3* plants. A–E, Plants were grown under 12 h light/12 h dark conditions and after five weeks 7–8 plants were collected and homogenized per line. Values are means of four technical replicates (A–C), and three technical parallels (D–E) ± SD, respectively. A, Starch content. B–C, Soluble sugar content. D–E, Sugar phosphate content. Asterisks denote the significance levels comparing *pgm2/3* mutants to Co1-0: * $p \leq 0.01$;** $p \leq 0.05$.

above). The resulting pellets were completely destained by washing with acetone followed by water. Then pellets were resolved in 0.1 M sodium acetate buffer (pH 5.0) and incubated for 20 min at 80°C. The suspension was cooled to RT and residual starch was removed by treatment with 25 U of α-amylase (from *Basillus sp.* Typ II-A, Sigma-Aldrich, Germany) and 7 U pullulanase (from *Klebsiella planticola*, Macerozyme, Ireland) as described elsewhere [32]. The residual pellet was washed at least five times with water and subjected to TFA hydrolysis (2 M final concentration) for 3 h at 100°C. After that samples were centrifuged and the supernatants were collected. Pellets were washed two times with water and supernatants pooled together. Collected supernatant represents matrix polysaccharides of the cell wall. Following lyophilization, samples were dissolved in water and monomer content was estimated [33] (glucose was used as a standard). Aliquots were subjected to HPAEC-PAD for monosaccharide separation (as described elsewhere [12]).

Isolation and quantification of crystalline cellulose

Residual pellets from cell wall matrix isolation were subjected to hydrolysis in Updegraff reagent (8:1:2 of concentrated acetic acid:concentrated nitric acid:water) [34] for 30 min at 100°C. Crystalline cellulose was separated, completely hydrolyzed into glucose, and determined as described elsewhere [35].

Metabolic Profiling

For GC-MS analyses, Col-0 and transgenic lines were grown in 12 h light/12 h dark regime and harvested at the end of the light and at the end of the dark. Plants were five-week-old. Leaves from several plants per line were pooled together and processed as previously described [36].

Trypan blue staining

Trypan blue (Sigma-Aldrich, Germany) staining was performed as described [37]. Leaves were boiled 1 min at 100°C with lactophenol-trypan blue solution (10 mL lactic acid, 10 mL glycerol, 10 g phenol, 10 mL 0.1% [w/v] trypan blue solution) and decolorized with chloral hydrate (2.5 g mL^{-1} distilled water) overnight.

Statistical analysis

Statistical analysis (Student's t-test [two-sided]) was performed using MS Excel 2010 (Microsoft Corporation, Washington, USA).

Results

Elimination of one cPGM isoform in Arabidopsis has no significant effect on starch metabolism

In native PAGE the total PGM activity was resolved in three distinct bands of activity, the fastest moving band represented the plastidial PGM (PGM1), whereas the slowest moving band represented PGM3 (At1g23190) and the intermediate band PGM2 (At1g70730). Both PGM2 and PGM3 are cytosolic isoforms [23,24]. The localization of the three isoforms was further confirmed by non-aqueous fractionation [38]. All three

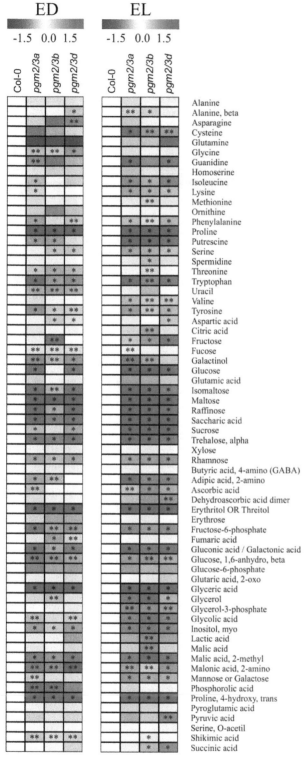

Figure 3. Overlay heat map of the metabolite changes in *pgm2/3* mutants in comparison with control (Col-0) using false-color scale. Red or blue indicate that the metabolite content is increased or decreased, respectively. Five-week-old plants were grown under 12 h light/12 h dark conditions and harvested at the end of light phase (EL) and dark phase (ED), and three replicates represented 3–4 plants were analyzed (two technical replicates each); asterisks denote the significance levels as comparing *pgm2/3* mutants to Co1-0 : * $p \leq 0.01$;** $p \leq 0.05$.

isoforms were detected in various organs (Fig. S1A in File S1). PGM activity was analyzed in leaves of different Arabidopsis accessions (Fig. S1B in File S1). Results indicate a wide diversity of cytosolic PGM isoforms. Consistent with previously published data [24], Cvi-0 was the single accession which displayed only one cytosolic isoform.

Two mutants lacking an isoform of cytosolic PGM (*pgm2*, *pgm3*) were previously analyzed [24]. No substantial differences compared to the wild type were observed even when various parameters like starch and soluble sugar content as well as root and shoot growth were examined. However, we here generated independent homozygous T-DNA mutant lines. The total reduction in PGM activity was determined to be 23% in *pgm3* plants and 35% in *pgm2* plants compared to control Col-0. These results were consistent with the PGM activity staining analysis (Fig. S1B in File S1), since the PGM2 band had a higher intensity than PGM3.

Additionally, PGM2 and PGM3 proteins from *A. thaliana* have previously been cloned and expressed in *Escherichia coli* and the recombinant proteins were analyzed for substrate specificity and affinity. However, no differences between PGM2 and PGM3 were observed [39].

In order to analyze the influence of different growth conditions on *pgm2* and *pgm3* mutants, plants were cultivated under various light/dark conditions (light phase: 7 h, 8 h, 10 h or 14 h). Still both mutants revealed a similar growth phenotype (data not shown) and starch content compared to the Col-0. The *pgm2* plants displayed an increased level of sucrose under different growth conditions but this was not observed for *pgm3* (Table 1). Most likely, PGM2 has a higher impact on glucose-phosphate turnover. However, no significant differences in steady-state levels of sugar phosphate contents (F6P, G1P, G6P) were observed (data not shown).

As the cytosolic pools of sugar phosphates are linked to starch metabolism via the action of two transglucosidases (DPE2 and AtPHS2), the activity of both enzymes and the composition of soluble heteroglycans (SHG$_L$) were analyzed. However, neither differences in enzyme activities nor composition of SHG$_L$ were observed (Fig. S2 in File S1).

Thus, it seems likely that PGM2 and PGM3 could substitute for one another since the residual PGM activity in either mutant is relatively abundant.

Simultaneous reduction of PGM2 and PGM3 activities affect plants growth and carbohydrate partitioning

Given that single *pgm2* and *pgm3* mutants do not reveal significant changes in e.g. starch metabolism, generation of double mutants is essential to clarify the role of cPGM for plant metabolism. An *amiRNA* cPGM construct was therefore transformed into Col-0 plants and four independent lines were generated. Transgenic *pgm2/3* lines were strongly retarded in growth and revealed diminished fresh weight compared to Col-0 (Fig. 1A–B). Additionally, *pgm2/3* leaves revealed small and abnormally curled leaves (Fig. 1C) and slightly elevated chlorophyll levels (Table S2 in File S1). Protein crude extracts of Col-0 and *pgm2/3* leaves were subjected to native PAGE and PGM activity staining (Fig. 1D). In all *pgm2/3* lines the two bands of cPGM activity were below the limit of detection (cPGM activity was not observed, even if 75 μg of protein crude extracts were loaded on the gel; data not shown). In addition, PGM activities in protein crude extracts were measured (Fig. S3A in File S1). In all three transgenic lines a strong reduction in total PGM activity was observed (residual activity 30–34%, [wt = 100%]). Furthermore, analyses of gene expression revealed that PGM2 and PGM3 were

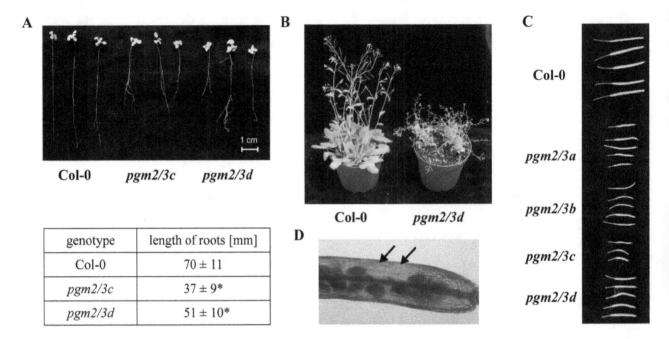

Figure 4. Roots and siliques of Col-0 and *pgm2/3* plants. A, Root length and morphology of Col-0 and *pgm2/3* lines. Plants were grown on vertical MS plates without any external sugar and antibiotics under long day conditions (16 h light/8 h dark). Plants were two-week-old. Length of central roots was measured. Values are means ± SD (n = 26−35). Asterisks indicate significant difference from Col-0 (Student Test, P≤0.01) B, Mature Col-0 and *pgm2/3d* plants. Col-0 and *pgm2/3* plants were six and 11- week-old, respectively. C, Morphology of siliques of Col-0 and *pgm2/3* lines. D, *pgm2/3d* silique. Siliques were destained in chloral hydrate solution (2.5 g in 1 mL distilled water). Black arrows indicate absence of seeds. C–D, Plants were grown under 14 h light/10 h dark regime.

strongly down-regulated in *pgm2/3* lines. In contrast PGM1 was somewhat up-regulated (Fig. S3B in File S1).

However, transgenic *pgm2/3* plants grown under prolonged day conditions (14 h light/10 h dark) revealed similar results with transgenic plants being significantly smaller than Col-0, but larger as compared to the 12 h light/12 h dark grown plants (Fig. S3C in File S1).

With respect to metabolites all *pgm2/3* lines showed increased starch content at the end of the dark phase compared to Col-0 (Fig. 2A). The increased starch content was also detected at the end of the light phase except for *pgm2/3a*. Similarly, starch content was significantly increased in *pgm2/3* lines compared to Col-0 when grown in 14 h light/10 h dark regime (data not shown). Transgenic *pgm2/3* lines displayed increased levels of glucose and sucrose on a fresh weight basis. In contrast the amount of fructose was comparable in the transgenic lines and Col-0 (Fig. 2B–C). Similar results were also obtained, if metabolite content was evaluated on a dry weight basis (data not shown).

Given that PGMs catalyze the interconversion of G1P and G6P, levels of sugar phosphates were determined. The *pgm2/3* plants displayed increased levels of G6P and fructose 6-phosphate (F6P) but G1P levels were similar to those in Col-0 (Fig. 2D–E). Nevertheless, further enzymes involved in the metabolism (DPE2 and phosphorylases) were not affected (Fig. S3D in File S1). In addition metabolic profiling was performed, revealing that numerous metabolites were increased both at the end of light and dark phase. At the end of the light period clear increases were seen in a range of sugars including maltose, glucose, trehalose, isomaltose and raffinose as well as the sugar alcohols galactinol, inositol and erythritol or threitol but fructose was unchanged or even decreased. Similarly, a large number of amino and organic acids were increased in the transgenic lines including tryptophan, proline, galacturonic acid, malate and shikimate (Fig. 3, Table S3 in File S1). By contrast, relatively few metabolites were consistently decreased in the transgenic lines at this time point those that were included were ornithine, phosphoric acid, asparagine, glutamine, and malonate. Consistent with these global effects on the primary

Table 2. Amount of crystalline cellulose and of cell wall matrix in Col-0 and *pgm2/3*.

genotype	crystalline cellulose [mg/g FW]	cell wall matrix [mg/g FW]
Col-0	5.17±0.42	4.73±0.01
pgm2/3a	6.24±0.11*	7.42±0.85*
pgm2/3b	5.80±0.06**	6.28±0.33*
pgm2/3c	5.43±0.24	6.63±0.58*

Plants were grown under 12 h light/12 h dark regime and harvested at the end of the light phase (six-week-old). Values are means of four replicates representing a mix of 7–10 plants ± SD. Asterisks denote the significance levels as comparing *pgm2/3* mutants to Co1-0 : * $p≤0.01$;** $p≤0.05$.

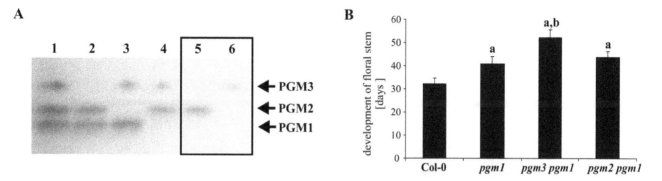

Figure 5. Characterization of knock-out mutants lacking one cytosolic and the plastidial PGM. A, Analysis of PGM activity in the Col-0 and *pgm3 pgm1* and *pgm2 pgm1* mutants using native PAGE and PGM activity staining. Separation gel 7.5% [T]. 35 µg proteins were loaded per lane. 1– Col-0, 2– *pgm3*, 3– *pgm2*, 4– *pgm1*, 5– *pgm3 pgm1*, 6– *pgm2 pgm1*. B, Analysis of floral stems development in Col-0 and different PGM knock-out plants. Plants were grown under long day conditions (14 h light/10 h dark). Days after germination were registered, when plants developed floral stems 1 cm long. Values are means ± SD (n = 24). a - significant difference from Col-0 (Student Test, $p \leq 0.01$), b - significant difference from *pgm1* (Student Test, $p \leq 0.01$).

metabolome being strongly influenced by the sugar status and more specifically by a likely inhibition of sucrose export, they became considerably stronger and more consistent by the end of the night. At this time point all three transgenic lines display alterations including maltose, glucose, trehalose, isomaltose, raffinose, galactinol, inositol, and erythritol or threitol, fructose 6-phosphates, tryptophan, proline, galacturonic acid, malate, and shikimate, which were also elevated in the day. Additionally, the levels of amino adipic acid, guanadine, glutamate, glycolate, lactate, and the branched chain amino acid increased in the dark. As for the situation observed in the light this is most likely the result of inhibition of sucrose export from the leaves. By contrast, at the end of the night the levels of malonate, pyruvate, glutamine and to a lesser extent succinate were significantly decreased in the transgenic lines. The exact reasons underlying these decreases are, however, unclear from the current study.

As G1P is strictly connected with formation of UDP-glucose in the cytosol, which acts as a major substrate for synthesis of cell wall constituents [40], crystalline cellulose and matrix component were analyzed. The *pgm2/3* lines displayed increased amounts of cell wall matrix components and in two of the lines the crystalline cellulose amount was altered (Table 2). Additionally, samples of cell wall matrix were hydrolyzed and the monomer composition was analyzed using HPAEC-PAD. The transgenic lines were characterized by an increased amount of all analyzed monosaccharides and changes in the arabinose/galactose ratio in comparison to Col-0 (Fig. S3E in File S1). For analyses of the impact of cPGM on roots Col-0 and two *pgm2/3* lines were grown on vertical MS plates. *amiRNA pgm2/3* plants carry antibiotic resistance markers, kanamycin and hygromycin. However, it was reported that hygromycin is toxic even to resistant plants during long exposure, which may cause their abnormal development [41]. Indeed, when *pgm2/3* plants were grown in the presence of antibiotics, roots of *pgm2/3* transgenic lines were much shorter and more branched as compared to Col-0 cultivated without antibiotics (data not shown). To avoid such effects, Col-0 and *pgm2/3* seeds were sown on vertical MS plates without antibiotics. After two weeks plants were gently removed from plates and the length of main root was measured (Fig. 4A). Additionally, the lack of cytosolic PGM activity was confirmed in these plants using native PAGE. The root length of transgenic plants was increased on plates without antibiotics (compared to MS plates containing antibiotics), which confirmed that the antibiotics might affect the

root growth of the transgenic plants. However, even without antibiotics the root length of transgenic plants was significantly decreased in comparison to Col-0 (Fig. 4A).

Furthermore, it was observed that *pgm2/3* lines were delayed in silique development, as compared to Col-0, independent of growth conditions (short day, long day) (Fig. 4B). The *pgm2/3* transgenic lines develop mature siliques approximately after 10–11 weeks under long day conditions (14 h light/10 h dark regime), whereas Col-0 achieves this after five to six weeks. Siliques from *pgm2/3* lines are much smaller (Fig. 4C) and possess a lower number of seeds compared to Col-0 (data not shown). In addition missing seeds were observed in the siliques of the transgenics (Fig. 4D).

Impact of simultaneous reduction of cytosolic and plastidial phosphoglucomutase activities on Arabidopsis plants

Action of the plastidial phosphoglucomutase (PGM1) is an essential step in starch synthesis. Arabidopsis mutants lacking PGM1 are strongly reduced in starch content [1,2]. In order to analyze the influence of single PGM2 or PGM3 mutation in the *pgm1* background, *pgm2* and *pgm3* mutants were crossed with *pgm1*. Both *pgm2 pgm1* and *pgm3 pgm1* are similar in growth compared to *pgm1*, under long day conditions (Fig. S4 in File S1). Crude extracts from double mutants were subjected to native PAGE and PGM activity staining (Fig. 5A). Both double mutants possess one band of cPGM activity each. Total PGM activity was reduced to 38±2% for *pgm3 pgm1* mutants and 36±2% for *pgm2 pgm1* plants (wt = 100%; n = 3).

Both double mutants possess very low yet still detectable amounts of starch (Table 3). *pgm3 pgm1* mutants revealed an elevated starch amount both in the light and in the dark compared to *pgm1*. However, when plants were grown under 12 h light/12 h dark or 16 h light/8 h dark, these results were not reproduced, as starch content was similar in *pgm1* and both double mutants under these photoperiod regimes (data not shown).

Furthermore, *pgm1* and both double mutants displayed elevated levels of soluble sugar compared to Col-0 (Table 3). Additionally, it was consistently observed that the double knock-out mutants flowered significantly later compared to Col-0 (data not shown). Therefore, floral stem development was investigated. *pgm1* mutants were delayed in floral stem development compared to Col-0, which is consistent with a previous report [42]. The *pgm2 pgm1* mutant displayed a floral stem development time similar to

Figure 6. Growth phenotype of *cp-pgm* plants. A, Seeds were sowed on MS medium containing sucrose and antibiotics (kanamycin [50 μg/mL], hygromycin [50 μg/mL]). Plants were grown under long day conditions (16 h light/8 h dark) and were two-week-old. Bar = 1 cm. B, *cp-pgm* plant before trypan blue staining. C, Col-0 and *cp-pgm* plants after trypan blue staining. The *cp-pgm* plant was five- week-old, germinated on MS plate (as above) and the two last weeks grown under continuous illumination. Leave of Col-0 from three-week-old plant grown under 12 h light/12 h dark conditions. Bars = 1 cm. D–F, Phenotype of cp-*pgm* plants under continuous illumination. Seeds were germinated on MS medium containing sucrose with antibiotics (kanamycin [50 μg/mL], hygromycin [50 μg/mL]). After four weeks plants were transferred to soil and grown further under continuous illumination. D, Plant was six-week-old. Bar = 1 cm. E–F, Flower buds of *cp-pgm* transgenic plants. Plant was six-week-old (E) and seven-week-old (F). Bars = 1 mm.

that of *pgm1*, by contrast *pgm3 pgm1* plants were significantly delayed (Fig. 5B). Although, *pgm1*, *pgm2 pgm1*, and *pgm3 pgm1* plants contained very low amounts of starch, they were not strongly compromised in growth under long day conditions and were able to develop normal flowers and seeds. By contrast, plants with reduced cPGM activity are strongly diminished in growth and seed development (Fig. 4). Therefore, transgenic Arabidopsis lines with a substantial reduction of total PGM were generated by introducing the cPGM *amiRNA* construct into *pgm1* mutants by Agrobacterium mediated transformation (*cp-pgm* plants). Seeds were germinated on MS medium supplemented with sucrose and antibiotics and transformants with well developed leaves and roots were identified (Fig. 6A). It was noted that sucrose is essential for

cp-pgm seed germination, as seeds sown on sucrose-free MS medium with appropriate antibiotics were not able to germinate.

In order to prove that the transgenic lines are strongly reduced in total PGM activity, protein crude extracts were subjected to native PAGE and PGM activity staining. The *cp-pgm* plants did not display any residual PGM activity (Fig. S5 in File S1). As a control the same crude extracts were used for phosphorylase activity staining, revealing activities comparable to Col-0 for both the cytosolic and plastidial phosphorylase isoforms (data not shown). After approximately three weeks *cp-pgm* plants were transferred to soil at different light/dark conditions: 12 h light/12 h dark, 14 h light/10 h dark and continuous illumination. Independent of growth conditions, plants were very tiny and

Table 3. Starch and soluble sugar content in Col-0 and PGM knock-out mutants.

genotype	starch content [mg glc equiv./g FW]		soluble sugars content (7 h in the light) [μmol/g FW]		
	7 h in the light	3.5 h in the dark	glucose	fructose	sucrose
Col-0	2.930±0.303	3.738±0.196	1.03±0.20	0.28±0.03	1.88±0.28
pgm1	0.012±0.003	0.010±0.001	4.23±0.65	1.04±0.21	2.69±0.11
pgm3 pgm1	0.025±0.005*	0.023±0.004*	4.91±0.59	0.94±0.04	2.70±0.17
pgm2 pgm1	0.015±0.003	0.016±0.003	4.67±0.51	0.87±0.11	2.74±0.31

Plants were grown under long day conditions (14 h light/10 h dark). Plants were five-week-old. Values are means of three biological replicates (two technical replicates each) ± SD. Asterisks indicate values significantly different from pgm1 and pgm2 pgm1 (Student Test, $p \leq 0.05$).

rapidly became chlorotic and dry (Fig. 6B). However, under prolonged light conditions and continuous illuminations plants stayed green longer. Nevertheless, trypan blue which selectively stains dead tissue revealed that the plants are not longer vital (Fig. 6C; [37]). That said, some transgenic cp-pgm plants were even able to develop normal looking flowering buds under continuous illumination (Fig. 6D–E), but further development of flowers failed as buds shriveled within one week (Fig. 6F). Even if plants were supplied for the entire growth period with exogenous sugars (MS medium+sucrose) they failed to grow to maturity (data not shown). Thus, significant reduction of total PGM activity leads to a dramatic dwarf phenotype and inability to develop functional flowers and seeds. Therefore, cp-pgm plants showed a more severe phenotype compared with transgenic potato plants reduced in total PGM activity [24]. Additionally, the phenotype exhibited by the lack of total PGM activity was corroborated by crossing pgm2/3d with pgm1 (named pgm2/3d pgm1 plants) which displayed the same phenotype as cp-pgm plants (data not shown). Despite of the tiny amount of available leaf material, initial analysis of the starch content in pgm2/3d pgm1 was performed revealing that pgm2/3d pgm1 plants possess very low amounts of starch (0.21±0.02 μmol glc. equiv./g FW), similar to pgm1 (0.25±0.06 μmol glc. equiv./g FW) at the middle of the day.

Discussion

Analyses of single knock-out mutants of both cytosolic phosphoglucomutase isoforms (pgm2 and pgm3) confirmed that the isoforms are redundant and expressed at a sufficient abundance to substitute for one another. Even the additional lack of PGM1 reveals only small alterations in metabolism and development in comparison to pgm1 (Table 3, Fig. S4 in File S1). Furthermore, investigations with purified recombinant Arabidopsis enzymes, reveal that the kinetic properties of both cytosolic isoforms are very similar (for example the K_M·s, using G1P as substrate, are PGM3 3.7±0.5 mM; PGM2 4.2±0.4 mM; [39]). The possible difference in substrate specificity observed for phosphoglucomutases of *Pseudomonas aeruginosa* [43] or *Giardia lamblia* [44], which show the additional interconversion of mannose 1-phosphate to mannose 6-phosphate, was not observed here. In competition experiments, where recombinant PGM2 or PGM3 were used with different amounts of mannose 1-phosphate in the presence of glucose 1-phosphate, no alteration in glucose 6-phosphate formation were observed. Furthermore, no formation of mannose 6-phosphate was detectable using HPAEC-PAD [39].

In contrast the pgm2/3 lines reveal a very considerable phenotype. Even when the cytosolic phosphoglucomutase activity was below the detection limit, there was still a slight residual expression of both cytosolic isoforms (Fig. S3B in File S1). This is

likely the reason for the severe yet not lethal phenotype. Thus, also the formation of seeds, albeit reduced or in some cases completely inhibited, could be explained and is in agreement with previous reports [24]. Furthermore, pgm2/3 reveals alterations in cell wall composition, which were not previously detected in transgenic potato plants with strong reduction of cPGM [22].

Surprisingly in the pgm2/3 lines a strong increase in sucrose, as well as the starch breakdown derived maltose, was observed. A significant increase in sucrose was additionally detected in the single knock-out line pgm2 (Table 1). The formation of sucrose in the light is dependent on cPGM activity, as G1P is essential for the formation of UDPglucose via both routes of sucrose synthesis. However, several pathways for formation of G1P and thereby sucrose remain in the cytosol of pgm2/3 plants: (i), the conversion via the mentioned residual cPGM activity in the plants, (ii), the formation of G1P in the night by the pathway of starch derived maltose, disproportionating enzyme 2, cytosolic heteroglycans, and the cytosolic phosphorylase [12,13], (iii), the direct transport of G1P from the chloroplasts into the cytosol as demonstrated from isolated chloroplasts [1]. That said on the basis of our results flux through all of these routes can be anticipated to be relatively minor since formation via the starch degradation pathway is restricted to the night period, and it was shown that the G1P transport rate across the chloroplast membrane is minor in Arabidopsis in comparison to situation observed in potato [27]. Furthermore, it has been demonstrated that G1P that is taken up by the Arabidopsis chloroplast is directly converted into starch via ADPglucose pyrophosphorylase pathway, indicating that free G1P is immediately metabolized thus reducing the possibility of the G1P export [1]. It is possible that the observed elevation of the expression of PGM1 (Fig. S3B in File S1) in the transgenic lines is an effort to overcome this limitation. Additionally, preliminary experiments point to an increased G1P transport rate in pgm2/3 plants compared to Col-0 (more than 20%) when measuring G1P uptake with isolated chloroplasts (data not shown).

However, it is not possible to explain the increase of sucrose in pgm2/3 compared to Col-0 merely in terms of its rate of synthesis. It would seem more likely to be the consequence of the reduced sink capacity in the heterotrophic tissues and, therefore, a reduced export from the leaves of these lines. When sink capacity is reduced, feedback to the autotrophic tissues occurs culminating in the high starch and maltose levels observed in these lines. Moreover, metabolic profiling reveals a massive effect on the entire plant metabolism. Furthermore, taking into account the carbohydrate partitioning between sucrose and starch, the increase in both is not unexpected. Sucrose is catabolized either by sucrose synthase or invertase. It is proposed that invertase rather than sucrose synthase might be the dominant route for sucrose catabolism in *A. thaliana* [45]. Consequently, products of sucrose catabolism would enter the hexose phosphate pool as G6P or F6P but not as G1P. Thus, it would appear that cPGM is essential for G1P production.

A strong reduction of G1P is also anticipated to affect the entire nucleotide sugar metabolism [40], resulting in reduced growth and altered cell wall formation. As shown for *pgm2/3* the composition of the cell wall is altered and the root length is reduced. This phenotype was also observed for plants deficient in cytosolic invertase (*cinv1*) revealing reduced cell wall flexibility, inhibited root cell elongation and shorter roots [46]. Furthermore, mutants lacking two isoforms of cytosolic invertase (*cinv/cinv2*) are drastically reduced in root growth [45].

Additionally, a development of curly leaves was described in plants exhibiting reduced expression of SUT1 [47,48] or plants expressed yeast derived invertase [49,50,51]. This leaf phenotype was postulated to be due to osmotic problems associated with carbohydrate accumulation, which is similar to the situation observed for *pgm2/3*. However, it is important to note that in some cases plants with alteration in cell wall synthesis, downstream of G1P, also display such curled leaves [52].

The tiny *cp-pgm* plants reveal an even more severe phenotype. Indeed under normal growth conditions these perturbations are lethal. Germination was only observed, when sucrose was supplemented, but also under these conditions complete formation of inflorescence and seeds were inhibited. As the expected residual cPGM activity is similar to the parental *pgm2/3* lines (not detectable), this is a strong indication that the glucose-phosphate interconversion via PGM1 and formation of G1P via the starch degradation pathway are essential in *pgm2/3* plants for the creation of the residual levels of G1P. The observed phenotype is much more severe than that observed for transgenic potato lines lacking both cPGM and pPGM activities [25]. The strongest reduced line was reported to have decreased leaf fresh weight of up to 33 percent. One explanation for the less distinct phenotype for potato is that in these plants a residual activity of both the pPGM and cPGM was still detectable (both 4%, [26]). However, also a second point is to mention, that the transport rate for G1P over the plastidial membranes seems to be much higher in potato compared to Arabidopsis [1,27]. Thus, the possible bypass of the

PGM lack via G1P transport is minor in Arabidopsis and therefore results in the observed more pronounced phenotype. Nevertheless, the higher transport rate of G1P observed for potato tuber is insufficient to completely overcome the limitations by lacking PGMs, especially in heterotrophic tissues, as the reduction in tuber fresh weight is far more pronounced with up to 75% reduction [25]. Overall, this points to a more flexible metabolism related to alternative carbon fluxes in potato then in Arabidopsis in respect to starch/sucrose turn-over.

Supporting Information

File S1   Supporting Information containing Tables S1–S3 and Figures S1–S5. Table S1. Primers used for PCR and qPCR analysis. **Table S2.** Chlorophyll content of Col-0 and *pgm2/3* plants. **Table S3.** Values of the metabolic profiling used for the generation of the heat map. **Figure S1.** Phosphoglucomutase activity in Arabidopsis leaves. **Figure S2.** Analysis of single knock-out lines *pgm2* and *pgm3* and Col-0 under long day conditions (14 h light/10 h dark). **Figure S3.** Characterization of Col-0 and *pgm2/3* plants. **Figure S4.** Growth phenotypes of Col-0 and PGM knock-out mutants. **Figure S5.** Phosphoglucomutase activity in Col-0 and PGM transgenic plants.

Acknowledgments

The authors gratefully thank Ulrike Matthes and Jessica Alpers for excellent technical assistants and Tom Orawetz for help screening the various transgenic lines and Sebastian Mahlow for help during preparation of the figures (all University of Potsdam). The authors also thank Julia Vogt and Anke Koch (both University of Potsdam) for help performing the qPCR experiments.

Author Contributions

Conceived and designed the experiments: IM HHK MG JF. Performed the experiments: IM HHK SA KH JF. Analyzed the data: IM HHK SA KH MG ARF JF. Contributed reagents/materials/analysis tools: IM HHK SA KH MG ARF JF. Contributed to the writing of the manuscript: IM HHK MG ARF JF.

References

1. Fettke J, Malinova I, Albrecht T, Hejazi M, Steup M (2011) Glucose-1-phosphate transport into protoplasts and chloroplasts from leaves of Arabidopsis. Plant Physiol 155: 1723–1734.

2. Streb S, Egli B, Eicke S, Zeeman SC (2009) The debate on the pathway of starch synthesis: a closer look at low-starch mutants lacking plastidial phosphoglucomutase supports the chloroplast-localized pathway. Plant Physiol 151: 1769–1772.

3. Kofler H, Häusler RE, Schulz B, Gröner F, Flügge U-I, et al. (2000) Molecular characterization of a new mutant allele plastid phosphoglucomutase in Arabidopsis, and complementation of the mutant with the wild-type cDNA. Mol Gen Genet 263: 978–986.

4. Gibon Y, Bläsing OE, Palacios-Rojas N, Pankovic D, Hendriks JHM, et al. (2004) Adjustment of diurnal starch turnover to short days: depletion of sugar during the night leads to a temporary inhibition of carbohydrate utilization, accumulation of sugars and post-translational activation of ADP-glucose pyrophosphorylase in the following light period. Plant J 39: 847–862.

5. Fettke J, Hejazi M, Smirnova J, Höchel E, Stage M, et al. (2009) Eukaryotic starch degradation: integration of plastidial and cytosolic pathways. J Exp Bot 60: 2907–2922.

6. Fettke J, Fernie AR, Steup M (2012) Transitory starch and its degradation in higher plant cells. In: Tetlow IJ, editor. Essential reviews in experimental biology: starch: origins, structure and metabolism, Vol. 5. London, UK: SEB. 309–372.

7. Weise SE, Weber APM, Sharkey TD (2004) Maltose is the major form of carbon exported from the chloroplast at night. Planta 218: 474–482.

8. Stitt M, Zeeman SC (2012) Starch turnover: pathways, regulation and role in growth. Curr Opin Plant Biol 15: 282–292.

9. Niittylä T, Comparot-Moss S, Lue WL, Messerli G, Trevisan M, et al.(2006) Similar protein phosphatases control starch metabolism in plants and glycogen metabolism in mammals. J Biol Chem 281: 11815–11818.

10. Lu Y, Sharkey TD (2004) The role of amylomaltase in maltose metabolism in the cytosol of photosynthetic cells. Planta 218: 466–473.

11. Chia T, Thorneycroft D, Chapple A, Messerli G, Chen J, et al. (2004) A cytosolic glucosyltransferase is required for conversion of starch in Arabidopsis leaves at night. Plant J 37: 853–863.

12. Fettke J, Chia T, Eckermann N, Smith A, Steup M (2006) A transglucosidase necessary for starch degradation and maltose metabolism in leaves at night acts on cytosolic heteroglycans (SHG). Plant J 46: 668–684.

13. Fettke J, Eckermann N, Poeste S, Pauly M, Steup M (2004) The glycan substrate of the cytosolic (Pho 2) phosphorylase isoform from *Pisum sativum* L.: identification, linkage analysis, and subcellular localization. Plant J 39: 933–946.

14. Fettke J, Poeste S, Eckermann N, Tiessen A, Pauly M, et al. (2005) Analysis of heteroglycans from leaves of transgenic potato (*Solanum tuberosum* L.) plants that under- or overexpress the Pho2 phosphorylase isozyme. Plant Cell Physiol 46: 1987–2004.

15. Cho MH, Lim H, Shin DH, Jeon JS, Bhoo SH, et al. (2011) Role of the plastidic glucose transporter in the export of starch degradation products from the chloroplasts in *Arabidopsis thaliana*. New Phytol 190: 101–112.

16. Weber A, Servaites JC, Geiger DR, Kofler H, Hille D, et al. (2000) Identification, purification, and molecular cloning of a putative plastidic glucose translocator. Plant Cell 12: 787–801.

17. Caspar T, Huber SC, Somerville C (1985) Alterations in growth, photosynthesis, and respiration in a starchless mutant of *Arabidopsis thaliana* (L.) deficient in chloroplast phosphoglucomutase activity. Plant Physiol 79: 11–17.

18. Hanson KR, McHale NA (1988) A starchless mutant of *Nicotiana sylvestris* containing a modified plastid phosphoglucomutase. Plant Physiol 88: 838–844.

19. Manjunath S, Kenneth LCH, Winkle PV, Bailey-Serres J (1998) Molecular and biochemical characterisation of cytosolic phosphoglucomutase in maize. Expression during development and in response to oxygen deprivation. Plant Physiol 117: 997–1006.

20. Fernie AR, Tauberger E, Lytovchenko A, Roessner U, Willmitzer L, et al. (2002) Antisense repression of cytosolic phosphoglucomutase in potato (*Solanum tuberosum*) results in severe growth retardation, reduction in tuber number and altered carbon metabolism. Planta 214: 510–520.

21. Mühlbach H, Schnarrenberger C (1978) Properties and intracellular distribution of two phosphoglucomutases from spinach leaves. Planta 141: 65–70.

22. Lytovchenko A, Sweetlove L, Pauly M, Fernie AR (2002) The influence of cytosolic phosphoglucomutase on photosynthetic carbohydrate metabolism. Planta 215: 1013–1021.

23. Fettke J, Nunes-Nesi A, Alpers J, Szkop M, Fernie AR, et al. (2008) Alterations in cytosolic glucose-phosphate metabolism affect structural features and biochemical properties of starch-related heteroglycans. Plant Physiol 148: 1614–1629.

24. Egli B, Kölling K, Köhler C, Zeeman SC, Streb S (2010) Loss of cytosolic phosphoglucomutase compromises gametophyte development in Arabidopsis. Plant Physiol 154: 1659–1671.

25. Lytovchenko A, Fernie AR (2003) Photosynthetic metabolism is severely impaired on the parallel reduction of plastidial and cytosolic isoforms of phosphoglucomutase. Plant Phys and Biochemistry 41: 193–200.

26. Fernie AR, Swiedrych A, Tauberger E, Lytovchenko A, Trethewey RN, et al. (2002) Potato plants exhibiting combined antisense repression of cytosolic and plastidial isoforms of phosphoglucomutase surprisingly approximate wild type with respect to the rate of starch synthesis. Plant Physiol Biochem 40: 921–927.

27. Fettke J, Albrecht T, Hejazi M, Mahlow S, Nakamura Y, et al. (2010) Glucose 1-phosphate is efficiently taken up by potato (*Solanum tuberosum*) tuber parenchyma cells and converted to reserve starch granules. New Phytol 185: 663–675.

28. Schwab R, Ossowski S, Riester M, Warthmann N, Weigel D (2006) Highly specific gene silencing by artificial microRNAs in Arabidopsis. Plant Cell 18: 1121–1133.

29. Nakagawa T, Kurose T, Hino T, Tanaka K, Kawamukai M, et al. (2007) Development of series of gateway binary vectors, pGWBs, for realizing efficient construction of fusion genes for plant transformation. J Biosci Bioeng 104: 34–41.

30. Zhang X, Henriques R, Lin SS, Niu QW, Chua NH (2006) *Agrobacterium*-mediated transformation of *Arabidopsis thaliana* using the floral dip method. Nature Protocols. 1: 1–6.

31. Stitt M, McC Lilley R, Gerhardt R, Heldt HW (1989) Metabolite levels in specific cells and subcellular compartments of plant leaves. Methods in Enzymology 174: 518–552.

32. Foster CE, Martin TM, Paul M (2010) Comprehensive compositional analysis of plant cell walls (Lignocellulosic biomass) Part II: Carbohydrates. Journal of Visualized Experiments. 37. Available : http://www.jove.com/index/Details. stp?ID=1837, doi: 10.3791/1837.

33. Waffenschmidt S, Jaenicke L (1987) Assay of reducing sugars in the nanomole range with 2,2'-bicinchoninate. Anal Biochem 165: 337–340.

34. Updegraf DM (1969) Semimicro determination of cellulose in biological materials. Anal Biochem 32: 420–424.

35. Fettke J, Eckermann N, Tiessen A, Geigenberger P, Steup M (2005) Identification, subcellular localization and biochemical characterization of

36. Fettke J, Nunes-Nesi A, Fernie AR, Steup M (2011) Identification of a novel heteroglycan-interacting protein, HIP 1.3, from *Arabidopsis thaliana*. J Plant Physiol 168: 1415–1425.

37. Koch E, Slusarenko A (1990) Arabidopsis is susceptible to infection by a downy mildew fungus. Plant Cell 2: 437–445.

38. Fettke J (2006) Stärkerelevante cytosolische Heteroglycane: Identifizierung und funktionelle Analyse. PhD thesis, University of Potsdam, Potsdam, Germany.

39. Herbst K (2011) Cytosolische Phosphoglucomutase in *Arabidopsis thaliana*: Heterologe Expression and Analyse. Diploma thesis. University of Potsdam, Potsdam, Germany.

40. Seifert GJ (2004) Nucleotide sugar interconversions and cell wall biosynthesis: how to bring the inside to outside. Curr Opin Plant Biol 7: 277–284.

41. Matsui M, Nakazawa M (2003) Selection of hygromycin-resistant Arabidopsis seedling. Biotechniques 34: 28–30.

42. Yu TS, Lue WL, Wang SM, Chen J (2000) Mutation of Arabidopsis plastid phosphoglucose isomerase affect leaf starch synthesis and floral initiation. Plant Physiol 123: 319–325.

43. Ye RW, Zielinski NA, Chakrabarty AM (1994) Purification and characterization of phosphomannomutase/phosphoglucomutase from *Pseudomonas aeruginosa* involved in biosynthesis of both alginate and lipopolysaccharide. Journal Bacteriol. 176: 4851–4857.

44. Mitra S, Cui J, Robbins PW, Samuelson J (2010) A deeply divergent phosphoglucomutase (PGM) of *Giardia lamblia* has both PGM and phospho-mannomutase activities. Glycobiology 20: 1233–1240.

45. Barratt DH, Derbyshire P, Findlay K, Pike M, Wellner N, et al. (2009) Normal growth of Arabidopsis requires cytosolic invertase but not sucrose synthase. Proc Natl Acad Sci USA 106: 13124–13129.

46. Lou Y, Gou JY, Xue HW (2007) PIP5K9, an Arabidopsis phosphatidylinositol monophosphate kinase, interacts with a cytosolic invertase to negatively regulate sugar-mediated root growth. Plant Cell 19: 163–181.

47. Riesmeier JW, Wilmitzer L, Frommer W (1994) Evidence for a role of the sucrose transporter in phloem loading and assimilate partitioning. EMBO J 13: 1–7.

48. Bürkle L, Hibberd JM, Quick WP, Kuehn C, Hirner B, et al. (1998) The H+-sucrose co-transporter NtSUT1 is essential for sugar export from tobacco leaves. Plant Physiol 118: 59–68.

49. Heineke D, Sonnewald U, Büssis D, Günter G, Leidreiter K, et al. (1992) Apoplastic expression of yeast-derived invertase in potato: effects on photosynthesis, leaf solute composition, water relations, and tuber composition. Plant Physiol 100: 301–308.

50. Dickinson C, Altabella T, Chrispeels M (1991) Slow-growth phenotype of transgenic tomato expressing apoplastic invertase. Plant Physiol 95: 420–425.

51. von Schaewen A, Stitt M, Schmidt R, Sonnewald U, Willmitzer L (1990) Expression of a yeast-derived invertase in the cell wall of tobacco and Arabidopsis plants leads to accumulation of carbohydrate and inhibition of photosynthesis and strongly influences growth and phenotype of transgenic tobacco plants. EMBO J. 9: 3033–3044.

52. Reiter WD, Chapple C, Somerville CR (1997) Mutants of *Arabidopsis thaliana* with altered cell wall polysaccharide composition. Plant J 12: 335–345.

water-soluble heteroglycans (SHG) in the leaves of *Arabidopsis thaliana* L.: distinct heteroglycans reside in the cytosol and in the apoplast. Plant J 43: 568–585.

13

The *SbMT-2* Gene from a Halophyte Confers Abiotic Stress Tolerance and Modulates ROS Scavenging in Transgenic Tobacco

Amit Kumar Chaturvedi, Manish Kumar Patel, Avinash Mishra*, Vivekanand Tiwari, Bhavanath Jha*

Discipline of Marine Biotechnology and Ecology, CSIR-Central Salt and Marine Chemicals Research Institute, Bhavnagar, Gujarat, India

Abstract

Heavy metals are common pollutants of the coastal saline area and *Salicornia brachiata* an extreme halophyte is frequently exposed to various abiotic stresses including heavy metals. The *SbMT-2* gene was cloned and transformed to tobacco for the functional validation. Transgenic tobacco lines (L2, L4, L6 and L13) showed significantly enhanced salt (NaCl), osmotic (PEG) and metals (Zn^{++}, Cu^{++} and Cd^{++}) tolerance compared to WT plants. Transgenic lines did not show any morphological variation and had enhanced growth parameters viz. shoot length, root length, fresh weight and dry weight. High seed germination percentage, chlorophyll content, relative water content, electrolytic leakage and membrane stability index confirmed that transgenic lines performed better under salt (NaCl), osmotic (PEG) and metals (Zn^{++}, Cu^{++} and Cd^{++}) stress conditions compared to WT plants. Proline, H_2O_2 and lipid peroxidation (MDA) analyses suggested the role of *SbMT-2* in cellular homeostasis and H_2O_2 detoxification. Furthermore *in vivo* localization of H_2O_2 and O_2^-; and elevated expression of key antioxidant enzyme encoding genes, *SOD*, *POD* and *APX* evident the possible role of *SbMT-2* in ROS scavenging/detoxification mechanism. Transgenic lines showed accumulation of Cu^{++} and Cd^{++} in root while Zn^{++} in stem under stress condition. Under control (unstressed) condition, Zn^{++} was accumulated more in root but accumulation of Zn^{++} in stem under stress condition suggested that *SbMT-2* may involve in the selective translocation of Zn^{++} from root to stem. This observation was further supported by the up-regulation of zinc transporter encoding genes *NtZIP1* and *NtHMA-A* under metal ion stress condition. The study suggested that *SbMT-2* modulates ROS scavenging and is a potential candidate to be used for phytoremediation and imparting stress tolerance.

Editor: Keqiang Wu, National Taiwan University, Taiwan

Funding: This study was supported by the Council of Scientific and Industrial Research (CSIR, www.csir.res.in), Government of India, New Delhi (BSC0109–SIMPLE; Senior Research Fellowship to AKC and VT). The funders had no role in study design, data collection and analysis, decision to publish, or preparation of the manuscript.

Competing Interests: The authors have declared that no competing interests exist.

* Email: avinash@csmcri.org (AM); bjha@csmcri.org (BJ)

Introduction

Abiotic stresses such as salinity, drought, temperature and heavy metals have significant effect on the agricultural production over the years [1]. Most often, plants may encounter these abiotic stresses simultaneously, resulting in to the substantial loss in agricultural productivity. Over the past two centuries, increased industrial and anthropogenic activities viz. mining, irrigation with sewage effluents/waste waters, use of phosphate fertilizers are the major sources of metal contamination to soil [2–3]. Salinity, drought and heavy metals have similar consequences ensuing oxidative stress, disturbances in ionic homeostasis and generation of reactive oxygen species (ROS) [1,4].

Metallothioneins (MTs) are the group of polypeptides which can bind with heavy metals through their thiols group via chelation and involved in the homeostasis of essential metals (Cu and Zn) and cellular detoxification of nonessential metals (Cd and Hg) [5,6]. Characteristics of metallothioneins are the cysteine (Cys) residue which is the basis of its classification. Based on the arrangement of Cys residues MTs are divided in two classes [7]. Plant MTs belong to Class II and further classified into four types (type 1 to 4) based on the position and allocation of cysteine

residues [7]. Spatial expression of all four MT type has been reported which is localized to root, stem, leaves and developing seeds [7,8] and show differential tolerance to different metals. Although there are several reports regarding the possible role of MTs in plants but its physiological role is still not fully conclusive because of difficulties in its isolation and stability [7,9]. Over-expression of MTs in various model systems like Arabidopsis, tobacco, yeast and *E. coli* established its functional role in homeostasis and tolerance to Cu, Zn and Cd [10,11,12,13], high salinity, drought, low temperature, heavy metal ions, abscisic acid (ABA) and ethylene [12,14]. Besides above, MTs are reported in the inhibition of root elongation [15], fruit ripening, seed development [7] and provide disease resistance against pathogen attack [16].

Heavy metals affect the plant at cellular, biochemical and molecular level causing the oxidative stress. These toxic metals generate free radicals and reactive oxygen species (ROS) which damages the cell membrane, nucleic acids and photosynthetic pigments [17]. The equilibrium between ROS generation and quenching is prerequisite for the cell survival which is maintained by the intricate anti-oxidative system comprising of enzymatic [superoxide dismutase (SOD), catalase (CAT), peroxidase (POD)

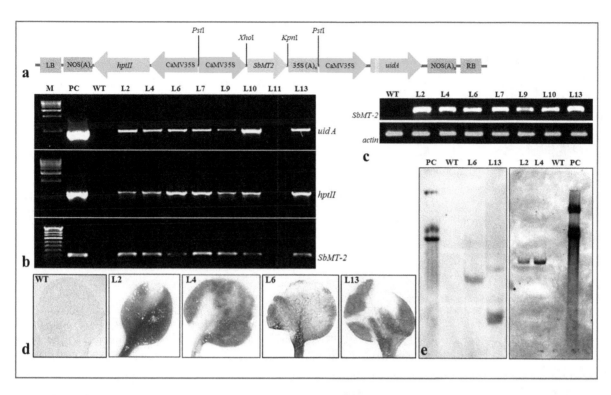

Figure 1. Analyses of transgenic tobacco plants. (a) Schematic map of plant expression gene construct pCAMBIA1301-*SbMT-2*. (b) Confirmation of transgenic lines by PCR amplification of *uidA*, *hptII* and *SbMT-2* genes (Lane M is molecular weight markers, Lane PC is positive control of PCR using plasmid *pCAMBIA1301-SbMT-2* as template, Lane WT is wild type control plant (non-transgenic) and Lane L is transgenic lines). (c) Over-expression of the *SbMT-2* gene in transgenic lines compared to wild-type control plants (Lane WT is wild type control plant and Lane L is transgenic lines) analyzed by semi-quantitative Rt-PCR, where *actin* was used as internal gene control. (d) Histochemical GUS staining of WT and transgenic plants. (e) Southern blot analysis (Lane PC is positive control plasmid *pCAMBIA1301-SbMT-2*, Lane WT is wild type control plant (non-transgenic) and Lane L is transgenic lines).

and non-enzymatic (ascorbate, glutathione and phenolic compounds) systems [4,12,17,18]. Under stress conditions the activity of this antioxidative system gets increased due to the increase of free radical formation [17].

Though MTs are known to be involved in abiotic/metal stresses and have been reported from several plant species, there are a few reports on halophytes under heavy metal stress [19]. *Salicornia brachiata* is an extreme halophyte and frequently exposed to heavy metals in coastal areas. The plant has nutritional value, unique oligosaccharide profile and requires NaCl for *in-vitro* regeneration [20–22]. The plant is considered as a model for the study of tolerance mechanism and several stress responsive genes have been isolated and characterized [13,23–28]. In our previous study, *SbMT-2* gene was isolated and physiological role was determined in *E. coli* which showed its role in the homeostasis and detoxification of Zn, Cu and Cd ions [13]. The gene (*SbMT-2*) was considered as a potential candidate to be utilized for the genetic engineering of plants for phytoremediation of heavy metals and stress tolerance. Therefore in the present study *SbMT-2* gene was transformed into tobacco for the functional validation. Biochemical, physiological and morphological responses of transgenic plants over-expressing *SbMT-2* gene under different abiotic stresses (metals- Zn, Cu & Cd; NaCl and osmotic stresses) were studied.

Results

Over-expression of the *SbMT-2* gene

A plant expression vector (Figure 1a), harboring *SbMT-2* gene driven by CaMV35S promoter was constructed and transformed

to tobacco plants. Fourteen independent transgenic lines (T_0) were raised and preliminarily screened by PCR using gene specific primers (data not shown). These lines were grown in containment facility and seeds (T_1) were collected. T_1 Seeds were germinated on hygromycin containing media and selected T_1 transgenic lines were confirmed by PCR using gene specific primers (Figure 1b). Expected size of amplicon was found in all lines except L11. Based on *SbMT-2* (Figure 1c) gene expression level, four independent T_1 transgenic lines; L2, L4, L6 and L13 were selected for the further morpho-physio-biochemical analyses. All transgenic lines showed high *gus* and *SbMT-2* gene expression (Figure 1c and 1d). Southern blot confirmed single and double copy gene integration to L2, L4, L6 and L13 lines, respectively (Figure 1e).

Growth parameters in T_1 transgenic lines

Transgenic lines L2, L4, L6 and L13 showed high percentage of seed germination in all stress treatments (metals- Zn, Cu and Cd; NaCl and PEG) compared to wild type (WT) plants (Figure S1). Transgenic lines grown under different stress treatments were comparatively healthier than WT plants (Figure 2) and showed enhanced growth parameters under stress conditions (Figure 3). Shoot and root lengths were found significantly higher in transgenic lines compared to WT plants in all stress treatments. Similarly, fresh and dry weights of transgenic lines were significantly higher than WT plants. Among all stress treatments studied, transgenic lines showed better growth in osmotic stress followed by metal and NaCl stress.

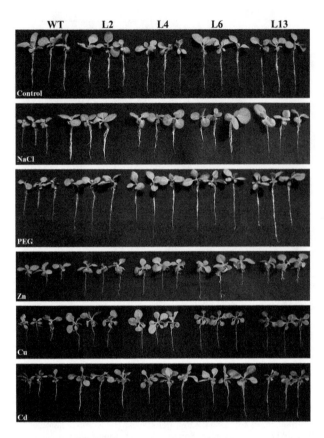

Figure 2. Comparative morphology of WT and T₁ transgenic lines under different stress condition. WT and four independent T₁ transgenic tobacco lines (L2, L4, L6 and L13) grown under control, salt (200 mM NaCl), osmotic (10% PEG), Zn (5 mM), Cu (0.2 mM) and Cd (0.2 mM) stress for 3 weeks.

Leaf senescence assay and chlorophyll content

Stress tolerance of T_1 transgenic lines was studied by leaf disc senescence assay and chlorophyll estimation. In leaf discs assay, stress-induced necrosis resulted in the decrease of chlorophyll content was lower in the *SbMT-2* over-expressing lines compared to WT plants (Figure 4). The damage caused by stress treatments was visualized by the degree of bleaching of leaf tissues and it was evident that the transgenic plants (L2, L4, L6 and L13) had a better ability to tolerate osmotic stress followed by metal and NaCl stress (Figure 4). The chlorophyll content in the WT plants reduced significantly with stress treatments while the transgenic lines (L2, L4, L6 and L13) retained higher chlorophyll contents than WT plants (Figure 5a).

Relative water content, Electrolyte leakage and membrane stability index

The relative water content (RWC) was found almost similar (insignificant difference) in transgenic and WT plants under unstressed condition (control), however RWC was significantly higher in transgenic lines compared to WT plants under all stress treatments (Figure 5b). Membrane permeability measured by electrolyte leakage (EL), was found significantly stable in transgenic lines, which showed reduced electrolyte leakage compared to WT plants under stress condition (Figure 5c). Similarly, membrane stability index (MSI) of transgenic lines was significantly

higher compared to WT under stress treatments (Figure 5d). RWC, EL and MSI evident that transgenic lines are thriving well in stress conditions compared to WT plants.

Localization of O_2^- & H_2O_2 and H_2O_2, proline & MDA content analysis

Leaves of transgenic lines and WT plants, subjected to different stress treatments showed *in vivo* localization of O_2^- and H_2O_2 (Figure 6). It was observed that leaves of WT plants showed more accumulation of O_2^- and H_2O_2 content compared to transgenic lines under various stress conditions. Transgenic lines showed significantly lower accumulation of H_2O_2, proline and MDA contents under stress condition compared to WT plants (Figure 7). Lower content of H_2O_2 exhibited the better ROS system in transgenic lines compared to WT plants. Lower accumulation of proline and MDA revealed that transgenic lines have higher osmoprotectants and lower lipid peroxidation, respectively compared to WT plants under stress condition.

Ion content analysis

Metallothioneins play a vital role in ion homeostasis and heavy metal binding. Metal ion sequestration was analyzed by ICP, which revealed high accumulation of metal ions in transgenic lines compared to WT plants (Figure 8). It was observed that metal ion contents were approximately same (insignificant difference) in WT and transgenic lines at control (unstressed) condition (Figure 8a). Under stress condition, metal ion accumulation was significantly higher in transgenic lines compared to WT plants (Figure 8b). Among different tissues, Zn accumulation was higher in shoot followed by root and leaves. However, high Cu and Cd contents were detected in roots followed by stem and leaves. Furthermore high affinity of *SbMT-2* was observed with Zn ions. Compared to control condition, Na^+ contents increased in transgenic as well as WT plants under NaCl stress treatment, however K^+ contents were decreased (Figure S2a). Similarly, transgenic lines showed higher K^+/Na^+ ratio compared to WT plants under NaCl stress condition (Figure S2b).

Transcript analysis of genes encoding metal transporters and antioxidative enzymes

Ion content analysis suggested that *SbMT-2* may involve in the selective translocation of Zn^{++} from root to stem. To further support this observation, expression analysis of zinc and heavy metal transporters encoding genes (*NtZIP1* and *NtHMA-A*) were studied under the metal stress (Zn, Cu and Cd) treatments. Expression of genes *NtZIP1* and *NtHMA-A* involved in heavy metal (Zn) translocation were up-regulated in transgenic lines compared to WT plants (Figure 9). Among different metal stress, the *NtZIP1* gene showed maximum up-regulation under Zn stress (Figure 9a). Transgenic lines, L2, L4, L6 and L13 showed about 10-, 13-, 8- and 6-fold expression compared to their respective control plants. However, *NtZIP1* expression was also up-regulated in WT plant. Similarly, the *NtHMA-A* gene was also up-regulated under metal stress compared to WT and control condition (Figure 9b). In contrast to other metal stress, inconsistent expression was observed among transgenic lines and WT under Cu stress treatment.

Common antioxidant enzymes involved in ROS scavenging mechanism are superoxide dismutase (SODs), peroxidase (POD) and ascorbate peroxidase (APX). Transcript analysis reveals that expression of the *NtSOD* gene increased concomitantly with stress treatments (except NaCl) in transgenic lines compared to WT plants (Figure 10a). About 3.5-fold expression was observed in L4

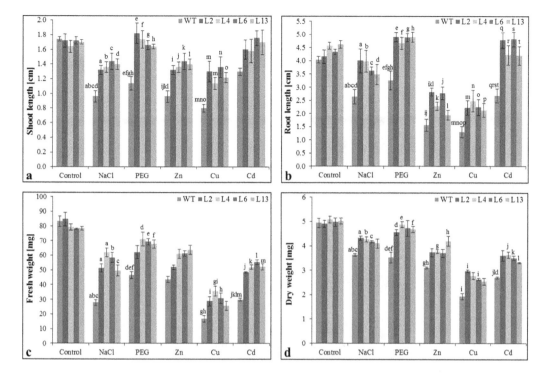

Figure 3. Analysis of *SbMT-2* transgenic tobacco lines under different stress condition. Comparison of (a) shoot length (cm), (b) root length (cm), (c) fresh weight (mg) and (d) dry weight (mg) of wild type (WT) and T_1 transgenic lines (L2, L4, L6 and L13) grown under control, salt (200 mM NaCl), osmotic (10% PEG), Zn (5 mM), Cu (0.2 mM) and Cd (0.2 mM) stress for 3 weeks. Graph represents the mean ± SE (of three replicates; n = 3) followed by similar letters are significantly different according to Tukey HSD at *P<0.05*.

line under Zn and osmotic stress, while about 2-fold expression was detected under Cu and Cd stress compared to control condition (untreated plants). Remaining L2, L6 and L13 lines showed higher relative expression of *NtSOD* gene under different stress conditions compare to WT plants. Surprisingly down-regulation of gene was found under NaCl stress in transgenic as well as WT plants compared to control (untreated) plants. Relative fold expression of *NtPOD* gene was found higher in transgenic lines compared to WT plants under metal stress treatments (Figure 10b). Transgenic line L4 showed maximum gene expression about 10 -fold under Zn, Cu and Cd stress. However, *NtPOD* gene expression was down-regulated in WT and transgenic lines under NaCl and PEG stress treatments. Elevated *NtAPX* gene expression was observed in transgenic lines (except L2 under NaCl and PEG) compared to WT plants under NaCl, osmotic and metal stress conditions (Figure 10c).

Discussion

Our previous study revealed that heterologous expression of *SbMT–2* gene not only augments zinc and copper tolerance but also increases metal ion sequestration in *E. coli* cells [13]. Further, in this study, *in planta* functional validation of *SbMT–2* gene has been elucidated. The *SbMT–2* gene was over-expressed in tobacco and its functional role was studied in different abiotic stresses *viz.* salt (200 mM NaCl), osmotic (10% PEG) and metals; Zn (1 mM), Cu (0.5 mM) and Cd (0.5 mM). Morphological variation was not observed in transgenic lines over-expressing *SbMT*-2 gene compared to WT plants (Figure 2). Transgenic lines showed higher seed germination percentage, increased root length, shoot length, fresh weight (FW), dry weight (DW) and chlorophyll content under different stress conditions compared to WT plants (Figure 3 and 5a) which reveal that *SbMT-2* leads to overcome the

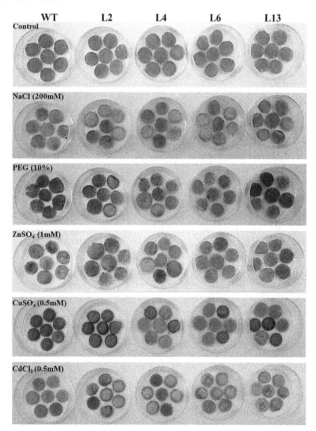

Figure 4. Leaf disc assay of transgenic tobacco lines (T₁) for different stress tolerance. Leaf discs of WT, L2, L4, L6 and L13 transgenic lines respectively were floated in different stress solution for 8 days.

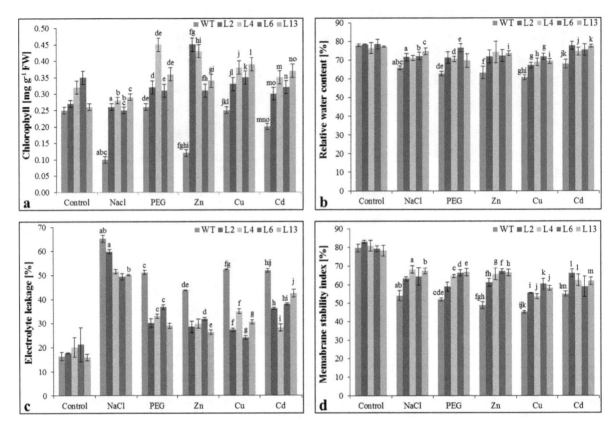

Figure 5. Analysis of *SbMT-2* transgenic tobacco lines under different stress condition. Comparison of (a) Chlorophyll content, (b) Relative water content, (c) Electrolyte leakage and (d) Membrane stability index of wild type (WT) and T_1 transgenic lines (L2, L4, L6 and L13) grown in hydroponic (control) and treated with salt (200 mM NaCl), osmotic (10% PEG), Zn (1 mM), Cu (0.5 mM) and Cd (0.5 mM) stress. Graph represents the mean ± SD (of three replicates; n = 3) followed by similar letters are significantly different according to Tukey HSD at *P<0.05*.

deleterious effect of salt, drought and metal stresses. Previously, ectopic expression of *BjMT2* gene conferred the increase in percentage seed germination, fresh weight (FW), dry weight (DW) and chlorophyll content in *Arabidopsis* under Cu and Cd stress [15]. Similar results were also observed under salt, drought, Zn and Cu stress in transgenic tobacco plants over-expressing clustered *OsMT1e-P* metallothionein gene [29].

Salt, drought and heavy metal stress are multigenic in nature, causing osmotic stress and thus create physiological drought conditions for plants. Transgenic lines displayed increase in relative water content under stress (Figure 5b) which indicates that *SbMT-2* gene may help in water retention to counteract the osmotic shock. Abiotic stress causes the perturbation in metabolic balance of the cell, resulting in the enhanced production of ROS, which in-turn damages cell membranes, nucleic acids and chloroplast pigments [15,30]. Electrolyte leakage (EL) and membrane stability index (MSI) is the indicator of cell membrane stability. In the present study electrolyte leakage was lower in transgenic lines while MSI was higher under stress condition (Figure 5c-d) which suggests the role of *SbMT-2* gene in cell membrane protection. It was further supported by lipid peroxidation analysis, H_2O_2 and proline quantification, in which high MDA, H_2O_2 and proline contents were found in WT plants compared to transgenic lines under stress condition (Figure 7). These result evident the possible role of *Sb*MT-2 in ROS scavenging/detoxification and maintaining the cellular homeostasis during stress condition. Furthermore, *in vivo* localization of

H_2O_2 and O_2^- under different abiotic stress (Figure 6), confirmed the role of *SbMT*-2 gene in ROS scavenging.

Although role of metallothionein genes in ROS scavenging and detoxification have studied [12,14,16,29,31] but the molecular mechanism of ROS detoxification/scavenging is still unknown. ROS, being a signaling molecule play an important role in development and regulation of different metabolic processes whereas ROS toxicity resulted into the oxidative damage to the cell membrane and its components [30]. Plant cell maintains a stringent regulation over its production and scavenging. In *Arabidopsis* about 150 genes are involved in the homeostasis of ROS which comprised of ROS scavenging enzymes and ROS producing proteins [30]. Superoxide dismutase (SOD), peroxidase (POD) and ascorbate peroxidase (APX) are important ROS scavenging enzymes activated under different stress conditions to maintain the ROS homeostasis. Expression of genes encoding these ROS scavenging enzymes was studied in transgenic lines and compared with WT plants (Figure 10) to confirm the role of the *SbMT*-2 gene in ROS scavenging and detoxification. Higher relative fold expression of *NtSOD* and *NtPOD* genes were found in transgenic lines under stress conditions, however similar result was also observed with the *NtAPX* gene except under salt stress. Moreover, high expression was detected in metal stress compared to salt and osmotic stress, which may be because of availability of cofactors of the enzymes.

Metallothioniens are involved in essential-metal homeostasis and impart protection against heavy metal toxicity by sequestra-

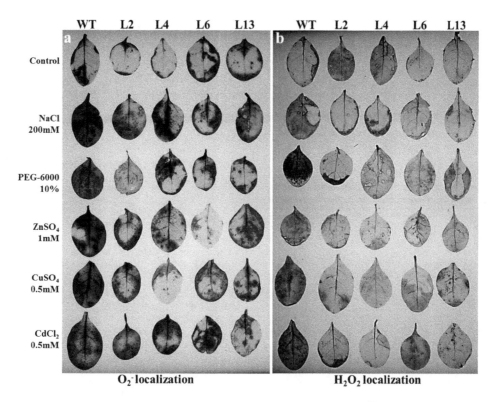

Figure 6. *In vivo* **localization of O₂⁻ and H₂O₂ in WT and transgenic lines under different stress condition.** Comparison of (a) O₂⁻ and (b) H₂O₂ localization in wild type (WT) and T₁ transgenic lines (L2, L4, L6 and L13) grown in hydroponic (control) and treated with salt (200 mM NaCl), osmotic (10% PEG), Zn (1 mM), Cu (0.5 mM) and Cd (0.5 mM) stress.

tion [6,9]. Plants maintain a high K^+/Na^+ ratio to combat with the deleterious effect of salinity. In present study transgenic lines over-expressing *SbMT-2* gene showed a high K^+/Na^+ ratio under stress compared to WT plants and control condition (Figure S2). It provides further evidence that *SbMT-2* may have role in ionic homeostasis and detoxification of H_2O_2 and thus impart salt tolerance. Transgenic lines showed accumulation of Cu^{++} and Cd^{++} in root compared to stem and leaves under stress condition. However, the *SbMT-2* gene over-expression leads to the accumulation of Zn^{++} in stem compared to control condition where it was accumulated more in roots (Figure 8). Zinc ion accumulation and tolerance varies among plant species and depends on MT type. The result exhibits *SbMT-2* may involve in the selective translocation of Zn^{++} from root to stem. Furthermore, the result also suggests that *SbMT-2* is involved in Cu and Cd binding and accumulation rather than translocation, as observed with Zn ions. Previously, it was reported that *SbMT2* protein exhibited high binding affinity and sequestration for Zn^{++} compared to Cu and Cd ions [13].

In order to support this observation, expression of type 1B heavy metal–transporting P-type ATPases (P_{1B} ATPase) transporter and zinc specific transporter encoding gene *NtHMA-A* and *NtZIP1* were analyzed under metal stress treatments (Figure 9). Up-regulation of genes in transgenic lines under metal stress especially, zinc compared to control and WT plants provides a supporting evidence that expression of *SbMT-2* gene may influence metal transporters and thus translocation of Zn was observed in the study. The qPCR analysis revealed higher up-regulation of these genes in transgenic lines compared to WT plants (about 6 to 10 fold up-regulation of *NtZIP1* gene in transgenic lines compared to about 2.5 fold of WT plants) under

metal stress condition and it might be due to introgression of *SbMT-2* gene. Furthermore, higher down-regulation of these genes (*NtZIP1*and *NtHMA-A*) in transgenic plants compared to WT was observed under de-stress treatments, performed by re-culturing the plants in un-stressed conditions (Table S1). Results further suggest that the *SbMT2* gene may influence the expression of these genes.

It was observed that expression of MT type 2 gene *PsMT(A1)* of *Pisum sativum* enhanced metal tolerance in white poplar and accumulation of zinc and copper (in leaves and roots) respectively [11]. Expression of seed specific MT gene *MT4a* in *Arabidopsis* increased Cu^{++} accumulation but did not show any response under Zn stress [10]. The *OsMT1a*, a type 1 metallothionein gene, involved in the Zn^{++} accumulation and thus provides tolerance to the transgenic rice [12]. *Elsholtzia haichowensis* metallothionein 1 (*Eh*MT1) over-expression in tobacco plants enhances copper tolerance and accumulation in root cytoplasm [32] however, expression of *BjMT2* gene in *A. thaliana* showed copper and cadmium tolerance [15]. Sequestration, translocation and thereby accumulation are the important mechanism used by plants for the phytoextraction [33] and therefore *SbMT-2* gene may be utilized for the phytoremediation.

Conclusion

In conclusion, the present study provides an useful insight that *SbMT-2* may involve in maintaining the cellular homeostasis by modulating ROS scavenging/detoxification during stress conditions and thus impart tolerance to salt and osmotic stress. It was observed that *SbMT-2* provides protection against heavy metal toxicity by metal ions accumulation and may be involved in the

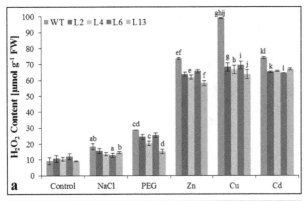

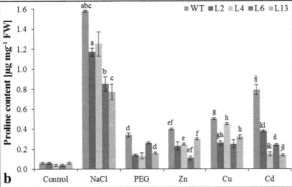

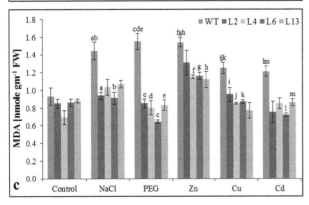

Figure 7. Estimation of H$_2$O$_2$, proline and MDA in transgenic tobacco lines under different stress condition. Comparison of (a) H$_2$O$_2$, (b) Proline and (c) MDA content of wild type (WT) and T$_1$ transgenic lines (L2, L4, L6 and L13) grown in hydroponic (control) and treated with salt (200 mM NaCl), osmotic (10% PEG), Zn (1 mM), Cu (0.5 mM) and Cd (0.5 mM) stress. Graph represents the mean $\pm$ SD (of three replicates; n = 3) followed by similar letters are significantly different according to Tukey HSD at $P<0.05$.

selective translocation of Zn^{++} from roots to stem (Figure 11). It is speculated that *SbMT-2* gene is a potential candidate for introgression to crop plants for imparting stress tolerance and phytoremediation.

Methods

Construction of plant transformation vector and tobacco transformation

The *SbMT-2* cDNA was amplified using forward (MT-2F: 5'-CTCGAGATGTCTCTTGCTTGTGGTGGTAAC-3') and re-

verse MT-2R: 5'- GGTACCTCATTTTGCAAGTG-CAAGGGTTG -3') primers containing *Xho*I and *Kpn*I sites respectively. The amplicon was digested with *Xho*I/*Kpn*I and cloned in pRT101 vector [34]. Thereafter, the gene cassette "CaMV35S-*SbMT-2*" was excised with *Pst*I and cloned into the pCAMBIA1301 vector. The resulting vector was mobilised into *Agrobacterium tumefaciens* (EHA105) and further transformed to tobacco (*Nicotiana tabacum* cv. xanthii) plants according to the standard protocol [35].

Confirmation of transgenic lines and expression of transgene

Total genomic DNA was isolated from leaves of transgenic and wild type (WT; control/untransformed) tobacco plants (T$_1$) using Qiagen DNeasy plant mini kit and quantified by Nanodrop Spectrophotometer (ND1000, Wilmington, USA). The integration of transgene in different lines was confirmed by PCR analyses using the *SbMT-2* gene, reporter gene *gus* and selection marker gene *hptII* specific primers (Table 1). Presence and expression of the *SbMT-2* gene in selected lines (L2, L4, L6 and L13) were confirmed by Southern blot and semi-quantitative Rt-PCR, respectively.

Genomic DNA (20 μg) from WT and transgenic lines L2, L4, L6 and L13 was digested with *Hind*III, separated on agarose (0.7%) by electrophoresis and transferred to a Hybond N$^+$ membrane (Amersham Pharmacia, UK) using alkaline transfer buffer (0.4 N NaOH with 1 M NaCl). DNA blot was hybridized with PCR-generated probe for the *uidA* gene labeled with DIG-11-dUTP following pre-hybridization and hybridization carried out at 42°C overnight in DIG EasyHyb buffer solution [36–39]. The hybridized membrane was detected by using CDP-Star chemiluminescent as substrate, following manufacturer user guide (Roche, Germany) and signals were visualized on X-ray film after 30 min.

Total RNA was extracted from transgenic tobacco (T$_1$) and wild-type plants using Qiagen RNeasy plant mini kit and the cDNA was made by reverse transcription with Superscript II RT (Invitrogen, USA). The actin gene was used as an internal control and both genes (*SbMT-2* and actin) were amplified using gene specific primers (Table 1). Histochemical GUS assay was performed with leaves as described by Jefferson [40].

Analyses of transgenic plants under different abiotic stress

Transgenic and wild type tobacco lines were maintained under controlled containment facility. Further morphological and physio-biochemical analyses were performed with T$_1$ transgenic lines under different abiotic stress treatments.

Growth parameters

Seeds of transgenic lines (L2, L4, L6 and L13) and WT plants were germinated on MS medium supplemented with 200 mM NaCl, 10% PEG 6000, 5 mM ZnSO$_4$, 0.2 mM CuSO$_4$ or 0.2 mM CdCl$_2$ under culture room conditions and the percentage of seed germination was calculated after 20 days.

Different growth parameters were measured under same stress conditions using T$_1$ seedlings. Seeds of transgenic lines were germinated on the MS media supplemented with 20 mg/l hygromycin, while seeds of WT plant were germinated on MS media only. WT and hygromycin positive T$_1$ seedlings were transferred to MS medium supplemented with 200 mM NaCl, 10% PEG, 5 mM ZnSO$_4$, 0.2 mM CuSO$_4$ or 0.2 mM CdCl$_2$ after eight days and grown for further 21 days. Growth

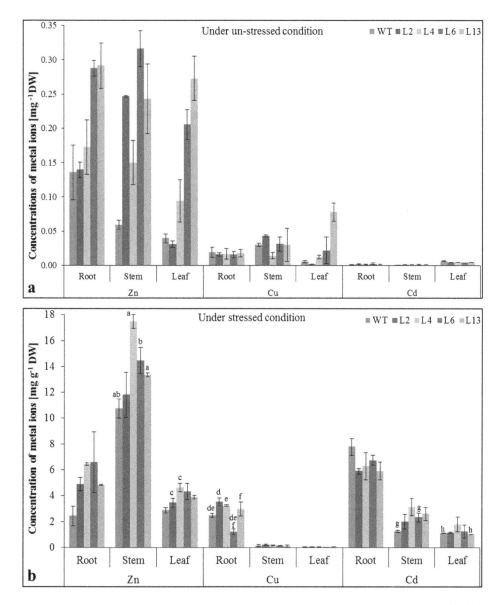

Figure 8. Metal ions content in transgenic tobacco lines under different stress condition. Comparison of Zn^{++}, Cu^{++} and Cd^{++} content in WT and transgenic lines in root, stem and leaf under (a) control and (b) stress condition (1 mM Zn, 0.5 mM Cu and Cd). Graph represents the mean $\pm$ SD (of three replicates; n = 3) followed by similar letters are significantly different according to Tukey HSD at $P<0.05$.

parameters; shoot length, root length, fresh weight and dry weight were recorded and compared with wild type plants.

Leaf senescence assay and chlorophyll estimation

Leaf discs from 30 days old WT and transgenic plants (L2, L4, L6 and L13) were used for the stress tolerance assay. Healthy leaves of similar age were detached and leaf discs (of 5 mm in diameter) were punched out. About 8 discs of each plant were floated in 5 ml ½ Hoagland media (control) supplemented with 200 mM NaCl, 10% PEG, 1 mM $ZnSO_4$, 0.5 mM $CuSO_4$ or 0.5 mM $CdCl_2$ for 7 days. The effect of these treatments on leaf discs were assessed by observing phenotypic changes.

Leaf discs (control and treated) were further subjected to chlorophyll isolation. Leaf discs were thoroughly homogenized in chilled N, N-dimethylformamide (DMF) at 4°C and thereafter

centrifuged at 10,000 g for 10 min. Supernatant was aspirated out and O.D. was recorded at 664 and 647 nm. Total chlorophyll content was calculated per gram fresh weight of tissue according to Porra et al. [41].

Relative water content, Electrolyte leakage and Membrane stability index

WT and T_1 seedlings (L2, L4, L6 and L13) were transferred to ½ Hoagland hydroponics culture and maintained for 20 days. Healthy young plants of same age and size were collected for each treatment (24 h for 200 mM NaCl and 0.5 mM $CuSO_4$; 12 h for 10% PEG, 1 mM $ZnSO_4$ and 0.5 mM $CdCl_2$).

About 100 mg (FW) leaves of control and treated plants were submerged into the deionised water and after 12 h turgid weight (TW) were recorded. Samples were further kept at 80°C for 48 h

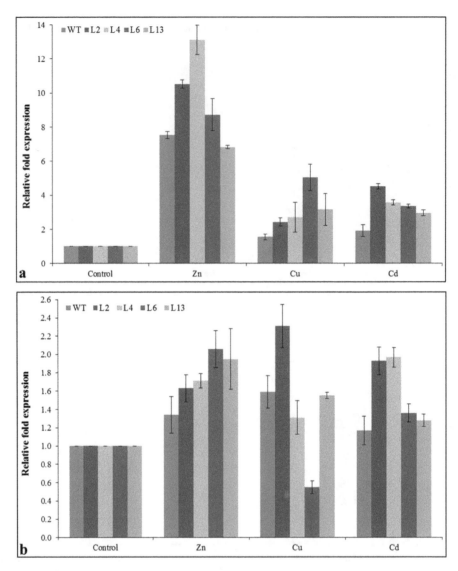

Figure 9. Expression analysis of zinc transporter encoding genes in transgenic tobacco lines under different metal ion stress condition. Comparison of relative fold expression of (a) *NtZIP1* and (b) *NtHMA-A* gene under Zn (1 mM), Cu (0.5 mM) and Cd (0.5 mM) stress.

to record the dry weight (DW). The relative water content was calculated as: $RWC~(\%) = (FW\text{-}DW/TW\text{-}DW) \times 100$.

Collected leaves were washed thoroughly with deionized water to remove surface-adhered electrolytes. Samples were kept in closed vials containing 10 ml deionised water and incubated at 25°C on a rotary shaker for 24 h. Subsequently, the electrical conductivity (EC) of the solution (L_t) was determined using conductivity meter (SevenEasy, Mettler Toledo AG 8603, Switzerland). Samples were autoclaved at 120°C for 20 min, cooled up to 25°C and electrical conductivity (L_0) was determined. The electrolyte leakage was determined: *Electrolyte leakage (%)* $= (L_t/L_0) \times 100$.

Healthy young leaves of stress-treated plants were taken and membrane stability index (MSI) was determined [42]. Leaves (200 mg) were kept in close vials containing 10 ml deionized water. A set of vials was incubated at 40°C for 30 min while second set of vials was incubated at 100°C for 10 min. Electrical conductivity was recorded for both sets (L1 for 40°C while L2 for

100°C) and MSI was calculated using the formula as: $MSI = [1\text{-}(L1/L2)] \times 100$.

In vivo localization of O_2^- and H_2O_2

Histochemical staining was performed for the *in vivo* detection of O_2^- and H_2O_2 using nitro-blue tetrazolium (NBT) and 3, 3-diaminobenzidine (DAB), respectively [43]. The presence of O_2^- in transgenic and WT leaves exposed to different stresses (as described above) was detected by immersing the leaf samples in NBT solution (1 mg ml^{-1} in 10 mM phosphate buffer; pH 7.8) at room temperature for 2 h and then illuminatedfor 12 h in light until blue spots appeared. For the localization of H_2O_2, treated leaves were incubated in DAB solution (1 mg ml^{-1} in 10 mM phosphate buffer; pH 3.8) at room temperature for 6 h in dark, thereafter exposed to the light until brown spots were appeared. Leaf samples were treated with absolute ethanol (for bleaching chlorophyll contents) before the documentation.

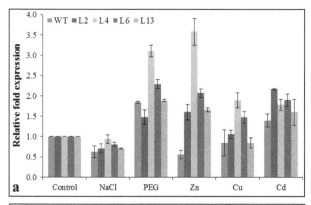

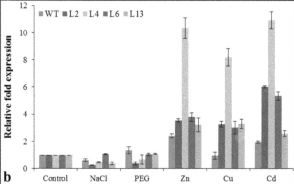

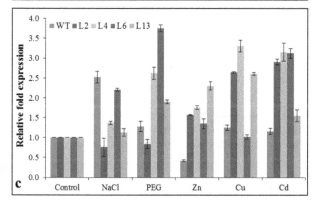

Figure 10. Expression analysis of antioxidant enzyme encoding genes in transgenic tobacco lines under different stress condition. Comparison of relative fold expression of (a) *NtSOD* gene, (b) *NtPOD* gene and (c) *NtAPX* gene under salt (200 mM NaCl), osmotic (10% PEG), Zn (1mM), Cu (0.5 mM) and Cd (0.5 mM) stress.

Quantification of H₂O₂, proline and lipid peroxidation

The H_2O_2 and free proline content in leaf samples (WT and transgenic lines under different stress conditions) were measured [44,45]. Proline and H_2O_2 levels were calculated by the standard curve, prepared against known concentration of proline or H_2O_2 measured at 520 and 560 nm absorbance respectively. Lipid peroxidation was estimated by determining the concentration of malondialdehyde (MDA) produced by thiobarbituric acid (TBA) reaction [46]. Leaf samples were extracted in 2 ml 0.1% trichloroacetic acid (TCA) and 0.5 ml extract was reacted with 2.0 ml of TBA reagent followed by boiling at 95°C for 30 min. Samples were cooled at ice, centrifuged at 10000 g for 5 min and absorbance of the supernatants was measured at 440 nm, 532 nm and 600 nm.

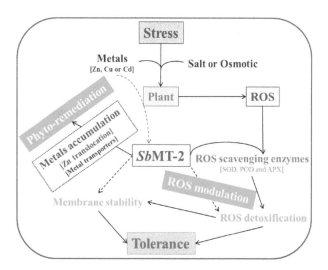

Figure 11. A hypothetical model for the role of *Sb*MT-2 in phyto-remediation and ROS modulation under abiotic stress.

Ion content analysis

For ion content analysis, tissues (root, shoot and leaves) from 4-week-old plants grown in hydroponic medium and treated with 200 mM NaCl, 10% PEG 6000, 1 mM ZnSO₄, 0.5 mM CuSO₄ or 0.5 mM CdCl₂ were washed with deionised water, dried in hot air oven for 48 h at 70°C and digested with 4 ml perchloric acid-nitric acid solution (3:1). The solution was heated to dry and further diluted to 25 ml with deionised water. Ion contents were measured by inductively coupled plasma optical emission spectrometer (Optima2000DV, PerkinElmer, Germany).

Transcriptional regulation of metal transporters and antioxidative enzymes

Total RNA was isolated from the stressed and control plants by using RNeasy plant mini kit (Qiagen) following the manufacturer's instructions and quantified with Nanodrop spectrophotometer (NanoDrop, USA). The cDNA was prepared using 2 µg of total RNA with a SuperScript RT III first-strand cDNA synthesis kit (Invitrogen, USA). The expression pattern of genes; *NtZIP1*, *NtHMA-A*, encoding metal transporters [47] and *NtSOD*, *NtPOX* and *NtAPX*, encoding antioxidant enzymes [48] which involved in heavy metal transportation and ROS scavenging, respectively were analyzed under different stress by using gene-specific primer pairs (Table 1), while the gene actin was used as an internal control. The quantitative real time PCR (qRT-PCR) was performed in a Bio-Rad IQ5 detection system (Bio-Rad, USA) with QuantiFast Kit (Qiagen, USA). Fold expression was calculated by the method described by Livak and Schmittgen [49] and specificity of qRT-PCR was monitored by melt curve analysis.

Statistical analysis

Each experiment was carried out in three replicates. All data was expressed as mean ± SD and subjected to analysis of variance (ANOVA) to determine the significance of difference between the means of WT and transgenic plants of each treatment group. A Tukey HSD multiple comparison of mean test was used, significant differences were considered at $P<0.05$ and indicated by similar letters.

Table 1. Primer sets used for the confirmation and transcript analysis of transgenic lines.

Genes	Primer sequences (5′ to 3′)	PCR conditions
uidA	F: GATCGCGAAAACTGTGGAAT	95°C, 5min; 34 Cycles: 95°C, 1min; 60°C, 1min; 72°C, 2 min; Extension: 72°C, 7 min
	R: TGAGCGTCGCAGAACATTAC	
hptII	F: TTCTTTGCCCTCGGACGAGTG	95°C, 5min; 34 Cycles: 95°C, 1min; 55°C, 1min; 72°C, 2 min; Extension: 72°C, 7 min
	R: ACAGCGTCTCCGACCTGATG	
SbMT-2	F: AGGTTGCAAGATGTTCC TG	95°C, 5min; 35 Cycles: 95°C, 30sec; 60°C, 45sec; 72°C, min; Melt curve: 60–95°C
	R: TCCATCGTTCTCAGCTGCTAC	with 0.5°C/cycle increment
Actin	F: CGTTTGGATCTTGCTGGTCGT	
	R: CAGCAATGCCAGGGAACATAG	
NtZIP1	F: TGGTGGCTCAGTCTGGAGAT	
	R: CGAAGGAGCTCAGAACTGGAA	
NtHMA-A	F: ACAAAGTGCTCGGACACCAA	
	R: CTTCTCGGTTGCAGAGTCCT	
NtSOD	F: AGCTACATGACGCCATTTCC	
	R: CCCTGTAAAGCAGCACCTTC	
NtPOD	F: CTTGGAACACGACGTTCCTT	
	R: TCGCTATCGCCATTCTTTCT	
NtAPX	F: CAAATGTAAGAGGAAACTCAGAGGA	
	R: CAGCCTTGAGCCTCATGGTACCG	

Supporting Information

Figure S1   Percentage of seed germination under different stress condition. Seeds of WT and transgenic lines were germinated on MS media supplemented with 200 mM NaCl, 10% PEG, 5 mM Zn, 0.2 mM Cu or 0.2 mM Cd. Graph represents the mean ± SD (of three replicates; n = 3) followed by similar letters are significantly different according to Tukey HSD at $P < 0.05$.

Figure S2   Sodium and potassium ion content in transgenic tobacco lines under different stress condition. Comparison of (a) Na^+ and K^+ content and (b) K^+/Na^+ ratio in WT and transgenic lines in root, stem and leaf under stressed (200 mM NaCl) and un-stressed conditions. Graph represents the mean ± SD (of three replicates; n = 3) followed

by similar letters are significantly different according to Tukey HSD at $P < 0.05$.

Table S1   Comparison of relative fold expression (down–regulation) of zinc transporter encoding genes (NtZIP1 and NtHMA–A) in transgenic tobacco lines and wild type plants under de–stress condition.

Acknowledgments

CSIR-CSMCRI Communication No. PRIS-144/2013.

Author Contributions

Conceived and designed the experiments: AM BJ. Performed the experiments: AKC MKP VT. Analyzed the data: AM AKC. Wrote the paper: AKC AM.

References

1. Sreenivasulu N, Sopory SK, Kavi Kishor PB (2007) Deciphering the regulatory mechanisms of abiotic stress tolerance in plants by genomic approaches. Gene 388: 1–13.
2. Seth CS, Remans T, Keunen E, Jozefczak M, Gielen H, et al. (2012) Phytoextraction of toxic metals: a central role for glutathione. Plant Cell Environ 35: 334–346.
3. Chary NS, Kamala CT, Raj DSS (2008) Assessing risk of heavy metals from consuming food grown on sewage irrigated soils and food chain transfer. Ecotoxicol Environ Safety 69: 513–524.
4. Tewari RK, Kumar P, Sharma PN, Bisht S (2002) Modulation of oxidative stress responsive enzymes by excess cobalt. Plant Sci 162: 381–388.
5. Hamer DH (1986) Metallothionein. Annu Rev Biochem 55: 913–951.
6. Huang GY, Wang YS (2010) Expression and characterization analysis of type 2 metallothionein from grey mangrove species (Avicennia marina) in response to metal stress. Aquat Toxicol 99: 86–92.
7. Cobbett C, Goldsbrough P (2002) Phytochelatins and metallothioneins: Roles in heavy metal detoxification and homeostasis. Annu Rev Plant Biol 53: 159–182.
8. Zhou JM, Goldsbrough PB (1994) Functional homologs of fungal metallothionein genes from Arabidopsis. Plant Cell 6: 875–884.
9. Hassinen VH, Tuomainen M, Peräniemi S, Schat H, Kärenlampi SO, et al. (2009) Metallothioneins 2 and 3 contribute to the metal-adapted phenotype but are not directly linked to Zn accumulation in the metal hyperaccumulator, Thlaspi caerulescens. J Exp Bot 60: 187–196.
10. Rodriguez Lloriente ID, Perez-Palacios P, Doukkali B, Caviedes MA, Pajuelo E (2010) Expression of the seed-specific metallothionein MT4a in plant vegetative tissues increase Cu and Zn tolerance. Plant Sci 178: 327–332.
11. Turchia A, Tamantini I, Camussia AM, Racchi ML (2012) Expression of a metallothionein A1 gene of Pisum sativum in white poplar enhances tolerance and accumulation of zinc and copper. Plant Sci 183: 50–56.
12. Yang Z, Wu Y, Li Y, Ling HQ, Chu C (2009) OsMT1a, a type 1 metallothionein, plays the pivotal role in zinc homeostasis and drought tolerance in rice. Plant Mol Biol 70: 219–229.
13. Chaturvedi AK, Mishra A, Tiwari V, Jha B (2012) Cloning and transcript analysis of type 2 metallothionein gene (SbMT–2) from extreme halophyte

Salicornia brachiata and its heterologous expression in *E. coli*. Gene 499: 280–287.

14. Xue T, Li X, Zhu W, Wu C, Yang G, et al. (2009) Cotton metallothionein *GhMT3a*, a reactive oxygen species scavenger, increased tolerance against abiotic stress in transgenic tobacco and yeast. J Exp Bot 60: 339–349.

15. Zhigang A, Cuijie L, Yuangang Z, Yejie D, Wachter A, et al. (2006) Expression of *BjMT2*, a metallothionein 2 from *Brassica juncea*, increases copper and cadmium tolerance in *Escherichia coli* and *Arabidopsis thaliana*, but inhibits root elongation in *Arabidopsis thaliana* seedlings. J Exp Bot 57: 3575–3582.

16. Wong HL, Sakamoto T, Kawasaki T, Umemura K, Shimamoto K (2004) Down-regulation of metallothionein, a reactive oxygen scavenger, by the small GTPase *OsRac1* in rice. Plant Physiol 135: 1447–1456.

17. Zhang FQ, Wang YS, Lou ZP, Dong JD (2007) Effect of heavy metal stress on antioxidative enzymes and lipid peroxidation in leaves and roots of two mangrove plant seedlings (*Kandelia candel* and *Bruguiera gymnorrhiza*). Chemosphere 67: 44–50.

18. Schutzenduble A, Polle A (2002) Plant responses to abiotic stresses: heavy metal-induced oxidative stress and protection by mycorrhization. J Exp Bot 53: 1351–1365.

19. Kholodova V, Volkov K, Kuznetsov V (2010) Plants under heavy metal stress in saline environments, in: Sherameti, I., Varma, A. (Eds.), Soil heavy metals, Soil Biology vol. 19. Springer Berlin Heidelberg. pp. 163–183.

20. Joshi M, Mishra A, Jha B (2012) NaCl plays a key role for *in vitro* micropropagation of *Salicornia brachiata*, an extreme halophyte. Ind Crops Prod 35: 313–316.

21. Jha B, Singh N, Mishra A (2012a) Proteome profiling of seed storage proteins reveals the nutritional potential of *Salicornia brachiata* Roxb., an extreme halophyte. J Agric Food Chem 60: 4320–4326.

22. Mishra A, Joshi M, Jha B (2013) Oligosaccharide mass profiling of nutritionally important *Salicornia brachiata*, an extreme halophyte. Carbohydr Polym 92: 1942–1945.

23. Jha B, Sharma A, Mishra A (2011) Expression of *SbGSTU* (tau class glutathione S-transferase) gene isolated from *Salicornia brachiata* in tobacco for salt tolerance. Mol Biol Rep 38: 4823–4832.

24. Singh N, Mishra A, Jha B (2014) Ectopic over-expression of peroxisomal ascorbate peroxidase (*SbpAPX*) gene confers salt stress tolerance in transgenic peanut (*Arachis hypogaea*). Gene On-line: 10.1016/j.gene.2014.06.037.

25. Joshi M, Jha A, Mishra A, Jha B (2013) Developing transgenic *Jatropha* using the *SbNHX1* gene from an extreme halophyte for cultivation in saline wasteland. PLoS ONE 8: e71136.

26. Singh N, Mishra A, Jha B (2014) Over-expression of the peroxisomal ascorbate peroxidase (*SbpAPX*) gene cloned from halophyte *Salicornia brachiata* confers salt and drought stress tolerance in transgenic tobacco. Mar Biotechnol 16: 321–332.

27. Udawat P, Mishra A, Jha B (2014) Heterologous expression of an uncharacterized universal stress protein gene (*SbUSP*) from the extreme halophyte, *Salicornia brachiata*, which confers salt and osmotic tolerance to *E. coli*. Gene 536: 163–170.

28. Tiwari V, Chaturvedi AK, Mishra A, Jha B (2014) The transcriptional regulatory mechanism of the peroxisomal ascorbate peroxidase (*pAPX*) gene cloned from an extreme halophyte, *Salicornia brachiata*. Plant Cell Physiol 55: 201–217.

29. Kumar G, Rituraj H, Kushwaha V, Sabharwal P, Kumari S, et al. (2012) Clustered metallothionein genes are co-regulated in rice and ectopic expression of *OsMT1e-P* confers multiple abiotic stress tolerance in tobacco via ROS scavenging. BMC Plant Biol 12: 107.

30. Mittler R (2002) Oxidative stress, antioxidants and stress tolerance. Trends Plant Sci 7: 405–410.

31. Zhu J, Zhang Q, Wua R, Zhang Z (2010) *Hb*MT2, an ethephon-induced metallothionein gene from *Hevea brasiliensis* responds to H_2O_2 stress. Plant Physiol Biochem 48: 710–715.

32. Xia Y, Qic Y, Yuana Y, Wanga G, Cuia J, et al. (2012) Overexpression of *Elsholtzia haichowensis* metallothionein 1 (*Eh*MT1) in tobacco plants enhances copper tolerance and accumulation in root cytoplasm and decreases hydrogen peroxide production. J Hazard Mater 233–234: 65–71.

33. Cherian S, Oliveira MM (2005) Transgenic Plants in Phytoremediation: Recent Advances and New Possibilities. Environ Sci Technol 39: 9377–9390.

34. Töpfer R, Matzeit V, Gronenborn B, Schell J, Steinbiss HH (1987) A set of plant expression vectors for transcriptional and translational fusions. Nucleic Acids Res 15: 5890.

35. Horsch RB, Fry JE, Hoffmann NL, Eichholtz D, Rogers SG, et al. (1985) A simple and general method for transferring genes into plants. Science 227: 1229–1231.

36. Joshi M, Mishra A, Jha B (2011) Efficient genetic transformation of *Jatropha curcas* L. by microprojectile bombardment using embryo axes. Ind Crop Prod 33: 67–77.

37. Singh N, Mishra A, Joshi M, Jha B (2010) Microprojectile bombardment mediated genetic transformation of embryo axes and plant regeneration in cumin (*Cuminum cyminum* L.). Plant Cell Tiss Organ Cult 103: 1–6.

38. Pandey S, Mishra A, Patel MK, Jha B (2013) An efficient method for *Agrobacterium*-mediated genetic transformation and plant regeneration in cumin (*Cuminum cyminum* L.). Appl Biochem Biotechnol 171: 1–9.

39. Tiwari V, Chaturvedi AK, Mishra A, Jha B (2014) An efficient method of *Agrobacterium*-mediated genetic transformation and regeneration in local Indian cultivar of groundnut (*Arachis hypogaea*) using grafting. Appl Biochem Biotechnol In-press, doi: 10.1007/s12010-014-1286-3.

40. Jefferson RA (1987) Assaying chimeric genes in plants: the GUS fusion system. Plant Mol Biol Rep 5: 387–405.

41. Porra RJ, Thompson WA, Kriedemann PE (1989) Determination of accurate extinction coefficients and simultaneous equations for assaying chlorophylls a and b extracted with four different solvents: verification of the concentration of chlorophyll standards by atomic absorption spectroscopy. Biochemica et Biophysica Acta 975: 384–391.

42. Sairam RK (1994) Effect of homobrassinolide application on metabolism and grain yield under irrigated and moisture stress conditions of two wheat varieties. Plant Growth Regul 14: 173–181.

43. Shi J, Fu XZ, Peng T, Huang XS, Fan QJ, et al. (2010) Spermine pretreatment confers dehydration tolerance of citrus in vitro plants via modulation of antioxidative capacity and stomatal response. Tree Physiol 30: 914–922.

44. He Z, Wang ZY, Li J, Zhu Q, Lamb C, et al. (2000) Perception of brassinosteroids by the extracellular domain of the receptor kinase BRI. Science 288: 2360–2363.

45. Bates LS, Waldern R, Teare ID (1973) Rapid determination of free proline for water stress studies. Plant Soil 39: 205–207.

46. Draper HH, Hadley M (1990) Malondialdehyde determination as index of lipid peroxidation. Methods in Enzymology 186: 421–431.

47. Siemianowski O, Barabasz A, Kendziorek M, Ruszczyńska A, Bulska E, et al. (2014) *HMA4* expression in tobacco reduces Cd accumulation due to the induction of the apoplastic barrier. J Exp Bot 65: 1125–1139.

48. Huang XS, Luo T, Fu XZ, Fan QJ, Liu JH (2011) Cloning and molecular characterization of a mitogen-activated protein kinase gene from *Poncirus trifoliata* whose ectopic expression confers dehydration/drought tolerance in transgenic tobacco. J Exp Bot 62: 5191–5206.

49. Livak KJ, Schmittgen TD (2001) Analysis of relative gene expression data using real-time quantitative PCR and the $2^{-\Delta\Delta C_T}$ method. Methods 25: 402–408.

14

BraLTP1, a Lipid Transfer Protein Gene Involved in Epicuticular Wax Deposition, Cell Proliferation and Flower Development in Brassica napus

Fang Liu[1], Xiaojuan Xiong[1], Lei Wu[1], Donghui Fu[2], Alice Hayward[3], Xinhua Zeng[1], Yinglong Cao[1], Yuhua Wu[1], Yunjing Li[1], Gang Wu[1]*

1 Key Laboratory of Oil Crop Biology of the Ministry of Agriculture, Oil Crops Research Institute, Chinese Academy of Agricultural Sciences, Wuhan, China, 2 The Key Laboratory of Crop Physiology, Ecology and Genetic Breeding, Ministry of Education, Agronomy College, Jiangxi Agricultural University, Nanchang, China, 3 Queensland Alliance for Agriculture and Food Innovation, The University of Queensland, Queensland, Australia

Abstract

Plant non-specific lipid transfer proteins (nsLTPs) constitute large multigene families that possess complex physiological functions, many of which remain unclear. This study isolated and characterized the function of a lipid transfer protein gene, BraLTP1 from Brassica rapa, in the important oilseed crops Brassica napus. BraLTP1 encodes a predicted secretory protein, in the little known VI Class of nsLTP families. Overexpression of BnaLTP1 in B. napus caused abnormal green coloration and reduced wax deposition on leaves and detailed wax analysis revealed 17–80% reduction in various major wax components, which resulted in significant water-loss relative to wild type. BnaLTP1 overexpressing leaves exhibited morphological disfiguration and abaxially curled leaf edges, and leaf cross-sections revealed cell overproliferation that was correlated to increased cytokinin levels (tZ, tZR, iP, and iPR) in leaves and high expression of the cytokinin biosynthsis gene IPT3. BnaLTP1-overexpressing plants also displayed morphological disfiguration of flowers, with early-onset and elongated carpel development and outwardly curled stamen. This was consistent with altered expression of a a number of ABC model genes related to flower development. Together, these results suggest that BraLTP1 is a new nsLTP gene involved in wax production or deposition, with additional direct or indirect effects on cell division and flower development.

Editor: Xianlong Zhang, National Key Laboratory of Crop Genetic Improvement, China

Funding: This work was supported by Major Research Project of CAAS Science and Technology Innovation Program, the National Natural Science Foundation of China (grant numbers 31400243 and 31201152) and the Natural Science Foundation of Hubei Province (grant number 2013CFB423). The funders had no role in study design, data collection and analysis, decision to publish, or preparation of the manuscript.

Competing Interests: The authors have declared that no competing interests exist.

* Email: wugang@caas.cn

Introduction

Plant non-specific lipid-transfer proteins (nsLTPs) are small, abundant, basic, secreted proteins in higher plants [1,2]. nsLTPs contain an 8 cysteine motif (8 CM) structure comprising eight cysteine residues linked with four disulphide bonds that stabilize a hydrophobic cavity that allows for the in vitro loading of a broad variety of lipid compounds [3,4].

nsLTPs are encoded by multigene families that were originally subdivided into type I (9 kDa) and type II (7 kDa) on the basis of molecular mass. More recently, several anther specific proteins in Maize (Zea mays) and rice (Oryza sativa) that displayed considerable homology with nsLTPs that have been proposed as a third type (type III) [5,6], which differ in the number of amino acid residues interleaved in the 8 CM structure. Recently, fifty two rice nsLTP genes, 49 Arabidopsis nsLTP genes and 156 putative wheat nsLTP genes were identified through genome-wide analyses [7]. Phylogenetic analysis revealed that the rice and Arabidopsis nsLTPs cluster into nine different clades, distinguished by a variable number of inter-cysteine amino acid residues [8]. Most studies to date have concentrated on type I, II and III family

members, with limited functional analysis of these other six structural types of nsLTPs.

Characterised nsLTPs have been implicated in variable and complex physiological functions, mainly related to stress resistance and development, including cuticular wax synthesis [8,9,10,11], abiotic stress [12,13,14], disease resistance [9,15,16,17,18], male reproductive development [19,20,21,22,23,24], and cell development [25,26,27].

One nsLTP family member, a glycosylphosphatidylinositol-anchored lipid transfer protein LTPG, was reported to function either directly or indirectly in cuticular lipid deposition, and mutant plant lines with decreased LTPG expression had reduced wax load on the stem surface [10]. Lee et al [9] reported that disruption of LTPG1 gene altered cuticular lipid composition, but not total wax and cutin monomer loads, and caused increased susceptibility to the fungus Alternaria brassicicola. Another gene, LTPG2, functionally overlaps with LTPG/LTPG1 during cuticular wax export or accumulation, and the total cuticular wax load was reduced in both ltpg2 and ltpg1 ltpg2 siliques [11]. These LTPG genes belong to type G nsLTPs classified by Edstam et al [28], and are not included in Boutrot's classification system [7].

Plant epidermal wax forms a hydrophobic layer covering aerial plant organs. This constitutes a barrier against nonstomatal water loss, as well as biotic stresses, and provides protection against pathogens [29]. Cuticular wax contains very-long-chain fatty acids (VLCFA) and their derivatives, such as alkanes and alcohols, with chain lengths of 20–34 carbons, and wax composition varies with species, organ, and developmental state [30]. Synthesis of VLCFA in the epidermis via acyl-CoA dehydratase PAS2 is essential for the proper control of cell proliferation in Arabidopsis [31]. Moreover, it was suggested that VLCFA, or its downstream derivatives or metabolites might function as signaling molecules to suppress cytokinin biosynthesis in the vasculature, thus fine-tuning cell division in internal tissue [31].

Despite progress in the functional analyses of these nsLTPs involved in cuticular wax deposition, the exact functions of most nsLTPs remain unclear, and complex expression profiles suggest disparate and unpredictable gene functions of unknown nsLTPs [22,23,27]. Most studies to date have focused on functions of type I, II and III nsLTP family members in many species, with few studies on type VI nsLTPs. Thus, the exploration of their roles may prove interesting, especially in non-model or crop species. In this study, we characterized a type VI nsLTP gene BraLTP1 from Brassica rapa, and investigated its biological function in the important allotetraploid oil crop, B. napus. Our results suggest that BraLTP1 has the basic characteristics of the nsLTPs gene family and is involved in wax deposition, cell proliferation and flower development. To our knowledge few reports linked nsLTPs to cell proliferation in plants, and nsLTP genes (excluding LTPG) involved in wax metabolisms in vivo have less been reported. This study will help to deepen our understanding of nsLTP family gene function and pave the way for the application of nsLTP gene in Brassica breeding.

Materials and Methods

Plant material

The plants used in this study were grown in pots containing mixture of moss peat (PINDSTRUP, Danmark) and field soil with the proportion of 3:1 in a plant growth room set to $20°C\pm2°C$ under a 16/8 h photo-period at a light intensity of 44 umol m^{-2} s^{-1} and 60–90% relative humidity.

Vector construction

The coding sequence of BraLTP1 was amplified from B. rapa accession Chiifu genomic DNA using primers designed to the published B. rapa sequence Bra011229 (http://brassicadb.org/brad/index.php) [32]. Primers were as follows: BraLTP1-F: 5′-GAGCTCACAACTTCCTTCAAAGCCACA-3′ and BraLTP1-R: 5′-GGATCCCAAACCTCATGGCACAATGTA-3′, containing 5′ restriction enzyme sites for SacI and BamHI respectively. PCR was carried out in 50 μL, with 50 ng DNA, 0.4 mM dNTPs, 0.2 μM each primer, 0.5 U LA Taq (TaKaRa, Japan) and 1×LA Taq buffer II (TaKaRa, Japan). Conditions were: 94°C for 3 min, 30 cycles at 94°C for 1 min, 55°C for 1 min and 72°C for 1 min. PCR product was checked by gel electrophoresis and target fragment was recovery and purified. The purified PCR product was cleaved using SacI and BamHI, and ligated between the CaMV 35S promoter and a terminal poly A sequence in the vector PBI121s (Fig. 1B), derived by modifying the multiple cloning site and deleting the GUS gene of PBI121 [33]. Positive clones by PCR using the above gene specific primers were chosen and sequenced (sangon company of Shanghai, China) to make sure they were correct. Standard molecular techniques [34,35] were used for DNA manipulation.

Genetic transformation

The 35S::BraLTP1 fragment in PBI121s was introduced into Agrobacterium tumefaciens GV3101 by electroporation, and positive clones were selected on on LB agar plates at 37°C, supplemented with appropriate concentration of antibiotics (gentamicin 50 mg L^{-1}, rifampicin 50 mg L^{-1} and kanamycin 50 mg L^{-1}) and PCR verified. A single positive colony was used to transform B. napus cv. Zhongshuang 6, an elite Chinese cultivar in China, as follows: Seeds of Zhongshuang 6 were soaked in 75% ethanol for 1 min and for 10–15 min in a 1.5% mercuric chloride solution. Five to six days after germination under darkness, etiolated hypocotyls were cut in 7 mm segments and mixed with 50 mL Agrobacterium in liquid DM media (MS+30 g L^{-1} sucrose+100 μM acetosyringone, pH 5.8) (OD ~0.3) for 0.5 h. Surface air dried hypocotyls were then transferred to co-cultured medium (MS+30 g L^{-1} sucrose+18 g L^{-1} manitol+1 mg L^{-1} 2, 4-D+0.3 mg L^{-1} kinetin+100 μM acetosyringone+8.5 g agrose, pH 5.8) for 2 days and then to a selection medium (MS+30 g L^{-1} sucrose+18 g L^{-1} manitol+1 mg L^{-1} 2, 4-D+0.3 mg L^{-1} kinetin+20 mg L^{-1} AgNO$_3$+8.5 g L^{-1} agrose+25 mg L^{-1} kanamycin+250 mg L^{-1} carbenicillin pH 5.8) for proliferation. After 3 weeks, hypocotyl callus was transferred to regeneration medium (MS+10 g L^{-1} glucose+0.25 g L^{-1} xylose+0.6 g L^{-1} MES hydrate+2 mg L^{-1} zeatin+0.1 mg L^{-1} indole-3-acetic acid+8.5 g L^{-1} agrose+25 mg L^{-1} kanamycin+250 mg L^{-1} carbenicillin, pH 5.8) for 2 weeks. Hypocotyls were transferred to new regeneration media every 2 weeks for 3~4 regeneration cycles before transfer to radication medium (MS+10 g L^{-1} sucrose+10 g L^{-1} agar, pH 5.8) for rooting (about 3 weeks). Transformed plants with roots were transplanted into pots and grown as described. For the construct, more than 60 independent 35S::BraLTP1 T$_0$ transgenic plants were generated, and more than 85% were positive transformants as detected using a forward primer designed to the CaMV 35S sequence (35S-F: 5′-AGGACACGCTGAAAT-CACCA-3) and a reverse primer designed to BraLTP1 (D-BraLTP1-R: 5′-GGATCCCAAACCTCATGGCACAATGTA-3′). T$_1$ seeds of PCR-positive transformants were harvested and grown to T$_2$ generation for phenotype identification.

Protein sequence analysis

BraLTP1 was aligned to homologous amino acid sequences from several cruciferae including Arabidopsis, B. rapa, B. napus and B. oleracea, using Align X multiple sequence alignment software (Vector NTI Advance 11.0, 2008 Invitrogen corporation). Homology search were conducted using BLAST 2.0 program of the National Center of Biotechnology Information (NCBI). Conserved domains were identified using CDD (http://www.ncbi.nlm.nih.gov/cdd/) and the signal peptide was determined by SignalP (http://www.cbs.dtu.dk/services/SignalP/).

Wax analysis

Epicuticular waxes were extracted from ~100 mg B. napus leaf disks from the fourth fully-expanded leaf from the apex for each plant by mixing in chloroform for 1 min, with 150 μl 100 μg L^{-1} triacontane added as an internal standard. The chloroform was then evaporated under gaseous N2, and the following steps for wax analysis was as described previously [36].

Water loss determination

For water loss analysis of detached leaves, leaves were detached from 8-week-old plants, placed on a petri dish for 1, 2, 3, 4, 5, and 6 h, and weighed [37]. Three independent experiments were performed, with 5 plants for each line in each experiment.

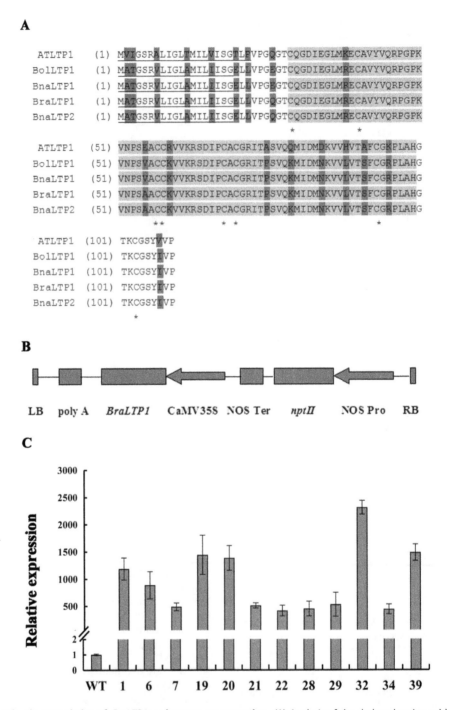

Figure 1. Basic protein characteristics of *BraLTP1* and vector construction. (A) Analysis of the deduced amino acid sequences of BraLTP1 with its homologous sequences in other cruciferae; variable sites (dark grey)the nsLTP-like conserved 8 CM domain (light gray) with conserved cysteine residues (asterisks) and putative extracellular secretory signals (underlined). Sequences are from *Arabidopsis thaliana* AtLTP1 (AT4g30880), *B. rapa BraLTP1* (Bra011229) [32], *B. oleracea* BolLTP1 (Bol018048) [60], *B. napus* BnaLTP1 (AY208878), and *B. napus* BnaLTP2 (KM062522). (B) T-DNA region of the *BnaLTP1* overexpression construct containing *BnaLTP1* driven by the CaMV 35S promoter. LB = Left border, RB = Right border, poly A = poly A terminator, *nptII* = kanamycin resistance, NOS Ter = nopaline synthase terminator, NOS Pro = nopaline synthase promoter. (C) Analysis of *BraLTP1* mRNA levels in 10-week-old wild type (WT) and T0 *35S::BraLTP1* transgenic plants by qRTPCR. Transgenic plants include *BraLTP1-1, -6, -7, -19, -20, -21, -22, -28, -29, -32, -34 and -39*. Standard errors were derived from three repeated experiment for the expression levels of each plants.

Leaf phenotype analysis

B. napus leaves were fixed for 24 h in 4% paraformaldehyde. After dehydration using an ethanol series (75% for 4 h, 85% for 2 h, 90% for 2 h, 95% for 1 h and 100% for 30 min twice) the leaves were cleared twice with xylene for 2 h each. The leaves were infiltrated and subsequently embedded in paraffin wax according to the method of Hu *et al* [38]. Seven micromolar sections were obtained using a Leica RM 2016 microtome (Leica, Nanterre Cedex, France) and stained with 1% safranin for 6 h. After dehydration using an ethanol series (50%, 70% and 80% for

3 min each), sections were stained with 0.5% fast green for 30 min and then dehydrated with 100% ethanol for 5 min. Observations were made and images were acquired with a Leica DM 2500 (Leica microsystems, DFC420C). Six leaf disks with 1.5 cm diameter were hole-punched from the 4th leaf of 8-week-old plants and weighed to estimate weight/unit leaf area. For quantification of cytokinin, sampling of ~100 mg fresh leaves from 4-week-old seedlings was repeated three times, and extraction and determination of cytokinin content were conducted as per Nobusawa et al [31].

Real-Time PCR

RNA was extracted using a TIANGEN RNAprep Pure Plant Kit (DP 432) according to the manufacturer's instructions. First-strand cDNAs were synthesized from DNaseI-treated total RNA using a TIANGEN FastQuant RT Kit (with gDNase) (KR106) according to the manufacturer's instructions. For transgene expression level and co-segregation experiments, real-time PCR was done using PCR SuperReal PreMix Plus (probe) (TIANGEN). The reaction system and process were followed by the manufac-turer's instructions, and four replicates were performed for each cDNA sample. PCR primers and TaqMan probes were designed on the basis of the BraLTP1 cDNA sequences as follows: sense: RT-BraLTP1-F: 5′-ATCGGTCTAGCAATGATC-3′, RT-BraLTP1-R: anti-sense: 5′-AGCACATTCTCTCATCAG-3′, RT-BraLTP1-probe: 5′-CTCGATGTCTCCTTGGCACG-3′. Specific primers for the B. napus Actin gene (GenBank accession number: AF111812.1) were used as an internal control (Actin-F: 5′-CACAGGAAATGCTTCTAAG-3′, Actin-R: 5′-GGATGGA-TATAGATCGTACC-3′, Actin-probe: 5′-ACTCACCACCAC-GAACCAGAA-3′). For transcriptional profiling of cell division and flower related genes, real-time PCR was done using PCR SuperReal PreMix Plus (SYBR Green) (TIANGEN). The reaction system and process followed the manufacturer's instructions, as above in four replicates for each cDNA sample. Table 1 provides information about the genes and primers used for the real-time PCR. Specially, IPT1, IPT4, IPT6 in Arabidopsis all correspond to the one Brassica napus IPT gene. Real-time PCR was performed in an optical 96-well plate with a Bio Rad CFX96 Real-Time System (C1000 Thermal Cycler) (Applied Biosystems, Hercules City, CA, USA).

Results

Protein sequence characterization

The B. rapa LTP1 coding region is predicted to encode a protein of 109 amino acid residues (Fig. 1A). Similarity searches revealed that BraLTP1 has 84% overall amino acid identity with AtLTP1 of Arabidopsis, 98% identity with both BolLTP1 from Brassica oleracea and BnaLTP1 from B. napus (which are 100% similar to each other), and 100% identity with BnaLTP2 of B. Napus (while gene sequences were 98% similar). B. napus (Brassica AC genome) is an allotetraploid species resulting from a cross between B. rapa (A genome) and B. oleracea (C genome) and thus it is not surprising that it contains LTP protein copies corresponding 100% to the A and C genomes derived from both of these diploid species (Nagaharu., 1935). All species possessed highest levels of sequence similarity in the nsLTP-like domain regions (light shaded amino acids 29 to 99) (Fig. 1A). A putative extracellular secretary signal (amino acids 1 to 22 in Arabidopsis and 1–19 in the other Brassica species) was 100% conserved in the Brassica species (Fig. 1A). Nine amino acid substitutions exist in the nsLTP-like domains between AtLTP1 and the Brassica genes. The eight strictly conserved cysteine residues in all plant LTPs that

form four intrachain disulfide bridges, were also 100% conserved among all aligned sequences.

In Arabidopsis, nine classes of nsLTP were identified that contain a variable number of inter-cysteine amino acid residues [7]. The number of amino acids between the eight cysteines in the 8 CM motif of BrLTP1 and it's aligned sequences are 10, 16, 0, 9, 1, 22 and 9, with a methionine and a valine residue present 10 and 4 aa before Cys7. These characteristics place the BraLTP1, AtLTP1, BolLTP1, and BnaLTP1/2 proteins as secreted LTPs in class VI of plant nsLTPs.

Overexpression of BraLTP1 in B. napus

Given the 100% similarity between the B. rapa LTP1 and B. napus LTP2 proteins, and the relative agronomic importance of B. napus, we carried out functional analysis of BraLTP1 in B. napus. Overexpression of BraLTP1 driven by the CaMV 35S promoter in 12 independent transformants of the B. napus L. cultivar Zhongshuang 6 showed expression levels 415 times to 2314 times higher than in wild type using qRTPCR (our primers could analyse BraLTP1 from Brassica rapa and its homologous gene from Brassica napus without difference) (Fig. 1C). The most visually striking feature of the BraLTP1 overexpression lines was a distinctly green leaf phenotype, with no waxy surface visible on either the abaxial or adaxial surfaces compared with the wild type (Fig. 2). Additional morphological differences in BraLTP1 over-expressing plants included humpy and wrinkled leaf surfaces, abaxially curvature of leaf edges, and a significant reduction in total plant size relative to the wild type and negative segregants (Fig. 2 and Fig. 3A). These phenotypes were widespread in most lines, and we chose two 35S::BraLTP1 transgenic lines (BraLTP1-20 and BraLTP1-22) for further study due to their low copy numbers, morphologically clear phenotype and high transgene expression, enabling easy generation and comparison to negative segregates as controls.

To confirm that these phenotypes resulted from specific over-expression of BraLTP1 rather than tissue culture or vector insertion effects, we performed co-segregation analysis. Leaf samples of positive and negative segregants of BraLTP1-20 and BraLTP1-22 lines, together with wild type, were selected based on genomic PCR for the insert (Fig. 3B) and real-time PCR of BraLTP1 expression (Fig. 3C). Plants of BraLTP1-20 and BraLTP1-22 that were positive for the insert and had high BraLTP1 transcript level also had the phenotypes described above, while negative segregants of these lines showed the same phenotype as the wild type controls (Fig. 3A, B and C). Therefore, 35S::BraLTP1 transgenic phenotypes perfectly cosegregated with overexpression of the BraLTP1 gene.

Overexpressing BraLTP1 reduces cuticular wax in leaves

Mutant analyses have implicated a role for nsLTPs in the transport of waxes or cutin monomers [8,9,10,11], however overexpression of nsLTPs has seldom been reported. To determine whether the visible leaf phenotypes of the B. napus BraLTP1 overexpressor lines resulted from decreased cuticle wax, the density of wax crystals on the leaf surface was assessed by scanning electron microscopy (SEM). A clear reduction in wax crystal density was observed on leaves of the BraLTP1-22 plants, accompanied by altered crystal shape and form (Fig. 4). This suggested that aletred epicuticular wax might lead to the visible phenotypes of BraLTP1 overexpressing leaves.

To determine the chemical composition of BraLTP1 overex-pressing leaves in greater detail, gas chromatography mass spectrometry (GC-MS) analyses were performed. A 78% reduction in levels of the C31 alkane, hentriacontane, a major component of

Table 1. Primers for real-time PCR checking genes related to cytokinin synthesis and flower development.

Gene name	Primer name	Sequence(5′→3′)
IPT1,4,6	Fna-75600-F	GAGGAGGCAAGTATGGAAGATAG
	Fna-75600-R	CGACGAACTCGAACTCATCATA
IPT2	Fna-41524-F	CAAACCAGGAGCTGACTATACC
	Fna-41524-R	AGCGGAGCTATTTGTGTCTG
IPT3	Fna-06414-F	TCAGGAATGAGCCGTTCTTAAA
	Fna-06414-R	GTTTGCAAGCTAACCCGAAAG
IPT5	Fna-36855-F	GAGCGGAGAAGCGTGATTAT
	Fna-36855-R	CCGACATGCAAGCAAACAG
IPT7	Fna-44012-F	TTGGGTCGACGTTTCCTTAC
	Fna-44012-R	GCTTTCGGATCGTGTACTTCT
IPT8	Fna-63259-F	GCTTGCCAAGAAGCAGATAGA
	Fna-63259-R	CTCTCTTGACGATGCCCTTAAC
IPT9	Fna-63954-F	GCCGTAGACAAAGAGGTGTAAG
	Fna-63954-R	CATTGAGCCAGTGGTACATAGG
CYP735A1	Fna-68846-F	GAAACTACCGCACTCCTTCTC
	Fna-68846-R	GCAGCCACATACCTCTCTAATC
CYP735A2	Fna-08361-F	CCTCATGCTCCTTGCTCATAA
	Fna-08361-R	TTGCTCAACGGAAGGGATAC
URH1	Fna-72547-F	GGGTGGAGACTCAAGGAATATG
	Fna-72547-R	CATGCCACTGATACTGGTGAA
AP1	Fna-16396-F	TTCTTAGGGCACAGCAAGAG
	Fna-16396-R	GCATGTATGGATGCTGGATTTG
AP2	Fna-27636-F	GCAGATGACGAATTTAACGAAGG
	Fna-27636-R	CTTCCCAACGACCACACTTAT
AP3	Fna-72759-F	ATCGAAGGATCACGTGCTTAC
	Fna-72759-R	AATGATGTCAGAGGCAGATGG
PI	Fna-60439-F	AAATGTTGGCGGAGGAGAA
	Fna-60439-R	GAATCGGCTGGACTCTGTATC
AG	Fna-37522-F	CTGATGCCAGGAGGAACTAAC
	Fna-37522-R	ATGCCGCGACTTGGAAATA
CRC	Fna-16399-F	AAGAGTGCCAATCCGGAAATA
	Fna-16399-R	GCTCCGGAAGTAATGGAAGTAG
SPT	Fna-54144-F	CTTTGGACCTTTCCCTCACTT
	Fna-54144-R	CATCAAACGCAGCATGTTCTC
LEUNIG	Fna-27110-F	ACAGGCAGTGAAGGAGAATG
	Fna-27110-R	CACCCAATTACGAGTAGGGAAG
AINTEGUMENTA	Fna-09982-F	AGACACAGATGGACTGGTAGA
	Fna-09982-R	GAGCAGCTTTCTCCTCCATATC
Actin	BnActin 88-F	GCTGACCGTATGAGCAAAGA
	BnActin 88-R	AGATGGATCCTCCAATCCAAAC

cuticular waxes, was seen in *BraLTP1*-22 leaves (49.6 µg g^{-1}) relative to wild type (221.3 µg g^{-1}). A second major wax constitutent, C29 alkane (nonacosane), was decreased by 44% in the overexpressor line relative to wild type; from 1431.7 µg g^{-1} to 82.8 µg g^{-1}. Other wax components were similarly reduced, ranging from 17% to 80% reductions (Fig. 5). Despite these defects in cuticular wax, *BraLTP1*-22 overexpressors did not display any organ fusions, unlike some other mutants with cuticle defects [39,40]. Together these results suggest a broad-range, non-specific reduction in wax deposition in *B. napus*.plants overexpressing *BraLTP1*.

Distortion of the cuticular layer often results in an increased permeability of leaves [41,42,43]. To test this, water loss assays of detached leaves from *BraLTP1*-20, *BraLTP1*-22 and wild type were performed. A significantly higher rate of water loss occurred for the detached leaves of *BraLTP* overexpressing lines when compared with wild type (Fig. 6). This is consistent with the observed abnormal cuticular layer of overexpressor leaves

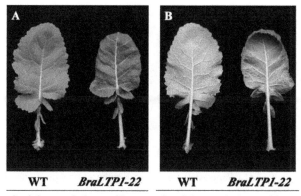

WT   *BraLTP1-22*   WT   *BraLTP1-22*

adaxial morphology of leaves   abaxial morphology of leaves

Figure 2. Leaf morphology of *35S::BraLTP1* overexpression line 22 (*BraLTP-22*). The adaxial (A) and abaxial (underside) (B) morphology of a representative wild type (WT) leaf (left) and *35S::BnaLPT1* (right) are shown. WT leaves appear glaucous with a smooth surface compared to *35S::BnaLTP1*, which has an unglaucous, dark green and bumpy leaf surface with abaxial edge-curling.

suggesting that wax defects can result in perturbed cuticle permeability. The result also hints that *BraLTP1* could be important for plant water relations and drought tolerance.

Overexpression of *BraLTP1* promotes cell overproliferation in leaf

In addition to the wax phenotype, the cellular morphology and weight of per unit leaf area was examined (Fig. 7A and B). *BraLTP1*-20 and *BraLTP1*-22 leaves were 0.033 g cm^{-2} and 0.031 g cm^{-2} respectively, significantly higher than negative segregants and wild type, which varied from 0.024 g cm^{-2} to 0.026 g cm^{-2} (P<0.01) (Fig. 7A). To examine changes at the cellular, structural level, paraffin-wax-embedded leaf cross-sections stained with Safranin and fast green were examined. This demonstrated that in *BraLTP1* overexpressing lines, the cellular layer of both palisade tissue and parenchyma tissue was increased; with palisade cells, parenchyma cells and epidermic cells smaller and more compact than negative segregants and wild type sections (Fig. 7B). Thus, increased cellular layering and a compact cell arrangement likely led to the increase in weight for per unit leaf area. This underlying change in leaf cell layer number and density likely contributes to the visible morphological defects including leaf curling.

To correlate these changes to cytokinin composition, we quantified cytokinin content, including the levels of the cytokinins isopentenyladenine (iP) and trans-zeatin (tZ), and of their riboside

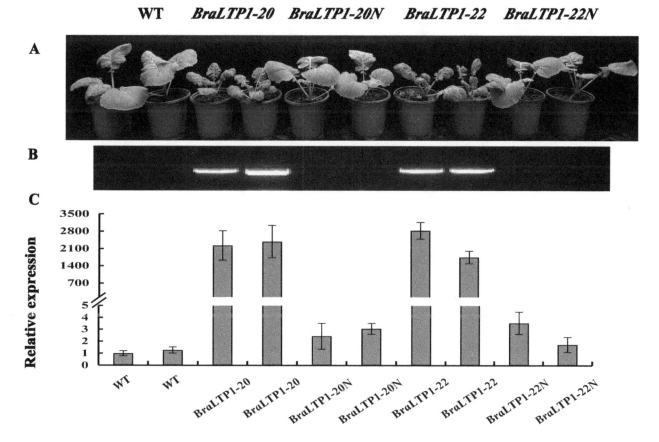

Figure 3. Cosegregation experiments of two *35S::BraLTP1* transgenic lines. (A) Phenotype of 5-week-old T1 plants segregating for *BraLTP1*-20 and *BraLTP1*-22. (B) PCR results using primers spanning the CaMV 35S promoter and *BraLTP1* in the vector, with DNA of corresponding plant in panel (A) as templates. (C) Transcript level of *BraLTP1* for each plant shown directly above in panel (A). PCR-positive transformants show high *BraLTP1*-expression and displayed the typical overexpression phenotypes including dark-green, wrinkled, curly leaves and smaller stature compared with their negative segregants (*BraLTP1-20N* and *BraLTP1-22N*).

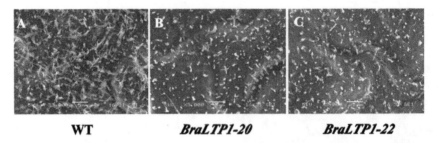

WT   **BraLTP1-20**   **BraLTP1-22**

Figure 4. Epicuticular wax and cuticle layer of air-dried adaxial leaf surfaces of wild type (WT), the *BraLTP1-20* **and** *BraLTP1-22* **overexpressor using scanning electron microscopy.** The experiments were repeated for 3 times with at least five plants for each time. (A) Wax crystals on the WT leaf were dense, with high proportion of tubular-like wax crystals. (B and C) Wax crystals are sparsely distributed on *BraLTP1-20 and BraLTP1-22 B. napus* leaves. Bar = 5 μm.

(iPR and tZR), As observed in *BraLTP1-22* line, the *BraLTP1* overexpressing leaves contained significantly higher amounts of tZ, tZR, iP, and iPR when compared with the wild type (P<0.01, Fig. 8), consistent with more cell accumulation in leaves (Fig. 7B). This indicates that active cytokinins are highly synthesized in leaves overexpressing *BraLTP1*.

BraLTP1 affects flower development

In the reproductive phase of the *B. napus* lifecycle, overexpressing *BraLTP1* resulted in early development of longer carpels and outward bending stamen in the flower. Early in flower development, carpels grew out of the apical flower buds, with the outward bending stamens clearly visible from the slightly opened sepals, giving an observably distinct phenotype from wild type flowers (Fig. 9A, B and C). Microscopic examination of the petals in the bud showed a variable severity of this phenotype in different lines. In the moderate version of this phenotype, the four petals

developed into nearly normal petals upon flower opening (Fig 9A, B and D), while in severe phenotype, poorly developed, shriveled petals were found resulting in only one to three petals in the opened flowers (data not show). These phenotypes occurred not only in primary inflorescences but also in other branched inflorescences. In addition, bumpy sepals and siliques with all distributed longitudinal ridges on the surface, similarly to the bumpy phenotype of leaves, were observed in *35S::BraLTP1* transgenic *B. napus* lines (Fig. 9D and E). Cross-section of sepals and siliques did not show a difference in cell proliferation between transgenic plants and wild type (data not show). Microscopic examination of the ovules within the carpels revealed developing embryos and endosperm, indicating self-compatibility and good fertilization; pollen viability identification revealed that pollen production was also normal (data not show). Silique filling occurred effectively, thus *BraLTP1* over-expression had few detrimental effects on plant fertility (Fig. 9E).

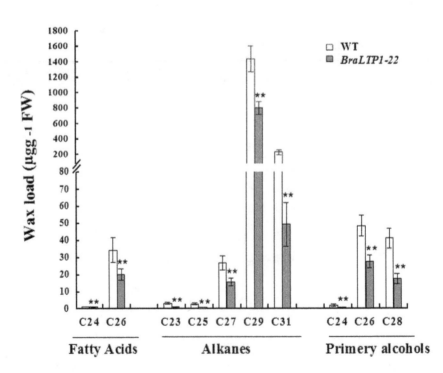

Figure 5. Cuticular wax composition and loads in leaves of *BnaLTP1-22* **overexpression line and wild type (WT)** *B. napus.* Error bars indicate SE of three 6 biological repeats (t-test: **P<0.01).

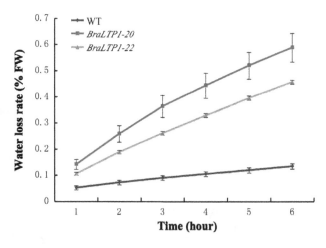

Figure 6. Water loss of detached leaves of wild type (WT), BraLTP1-20 and BraLTP1-22 plants. The data are the means ± SD of three replicates ($n=5$ for each experiment); $P<0.01$ (ANOVA, $P=6.53 \times 10^{-56}$).

Co-regulated genes to cytokinin synthesis and flower organ development

The transcript abundance of 10 genes involved in the cytokinin synthesis pathway, and 13 genes involved in floral organ development, were investigated by real-time PCR in the wrinkled leaves of *BraLTP1* overexpression lines as well as in wild type (Table 1).

In *Arabidopsis*, seven genes for adenosine phosphates isopentenyl transferase (*AtIPT1* and *AtIPT3* to *AtIPT8*) have been identified as cytokinin biosynthesis genes [44,45]. *IPT* and *CYP735A*, are responsible for the differential distribution of *de novo* synthesis pathways for isopentenyladenine (iP), trans zeatin (tZ) [46]. Recently, a novel uridine ribohydrolase, *URH1*, was characterized that degrades isopentenyladenosine in *Arabidopsis* [47]. In leaves of *BraLTP1* overexpression lines, the cytokinin biosynthesis genes *IPT3* were significantly increased (P<0.01), with 3.8 times of expression levels in wild type, while *IPT7*, *IPT9* and *CYP735A2* were decreased (P<0.01), and there was no significant change in *IPT1*, *IPT2*, *IPT4*, *IPT5*, *IPT6*, *IPT8*, *CYP735A1* and *URH1* (P>0.05) (Fig. 10A).

The ABC model of flower development describes how the combinatorial interaction of three classes of genes directs the development of four types of floral organs [48,49]. Here, in *BraLTP1* overexpression lines, the expression level of genes involved in flower development including *AP1*, *AP2*, *AP3*, *PI*, *CRC* were significantly decreased in the abnormally-developed flowers, while *AG*, *SPT* and *LEUNIG* were significantly increased. There was no significant change in *AINTEGUMENTA* (Fig. 10B).

The decreased transcript levels of *AP1*, *AP2*, *AP3*, *PI* are consistent with the phenotype of *BraLTP1* overexpressing flowers in *B. napus*, with short and humpy sepals and outwardly bent stamens. Furthermore, the increased expression levels of *AG*, *SPT* and *LEUNIG* were consistent with the over-developed longer carpels. These results, plus phenotypic data suggest that *BraLTP1* plays a role in flower development.

Discussion

Functional characterization of *BraLTP1*

In this study, we report the isolation and characterization of an *nsLTP*-like gene from *B. rapa*; *BraLTP1*. Sequence analysis

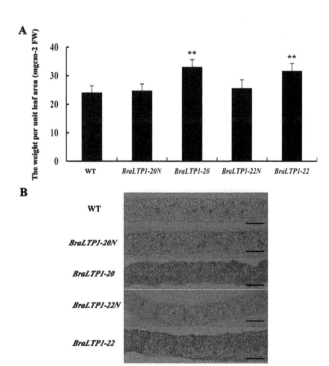

Figure 7. Overexpression of *BraLTP1* promotes cell proliferation. (A) The weight of per unit leaf area of wild type (WT), *BraLTP1*-20 and *BraLTP1*-22 and their segregated negative controls *BraLTP1*-20N and *BraLTP1*-22N. Leaf disks were collected from the fourth fully-expanded leaf from the apex taken from 8-week-old plants. Data are the mean±SD from three independent experiments using leaves of five plants. **Significant differences at the levels of P<0.01. (B) Representative leaf cross-sections of segregating *35S::BraLTP1* and WT plants. The fourth fully-expanded leaf from the apex were taken from 6-week old plants of WT, *35S::BraLTP1* transgenic plants (*BraLTP1*-20 and *BraLTP1*-22), and null segregates of these two lines (*BraLTP1*-20N and *BraLTP1*-22N). The experiments were repeated for 3 times. Bar = 200 μm.

showed that the *BraLTP1* protein is a single copy gene in the 'A' genome of *B. rapa*, with a homologous gene in *B. olercaea*, 'C' genome and two corresponding 'A' and 'C' genome copies in the amphidiploid 'AC' genome of *B. napus*. *BnaLTP1* was first reported by Dong [50] as a *B. napus* seed specific gene named *Bn15D18B* (genbank number: AY208878), which was differentially screened in a seed-cDNA library harvested 15 days after pollination (DAP). The amino acid similarity of *Arabidopsis*, *B. rapa*, *B. oleracea* and *B. napus* LTP1 copies is high and extends throughout the whole protein, with increasing divergence consistent with the older evolutionary relationship of Arabidopsis. The configuration of the 8CM domain and inter-cysteine amino acid residues places *BraLTP1* in class VI of *nsLTP*, which in *Arabidopsis* is composed predominantly of uncharacterized proteins including *At1g32280.1*, *At4g30880.1*, *At4g33550*, and *At5g56480.1* [7]. Until recently, all type VI *nsLTP* genes were less studied, with unknown functions, providing good opportunity to expose new physiological functions of this family in processes such as cell division, as shown herein.

In this study, we cloned and functionally analysed a type VI *nsLTP* in *B. napus*. Over-expression of the *BraLTP1* gene caused growth defects in the seedling and reproductive organs of *B. napus*. These included a distinct green, disorganized leaf surface with curled edges and abnormally developed flower. Decreased levels of epicuticular wax accumulated on the leaf epidermis, and cell layering and cell density were increased in the mesophyll cell

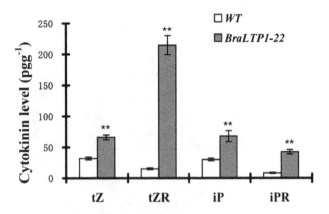

Figure 8. Cytokinin content in *BraLTP1* overexpressing leaves. Amounts of tZ, tZR, iP, and iPR were measured in leaves of 4-week-old *BraLTP1-22* and wild type (WT). Data are presented as mean±SD (n = 3).

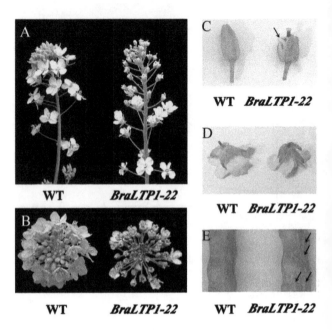

Figure 9. Morphological deformities of *BraLTP1* overexpressing flowers. (A) Lateral view of wild type (WT) and *35S::BraLTP1* (*BraLTP1-22*) primary inflorescences. (B) Top view of WT and *35S::BraLTP1* (*BraLTP1-22*) inflorescences. (C and D) elongated carpel, outward bending stamen (arrow indicated), and disorganized sepals of open (D) and budded (C) flowers of *35S::BraLTP1*(*BraLTP1-22*) vs normal development of WT plants. (E) Silique epidermis of *35S::BraLTP1* (*BraLTP1-22*) and WT plants; arrow indicates the longitudinal ridges that all disturbed in different places.

of *BraLTP1* overexpressing leaves. Increased cytokinin levels including of tZ, tZR, iP, and iPR and increased expression of the cytokinin-synthesis gene *IPT3* in *BraLTP1* overexpression lines correlated well with the enhanced cell proliferation phenotype. Overexpression of *BraLTP1* also led to the altered transcription of many important ABC model flower development genes, coinciding with the visible morphological and developmental perturbations in these lines. Overall, our experiments suggest that *BraLTP1* is an important *nsLTP* gene affecting wax deposition, cell proliferation, and leaf and flower morphology development in *B. napus*.

Overexpressing *BraLTP1* leads to a reduction of wax load

NsLTPs are proposed to play a role in the delivery of wax components during the assembly of the cuticle [26,51]. Previous studies suggested that in *nsLTP* mutants such as *ltpg1*, *ltpg2* and *ltpg1ltpg2*, wax load decreases with the reduced expression of *nsLTP* genes [10,11,12]. Therefore, we hypothesised that overexpressing *nsLTP1* in *B. napus* might lead to wax enrichment. However, in *BraLTP1* overexpressing plants, wax accumulation was decreased in seedling leaves. The amount of wax on *BraLTP1-22* leaves was significantly reduced with proportional deficiencies in the component aldehydes, alkanes, alcohols and acids compared with wild type. This universal reduction, with no specific component alteration, in *35S::BraLTP1* transgenic plants suggested that the effect of *BraLTP1* overactivity was not substrate specific.

There are a few possible explanations for this contradicting observation: (1) the genes above belong to glycosylphosphatidylinositol-anchored *LTPs*, which are different from the type VI *BraLTP1* and exercise different mechanisms *in vivo*; (2) overexpression of *BraLTP1* may lead to disordered or destructive secretion of wax out of cells, which is subsequently lost from the surface; (3) overexpression of *BraLTP1* gene somehow feedback inhibits, or competitively inhibits, other *nsLTPs* with important complimentary lipid synthesis or transport abilities. Until recently, functional analysis of wax-synthesis-related *nsLTP* genes has focused on mutants, while transgenic overexpression is seldom reported. To our knowledge the only prior overexpression study was on *LTP3* in *Arabidopsis*, overexpression of *LTP3* enhanced freezing and drought tolerance in *Arabidopsis* but with no change of cuticular wax seen [15]. Further investigation of the molecular mechanism of *BraLTP1* action will shed more light on its function in wax metabolism and feedback regulation.

Overexpressing *BraLTP1* leads to cell overproliferation

Besides wax load reduction, we also found that (1) *35S::BraLTP1* transgenic *B. napus* plants exhibit disorganized leaf patterning/morphology; (2) mesophyll cells were over proliferated, with increased cell layer number and cell density, and; (3) cytokinin levels were significantly increased and (4) the cytokinin-synthesis gene *IPT3* was increased 3.8 fold in transgenic leaves. It is known that wax is composed of VLCFA and their derivatives, thus our data is consistent with Nobusawa *et al* [31], wherein VLCFA synthesis in the epidermis confines cytokinin biosynthesis via *IPT3* to the vasculature and restricts cell proliferation. While it is also possible that *BraLTP1* itself plays a direct role in cell division by affecting related genes. However, the internal mechanism remains to be clarified.

The decreased expression of other cytokinin synthesis related genes seen here may be due to feedback downregulation through the pathway, or functionally differentiated roles of *IPTs* in response to environmental conditions. VLCVA, its derivatives, or VLCFA-related lipids, may function as signaling molecules to control cell division by affecting cytokinin related gene transcription [31,52,53,54]. Further studies are needed to examine such possibilities and explore the specific mediators or ligands which suppress cell proliferation in tissues. Mutant material in Nobusawa's study was difficult to observe macroscopically because of severe growth defects and in our experiment, overexpressing line for *BraLTP1* gene analysis produced a moderate, morphologically observable phenotype that produced fertile seed for continued research, shedding important light on the system for study of plant-enviroment interaction.

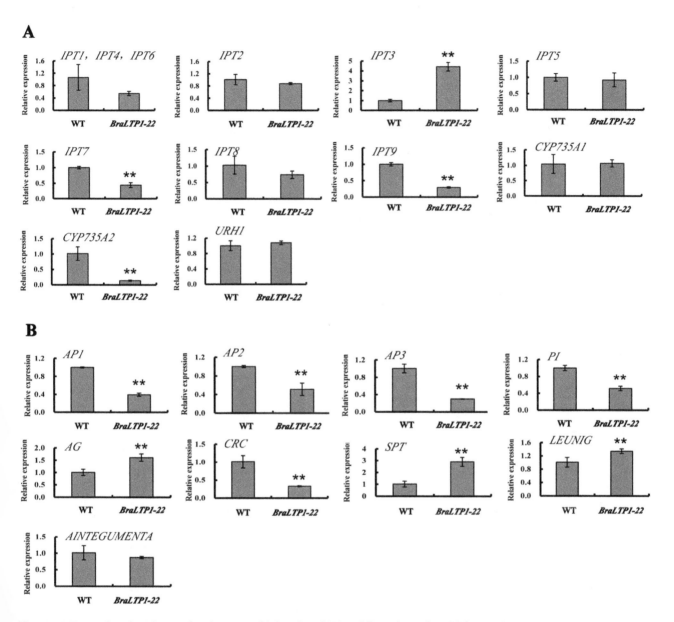

Figure 10. Transcript abundance of various cyctokinin-related (A) and flowering-related (B) genes in *35S::BraLTP1* **(***BraLTP1-22***) and wild type (WT) plants as determined by qRTPCR.** Data is the average of three plants with standard errors. Asterixes indicate significant differences to WT (** P<0.01).

Overexpressing *BraLTP1* led to altered flower morphology

Some studies have reported that *nsLTPs* are involved in flower development. For example, multiple *nsLTP* genes were identified to be differentially expressed in petals during different developmental periods in carnation flowers, suggesting their contributions to petal development [55]. *FIL1*, a non-specific lipid-transfer protein with an *nsLTP*-like domain was reported to be important in petal and stamen formation in *Antirrhinum* [56]. The identification of *Antirrhinum nsLTPs* as target genes of the class B MADS box transcription factors *DEFICIENS*, suggested a function during late petal and stamen development [57]. Kotilainen *et al* [58] reported that the *gltpl* gene in *Gerbera hybrida* var. Regina was expressed only in the corolla and carpels and was developmentally regulated during corolla development.

In the ABC model of flower development, the A-class genes *APETALA1* (*AP1*) and *AP2* confer sepal. Their activity overlaps with B-class genes *APETALA3* (*AP3*) and *PISTILLATA* (*PI*), which develops into petals. B-class genes and the C-class gene *AGAMOUS* (*AG*) specify stamen, while *AG* promotes carpel development [48,49]. Two newly characterized genes, *CRABS CLAW* (*CRC*) and *SPATULA* (*SPT*), function similarly to AG to promote carpel differentiation. *LEUNIG* and *AINTEGU-MENTA* are also putative genes affecting carpel development [59]. In our study, overexpressing *BraLTP1* led to altered expression levels of class ABC genes, which control flower organogenesis, with *AP1*, *AP2*, *AP3*, *PI* decreased. This is consistent with the morphological defects of sepals and stamens seen. The expression level of *AG*, *SPT* and *LEUNIG* were increased, which may result in the early development of longer carpels (Fig. 10B). Combined with previous studies of *nsLTP* on flower development, we speculate that *BraLTP1* may affect flower development through the regulation of morphologically important cellular components like the cell wall and cuticle, to affect flower-related genes.

Conclusion

This study identifies a novel *nsLTP* gene *BraLTP1* that influences wax deposition, cell proliferation and flower development when overexpressed in *B. napus*. Although the precise biological role is yet to be determined, we suggest that *BraLTP1* may link the metabolism of wax lipids and/or cell wall

components in the epicuticular, or internal plant interfaces, to the coordinated execution of developmental programs, including cell division and flower development. Therefore, *BraLTP1* likely plays important roles in different developmental periods in plants.

Author Contributions

Conceived and designed the experiments: FL DF GW. Performed the experiments: XX LW XZ YC YW. Analyzed the data: FL AH. Contributed reagents/materials/analysis tools: YL. Wrote the paper: FL DF AH GW.

References

1. Thoma S, Kaneko Y, Somerville C (1993) A non-specific lipid transfer protein from Arabidopsis is a cell wall protein. The Plant Journal 3: 427–436.
2. Kader JC (1996) Lipid-Transfer Proteins in Plants. Annu Rev Plant Physiol Plant Mol Biol 47: 627–654.
3. Vergnolle C, Arondel V, Jolliot A, Kader JC (1992) Phospholipid transfer proteins from higher plants. Methods Enzymol 209: 522–530.
4. Shin DH, Lee JY, Hwang KY, Kim KK, Suh SW (1995) High-resolution crystal structure of the non-specific lipid-transfer protein from maize seedlings. Structure 3: 189–199.
5. Lauga B, Charbonnel-Campaa L, Combes D (2000) Characterization of MZm3-3, a Zea mays tapetum-specific transcript. Plant Sci 157: 65–75.
6. Boutrot F, Guirao A, Alary R, Joudrier P, Gautier MF (2005) Wheat non-specific lipid transfer protein genes display a complex pattern of expression in developing seeds. Biochimica et Biophysica Acta (BBA)-Gene Structure and Expression 1730: 114–125.
7. Boutrot F, Chantret N, Gautier MF (2008) Genome-wide analysis of the rice and Arabidopsis non-specific lipid transfer protein (nsLtp) gene families and identification of wheat nsLtp genes by EST data mining. BMC Genomics 9: 86.
8. Pyee J, Yu H, Kolattukudy PE (1994) Identification of a lipid transfer protein as the major protein in the surface wax of broccoli (Brassica oleracea) leaves. Arch Biochem Biophys 311: 460–468.
9. Lee SB, Go YS, Bae HJ, Park JH, Cho SH, et al. (2009) Disruption of glycosylphosphatidylinositol-anchored lipid transfer protein gene altered cuticular lipid composition, increased plastoglobules, and enhanced susceptibility to infection by the fungal pathogen Alternaria brassicicola. Plant Physiol 150: 42–54.
10. Debono A, Yeats TH, Rose JK, Bird D, Jetter R, et al. (2009) Arabidopsis LTPG is a glycosylphosphatidylinositol-anchored lipid transfer protein required for export of lipids to the plant surface. Plant Cell 21: 1230–1238.
11. Kim H, Lee SB, Kim HJ, Min MK, Hwang I, et al. (2012) Characterization of glycosylphosphatidylinositol-anchored lipid transfer protein 2 (LTPG2) and overlapping function between LTPG/LTPG1 and LTPG2 in cuticular wax export or accumulation in Arabidopsis thaliana. Plant Cell Physiol 53: 1391–1403.
12. Hincha DK (2002) Cryoprotectin: a plant lipid-transfer protein homologue that stabilizes membranes during freezing. Philos Trans R Soc Lond B Biol Sci 357: 909–916.
13. Liu KH, Lin TY (2003) Cloning and characterization of two novel lipid transfer protein I genes in Vigna radiata. DNA Seq 14: 420–426.
14. Guo L, Yang H, Zhang X, Yang S (2013) Lipid transfer protein 3 as a target of MYB96 mediates freezing and drought stress in Arabidopsis. J Exp Bot 64: 1755–1767.
15. Carvalho Ade O, Gomes VM (2007) Role of plant lipid transfer proteins in plant cell physiology-a concise review. Peptides 28: 1144–1153.
16. Blein JP, Coutos-Thevenot P, Marion D, Ponchet M (2002) From elicitins to lipid-transfer proteins: a new insight in cell signalling involved in plant defence mechanisms. Trends Plant Sci 7: 293–296.
17. Maldonado AM, Doerner P, Dixon RA, Lamb CJ, Cameron RK (2002) A putative lipid transfer protein involved in systemic resistance signalling in Arabidopsis. Nature 419: 399–403.
18. Jung HW, Tschaplinski TJ, Wang L, Glazebrook J, Greenberg JT (2009) Priming in systemic plant immunity. Science 324: 89–91.
19. Foster GD, Robinson SW, Blundell RP, Roberts MR, Hodge R, et al. (1992) A Brassica napus mRNA encoding a protein homologous to phospholipid transfer proteins, is expressed specifically in the tapetum and developing microspores. Plant Science 84: 187–192.
20. Ariizumi T, Amagai M, Shibata D, Hatakeyama K, Watanabe M, et al. (2002) Comparative study of promoter activity of three anther-specific genes encoding lipid transfer protein, xyloglucan endotransglucosylase/hydrolase and polygalacturonase in transgenic Arabidopsis thaliana. Plant Cell Reports 21: 90–96.
21. Imin N, Kerim T, Weinman JJ, Rolfe BG (2006) Low temperature treatment at the young microspore stage induces protein changes in rice anthers. Mol Cell Proteomics 5: 274–292.
22. Park SY, Jauh GY, Mollet JC, Eckard KJ, Nothnagel EA, et al. (2000) A lipid transfer-like protein is necessary for lily pollen tube adhesion to an in vitro stylar matrix. Plant Cell 12: 151–164.
23. Chae K, Gonong BJ, Kim SC, Kieslich CA, Morikis D, et al. (2010) A multifaceted study of stigma/style cysteine-rich adhesin (SCA)-like Arabidopsis lipid transfer proteins (LTPs) suggests diversified roles for these LTPs in plant growth and reproduction. J Exp Bot 61: 4277–4290.
24. Chae K, Kieslich CA, Morikis D, Kim SC, Lord EM (2009) A gain-of-function mutation of Arabidopsis lipid transfer protein 5 disturbs pollen tube tip growth and fertilization. Plant Cell 21: 3902–3914.
25. Nieuwland J, Feron R, Huisman BA, Fasolino A, Hilbers CW, et al. (2005) Lipid transfer proteins enhance cell wall extension in tobacco. Plant Cell 17: 2009–2019.
26. Sterk P, Booij H, Schellekens GA, Van Kammen A, De Vries SC (1991) Cell-specific expression of the carrot EP2 lipid transfer protein gene. Plant Cell 3: 907–921.
27. Eklund DM, Edqvist J (2003) Localization of nonspecific lipid transfer proteins correlate with programmed cell death responses during endosperm degradation in Euphorbia lagascae seedlings. Plant Physiol 132: 1249–1259.
28. Edstam MM, Viitanen L, Salminen TA, Edqvist J (2011) Evolutionary history of the non-specific lipid transfer proteins. Mol plant 4: 947–964.
29. Riederer M, Muller C (2006) Biology of the Plant Cuticle. Oxford: UK: Blackwell: 11–125.
30. Samuels L, Kunst L, Jetter R (2008) Sealing plant surfaces: cuticular wax formation by epidermal cells. Annu Rev Plant Biol 59: 683–707.
31. Nobusawa T, Okushima Y, Nagata N, Kojima M, Sakakibara H, et al. (2013) Synthesis of very-long-chain fatty acids in the epidermis controls plant organ growth by restricting cell proliferation. PLoS Biol 11: e1001531.
32. Wang X, Wang H, Wang J, Sun R, Wu J, et al. (2011) The genome of the mesopolyploid crop species Brassica rapa. Nat Genet 43: 1035–1039.
33. Gleave A (1992) A versatile binary vector system with a T-DNA organisational structure conducive to efficient integration of cloned DNA into the plant genome. Plant Molecular Biology 20: 1203–1207.
34. Maniatis TA, Fritsch EF, Sambrook J (1992) Molecular cloning: a laboratory manual. Cold Spring Harbor: Cold Spring Harbor Laboratory Press.
35. Sambrook J, Fritsch EF, Maniatis T (1989) Molecular cloning: a laboratory manual, 2nd ed. Cold Spring Harbor: Cold Spring Harbor Laboratory Press.
36. Todd J, Post-Beittenmiller D, Jaworski JG (1999) KCS1 encodes a fatty acid elongase 3-ketoacyl-CoA synthase affecting wax biosynthesis in Arabidopsis thaliana. Plant J 17: 119–130.
37. Hua D, Wang C, He J, Liao H, Duan Y, et al. (2012) A plasma membrane receptor kinase, GHR1, mediates abscisic acid- and hydrogen peroxide-regulated stomatal movement in Arabidopsis. Plant Cell 24: 2546–2561.
38. Hu Z, Wang X, Zhan G, Liu G, Hua W, et al. (2009) Unusually large oilbodies are highly correlated with lower oil content in Brassica napus. Plant Cell Rep 28: 541–549.
39. Kurdyukov S, Faust A, Trenkamp S, Bar S, Franke R, et al. (2006) Genetic and biochemical evidence for involvement of HOTHEAD in the biosynthesis of long-chain alpha-,omega-dicarboxylic fatty acids and formation of extracellular matrix. Planta 224: 315–329.
40. Bird D, Beisson F, Brigham A, Shin J, Greer S, et al. (2007) Characterization of Arabidopsis ABCG11/WBC11, an ATP binding cassette (ABC) transporter that is required for cuticular lipid secretion. Plant J 52: 485–498.
41. Li C, Wang A, Ma X, Pourkheirandish M, Sakuma S, et al. (2013) An eceriferum locus, cer-zv, is associated with a defect in cutin responsible for water retention in barley (Hordeum vulgare) leaves. Theoretical and Applied Genetics 126: 637–646.
42. Tanaka T, Tanaka H, Machida C, Watanabe M, Machida Y (2004) A new method for rapid visualization of defects in leaf cuticle reveals five intrinsic patterns of surface defects in Arabidopsis. Plant J 37: 139–146.
43. Lu S, Zhao H, Parsons EP, Xu C, Kosma DK, et al. (2011) The glossyhead1 allele of ACC1 reveals a principal role for multidomain acetyl-coenzyme A carboxylase in the biosynthesis of cuticular waxes by Arabidopsis. Plant Physiol 157: 1079–1092.
44. Kakimoto T (2001) Identification of plant cytokinin biosynthetic enzymes as dimethylallyl diphosphate:ATP/ADP isopentenyltransferases. Plant Cell Physiol 42: 677–685.
45. Takei K, Sakakibara H, Taniguchi M, Sugiyama T (2001) Nitrogen-dependent accumulation of cytokinins in root and the translocation to leaf: implication of

cytokinin species that induces gene expression of maize response regulator. Plant Cell Physiol 42: 85–93.

46. Kudo T, Kiba T, Sakakibara H (2010) Metabolism and long-distance translocation of cytokinins. J Integr Plant Biol 52: 53–60.

47. Jung B, Florchinger M, Kunz HH, Traub M, Wartenberg R, et al. (2009) Uridine-ribohydrolase is a key regulator in the uridine degradation pathway of Arabidopsis. Plant Cell 21: 876–891.

48. Bowman JL, Smyth DR, Meyerowitz EM (1991) Genetic interactions among floral homeotic genes of Arabidopsis. Development 112: 1–20.

49. Coen ES, Meyerowitz EM (1991) The war of the whorls: genetic interactions controlling flower development. Nature 353: 31–37.

50. Dong J, Keller W, Yan W, Georges F (2004) Gene expression at early stages of Brassica napus seed development as revealed by transcript profiling of seed-abundant cDNAs. Planta 218: 483–491.

51. Yeats TH, Rose JK (2008) The biochemistry and biology of extracellular plant lipid-transfer proteins (LTPs). Protein Sci 17: 191–198.

52. Black PN, Faergeman NJ, DiRusso CC (2000) Long-chain acyl-CoA-dependent regulation of gene expression in bacteria, yeast and mammals. J Nutr 130: 305S–309S.

53. Worrall D, Ng CK, Hetherington AM (2003) Sphingolipids, new players in plant signaling. Trends Plant Sci 8: 317–320.

54. Savchenko T, Walley JW, Chehab EW, Xiao Y, Kaspi R, et al. (2010) Arachidonic acid: an evolutionarily conserved signaling molecule modulates plant stress signaling networks. Plant Cell 22: 3193–3205.

55. Harada T, Torii Y, Morita S, Masumura T, Satoh S (2010) Differential expression of genes identified by suppression subtractive hybridization in petals of opening carnation flowers. J Exp Bot 61: 2345–2354.

56. Nacken WK, Huijser P, Beltran JP, Saedler H, Sommer H (1991) Molecular characterization of two stamen-specific genes, tap1 and fil1, that are expressed in the wild type, but not in the deficiens mutant of Antirrhinum majus. Mol Gen Genet 229: 129–136.

57. Bey M, Stuber K, Fellenberg K, Schwarz-Sommer Z, Sommer H, et al. (2004) Characterization of antirrhinum petal development and identification of target genes of the class B MADS box gene DEFICIENS. Plant Cell 16: 3197–3215.

58. Kotilainen M, Helariutta Y, Elomaa P, Paulin L, Teeri TH (1994) A corolla- and carpel-abundant, non-specific lipid transfer protein gene is expressed in the epidermis and parenchyma of Gerbera hybrida var. Regina (Compositae). Plant Mol Biol 26: 971–978.

59. Liu Z, Franks RG, Klink VP (2000) Regulation of gynoecium marginal tissue formation by LEUNIG and AINTEGUMENTA. Plant Cell 12: 1879–1892.

60. Liu S, Liu Y, Yang X, Tong C, Edwards D, et al. (2014) The Brassica oleracea genome reveals the asymmetrical evolution of polyploid genomes. Nat Commun 5: 3930.

15

The Positive Regulatory Roles of the TIFY10 Proteins in Plant Responses to Alkaline Stress

Dan Zhu[1,3]**, Rongtian Li**[2]**, Xin Liu**[1]**, Mingzhe Sun**[3]**, Jing Wu**[3]**, Ning Zhang**[3]**, Yanming Zhu**[3]*****

1 College of Life Science, Qingdao Agricultural University, Qingdao, P.R. China, **2** Key Laboratory of Molecular Biology, College of Heilongjiang Province, Heilongjiang University, Harbin, P.R. China, **3** Plant Bioengineering Laboratory, Northeast Agricultural University, Harbin, P.R. China

Abstract

The TIFY family is a novel plant-specific protein family, and is characterized by a conserved TIFY motif (TIFF/YXG). Our previous studies indicated the potential roles of TIFY10/11 proteins in plant responses to alkaline stress. In the current study, we focused on the regulatory roles and possible physiological and molecular basis of the TIFY10 proteins in plant responses to alkaline stress. We demonstrated the positive function of TIFY10s in alkaline responses by using the AtTIFY10a and AtTIFY10b knockout Arabidopsis, as evidenced by the relatively lower germination rates of attify10a and attify10b mutant seeds under alkaline stress. We also revealed that ectopic expression of GsTIFY10a in Medicago sativa promoted plant growth, and increased the NADP-ME activity, citric acid content and free proline content but decreased the MDA content of transgenic plants under alkaline stress. Furthermore, expression levels of the stress responsive genes including NADP-ME, CS, H⁺-ppase and P5CS were also up-regulated in GsTIFY10a transgenic plants under alkaline stress. Interestingly, GsTIFY10a overexpression increased the jasmonate content of the transgenic alfalfa. In addition, we showed that neither GsTIFY10a nor GsTIFY10e exhibited transcriptional activity in yeast cells. However, through Y2H and BiFc assays, we demonstrated that GsTIFY10a, not GsTIFY10e, could form homodimers in yeast cells and in living plant cells. As expected, we also demonstrated that GsTIFY10a and GsTIFY10e could heterodimerize with each other in both yeast and plant cells. Taken together, our results provided direct evidence supporting the positive regulatory roles of the TIFY10 proteins in plant responses to alkaline stress.

Editor: Keqiang Wu, National Taiwan University, Taiwan

Funding: This work was supported by Heilongjiang Provincial Higher School Science and Technology Innovation Team Building Program (grant no. 2011TD005 to YMZ), National Natural Science Foundation of China (grant no. 31171578 to YMZ), National Major Project for Cultivation of Transgenic Crops (grant no. 2011ZX08004-002 to YMZ). The funders had no role in study design, data collection and analysis, decision to publish, or preparation of the manuscript.

Competing Interests: The authors have declared that no competing interests exist.

* Email: ymzhu@neau.edu.cn

Introduction

Salt-alkaline stress is one of the most severe environmental challenges and affects all aspects of plant physiological and metabolic processes [1,2]. Compared with neutral salt stress, soil alkalization leads to high pH stress, poor fertility, dispersed physical property and low water content, and thereby causes much stronger inhibition of plant growth and development [3–5]. In recent years, research on plant responses to salt stress have identified the molecular basis of stress signal transduction pathways and salt tolerance mechanisms, and have been at the forefront of plant stress biology [6]. Unfortunately, until now, little attention has been paid on the molecular mechanisms of plant responses to alkaline stress [7,8].

Alfalfa (Medicago sativa L.) is an important worldwide leguminous forage crop and distributes over a wide range of climatic conditions [9,10]. It has become one of the most important plants due to its high productivity, high feed value and potential roles in soil improvement and soil conservation [11,12]. However, alfalfa yield and symbiotic nitrogen-fixation capacity were severely restricted by adverse environmental stresses, especially soil salinity and alkalinity [10,13]. With the

global climate change and the global shrinkage of arable lands, a grimmer reality of soil salinity and alkalinity is painted. Therefore, it is of fundamental importance to explore salt/alkaline-tolerant alfalfa through rational breeding and genetic engineering strategies.

The TIFY family, a novel plant-specific protein family, is characterized by a conserved TIFY motif (TIFF/YXG), and comprises 18 members in Arabidopsis and 20 members in rice [14,15]. It has been well suggested that TIFY genes play important roles in the jasmonate (JA) signaling pathway [16,17], plant growth and development[18–21], and pathogen responses [22–24]. For example, in Arabidopsis, TIFY genes are suggested to negatively regulate the key transcriptional activator of JA responses [25], such as MYC2 [26], MYC3, MYC4 [26,27], MYB21, MYB24 [28], bHLH017 and bHLH003 [29]. Furthermore, AtTIFY4a and AtTIFY4b regulate lamina size and curvature [30], whereas AtTIFY1 plays a role in petiole and hypocotyl elongation [31]. Recent research in tomato and tobacco suggested that JAZ proteins regulated the progression of cell death during host and nonhost interactions [23].

Recently, several lines of direct evidence supported that TIFY genes also fulfilled important function in plant responses to

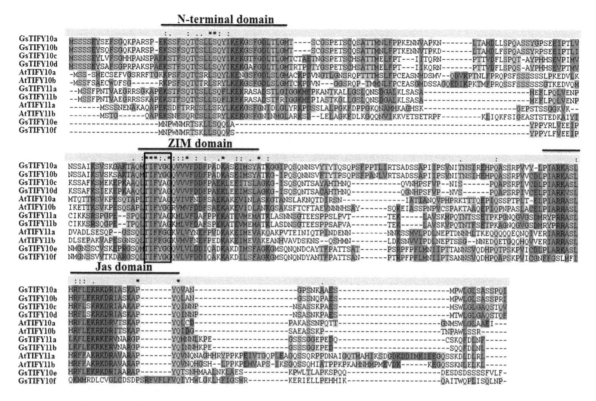

Figure 1. Sequence alignment of the Arabidopsis and wild soybean TIFY10/11 subgroup members based on the full-length amino acid sequences. The conserved N-terminal domain, ZIM domain and Jas domain were marked as solid lines. The TIFY motif (TIFF/YXG) was marked as a black solid box. Sequences were aligned by using ClustalX, and gaps were introduced to maximize alignment.

environmental challenges [32,33]. Overexpression of *OsTIFY11a* resulted in increased tolerance to salt and dehydration stresses [15]. Furthermore, OsTIFY3 acted as a transcriptional regulator of the OsbHLH148-mediated JA signaling pathway leading to drought tolerance [34]. Although a series of studies have demonstrated the biological function of TIFYs in salt and drought tolerance, little evidence is given on their roles in alkaline stress responses.

Glycine soja, is a wild soybean species and belongs to the same family *Leguminosae* with *Medicago sativa*. *Glycine soja* has extreme excellent tolerance to salt-alkaline stress [35], which makes it as an ideal candidate for exploring resistant genes and breeding of transgenic legume crops with superior salt-alkaline tolerance. In previous studies, we constructed a transcriptional profile of *Glycine soja* (G07256) roots in response to alkaline stress (50 mM NaHCO$_3$, pH8.5) [36], and identified three TIFY genes *GsTIFY10a*, *GsTIFY10e* (also named as *GsJAZ2*) and *GsTIFY11b* as alkaline stress responsive genes. We further demonstrated that overexpression of *GsTIFY10a* and *GsTIFY10e* in *Arabidopsis* improved plant alkaline tolerance [37,38]. In contrast, *GsTIFY11b* overexpression led to decreased salt tolerance [39]. A genomic analysis revealed 34 TIFY genes in *Glycine soja* genome and these GsTIFY proteins were clustered into two groups [40]. Group I comprises 9 members containing a GATA zinc-finger domain (GsTIFY1a, 1b, 1c, 1d, 2a, 2b, 2c, 2d, and 2e), and group II consists of 25 members without GATA zinc-finger domains. Among the group II TIFY proteins, GsTIFY10s (10a, 10b, 10c, 10d, 10e, and 10f) and GsTIFY11s (11a and 11b) were clustered together into one subgroup, here designated as GsTIFY10/11 subgroup. Transcriptional profiles revealed that all GsTIFY10/11 members were dramatically up-regulated at the early stage of

alkaline stress, indicating potential roles of GsTIFY10/11s in alkaline stress responses.

In this study, we aimed to identify the regulatory roles and possible physiological and molecular basis of the TIFY10 proteins in plant responses to alkaline stress. We verified the positive function of TIFY10s in alkaline responses by using the *AtTIFY10a* and *AtTIFY10b* knockout Arabidopsis. We also demonstrated the increased alkaline tolerance of transgenic alfalfa ectopically expressing *GsTIFY10a*, and investigated the physiological basis by which *GsTIFY10a* overexpression conferred to increased alkaline tolerance. Finally, we determined the transcriptional activity and dimerization characteristics of GsTIFY10a and GsTIFY10e in yeast and plant cells. Taken together, our results provided direct evidence that TIFY10 proteins positively regulated plant alkaline stress responses.

Results

The TIFY10/11 subgroup in Glycine soja and *Arabidopsis thaliana*

As shown in Fig. 1, the TIFY10/11 subgroup comprises 8 members in *Glycine soja* and 4 members in *Arabidopsis*. Protein sequence analyses showed that all TIFY10/11 proteins contained two highly conserved domains: a ZIM/TIFY domain which mediated homo- and hetero-dimerization, and a Jas domain which played a critical role in repression of JA signaling. It is worth noted that except for GsTIFY10e and GsTIFY10f, all other TIFY10/11 proteins included an N-terminal domain (Fig. 1, Fig. 2C). We further examined the phylogenetic relationship of TIFY10/11s in *Glycine soja* and *Arabidopsis thaliana*. As shown in Fig. 2A, TIFY10/11 proteins were divided into two branches (TIFY10 and

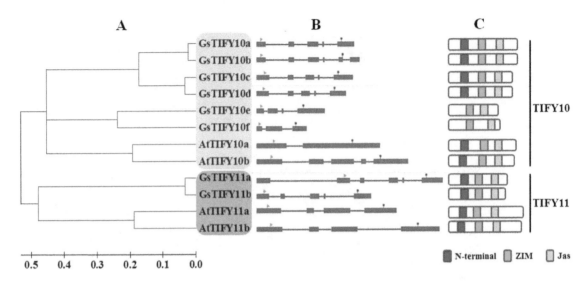

Figure 2. The Arabidopsis and wild soybean TIFY10/11 subgroup proteins. a. Phylogenetic analysis of the Arabidopsis and wild soybean TIFY10/11 subgroup proteins. A neighbor-joining tree was constructed with the full-length TIFY10/11 protein sequences by using MEGA 5.0. **b.** Exon/intron structures of the Arabidopsis and wild soybean TIFY10/11 genes. Exons were represented by blue boxes, and grey lines connecting two exons represented introns. Both the exons and introns were drawn to scale. **c.** The distribution of conserved domains within Arabidopsis and wild soybean TIFY10/11 proteins. The relative positions of each conserved domain within each protein were shown in color.

TIFY11). GsTIFY10a and GsTIFY10b, GsTIFY10c and GsTIFY10d, GsTIFY10e and GsTIFY10f were closely related with each other, respectively. Among the TIFY10 proteins, GsTIFY10a/b/c/d exhibited high similarity in exon distribution (Fig. 2B), amino acid sequence and domain architecture (Fig. 2C).

To get better understanding of TIFY10/11s sequence diversity, we analyzed the species and numbers of the cis-regulatory elements in their promoter regions. A series of typical elements related to environmental stress, hormone responsiveness and transcription factor (TF) binding sites were identified (Table 1), indicating the involvement of TIFY10/11s in hormone-dependent environmental stress responses and the regulation of TFs on TIFY10/11 expression or activity. Among the 33 elements listed in Table 1, several elements, including ABRELATERD1, ABRERATCAL, ACGTATERD1, ARR1AT, CACGTGMO-TIF, DPBFCOREDCDC3, GAREAT, GT1GMSCAM4 and MYCCONSENSUSAT, were found in all of the TIFY10/11s promoter regions. We also noticed several elements only existed in specific TIFY10/11 promoters, for example ABREATRD22, ABREOSRAB21, ABREMOTIFAOSOSEM, ACGTABREMO-TIFA2OSEM, ACGTCBOX and ACGTTBOX.

We further checked the responses of GsTIFY10/11s expression to alkaline stress based on the RNA-seq data. Our results revealed that expression of all the GsTIFY10/11 genes was rapidly and greatly induced by alkaline stress. They showed similar alkaline induced expression patterns, with a maximum point at 1 h after NaHCO$_3$ treatment (Fig. S1A). Among them, *GsTIFY10e* and *GsTIFY10f* exhibited the greatest alkaline stress induction, while *GsTIFY10a* and *GsTIFY10b* showed the highest expression levels (Fig. S1B). Taken together, these results strongly suggested the potential roles of GsTIFY10/11 proteins in alkaline stress responses.

The *AtTIFY10a/b* knockout decreased alkaline tolerance at the seed germination stage

To confirm the regulatory roles of TIFY10 proteins in alkaline stress responses, we adopted the T-DNA insertion mutant Arabidopsis of AtTIFY10a (*attify10a*, SALK_011957) and AtTI-

FY10b (*attify10a*, SALK_025279) (Fig. 3A). PCR-based analysis demonstrated the homozygous T-DNA insertion in the *attify10a* and *attify10b* mutants, and RT-PCR results confirmed that the *AtTIFY10a* and *AtTIFY10b* genes did not express in the *attify10a* and *attify10b* mutants, respectively (Fig. 3B).

The wild-type (WT) and mutant Arabidopsis seeds were germinated and grown on 1/2MS solid medium at pH5.8 (Control) or pH8.5 (Alkaline stress), respectively. As shown in Fig. 3C, WT and *AtTIFY10a/b* knockout Arabidopsis seedlings showed similar growth on 1/2MS solid medium at pH5.8, but growth of mutant seedlings was inhibited more severely than that of WT on 1/2MS solid medium at pH8.5. Under alkaline stress, WT seeds maintained relatively high germination rates (86.5%) on the 3rd day, but the germination rates of *attify10a* and *attify10b* seeds dropped to 54.2% and 77.0%, respectively (Fig. 3D). These results suggested that alkaline tolerance of *AtTIFY10a/b* knockout decreased at the seed germination stage, and further confirmed the positive roles of TIFY10 proteins in alkaline stress responses.

GsTIFY10a overexpression in alfalfa enhanced plant alkaline tolerance

In previous studies, we have demonstrated the involvement of three GsTIFY genes in salt-alkaline stress, among which *GsTIFY10a* could dramatically improve the alkaline tolerance. In an attempt to generate transgenic alfalfa with superior alkaline tolerance, we ectopically expressed *GsTIFY10a* in the wild type *Medicago sativa* through the *Agrobacterium tumefaciens*-mediated transformation strategy. The *GsTIFY10a* gene was under the control of the cauliflower mosaic virus (CaMV) 35S promoter, with the binding enhancers E12 and omega (Fig. 4A). After glufosinate selection, the regenerated alfalfa seedlings were analyzed by PCR and semi-quantitative RT-PCR assays. We identified a total of six transgenic lines (Fig. 4B), and three of them, with different expression levels (#12, #13 and #28), were used to examine the responses to alkaline stress.

We first compared the growth performance of WT and *GsTIFY10a* transgenic alfalfa plants under alkaline stress. As shown in Fig. 4C, under control conditions, transgenic lines

Table 1. Sequence diversity and cis-regulatory elements of soybean and Arabidopsis TIFYs promoters.

No.	Site Name	Signal Sequence	Description	GsTIFY10/11s								AtTIFY10/11s			
				10a	10b	10c	10d	10e	10f	11a	11b	10a	10b	11a	11b
1	ABREATRD22	RYACGTGGYR	ABRE in Arabidopsis dehydration-responsive gene rd22			1	1								
2	ABRELATERD1	ACGTG	ABRE-like sequence required for etiolation-induced expression of erd1 in *Arabidopsis*	3	4	2	3	4	5	2	2	5	7	5	6
3	ABREOSRAB21	ACGTSSSC	ABRE of wheat Em and rice rab21 genes		1										
4	ABREMOTIFAOSOSEM	TACGTGTC	ABRE-like sequence found in rice Osem gene promoter										1	1	
5	ABRERATCAL	MACGYGB	ABRE-related sequence in the upstream regions of Ca^{2+}-responsive upregulated genes	2	4	2	2	3	3	1	2	7	4	3	5
6	ACGTABREMOTIFA2OSEM	ACGTGKC	Sequence requirement of ACGT-core of motif A in ABRE of the rice gene, OSEM									1	1	1	2
7	ACGTATERD1	ACGT	ACGT sequence required for etiolation-induced expression of erd1 in *Arabidopsis*	4	6	6	4	6	8	4	2	12	18	12	8
8	ACGTCBOX	GACGTC	"C-box" according to the nomenclature of ACGT elements								2				
9	ACGTTBOX	AACGTT	"T-box" according to the nomenclature of ACGT elements											2	
10	ARFAT	TGTCTC	ARF binding site found in the promoters of primary/early auxin response genes; AuxRE; enriched in the 5'-flanking region of genes up-regulated by both IAA and BL	1	1					1		2	1		
11	ARR1AT	NGATT	"ARR1-binding element" found in Arabidopsis (CK)	11	4	8	5	6	5	7	11	13	13	10	6
12	ASF1MOTIFCAMV	TGACG	ASF-1 binding site; are found in many promoters and are involved in transcriptional activation of several genes by auxin and/or salicylic acid		1	1	1		2	1	2	3	1	2	2
13	BIHD1OS	TGTCA	Binding site of OsBIHD1, a rice BELL homeodomain transcription factor (disease)	4	2	3	1	4	4	1	1	2	1	2	5
14	CACGTGMOTIF	CACGTG	"CACGTG motif"; "G-box"; Binding site of Arabidopsis GBF4	2	2	2	2	2	2	2	2	2	4	2	4
15	CACGCAATGMGH3	CACGCAAT	Sequence found in D4 element in Soybean GH3 gene promoter; Confers auxin inducibility			1							2		
16	CATATGGMSAUR	CATATG	Sequence found in soybean SAUR15A promoter; Involved in auxin responses				2					2			
17	CBFHV	RYCGAC	Binding site of barley CBF1 and CBF2	1	1		1	1	1	2		2		1	
18	CGCGBOXAT	VCGCGB	"CGCG box" recognized by AtSR1-6 (Arabidopsis thaliana signal-responsive genes)		2							4	2	2	
19	CPBCSPOR	TATTAG	The sequence critical for Cytokinin-enhanced Protein Binding in vitro		1	3	3	2	2	1	2	2			3
20	CRTDREHVCBF2	GTCGAC	Preferred sequence for AP2 transcriptional activator HvCBF2; Core CRT/DRE motif							2		2			
21	DPBFCOREDCDC3	ACACNNG	A novel class of bZIP transcription factors binding core sequence (ABA)	1	1	1	1	1	1	1	2	3	5	2	3
22	DRE2COREZMRAB17	ACCGAC	"DRE2" core found in maize rab17 gene promoter; rab17 is induced by ABA	1								1		1	
23	DRECRTCOREAT	RCCGAC	Core motif of DRE/CRT cis-acting element in many genes in *Arabidopsis* and rice	1	1									1	1

Table 1. Cont.

No.	Site Name	Signal Sequence	Description	GsTIFY10/11s								AtTIFY10/11s			
				10a	10b	10c	10d	10e	10f	11a	11b	10a	10b	11a	11b
24	ERELEE4	AWTTCAAA	"ERE (ethylene responsive element)" of tomato E4 and carnation GST1 genes; ERE motifs mediate ethylene-induced activation of the U3 promoter region	1	1		1					1		2	
25	GADOWNAT	ACGTGTC	Sequence present in 24 genes in the GA-down regulated d1 cluster (106 genes) found in Arabidopsis seed germination; This motif is similar to ABRE										1	1	2
26	GARE1OSREP1	TAACAGA	"Gibberellin-responsive element (GARE)" found in rice cystein proteinase promoter			1						1	2		1
27	GAREAT	TAACAAR	GARE (GA-responsive element)	1	2	3	1	2	1	1	2	1	3	1	1
28	GT1GMSCAM4	GAAAAA	"GT-1 motif" found in the promoter of soybean CaM isoform, SCaM-4; Plays a role in pathogen- and salt-induced SCaM-4 gene expression	5	4	1	3	3	4	3	3	1	2	3	3
29	LTRECOREATCOR15	CCGAC	Core of low temperature responsive element of cor15a gene in Arabidopsis	1	1	1						2		2	3
30	MYB1AT	WAACCA	MYB recognition site found in the promoters of the dehydration-responsive gene rd22	2	3	1	3	3	2		1	3	3	2	2
31	MYB2CONSENSUSAT	YAACKG	MYB recognition site found in the promoters of the dehydration-responsive gene rd22	1				1	1	1		6	6	2	3
32	MYBCORE	CNGTTR	Binding site plant MYB proteins ATMYB1 and ATMYB2 from *Arabidopsis*	1		1	3	4	4	1		1	8	4	6
33	MYCCONSENSUSAT	CANNTG	MYC recognition site found in the promoters of the dehydration-responsive gene rd22	8	8	6	6	10	14	2	4	6	14	4	8

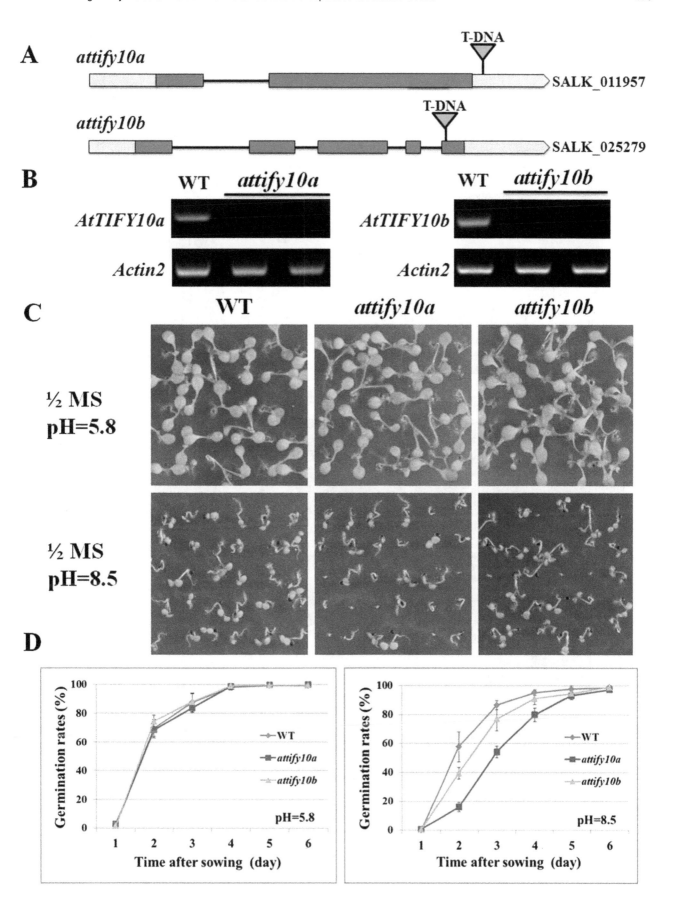

Figure 3. The *AtTIFY10a/b* knockout mutant *Arabidopsis* showed decreased alkaline tolerance at the seed germination stage. a. Schematic representation of the *AtTIFY10a/b* T-DNA insertion mutant lines. The exons and introns of the *AtTIFY10a/b* genes were showed as boxes and lines, and the T-DNA insertion sites were marked as triangles. **b.** RT-PCR analyses showing that *AtTIFY10a/b* did not expressed in the *attify10a/b* mutants. **c.** The growth performance of WT, *attify10a* and *attify10b* mutant Arabidopsis under alkaline stress. *Arabidopsis* seeds were germinated and grown on 1/2MS medium at pH5.8 or pH8.5. Photographs were taken 6 days after germination. **d.** Seed germination rates of WT and mutant lines. Seeds were considered to be germinated when the radicles completely penetrated the seed coats. A total of 90 seeds from each line were used for each experiment. Data are means (±S.E.) of three replicates.

showed no obvious differences in seedling growth compared with WT. After 100 or 150 mM NaHCO₃ treatment for 14 d, both WT and transgenic lines showed growth retardation in a dose-dependent manner. However, the growth inhibition of transgenic lines was less severe than that of WT. After 150 mM NaHCO₃ treatment, all the transgenic lines could maintain continous growth, whereas WT plants showed severe chlorosis (Fig. 4C). In details, the shoot length, ground fresh weight and dry weight of

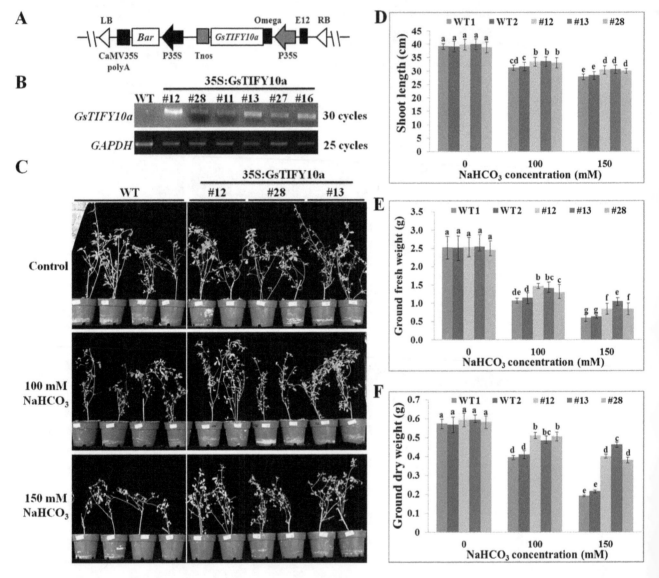

Figure 4. Overexpression of *GsTIFY10a* in alfalfa promoted plant growth under alkaline stress. a. Schematic representation of expression constructs to ectopically express *GsTIFY10a* in Medicago *sativa*. **b.** Semi-quantitative RT-PCR analysis showing the transcript levels of *GsTIFY10a* in transgenic alfalfa lines. **c.** Growth performance of WT and transgenic lines under control conditions or NaHCO₃ treatments. **d.** The shoot length of WT and transgenic plants. **e.** The ground fresh weight of WT and transgenic plants. **f.** The ground dry weight of WT and transgenic plants. For phenotypic analysis under alkaline stress, the propagated WT and *GsTIFY10a* transgenic plants with similar sizes (approximately 25 cm high) were treated with 1/8 Hoagland nutrient solution containing either 0, or 100, or 150 mM NaHCO₃ every 3 days for a total of 12 days. Photographs were taken 12 days after initial treatment. Thirty plants of each line were used for each experiment. Data are means (±SE) of three replicates. Significant differences were determined by one-way ANOVA (P<0.0001) statistical analysis. Different letters show significant differences between groups as indicated by Dunnett's posttests (P<0.05).

both WT and transgenic alfalfa plants were decreased gradually with increased $NaHCO_3$ concentration, but transgenic lines were much taller (Fig. 4D) and displayed more biomass accumulation (Fig. 4E and 4F) than WT. These results suggested that overexpression of *GsTIFY10a* in alfalfa promoted plant growth under alkaline stress.

To further elucidate the influence of *GsTIFY10a* overexpression in alfalfa, several alkaline stress-related physiological and biochemical parameters were analyzed in both WT and transgenic plants, respectively. It is generally accepted that alkaline stress is characterized by high pH value, and always causes much stronger inhibition of plant growth than salt stress [41]. Previous studies showed that NADP-ME could help to maintain the cytosolic pH homeostasis[42], and citric acid was an indicator of plant responses to pH challenge [43]. Therefore, we investigated the NADP-ME activity and citrate acid content of WT and transgenic plants, in an attempt to understand the physiological mechanisms responsible for the increased alkaline tolerance of *GsTIFY10a* transgenic alfalfa. As shown in Fig. 5A and 5B, alkaline stress obviously improved the NADP-ME activity and citrate acid content in both WT and transgenic lines, however, an obvious up-regulation was observed in the transgenic lines. These results indicated that the alleviation of high pH damage in *GsTIFY10a* transgenic alfalfa might partially result from the ability to maintain the cytosolic pH homeostasis through increased NADP-ME activity and citrate acid content.

As a type of compatible osmolyte, proline plays a critical role in protecting plants from environmental stresses [44]. Our results revealed that transgenic alfalfa plants accumulated more free proline than WT in the presence of 100 or 150 mM $NaHCO_3$ (Fig. 5C). To further test the cell membrane stability, we further determined the malon dialdehyde (MDA) content of WT and transgenic plants. Under control conditions, the MDA content of transgenic lines was similar to that of WT (Fig. 5D). After $NaHCO_3$ treatment for 14 d, the MDA content of WT were significantly higher than that of transgenic alfalfa. Collectively, these results demonstrated that the increased alkaline tolerance of *GsTIFY10a* transgenic alfalfa might be related to the elevated levels of NADP-ME activity, citrate content and proline content, as well as reduced MDA content.

GsTIFY10a overexpression up-regulated the expression levels of stress responsive genes

As described above, *GsTIFY10a* overexpression increased the NADP-ME activity, citrate content and proline content of transgenic alfalfa. To explore the molecular basis of *GsTIFY10a* in the cytoplasmic pH regulation and osmotic regulation under alkaline stress, we examined the expression levels of *NADP-ME*, *CS*, *H⁺-Ppase*, and *P5CS* in WT and two transgenic lines (#12 and #13). The real-time PCR results showed that their expression was greatly induced by alkaline stress in both WT and transgenic lines (Fig. 6). Expectedly, their expression levels in transgenic plants were significantly higher than that in WT, which explained

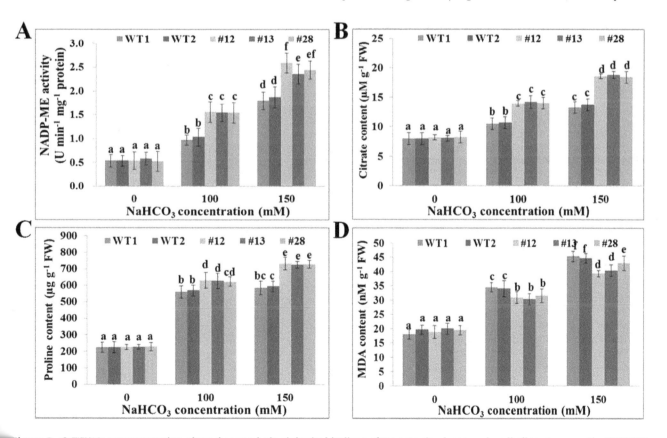

Figure 5. *GsTIFY10a* overexpression altered several physiological indices of transgenic plants under alkaline stress. a. The NADP-ME activity of WT and transgenic lines. **b**. The citric acid content of WT and transgenic lines. **c**. The free proline content of WT and transgenic lines. **d**. The MDA content of WT and transgenic lines. Thirty plants of each line were used for each experiment. Data are means (±SE) of three replicates. Significant differences were determined by one-way ANOVA (P<0.0001) statistical analysis. Different letters show significant differences between groups as indicated by Dunnett's posttests (P<0.05).

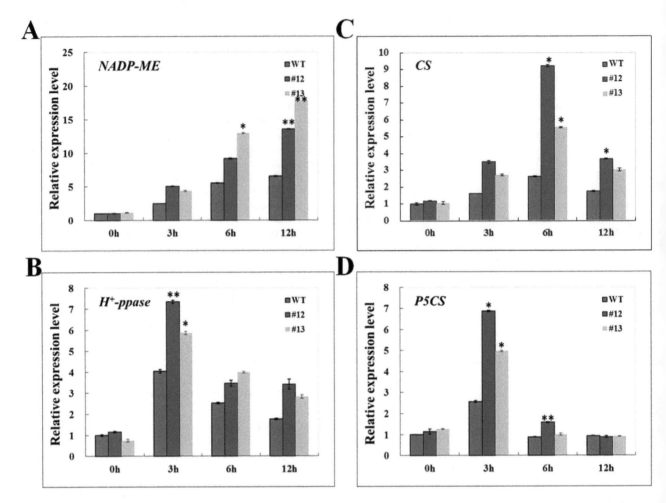

Figure 6. *GsTIFY10a* **overexpression up-regulated the expression levels of several stress responsive genes. a.** Increased expression levels of *NADP-ME* in transgenic plants under alkaline stress. **b.** Increased expression levels of *CS* in transgenic plants. **c.** Increased expression levels of *H⁺-ppase* in transgenic plants. **d.** Increased expression levels of *P5CS* in transgenic plants. To explore the expression patterns of stress-responsive genes, the 4-week-old WT and transgenic seedlings (line #12 and #13) after shoot cottage were treated with 1/8 Hoagland solution containing 50 mM $NaHCO_3$ (pH 8.5) for 0, 3, 6, and 12 h, respectively. Relative transcript levels were determined by quantitative real-time PCR with the *MtGAPDH* gene as an internal reference, and were normalized to WT at 0 h. Values represented the means of three independent biological replicates, and three technological replicates for each. *$P<0.05$; **$P<0.01$ by Student's t-test.

the up-regulation of the NADP-ME activity, citrate acid content and free proline content in transgenic lines. These results also implied that *GsTIFY10a* overexpression promoted the transcript accumulation levels of the stress responsive genes, which might be helpful for the intracellular pH homeostasis and osmotic regulation under alkaline stress.

Increased JA content in *GsTIFY10a* transgenic lines

TIFY/JAZ proteins are repressors of JA signaling in plants. Our previous studies revealed that *GsTIFY10a* overexpression in Arabidopsis repressed the transcription of JA responsive genes [38]. Hence, in this study, we determined the JA content in the transgenic alfalfa lines. As shown in Fig.7, JA contents were significantly increased in the transgenic alfalfa lines ($F_{3,8} = 581.042$, $P = 1.07*10^{-9}$).

Transcriptional activity assays of GsTIFY10a and GsTIFY10e

Our previous studies demonstrated that two of the *Glycine soja* TIFY10 proteins GsTIFY10a and GsTIFY10e positively regulated

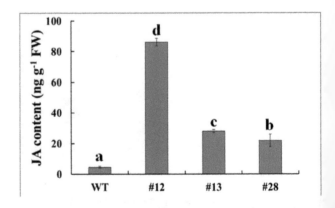

Figure 7. Increased JA content in *GsTIFY10a* transgenic plants. Leaves of WT and transgenic plants were harvested for JA extraction, and subjected for HPLC analysis to determine the content of endogenous JA. Each data point represents the mean (±SE) of three samples from independent sets of plants. Significant differences were found using one-way ANOVA analysis ($P = 1.07*10^{-9}$)

plant tolerance to alkaline stress. Both of them were found to localize at the nuclei of plant cells, indicating that they might act as transcriptional regulators. To further investigate the molecular basis of the TIFY10 proteins in alkaline responses, we identified the transcriptional activity of the GsTIFY10a and GsTIFY10e proteins. The full-length *GsTIFY10a* and *GsTIFY10e* genes were fused to the GAL4 DNA-binding domain in the pGBKT7 vector, and then introduced to the yeast reporter strain AH109. The AtbZIP1 transcription factor, which showed transcriptional activity in yeast cells, was used as a positive control [45,46]. LacZ activity was assessed by using the β-galactosidase filter lift assays. As shown in Fig.8, only the recombinant yeast cells carrying the AtbZIP1-BD vector displayed LacZ activity. These results implied that neither GsTIFY10a nor GsTIFY10e showed transcriptional activity in yeast cells.

Dimerization analyses between GsTIFY10a and GsTIFY10e

It has been suggested that the ZIM domains mediated homo- and heterodimerization of the TIFY proteins [47]. To verify if GsTIFY10a and GsTIFY10e could form homodimers or heterodimers with each other, we performed the yeast two hybrid analyses. The AtbZIP63 transcription factor, which formed homodimers in plant cells [48], was used as a positive control. As shown in Fig.9A, the yeast cells carrying GsTIFY10a-BD/GsTIFY10a-AD, GsTIFY10a-BD/GsTIFY10e-AD, GsTIFY10e-BD/GsTIFY10a-AD and AtbZIP63-BD/AtbZIP63-AD (positive control) were capable of growth on the both SD/-Trp-Lcu and SD/-Trp-Leu-Ade-His medium. However, the yeast cells harboring GsTIFY10e-BD/GsTIFY10e-AD, GsTIFY10a-BD/AD (negative control), GsTIFY10e-BD/AD (negative control) could not grow on the SD/-Trp-Leu-Ade-His medium. These results demonstrated that GsTIFY10a, not GsTIFY10e, could form homodimers, and GsTIFY10a and GsTIFY10e could heterodimerize with each other in yeast cells.

To further verify their physical interaction in living plant cells, we performed the bimolecular fluorescence complementation (BiFC) assays. To this end, we fused GsTIFY10a to the N-terminal YFP fragment and GsTIFY10a/e to the C-terminal YFP fragment, to generate GsTIFY10a-YFPN and GsTIFY10a/e-YFPC constructs, respectively. GsTIFY10a-YFPN/GsTIFY10a-YFPC and GsTIFY10a-YFPN/GsTIFY10e-YFPC were co-transformed into *Arabidopsis* protoplasts, respectively, and the empty YFPN/YFPC was used as a negative control. As shown in Fig. 9B, YFP fluorescence was observed from protoplasts co-transformed with GsTIFY10a-YFPN/GsTIFY10a-YFPC and GsTIFY10a-YFPN/GsTIFY10e-YFPC, but not from YFPN/YFPC. The BiFc results confirmed the homodimerization of GsTIFY10a, as well as the heterodimerization between GsTIFY10a and GsTIFY10e in plant cells.

Discussions

The TIFY protein family, characterized by a highly conserved TIFY motif, constitutes a particular class of plant-specific transcription factors with a broad range of biological functions. According to their distinct domain architectures, TIFY proteins could be classified into four subfamilies: the TIFY subfamily containing only the TIFY domain, the JAZ subfamily containing the TIFY domain and the Jas domain, the PPD subfamily containing the PPD, TIFY and a truncated Jas domain, and the ZML subfamily containing the TIFY, CCT and ZML domains[14]. Among them, the JAZ subfamily proteins could be further clustered into five groups (group I-V) [14]. The first group (group I) is composed of the TIFY10 and TIFY11 proteins. In this study, we focused on the group I TIFY proteins in Arabidopsis and wild soybean. All of the group I TIFY proteins contained the conserved TIFY and Jas domains (Fig. 1, 2), which was the canonical characteristic of JAZ subfamily. It is noteworthy that except for GsTIFY10e and GsTIFY10f, all other TIFY10/11s included an N-terminal domain (Fig. 1, 2).

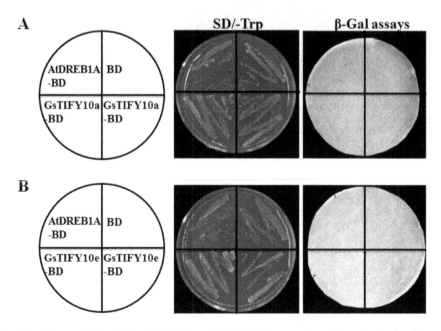

Figure 8. Transcriptional activity analysis of the GsTIFY10a and GsTIFY10e proteins. a. Transcriptional activity analysis of GsTIFY10a. **b**. Transcriptional activity analysis of GsTIFY10e. The pGBKT7-GsTIFY10a and pGBKT7-GsTIFY10e vectors were transformed into the yeast reporter strain AH109, and LacZ activity was assessed by using the β-galactosidase filter lift assays. The AtDREB1A transcription factor was used as a positive control.

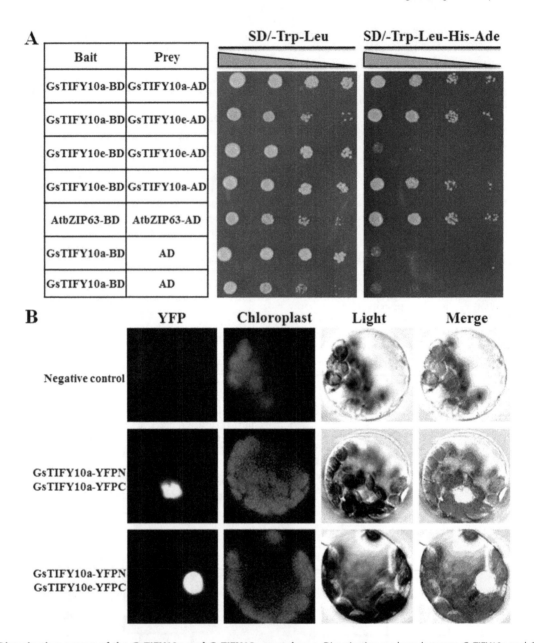

Figure 9. Dimerization assays of the GsTIFY10a and GsTIFY10e proteins. a. Dimerization analyses between GsTIFY10a and GsTIFY10e in yeast cells. Pictures showed the growth performance of recombinant yeast cells harboring different plasmids on SD/-Trp-Leu and SD/-Trp-Leu-Ade-His medium. The AtbZIP63-BD/AtbZIP63-AD combination was used as a positive control, and the GsTIFY10a-BD/AD and GsTIFY10e-BD/AD combinations were used as negative controls. **b**. Dimerization analyses between GsTIFY10a and GsTIFY10e in living plant cells. The YFPN/YFPC combination was used as a negative control. Pictures showed the YFP fluorescence, chlorophyll auto-fluorescence, light and overlay visions.

The best-characterized TIFY10 protein is the Arabidopsis TIFY10a/JAZ1, which was found to play a critical role in JA signaling. In the absence of JAs, JAZ1 interacts with and represses the downstream transcriptional activators, such as MYC2 [26], MYC3, MYC4 [27], MYB21, MYB24 [28], bHLH017 and bHLH003 [29], which control the expression of JA-responsive genes. In the presence of JAs, AtTIFY10a protein is recognized and degraded by the SCF (COI1) E3 ubiquitin ligase, releasing the downstream transcription factors [49]. Jasmonoyl isoleucine (JA-Ile) and coronatine could promote the physical interaction between JAZ1 and COI1, and the C-terminal Jas domain of JAZ1 is critical for JA-Ile/coronatine-dependent interaction with COI1 [49,50]. Recently, a transacting factor AtBBD1 was

suggested to interact with AtTIFY10a and bind to the JARE element upstream of the JA responsive gene AtJMT [51]. In addition to the negative regulatory role in JA signaling pathway, AtTIFY10a was also reported to be involved in phytochrome A [52] and auxin signaling [53]. Furthermore, AtTIFY10a physically interacted with ICE1 and ICE2 transcription factors, repressed the transcriptional function of ICE1, and thereby affected plant freezing stress responses [33]. However, no direct evidence supporting its role in alkaline stress responses was reported until now.

Recently, the systemic transcriptional analyses and several lines of genetic evidence suggested the potential regulatory roles of TIFY10/11s in plant responses to environmental challenges

[15,33]. Our previous research also revealed the alkaline stress induced expression of *Glycine soja* TIFY10/11 genes [36,40], which was further confirmed by the RNA-seq data (Fig. S1). Consistently, we also observed some cis-regulatory elements related to environmental stress response in the TIFY10/11s promoters (Table 1). In addition, we also found elements involved in hormone responsiveness, suggesting the crosstalk between environmental stress and hormone signal. Several elements were common to all TIFY10/11s, implying that TIFY10/11 proteins were involved in the same signal transduction pathway. Remarkably, several elements only existed in specific TIFY10/11 promoters, which might indicate the specific response of TIFY10/11s to environmental stress.

Among the eight GsTIFY10/11 genes, overexpression of *GsTIFY10a* and *GsTIFY10e* in *Arabidopsis* obviously improved plant alkaline tolerance [37,38], while *GsTIFY11b* overexpression led to decreased salt tolerance [39]. In this study, we suggested the positive roles of AtTIFY10a/b in plant tolerance to alkaline stress by using the plate germination assays (Fig. 3), further supporting the important roles of TIFY10s in plant alkaline stress responses. It is worth to note that, under the same alkaline stress treatment, *attify10a* (54.2% on the 3rd day) exhibited relatively lower germination rates than *attify10b* (77.0%), indicating a more important role of *AtTIFY10a* than *AtTIFY10b* in alkaline stress responses.

In the current study, we also transformed the *GsTIFY10a* gene into *Medicago sativa* in an attempt to obtain transgenic alfalfa with superior alkaline tolerance. Our results demonstrated that *GsTIFY10a* overexpression dramatically promoted growth of transgenic plants under alkaline stress. *GsTIFY10a* transgenic lines displayed much better at shoot height, ground fresh weight and dry weight than WT under alkaline stress (Fig. 4), which was in line with the better growth performance of *GsTIFY10a* transgenic Arabidopsis [38]. Furthermore, we also investigated the potential physiological and molecular basis of *GsTIFY10a* in response to alkaline stress (Fig. 5, 6). Firstly, *GsTIFY10a* overexpression could help plant to deal with the high pH damage by up-regulating the NADP-ME activity and citrate acid content (Fig. 5). The up-regulation of *NADP-ME* and *CS* gene expression in transgenic plants might be helpful to explain the increase of NADP-ME activity and citrate acid content (Fig. 6). Meanwhile, we also observed an obvious increase of H^+-*ppase* expression in transgenic lines under alkaline stress (Fig. 6), which was also helpful for maintaining the cytosolic pH homeostasis. These results were consistent with our previous observation that *GsTIFY10a* overexpression up-regulated the expression of *NADP-ME* and H^+-*ppase* in Arabidopsis [38]. Secondly, *GsTIFY10a* overexpression led to greater proline accumulation (Fig. 5) and up-regulated expression of the *P5CS* gene (Fig. 6), which encodes a key enzyme in the proline biosynthesis process. Proline serves as a compatible osmolyte, molecular chaperone and ROS scavenger in plant responses to environmental stress [54]. The increased accumulation of free proline in *GsTIFY10a* transgenic lines might be of great importance for the effective osmo-regulation and ROS scavenging of plant cells under alkaline stress. In addition, *GsTIFY10a* transgenic lines displayed lower levels of MDA content under alkaline stress (Fig. 5D). MDA is widely recognized as an indicator for lipid peroxidation resulted from the elevated ROS accumulation in plant cells [55]. The decreased MDA accumulation also indicated the more effective adaptation of transgenic plants to ROS damage caused by alkaline stress. Taken together, we speculated that *GsTIFY10a* overexpression is of fundamental importance for the cytosolic pH regulation, osmo-egulation and ROS scavenging by regulating the related gene expression and enzyme activity, and thereby promoted plant growth under alkaline stress.

A great number of studies have demonstrated that JA signaling is critical for plant stress responses and TIFY proteins are involved in plant tolerance to both biotic and abiotic stresses such as wounding [52,56], pathogen [23,57], drought stress [15,34], and salt stress [20,32]. Our studies suggested the involvement of TIFYs in alkaline stress responses and the crosstalk between alkaline stress and JA signaling [37,38,40]. All of the *GsTIFY10/11* genes were dramatically up-regulated at the early stage of alkaline stress treatment (Fig. S1). Overexpression of GsTIFY10e/JAZ2 significantly improved plant alkaline stress tolerance [37]. Knockout of AtTIFY10a and AtTIFY10b also inhibited seed germination under alkaline stress (Fig. 3). Furthermore, *GsTIFY10a* expression was greatly induced by both alkaline stress and MeJA treatment [38].

In addition to the increased alkaline tolerance, *GsTIFY10a* overexpression in Arabidopsis conferred MeJA insensitivity. *GsTIFY10a* repressed transcription of the JA responsive genes such as *PDF1.2* (Plant Defensin 1.2), *VSP2* (Vegetative Storage Protein 2), *AOS* (Allelen Oxide Synthase) and *LOX2* (Lipoxygenase 2), whose expression was also induced by alkaline stress. However, in the present study, we found that JA contents were significantly increased in the transgenic alfalfa lines (Fig.7). As we know, environmental stress could stimulate the biosynthesis of JAs. The elevated JAs levels promoted the interaction of TIFY/JAZ with SCF^{CoI1}, which led to degradation of TIFY/JAZ proteins, and subsequently release of the targeted TFs. Activation of TFs induced transcription of JA-responsive genes, and the newly synthesized TIFY/JAZ proteins could restore the repression of TFs. The *GsTIFY10a* transgenic lines always kept high levels of GsTIFY10a proteins, which repressed the TFs activity and JA signaling. Hence, plants need to synthesize more JAs to degrade the GsTIFY10a proteins. Collectively, these results strongly suggested the crosstalk between alkaline stress and JA signaling. Molecular basis of the hypothesis that TIFY10 proteins mediate alkaline stress responses through JA signaling need be further studied.

The JAZ proteins were suggested to act as transcriptional repressors in JA signaling because they do not contain a known DNA binding domain. Due to the nuclear localization of several JAZ subfamily members [25,58], it is proposed that they might exert their effect on gene expression through protein interaction with transcription factors. Our previous revealed that, like other JAZs, GsTIFY10a and GsTIFY10e localized in the nuclei of plant cells. In this study, we found that GsTIFY10a and GsTIFY10e showed no transcriptional activity in yeast cells (Fig.8), indicating them as transcriptional repressors. On the other hand, it has been well demonstrated that the ZIM domains of JAZ/TIFYs mediate formation of homo- and heteromeric complexes [59]. For example, the Arabidopsis JAZ1/TIFY10a and JAZ2/TIFY10b could form homodimers and heterodimers [59]. In this study, by using the Y2H and BiFc technologies, we found that GsTIFY10a could form homodimers, as well as heterodimers with GsTIFY10e, while GsTIFY10e could not homodimerize (Fig.9). Similarly, in case of Arabidopsis JAZs, JAZ7, JAZ8, JAZ9, JAZ11, and JAZ12 could not homodimerize, but JAZ9 and JAZ12 could form heterodimers with other JAZs [59].

In summary, here we provide novel insights into the regulatory roles and potential molecular basis of TIFY10 proteins in plant responses to alkaline stress. We obtained the transgenic alfalfa with superior alkaline tolerance by ectopically expressing *GsTIFY10a*, and investigated the possible physiological and molecular mechanism by which *GsTIFY10a* regulated plant alkaline tolerance.

For now, the *GsTIFY10a* transgenic alfalfa is at the stage of the "biosafety evaluation of genetically modified organisms", and will provide a bearing on future aims at salt-alkaline soil management.

Materials and Methods

Plant materials and growth conditions

Seeds of the wild type *Arabidopsis thaliana* (Columbia ecotype), *attify10a* (SALK_011957) and *attif10b* (SALK_025279) mutant lines were kindly provided by the European Arabidopsis Stock Centre (UK). Both WT and mutant Arabidopsis seeds were germinated and grown in standard nutrient solution as described by Tocquin et al. (2003). The Arabidopsis seedlings were maintained in a greenhouse under controlled environmental conditions (21–23°C, 100 µmol m^{-2} s^{-1}, 60% relative humidity, 16 h light/8 h dark cycles).

Alfalfa (*Medicago sativa* L. cv. Nongjing No. 1, a local species) was used in this study, and was kindly obtained from Heilongjiang Academy of Agricultural Sciences (Harbin, China). Alfalfa was grown in a greenhouse under controlled environmental conditions (24–26°C, 600 µmol m^{-2} s^{-1}, 80% relative humidity, 16 h light/8 h dark cycles).

Seed germination assays of *AtTIFY10a/b* knockout Arabidopsis

Homozygous T-DNA insertion mutant Arabidopsis was obtained by using a PCR-based method as described previously [60]. Briefly, the gene specific primers across the T-DNA insertion sites of *attify10a* and *attif10b* mutant (5′-TGCGATCCAGCCAA-AGCGTCT-3′ and 5′-GAGTATTTGATAGTATGGTTCGT-CAACAA-3′ for *attify10a*, 5′-TGAGAAGAGGAAGGATAGG-TAATGCAT-3′ and 5′-CAAGCTAATGTTGAGATCGGCA-GAG-3′ for *attify10b*) were used for homozygous identification. T-DNA insertion was confirmed by PCR amplification with the following primers (5′-TGCGATCCAGCCAAAGCGTCT-3′ and 5′-GATTTGGGTGATGGTTCACGTAGTG-3′ for *attify10a*, 5′-GAGAAGAGGAAGGATAGGTAATGCAT-3′ and 5′-GAT-TTGGGTGATGGTTCACGTAGTG-3′ for *attify10b*).

RT-PCR analysis was used to confirm the silence of *AtTIFY10a* and *AtTIFY10b* genes in the T-DNA insertion mutants. The following gene specific primers were used (5′-TGCGATCCAGC-CAAAGCGTCT-3′ and 5′-GAGTATTTGATAGTATGGTT-CGTCAACAA-3′ for *AtTIFY10a*, 5′-TGAGAAGAGGAAGGA-TAGGATCACATC-3′ and 5′-CAAGCTAATGTTGAGATCG-GCAGAG-3′ for *AtTIFY10b*).

For plate germination assays under alkaline stress, seeds of WT and mutant lines were surfaced-sterilized with 5% sodium hypochlorite (NaClO) for 6–8 min with shaking, washed with sterilized distilled water for 6–8 times, and kept at 4°C for 3 days to break seed dormancy. After that, WT and mutant *Arabidopsis* seeds were sown on 1/2 MS agar medium with 1% (w/v) sucrose and 0.8% (w/v) agar at pH5.8 (Control conditions) or pH8.5 (Alkaline stress). The germination percent was recorded for consecutive 6 days after sowing, and pictures were taken to show the growth performance of each line. Ninety seeds from each line were used for each experiment and the experiments were repeated for three times.

Generation of *GsTIFY10a* overexpression transgenic alfalfa

In order to investigate the effect of *GsTIFY10a* on alkaline stress tolerance, *GsTIFY10a* was cloned into the pBEOM plant expression vector, and was under the control of the cauliflower mosaic virus (CaMV) 35S promoter, with the binding enhancers

E12 and omega. The *Bar* gene was used as the selectable marker. The recombinant vector pBEOM-GsTIFY10a was introduced into *Agrobacterium tumefaciens* strain EHA105, and then introduced into *Medicago sativa* by using the cotyledonary node method as described previously [13]. Briefly, alfalfa seeds were surface sterilized with 70% ethanol for 1 min, 0.1% (v/v) HgCl$_2$ for 15 min, and then washed with sterilized water for 3–5 times. Sterilized seeds were germinated and grown on 1/2MS medium for 8 days, and then seedlings were aseptically excised at the cotyledonary node position as explants. The explants were infected with *A. tumefaciens* EHA105 for 15 min, and placed vertically into MS medium (pH 5.2) containing 1 mg L^{-1} 6-benzylamino-purine (6-BA) and 100 mM L^{-1} acetosyringone for 3 days. After that, explants were washed in liquid MS medium (pH 5.8) containing 1 mg L^{-1} 6-BA, 0.5 mg L^{-1} glufosinate and 100 mg L^{-1} amoxicillin. Then explants were transferred onto solid MS medium (pH 5.8) containing 1 mg L^{-1} 6-BA, 0.5 mg L^{-1} glufosinate and 100 mg L^{-1} amoxicillin, and grown for another 14 days to form the glufosinate-resistant regenerated shoots. The regenrated shoots were then transferred to new solid medium as described above, and grown for another 14 days. Elongated shoots were then transferred to new 1/2MS medium only containing 100 mg L^{-1} amoxicillin, until roots appeared.

The regenerated alfalfa plants were then transplanted into soil under controlled conditions and confirmed by PCR analysis using CaMV35S promoter specific forward primer and *Bar* gene specific reverse primer (5′-TGCACCATCGTCAACCACTA-CATCG-3′ and 5′-CCAGCTGCCAGAAACCCACGTCATG-3′). To eliminate the potential existence of the chimeric transgenic plants, new-born leaves from different branches of each regenerated plants were used for PCR identification. The transcript levels of *GsTIFY10a* in the PCR-positive plants were further analyzed by semi-quantitative RT-PCR analyses with gene specific primers (5′-ACAGAGCCAGCCTTCATTTCC-3′ and 5′-CGAACCCG-ACTCACGAAGAAG-3′). The alfalfa glyceraldehyde-3-phosphate dehydrogenase gene (*MtGAPDH*, Accession: Medtr3g085850) was used as an internal control, and PCR amplified with the following primer pair (5′-GTGGTGCCAAGAAGGTTGTTAT-3′ and 5′-CTGGGAATGATGTTGAAGGAAG-3′).

Phenotypic analyses of transgenic alfalfa under alkaline stress

For phenotypic analyses under alkaline stress, the lignified WT and transgenic alfalfa plants were used for vegetative propagation through stem cuttings. The propagated seedlings were transplanted into plastic culture pots filled with a mixture of peat moss: soil (1:1; v/v), irrigated with 1/8 Hoagland nutrient solution and grown in a greenhouse under controlled conditions. To eliminate the potential existence of the chimeric transgenic plants, each propagated seedlings were subjected for PCR identification before stress treatment. PCR-positive plants with similar sizes (approximately 25 cm high) were then exposed to alkaline stress by irrigating with 1/8 Hoagland solution containing either 0, or 100, or 150 mM NaHCO$_3$ every 3 days for a total of 12 days. Photographs were taken on the 12th day. Thirty plants of each line were used for each experiment, and the experiments were repeated for at least three times.

The MDA content was determined according to the protocol described by Peever et al. [61]. The citric acid content was determined by using a spectrophotometer (UV-2550, Shimadzu Japan) at the absorbance of 490 nm according to the method of Zhu [62]. Free proline content was measured according to the method of Bates et al. [63]. NADP-ME activity was measured a described by Geer et al. [64].

Table 2. Gene-specific primers used for quantitative real-time PCR assays.

Gene name	Gene ID	Primer Sequence (5′ to 3′)
GsGAPDH	DQ355800	Forward: GACTGGTATGGCATTCCGTGT Reverse: GCCCTCTGATTCCTCCTTGA
GsTIFY10a		Forward: TCCAGCGCAATTAAGTCTGTGAG Reverse: TTTGGTGGCATAGGACATGATCT
MtGAPDH	Medtr3g085850	Forward: GTGGTGCCAAGAAGGTTGTTAT Reverse: CTGGGAATGATGTTGAAGGAAG
P5CS	EU371644.1	Forward: TCGGGGTCCAGTAGGAGTTG Reverse: AGTAGTTAGGTCTTTGTGGGTGTAGG
H⁺-ppase	XM_003609415.1	Forward: TCTCCACCGACGCATCTATCA Reverse: GGCATTATCCCAAGCACCG
NADP-ME	XM_003630679.1	Forward: TGGCGGTGTTGAGGATGTCT Reverse: CTGACAGTGGGAGGTAAGAGGC
Citrate Synthase	HM030734.1	Forward: TTAAAGCCAGGAAACGAAAGC Reverse: AAGGCAACAGCAACCTCAATC

All of the above numerical data were subjected to statistical analyses using EXCEL 2010 and/or IBM SPSS statistics 19, and analyzed by Student's T-test and/or one-way ANOVA analysis.

Quantitative real-time PCR analyses

To investigate the expression patterns of stress-responsive genes, the 4-week-old seedlings after shoot cottage were treated with 1/8 Hoagland solution containing 50 mM $NaHCO_3$ (pH 8.5) for 0, 3, 6, and 12 h, respectively. Equal amounts of leaves were harvested and stored at $-80°C$ after snap-frozen in liquid nitrogen.

Total RNA was extracted by using RNeasy Plant Mini Kit (Qiagen, Valencia, CA, USA), and subjected to cDNA synthesis by using SuperScript^{TM} III Reverse Transcriptase kit (Invitrogen, Carlsbad, CA, USA). Quantitative real-time PCR was performed using a Stratagene MX3000P real-time PCR instrument and the SYBR Select Master Mix (Applied Biosystems, USA). The MtGAPDH gene was used as an internal reference. Expression levels of all candidate genes were calculated by using the $2^{-\Delta\Delta CT}$ method, and the relative intensities were calculated and normalized as described previously 65. Three independent biological replicates were carried out and subjected to real-time PCR in triplicate. Primers used for quantitative real-time PCR were designed by using Primer 5 software and listed in Table 2.

Transcriptional activation assays

The full-length coding regions of GsTIFY10a and GsTIFY10e were amplified by using the following primer pairs: 5′-CTTGAATTCTCGAGCTCATCGG-3′ and 5′-GAAGTCGA-CGATTTGAGGTGAA-3′ (for GsTIFY10a), 5′-CGGGAATT-CAATCCATGGAAC-3′ and 5′-GCCGTCGACTAACACAAA-GCTGG-3′ (for GsTIFY10e). The PCR products were cloned to the pGBKT7 vector to express the GsTIFY10a-BD and GsTIFY10e-BD fused proteins.

For the transcriptional activation assays, the corresponding vectors pGBKT7 (negative control), pGBKT7-GsTIFY10a, pGBKT7-GsTIFY10e and pGBKT7-AtbZIP1 (positive control) were transformed into yeast strain AH109. The transformants were selected on SD/-Trp medium. LacZ activity was analyzed by using the β-galactosidase filter lift assay.

Dimerization analyses of GsTIFY10a and GsTIFY10b in yeast cells

The full-length coding regions of GsTIFY10a and GsTIFY10e were amplified by using the following primer pairs: 5′-GCTCA-TATGTCGAGCTCATCGG-3′ and 5′-CTTGAATTCGATT-TGAGGTGAA-3′ (for GsTIFY10a), 5′- GCCCATATGACCT-CAGTGGAG-3′ and 5′-GTTGGATCCTAACACAAAGCTG-G-3′ (for GsTIFY10e). The PCR products were cloned to the

pGADT7 vector to express the GsTIFY10a-AD and GsTIFY10e-AD fused proteins.

For homodimerization analyses of GsTIFY10a and GsTI-FY10e, the combinations pGBKT7-GsTIFY10a/pGADT7-GsTI-FY10a and pGBKT7-GsTIFY10e/pGADT7-GsTIFY10e were co-transformed into yeast strain AH109. For heterodimerization analyses between GsTIFY10a and GsTIFY10e, the combinations pGBKT7-GsTIFY10a/pGADT7-GsTIFY10e and pGBKT7-GsTIFY10e/pGADT7-GsTIFY10a were co-transformed into yeast. The pGBKT7-GsTIFY10a/pGADT7 and pGBKT7-GsTI-FY10e/pGADT7 combinations were used as negative controls, and the pGBKT7-AtbZIP63/pGADT7-AtbZIP63 combination was used as a positive control. The transformants were selected on SD/-Trp-Leu and SD/-Trp-Leu-His-Ade medium, respectively.

Bimolecular fluorescence complementation (BiFc) assays

The full-length coding region of GsTIFY10a was PCR amplified and fused into the N-terminus of the amino-terminal half of YFP protein (YFPN), to generate the GsTIFY10a-YFPN construct. The coding regions of GsTIFY10a and GsTIFY10e were fused into the N-terminus of the carboxyl-terminal half of YFP protein (YFPC), to generate the GsTIFY10a-YFPC and GsTIFY10e-YFPC construct.

BiFc assays were conducted by transient transformation of Arabidopsis protoplasts prepared as described [66]. Yellow fluorescence was measured by excitation at 514 nm and emission at 527 nm, by using confocal laser-scanning microscope Leica SP2 (Leica, Wetzlar, Germany). YFP fluorescence, chlorophyll auto-fluorescence and light visions were recorded in separate channels and then merged into an overlay image.

Supporting Information

Figure S1 Expression patterns of the Arabidopsis and wild soybean TIFY10/11 subgroup genes under 50 mM $NaHCO_3$ (pH 8.5) treatment based on the RNA-seq data. **a.** Relative expression levels of the TIFY10/11 genes under alkaline stress. The expression levels at 0 h were considered as 1. **b.** Expression levels of the TIFY10/11 genes under alkaline stress.

Acknowledgments

We thank members of the lab for discussions and comments on this manuscript. We gratefully acknowledge support from Key Laboratory of Agricultural Biological Functional Genes - College of Heilongjiang Province.

Author Contributions

Conceived and designed the experiments: DZ MS YZ. Performed the experiments: DZ JW NZ. Analyzed the data: MS RL. Contributed reagents/materials/analysis tools: YZ RL XL. Wrote the paper: DZ MS.

References

1. Xu J, Ji P, Wang B, Zhao L, Wang J, et al. (2013) Transcriptome sequencing and analysis of wild Amur Ide (*Leuciscus waleckii*) inhabiting an extreme alkaline-saline lake reveals insights into stress adaptation. PLoS ONE 8: e59703.

2. Sun Y, Wang F, Wang N, Dong Y, Liu Q, et al. (2013) Transcriptome exploration in Leymus chinensis under saline-alkaline treatment using 454 pyrosequencing. PLoS ONE 8: e53632.

3. Zhang WJ, Niu Y, Bu SH, Li M, Feng JY, et al. (2014) Epistatic association mapping for alkaline and salinity tolerance traits in the soybean germination stage. PLoS ONE 9: e84750.

4. Han X, Cheng Z, Meng H (2012) Soil properties, nutrient dynamics, and soil enzyme activities associated with garlic stalk decomposition under various conditions. PLoS ONE 7: e50868.

5. Wang H, Wu Z, Han J, Zheng W, Yang C (2012) Comparison of ion balance and nitrogen metabolism in old and young leaves of alkali-stressed rice plants. PLoS ONE 7: e37817.

6. Guan Q, Wu J, Yue X, Zhang Y, Zhu J (2013) A nuclear calcium-sensing pathway is critical for gene regulation and salt stress tolerance in Arabidopsis. PLoS Genet 9: e1003755.

7. Xu W, Jia L, Shi W, Baluska F, Kronzucker HJ, et al. (2013) The tomato 14-3-3 protein TFT4 modulates H^+ efflux, basipetal auxin transport, and the PKS5-J3 pathway in the root growth response to alkaline stress. Plant Physiol 163: 1817–1828.

8. Liu J, Guo Y (2011) The alkaline tolerance in Arabidopsis requires stabilizing microfilament partially through inactivation of PKS5 kinase. J Genet Genomics 38: 307–313.

9. Tong Z, Xie C, Ma L, Liu L, Jin Y, et al. (2014) Co-expression of bacterial aspartate kinase and adenylylsulfate reductase genes substantially increases sulfur amino acid levels in transgenic alfalfa (*Medicago sativa* L.). PLoS ONE 9: e88310.

10. Li W, Wei Z, Qiao Z, Wu Z, Cheng L, et al. (2013) Proteomics analysis of alfalfa response to heat stress. PLoS ONE 8: e82725.

11. Shi Y, Guo R, Wang X, Yuan D, Zhang S, et al. (2014) The regulation of alfalfa saponin extract on key genes involved in hepatic cholesterol metabolism in hyperlipidemic rats. PLoS ONE 9: e88282.

12. Bogino P, Abod A, Nievas F, Giordano W (2013) Water-limiting conditions alter the structure and biofilm-forming ability of bacterial multispecies communities in the alfalfa rhizosphere. PLoS ONE 8: e79614.

13. Sun M, Sun S, Zhao Y, Zhao C, Duanmu H, et al. (2014) Ectopic expression of GsPPCK3 and SCMRP in Medicago sativa enhances plant alkaline stress tolerance and methionine content. PLoS ONE 9: e89578.

14. Bai Y, Meng Y, Huang D, Qi Y, Chen M (2011) Origin and evolutionary analysis of the plant-specific TIFY transcription factor family. Genomics 98: 128–136.

15. Ye H, Du H, Tang N, Li X, Xiong L (2009) Identification and expression profiling analysis of TIFY family genes involved in stress and phytohormone responses in rice. Plant Mol Biol 71: 291–305.

16. Qi T, Huang H, Wu D, Yan J, Qi Y, et al. (2014) Arabidopsis DELLA and JAZ proteins bind the WD-repeat/bHLH/MYB complex to modulate gibberellin and jasmonate signaling synergy. Plant Cell 26: 16.

17. Van der Does D, Leon-Reyes A, Koornneef A, Van Verk MC, Rodenburg N, et al. (2013) Salicylic acid suppresses jasmonic acid signaling downstream of SCFCOI1-JAZ by targeting GCC promoter motifs via transcription factor ORA59. Plant Cell 25: 744–761.

18. Oh Y, Baldwin IT, Galis I (2013) A jasmonate ZIM-domain protein NaJAZd regulates floral jasmonic acid levels and counteracts flower abscission in Nicotiana attenuata plants. PLoS ONE 8: e57868.

19. Toda Y, Yoshida M, Hattori T, Takeda S (2013) RICE SALT SENSITIVE3 binding to bHLH and JAZ factors mediates control of cell wall plasticity in the root apex. Plant Signal Behav 8.

20. Toda Y, Tanaka M, Ogawa D, Kurata K, Kurotani K, et al. (2013) RICE SALT SENSITIVE3 forms a ternary complex with JAZ and class-C bHLH factors and regulates jasmonate-induced gene expression and root cell elongation. Plant Cell 25: 1709–1725.

21. Hakata M, Kuroda M, Ohsumi A, Hirose T, Nakamura H, et al. (2012) Overexpression of a rice TIFY gene increases grain size through enhanced accumulation of carbohydrates in the stem. Biosci Biotechnol Biochem 76: 2129–2134.

22. Song S, Qi T, Fan M, Zhang X, Gao H, et al. (2013) The bHLH subgroup IIId factors negatively regulate jasmonate-mediated plant defense and development. PLoS ONE 8: e1003653.

23. Ishiga Y, Ishiga T, Uppalapati SR, Mysore KS (2013) Jasmonate ZIM-domain (JAZ) protein regulates host and nonhost pathogen-induced cell death in tomato and Nicotiana benthamiana. PLoS ONE 8: e75728.

24. Demianski AJ, Chung KM, Kunkel BN (2012) Analysis of Arabidopsis JAZ gene expression during Pseudomonas syringae pathogenesis. Mol Plant Pathol 13: 46–57.

25. Thines B, Katsir L, Melotto M, Niu Y, Mandaokar A, et al. (2007) JAZ repressor proteins are targets of the SCF(COI1) complex during jasmonate signalling. Nature 448: 661–665.

26. Fernandez-Calvo P, Chini A, Fernandez-Barbero G, Chico JM, Gimenez-Ibanez S, et al. (2011) The Arabidopsis bHLH transcription factors MYC3 and MYC4 are targets of JAZ repressors and act additively with MYC2 in the activation of jasmonate responses. Plant Cell 23: 701–715.

27. Niu Y, Figueroa P, Browse J (2011) Characterization of JAZ-interacting bHLH transcription factors that regulate jasmonate responses in Arabidopsis. J Exp Bot 62: 2143–2154.

28. Song S, Qi T, Huang H, Ren Q, Wu D, et al. (2011) The Jasmonate-ZIM domain proteins interact with the R2R3-MYB transcription factors MYB21 and MYB24 to affect Jasmonate-regulated stamen development in Arabidopsis. Plant Cell 23: 1000–1013.

29. Fonseca S, Fernandez-Calvo P, Fernandez GM, Diez-Diaz M, Gimenez-Ibanez S, et al. (2014) bHLH003, bHLH013 and bHLH017 Are New Targets of JAZ Repressors Negatively Regulating JA Responses. PLoS ONE 9: e86182.

30. White DW (2006) PEAPOD regulates lamina size and curvature in Arabidopsis. Proc Natl Acad Sci U S A 103: 13238–13243.

31. Shikata M, Matsuda Y, Ando K, Nishii A, Takemura M, et al. (2004) Characterization of Arabidopsis ZIM, a member of a novel plant-specific GATA factor gene family. J Exp Bot 55: 631–639.

32. Ismail A, Riemann M, Nick P (2012) The jasmonate pathway mediates salt tolerance in grapevines. J Exp Bot 63: 2127–2139.

33. Hu Y, Jiang L, Wang F, Yu D (2013) Jasmonate regulates the inducer of cbf expression-C-repeat binding factor/DRE binding factor1 cascade and freezing tolerance in Arabidopsis. Plant Cell 25: 2907–2924.

34. Seo JS, Joo J, Kim MJ, Kim YK, Nahm BH, et al. (2011) OsbHLH148, a basic helix-loop-helix protein, interacts with OsJAZ proteins in a jasmonate signaling pathway leading to drought tolerance in rice. Plant J 65: 907–921.

35. Chen P, Yan K, Shao H, Zhao S (2013) Physiological Mechanisms for High Salt Tolerance in Wild Soybean (Glycine soja) from Yellow River Delta, China: Photosynthesis, Osmotic Regulation, Ion Flux and antioxidant Capacity. PLoS ONE 8: e83227.

36. Ge Y, Li Y, Zhu YM, Bai X, Lv DK, et al. (2010) Global transcriptome profiling of wild soybean (Glycine soja) roots under NaHCO3 treatment. BMC Plant Biol 10: 153.

37. Zhu D, Cai H, Luo X, Bai X, Deyholos MK, et al. (2012) Over-expression of a novel JAZ family gene from Glycine soja, increases salt and alkali stress tolerance. Biochem Biophys Res Commun 426: 273–279.

38. Zhu D, Bai X, Chen C, Chen Q, Cai H, et al. (2011) GsTIFY10, a novel positive regulator of plant tolerance to bicarbonate stress and a repressor of jasmonate signaling. Plant Mol Biol 77: 285–297.

39. Zhu D, Bai X, Zhu YM, Cai H, Li Y, et al. (2012) [Isolation and functional analysis of GsTIFY11b relevant to salt and alkaline stress from Glycine soja]. Yi Chuan 34: 230–239.

40. Zhu D, Bai X, Luo X, Chen Q, Cai H, et al. (2013) Identification of wild soybean (Glycine soja) TIFY family genes and their expression profiling analysis under bicarbonate stress. Plant Cell Rep 32: 263–272.

41. Yang CW, Wang P, Li CY, Shi DC, Wang DL (2008) Comparison of effects of salt and alkali stresses on the growth and photosynthesis of wheat. Photosynthetica 46: 107–144.

42. Edwards GE, Andreo CS (1992) NADP-malic enzyme from plants. Phytochemistry 31: 1845–1857.

43. Garcia J, Torres N (2011) Mathematical modelling and assessment of the pH homeostasis mechanisms in Aspergillus niger while in citric acid producing conditions. J Theor Biol 282: 23–35.

44. Xiong H, Li J, Liu P, Duan J, Zhao Y, et al. (2014) Overexpression of OsMYB48-1, a Novel MYB-Related Transcription Factor, Enhances Drought and Salinity Tolerance in Rice. PLoS ONE 9: e92913.

45. Sun X, Li Y, Cai H, Bai X, Ji W, et al. (2012) The Arabidopsis AtbZIP1 transcription factor is a positive regulator of plant tolerance to salt, osmotic and drought stresses. J Plant Res 125: 429–438.

46. Weltmeier F, Rahmani F, Ehlert A, Dietrich K, Schutze K, et al. (2009) Expression patterns within the Arabidopsis C/S1 bZIP transcription factor network: availability of heterodimerization partners controls gene expression during stress response and development. Plant Mol Biol 69: 107–119.

47. Chini A, Fonseca S, Chico JM, Fernandez-Calvo P, Solano R (2009) The ZIM domain mediates homo- and heteromeric interactions between Arabidopsis JAZ proteins. Plant Journal 59: 77–87.

48. Siberil Y, Doireau P, Gantet P (2001) Plant bZIP G-box binding factors. Modular structure and activation mechanisms. Eur J Biochem 268: 5655–5666.

49. Melotto M, Mecey C, Niu Y, Chung HS, Katsir L, et al. (2008) A critical role of two positively charged amino acids in the Jas motif of Arabidopsis JAZ proteins in mediating coronatine- and jasmonoyl isoleucine-dependent interactions with the COI1 F-box protein. Plant J 55: 979–988.

50. Withers J, Yao J, Mecey C, Howe GA, Melotto M, et al. (2012) Transcription factor-dependent nuclear localization of a transcriptional repressor in jasmonate hormone signaling. Proc Natl Acad Sci U S A 109: 20148–20153.

51. Seo JS, Koo YJ, Jung C, Yeu SY, Song JT, et al. (2013) Identification of a novel jasmonate-responsive element in the AtJMT promoter and its binding protein for AtJMT repression. PLoS ONE 8: e55482.

52. Robson F, Okamoto H, Patrick E, Harris SR, Wasternack C, et al. (2010) Jasmonate and phytochrome A signaling in Arabidopsis wound and shade responses are integrated through JAZ1 stability. Plant Cell 22: 1143–1160.

53. Grunewald W, Vanholme B, Pauwels L, Plovie E, Inze D, et al. (2009) Expression of the Arabidopsis jasmonate signalling repressor JAZ1/TIFY10A is stimulated by auxin. EMBO Rep 10: 923–928.

54. Liu J, Zhu JK (1997) Proline accumulation and salt-stress-induced gene expression in a salt-hypersensitive mutant of Arabidopsis. Plant Physiol 114: 591–596.

55. Mittler R (2002) Oxidative stress, antioxidants and stress tolerance. Trends Plant Sci 7: 405–410.

56. Chung HS, Koo AJ, Gao X, Jayanty S, Thines B, et al. (2008) Regulation and function of Arabidopsis JASMONATE ZIM-domain genes in response to wounding and herbivory. Plant Physiol 146: 952–964.

57. Jiang S, Yao J, Ma KW, Zhou H, Song J, et al. (2013) Bacterial effector activates jasmonate signaling by directly targeting JAZ transcriptional repressors. PLoS Pathog 9: e1003715.

58. Yan Y, Stolz S, Chetelat A, Reymond P, Pagni M, et al. (2007) A downstream mediator in the growth repression limb of the jasmonate pathway. Plant Cell 19: 2470–2483.

59. Chung HS, Howe GA (2009) A critical role for the TIFY motif in repression of jasmonate signaling by a stabilized splice variant of the JASMONATE ZIM-domain protein JAZ10 in Arabidopsis. Plant Cell 21: 131–145.

60. Alonso JM, Stepanova AN, Leisse TJ, Kim CJ, Chen H, et al. (2003) Genome-wide insertional mutagenesis of Arabidopsis thaliana. Science 301: 653–657.

61. Peever TL, Higgins VJ (1989) Electrolyte leakage, lipoxygenase, and lipid peroxidation induced in tomato leaf tissue by specific and nonspecific elicitors from *Cladosporium fulvum*. Plant Physiol 90: 867–875.

62. de la Fuente JM, Ramirez-Rodriguez V, Cabrera-Ponce JL, Herrera-Estrella L (1997) Aluminum tolerance in transgenic plants by alteration of citrate synthesis. Science 276: 1566–1568.

63. Bates LS, Waldren RP, Teare ID (1973) Rapid determination of free proline for water-stress studies. Plant and Soil 39: 3.

64. Geer BW, Krochko D, Williamson JH (1979) Ontogeny, cell distribution, and the physiological role of NADP-malic enxyme in Drosophila melanogaster. Biochem Genet 17: 867–879.

65. Willems E, Leyns L, Vandesompele J (2008) Standardization of real-time PCR gene expression data from independent biological replicates. Anal Biochem 379: 127–129.

66. Yoo SD, Cho YH, Sheen J (2007) Arabidopsis mesophyll protoplasts: a versatile cell system for transient gene expression analysis. Nat Protoc 2: 1565–1572.

16

Split-Cre Complementation Restores Combination Activity on Transgene Excision in Hair Roots of Transgenic Tobacco

Mengling Wen[1,2], **Yuan Gao**[1,2], **Lijun Wang**[1,2], **Lingyu Ran**[1,2], **Jiahui Li**[1,2], **Keming Luo**[1,2,3]*

1 Key Laboratory of Eco-environments of Three Gorges Reservoir Region, Ministry of Education, Institute of Resources Botany, School of Life Sciences, Southwest University, Chongqing, China, **2** Chongqing Key Laboratory of Transgenic Plant and Safety Control, Southwest University, Chongqing, China, **3** Key Laboratory of Adaptation and Evolution of Plateau Biota, Northwest Institute of Plateau Biology, Chinese Academy of Sciences, Xining, China

Abstract

The Cre/loxP system is increasingly exploited for genetic manipulation of DNA *in vitro* and *in vivo*. It was previously reported that inactive "split-Cre" fragments could restore Cre activity in transgenic mice when overlapping co-expression was controlled by two different promoters. In this study, we analyzed recombination activities of split-Cre proteins, and found that no recombinase activity was detected in the *in vitro* recombination reaction in which only the N-terminal domain (NCre) of split-Cre protein was expressed, whereas recombination activity was obtained when the C-terminal (CCre) or both NCre and CCre fragments were supplied. We have also determined the recombination efficiency of split-Cre proteins which were co-expressed in hair roots of transgenic tobacco. No Cre recombination event was observed in hair roots of transgenic tobacco when the NCre or CCre genes were expressed alone. In contrast, an efficient recombination event was found in transgenic hairy roots co-expressing both inactive split-Cre genes. Moreover, the restored recombination efficiency of split-Cre proteins fused with the nuclear localization sequence (NLS) was higher than that of intact Cre in transgenic lines. Thus, DNA recombination mediated by split-Cre proteins provides an alternative method for spatial and temporal regulation of gene expression in transgenic plants.

Editor: Hong Luo, Clemson University, United States of America

Funding: This work was supported by the National Natural Science Foundation of China (31370672, 31171620), the National Key Project for Research on Transgenic Plant (2011ZX08010-003), 100 Talents Programme of The Chinese Academy of Sciences, the Natural Science Foundation Project of CQ CSTC (CSTC2013JJB8007), the program for New Century Excellent Talents in University (NCET-11-0700), the Fundamental Research Funds for the Central Universities (XDJK2014a005). The funders had no role in study design, data collection and analysis, decision to publish, or preparation of the manuscript.

Competing Interests: The authors have declared that no competing interests exist.

* Email: kemingl@swu.edu.cn

Introduction

The phage P1 Cre recombinase is a member of the tyrosine recombinase family and catalyzes site-specific DNA recombination between tandem 34-bp loxP DNA sequences [1,2]. If two loxP sites are introduced in the same orientation into a genomic locus, Cre-mediated recombination will result in the deletion of the loxP-flanked DNA sequences. The Cre/loxP recombination system is a sophisticated tool for general knockouts, conditional knockouts and reporter strains, and has been widely used in a variety of organisms, including yeasts [3,4], plants [5–9] and animals [1,9–12]. In general, Cre recombinase is expressed under the control of a cell-or tissue-specific promoter to achieve targeted gene knockout in a spatial-temporal fashion [13–15]. However, it is not always facile to find a gene-specific promoter to control expression of the Cre recombinase specifically in a desired cell type.

Active protein can be cleaved into two inactive fragments which can directly re-associate to restore activity [16–18]. Cre recombinase consists of 343 amino acids that form two distinct domains. The N-terminal domain encompasses residues 20–129 and

contains five α-helical segments linked by a series of short loops. The C-terminal domain contains amino acids 132–341 and harbors the active site of the enzyme [19]. Based on its protein structure, the Cre recombinase has previously been split into two complementation polypeptides at different break points such as Asn59/Asn60, Leu104/Arg106 and Gly190/Gly191[16,20,21,22,23], and the recombination activity could be reconstituted *in vivo*. In a previous report, Cre recombinase was divided into two independent polypeptides, a-NH$_2$ terminal with the amino acids 19–59 and b-COOH terminal with the amino acids 60–343 [24]. When two fragments with overlapping amino acid sequences of the Cre gene were co-expressed, recombinase activity was restored even without the addition of dimerization modules [24,25]. Maruo et al. (2008) systematically analyzed the efficiency of Cre complementation by screening multiple dimerization modules in Cos7 cells and primary neurons [26]. To improve the efficiency of split-Cre a-complementation, two inactive fragments were reconstituted by the leucine zipper domain dimerization [20]. However, reassembling split-Cre protein has not yet been reported in higher plants. In this study, we used the a-complementation approach to split Cre and introduce the two inactive fragments into transgenic tobacco (*Nicotiana tabacum* cv. Xanthi). Our experiments revealed that no recombination activity was detected in transgenic tobacco hair roots when individual N- or C-terminal fragments of Cre recombinase gene were expressed. While Cre enzyme activity was able to be restored *in vivo* when co-expressed these polypeptides. Therefore, we provide a new strategy for DNA recombination and gene expression regulation in plants.

Materials and Methods

Plant material and bacterial strains

Nicotiana tabacum cv. Xanthi was grown on Murashige and Skoog medium in a greenhouse under an 18/6h(light/dark)photoperiod at 25°C.

Escherichia coli strain DH5a was used as the recipient for transformation, genetic manipulation and production of plasmid DNA for sequencing. *E. coli* strain BL-21 (DE3) was used for protein expression. The disarmed *Agrobacterium rhizogenes* strain C58C1 was used for tobacco transformation.

Vector construction

The N- (amino acids 1–59) and C-terminal (aa 60–343) moieties of Cre recombinase [20,21] and full-length Cre were amplified by PCR using primers NCre-F, NCre-R, CCre-F and CCre-R (listed in Table 1) with *Eco*RI and *Hin*dIII restriction sites at their 5' ends. The NCre and CCre gene fragments were cloned into the multiple cloning sites of prokaryotic expression vector pMAL-C2X digested with the same enzymes, respectively. Cre gene fragment was cloned into prokaryotic expression vector pET-28a. All the recombinant plasmids were then transformed into host cells *E. coli* BL-21 (DE3).

To construct the plant binary vectors, we synthesized the loxP-nos-loxP fusion sequences by a commercial company (Huada, Shenzhen, China). Sequences were as follows:5'-CGGGATCCG-**CATAACTTCGTATAATGTATGCTATACGAAGTTAT**A-GATCTTCCGTTCAAACATTTGGCAATAAAGTTTCTTAA-*GATTGAATCCTGTTGCCGGTCTTGCGATGATTATCATA-TAATTTCTGTTGAATTACGTTAAGCATGTAATAATTAA-*

*CATGTAATGCATGACGTTATTTATGAGATGGGTTTT-TATGATTAGAGTCCCGCAATTATACATTTAATACGCGA-TAGAAAACAAAATATAGCGCGCAAACTAGGATAAAT-TATCGCGCGCGGTGTCATCTATGTTACTAGATCGGG***A-TAACTTCGTATAATGTATGCTATACGAAGTTAT**G-GATCCCG-3'. The bold letters represent the loxP site (34 bp) and the italic letters represent the NOS terminator sequence (253 bp). The underlined letters show the restriction enzyme sites: *Bam*HI, *Bgl*II and *Bam*HI, respectively. The *Bam*HI-digested loxP-nos-loxP fragment was ligated to the binary vector pCAMBIA1305.1 which was digested with *Bgl*II, producing the vector ploxP. The NCre, CCre and full-length Cre fragments were amplified by PCR and inserted respectively into the vector pCXSN [27], respectively. And then the 35S-NCre-nos, 35S-CCre-nos, 35S-Cre-nos fusion gene segments were excised from the resulting pCXSN vectors with *Eco*RI/*Hin*dIII digestion and then ligated into the corresponding sites of the ploxP vector, producing the vectors pCre, pNCre and pCCre. To add an extra nuclear localization signal (NLS) sequence to N-terminus of Cre, NCre and CCre, the oligos (nNCre-F and nCCre-F) (Table 1) were utilized. PCR fragments were cloned into the ploxP vector to produce pnCre, pnNCre and pnCCre, respectively. To construct the vectors pCCre-nNCre and pnCCre-nNCre, the loxP-nNCre-loxP fusion gene fragment was amplified by PCR and ligated into the pCCre and pnCCre by digesting with *Bgl*II, respectively. All the plant binary vectors were introduced into the *A. rhizogenes* strain C58C1 *via* a simple freeze/thaw transformation method [28].

Expression and purification of proteins

The *E. coli* strain BL-21 (DE3) was transformed with expression vectors containing NCre, CCre and full length Cre proteins. A colony of the transformed cells was cultured in LB medium with ampicillin (100 μg/mL) at 37°C with 180 rpm until $OD_{600} = 0.6$. Protein expression was induced by isopropyl β-D-thiogalactoside (IPTG) at 0.1 mM. Incubation was continued to culture at 25°C with 180 rpm for 4 h before the bacteria were harvested by centrifugation. The cells were resuspended in phosphate buffered saline (PBS) after washing. Clear lysate was obtained after

Table 1. DNA oligo sequences utilizes in this report.

Primer name	Primer sequence (5'-3')	Restriction enzyme site
NCre-F	cggaattcatgtccaatttactgaccgtac	*Eco*R I
NCre-R	cccaagcttctaattcaacttgcaccatgcc	*Hin*d III
CCre-F	cggaattcatgaaccggaaatggtttcccg	*Eco*R I
CCre-R	cccaagcttctaatcgccatcttccagca	*Hin*d III
pX6-NCre-F	gaagatctatgtccaatttactgaccgtac	*Bgl* II
pX6-NCre-R	catgggatccctaattcaacttgcaccatgcc	*Bam*H I
pX6-NLS-NCre-F	gaagatctcccaagaagaagaggaaggtgatg	*Bgl* II
	tccaatttactgaccgtac	
pX6-CCre-F	atgaaccggaaatggtttcccg	/
pX6-CCre-R	ctaatcgccatcttccagca	/
pX6-NLS-CCre-F	atgcccaagaagaagaggaaggtgaaccggaa	/
	atggtttcccg	
pCa-F	gatgacgcacaatcccactatcc	/
pCa-R	gtacagactagttcgtcggttctg	/
F1	cgggatccgaacgtgcaaaacaggctct	/

centrifugation because NCre, CCre and full length Cre proteins are expressed in the soluble fraction. The purity and relative concentrations of these proteins were examined by 12% SDS-PAGE [29,30]. All of the purified proteins were stored at -80°C after adding glycerol with a 1:1 ratio.

In vitro assays of recombination activity

In order to detect the recombination activity of purified proteins, including split-Cre (NCre and CCre) and full-length Cre, the plasmid ploxP-CCre629, in which a 1200-bp DNA fragment was flanked by two loxP recognition sites in the same orientation, was digested at 37°C for 1 h by the purified proteins. The reaction system was as follows: 1 μL 10 × Buffer L (TaKaRa, Dalian, China), 5 μL plasmid (90 ng/μL), 3 μL purified protein and 1 μL ddH$_2$O, total 10 μL. As a control, the plasmid ploxP-CCre629 was also digested with HindIII and BamHI at 37°C for 1 h. The digested product was used for DNA electrophoresis.

Transformation of tobacco plants

A. rhizogenes strain C58C1 with the plant binary vectors was incubated in liquid YEP medium supplemented with 50 mg/L kanamycin and 40 mg/L rifampicin at 28°C and 180 rpm until the cultures reached an optimal density of approximately 0.6–0.8 at OD$_{600}$ [31]. After centrifuged for 10 min at 4,000 rpm and 4°C, the cultures were resuspended with an equal volume of liquid MS medium (MS medium, 100 μmol/L acetosyringone; pH5.8) [32]. Tobacco transformation was performed using the leaf disc method as described previously [33]. After growing on the co-cultivation medium (MS medium, 100 μmol/L acetosyringone, 30 g/L sucrose, 6 g/L agar, pH5.8) in darkness at 25°C for 2 days, the leaf discs were transferred to a selective medium (MS medium, 10 mg/L hygromycin, 150 mg/L rifampicin, 30 g/L sucrose, 6 g/L agar, pH5.8) under a photoperiod of 16:8 (light:dark) h at 25°C.

GUS staining assay

Activity of β-glucuronidase (GUS) in transgenic hair roots was determined by a GUS histochemical staining assay [34]. Transgenic hair roots were placed in 1 mM X-gluc (5-bromo-4-choloro-3-indolyl-b-glucuronic acid) solution and incubated at 37°C overnight and was subsequently recorded photographically.

RNA extraction and reverse-transcriptase PCR (RT-PCR)

Total RNA of hair roots was extracted using TRIzol Reagent (Invitrogen, Beijing, China) according to the manufacturer's instructions. First-strand cDNA was synthesized from 1 μg of total RNA using PrimeScript RT reagent Kit with gDNA Eraser (TaKaRa, Dalian, China). RT-PCR was performed as previously described for genomic PCR using gene-specific primers (Table 1) for different genes. Reaction products were resolved by electrophoresis in 1.5% agarose gel. A pair of specific primers for 18S of N. tabacum [35] were used in a control reaction.

DNA extraction and molecular analysis of transgenic plants

Genomic DNA was extracted from transgenic and untransformed control hair roots using the modified CTAB extraction method as described previously [36]. Putative transgenic hair roots were screened preliminarily to confirm the presence of the transgenes by PCR method [37]. Two gene-specific primers pCa-F and pCa-R (Table 1), which flanked two loxP sites, were designed for detection of transgene excision. PCR was conducted at 94°C for 5 min, followed by 30 cycles of 94°C for 30 s, 56°C for

30 s, 72°C for 1 min, and a final extension step at 72°C for 10 min. The PCR products were loaded on 1% (w/v) agarose gel and visualized after ethidium bromide staining. The PCR fragment was cloned into pMD19 vector (TaKaRa, Dalian, China) and sequenced by Beijing Genomics Institute.

Results

In vitro assays for recombination activity of split-Cre complementation

To establish a split-Cre complementation system, the coding sequence of Cre recombinase was cleaved into two complementation-competent fragments, named NCre (amino acids residues 1-59) and CCre (amino acids residues 60-343) (Fig. 1A), according to previous reports [20,21]. These split-Cre- and Cre-genes were cloned into the expression vector pMAL-C2X and pET-28a (Novagen) and recombinase proteins were produced in reticulocyte lysates. Under the induction of isopropyl β-D-thiogalactoside (IPTG), Split-Cre and full-length Cre proteins were purified to detect the recombination activity (Fig. S1). In vitro excision recombination reactions were conducted using linear fragments from ploxP-CCre629 as substrates (Fig. 1B). The substrates were recombined equally well when full-length Cre or both NCre and CCre were supplied (Fig. 1C). Interestingly, successful recombination was also detected in the reaction when CCre protein was used alone. While there was no related reports stating the recombination activity of CCre protein in vitro, and no recombination activity was detected in vivo in previous studies. In contrast, no recombination activity was found when only NCre was added (Fig. 1C), consistent with a previous in vivo study in the brain of transgenic mice [20,21].

Functional complementation of split-Cre in transgenic tobacco hairy roots

We constructed a series of plant expression vectors for split-Cre complementation system (Fig. 2A). The plant expression vector pCAMBIA1305.1 [38], in which the E. coli gusA gene has been replaced by GUSPlus, served as an empty control. These recombinant plasmids carrying the split-Cre and full-length Cre genes were generated based on the pCAMBIA1305.1 vector. The gene cassette ploxP containing nos terminator sequences flanked by two 34-bp loxP sites in direct orientation, was used as a negative control. pCre and pnCre, containing full-length Cre and NLS-fused Cre driven by the CaMV 35S promoter served as positive controls. The schematic diagrams of all plant binary vectors were showed in Fig. 2A. The gene cassettes pNCre and pnNCre contained NCre and NLS-fused NCre, whereas pCCre and pnCCre contained CCre and NLS-fused CCre, respectively. The gene cassette pCCre-nNCre carried CCre and NLS-fused NCre. In the gene cassette pnCCre-nNCre, a NLS was fused into N terminus of the CCre and NCre genes, respectively.

All of the recombinant plasmids were introduced into tobacco plants by A. rhizogenes-mediated transformation. The hair roots of N. tabacum transformants with hygromycin resistance were subjected to GUS staining assay. To characterize the excision efficiency of each recombination event, we used the GUS-positive ratio to calculate the excision efficiency. Table 2 showed the total number of transgenic events analyzed for each gene cassette and the number of GUS-positive roots. As showed in Fig. 2B, no GUS activity was observed in transgenic lines harboring ploxP, pNCre, pnNCre, pCCre and pnCCre, indicating that each half (NCre and CCre) of split Cre alone, even fused with an extra NLS, did not have any recombinase activity in vivo. In contrast, all transgenic hair roots containing pCCre-nNCre and pnCCre-nNCre dis-

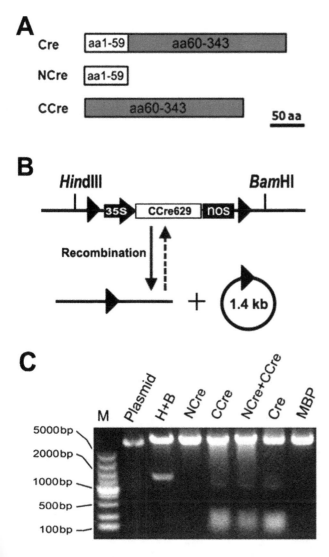

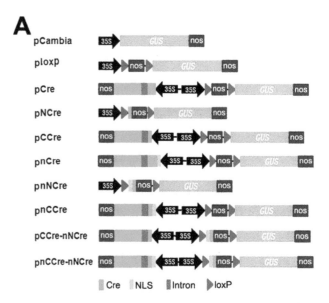

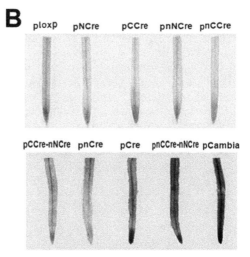

Figure 1. Digram of the split-Cre model and *in vitro* recombination of Split-Cre protein. A: Digram of the split-Cre model. The intact Cre was designed to be split at the 60th amino acid residue. Two molecules of split-Cre were named NCre and CCre respectively. **B: Structure of the substrate catalyzed by purified protein. C: Recombination assay of Split-Cre protein *in vitro*.** M: DL5000 Marker; Plasmid: 2μl plasmid (90 ng/μl) of pLoxp-lc-CCre629. The plasmid was respectively digested by *Hind*III and *Bam*HI (H+B), split protein NCre (NCre), split protein CCre (CCre), combination of split protein NCre and CCre (NCre + CCre), intact protein Cre (Cre) and MBP. Plasmid and MBP were used as negative control, H+B digestions were used as positive control. MBP tag was used to purify fusion proteins.

Figure 2. The *in vivo* recombination of split-Cre protein and the deletions determined by GUS activity. A: Digram of plant expression vectors. pCambia refer to vector of pCambia1305.1. **B: GUS staining of transgenic hair roots for each transformant.** "n" represents nuclear localization signal. The following are all the same.

played blue (Fig. 2B), indicating that recombination activity of Cre is present in these transgenic plants. Transgenic lines harboring the binary vectors pCAMBIA1305.1, pCre and pnCre also showed GUS activity as expected (Fig. 2B). The results demonstrated that recombination activity of intact Cre protein could be reconstituted *in vivo* when both N- and C-terminal fragments of Cre recombinase were co-expressed, whereas no recombination activity was observed when either NCre or CCre was expressed alone.

To determine whether the split-Cre genes were indeed expressed in the hairy roots of transgenic tobacco, we used gene-specific primers to perform RT-PCR analysis. The expected

DNA fragments of split-Cre recombinase were detected in these tested transformants containing pnNCre and pnCCre vectors (Fig. 3A). The *18S* rRNA complementary primers were used as an internal control. No transcripts were found in wild-type plants. Two specific PCR-amplified products for NCre and CCre were obtained in transgenic lines harboring pCCre-nNCre and pnCCre-nNCre (Fig. 3B), indicating that all the split-Cre genes transformed into transgenic tobacco plants were constitutively expressed, at least on the transcriptional level, resulting in the successful deletion of the transgene fragments flanked by two loxp sites.

Molecular characterization of site-specific DNA excision in transgenic hairy roots

The transgene excision from hairy roots of transgenic tobacco was confirmed by PCR analysis. The genomic DNA samples extracted from different transgenic lines were used as templates for

Table 2. GUS positive ratio of different transgenic tobacco hair roots.

Vectors	GUS (+) No. of roots (Blue)	GUS (-) No. Of roots (White)	Total	GUS positive ratio (%)
pCAMBIA1305.1	121	0	121	100
pCA-Cre	64	66	130	49.2
pCA-nCre	77	52	129	59.7
pCA-CCre-nNCre	63	74	137	46.0
pCA-nCCre-nNCre	90	44	134	67.2
pCA-LoxP	0	49	49	0
pCA-NCre	0	53	53	0
pCA-nNCre	0	46	46	0
pCA-CCre	0	57	57	0
pCA-nCCre	0	61	61	0

detecting the excision events. Transgenic lines carrying pCre, pnCre, pCCre-nNCre and pnCCre-nNCre vectors showed visible post-excision signals (369 bp amplification fragments) (Fig. 4B), compared to the pre-excision signals in transgenic lines pCre and pnCre (664 bp amplification fragments) and in transgenic lines pCCre-nNCre and pnCCre-nNCre (862 bp amplification fragments), respectively. No excision events were observed in transgenic lines harboring pNCre, pnNCre, pCCre, pnCCre and ploxP vectors (Fig. 4A).

Furthermore, DNA sequencing analysis revealed that the 369-bp amplification products consisted of a single loxP site and the junction T-DNA sequences located outside two loxP repeats (Fig. 4C). This result further confirmed that excision events did occur in transgenic lines pCCre-nNCre and pnCCre-nNCre, thus demonstrating that split-Cre fragments can rebuild recombination activity *in vivo* when co-expressed in plants.

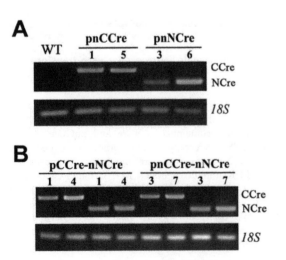

Figure 3. Analysis of CCre and NCre transcription in transgenic hairy roots. Semi-quantitative RT-PCR analysis of the transcription level of CCre or NCre in transformants carrying pnCCre or pnNCre (A) and pCCre-nNCre or pnCCre-nNCre (B). Tobacco 18S was used as an internal control. Total RNA was isolated from roots. Numbers represent the different lines of each recombinant.

Determination of excision efficiency in the transgenic events

To examine the excision efficiency of each recombinant, we analyzed the GUS positive ratio in transgenic hair roots. As shown in Fig. 5, transgenic plants containing pCAMBIA1305.1 showed strong GUS activity, whereas for transgenic lines hosting ploxP (no recombinase gene included), or pNCre, pnNCre, pCCre and pnCCre, in which each half (NCre and CCre) of split Cre was contained alone, we observed a negative GUS staining as expected. In contrast, transgenic plants hosting pCCre-nNCre (both the NCre and CCre genes expressed simultaneously) had average GUS-positive ratio of 46.0% based on a total of 137 independent transgenic events. The similar GUS-positive ratio (49.2%) was obtained in transgenic plants hosting pCre (containing the intact Cre gene). Furthermore, we found that a higher GUS positive ratio was generated in transgenic plants harboring pnCre (59.7%) and pnCCre-nNCre (67.2%) (Fig. 5), in which the nuclear localization signal (NLS) of the simian virus 40 large T antigen (SV40) was fused at the amino terminus of Cre recombinase, indicating that the NLS sequence can improve the localization of Cre recombinase to the nucleus, resulting in increasing excision efficiency by building the Cre cassette. The results demonstrated that co-expression of NCre and CCre leads to the efficient reconstitution of Cre recombinase from two inactive precursor fragments in transgenic plants.

Discussion

The Cre/loxP recombination system has been intensively used in genetic analysis of animals and higher plants [1,5-11]. One main challenge for this system is to control the expression of Cre gene in spatially and temporally desirable manners. To regulate Cre activity, in general, its expression is under the control of a cell-type specific promoter [7,8]. However, the expression pattern of a single promoter activity is often insufficient to achieve accurate results. To overcome this limitation, split-Cre systems based on the structure of Cre recombinase have been reported previously [19,22]. In these systems, Cre protein was generally cleaved into two complementation-competent fragments at the breakpoints in the N-terminal domain and each of these split-Cre proteins expressed alone had no enzymatic activity. But the inactive Cre moieties readily reconstituted into a functional enzyme with recombination activity when co-expressed in transgenic animals [20,21]. In this study, the split-Cre proteins were reassembled in transgenic plants when co-expressed (Fig. 2 and 4) and the

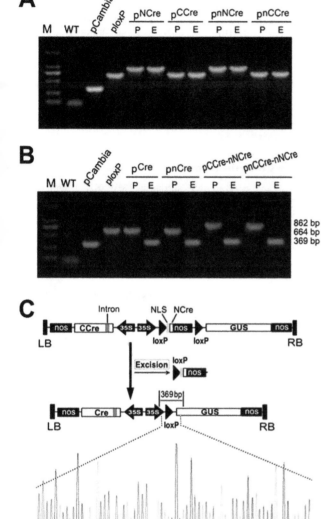

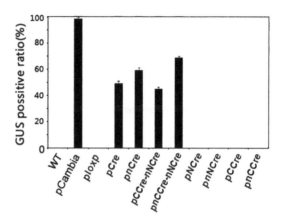

Figure 5. Excision ratio in the transgenic hairy roots determined by GUS staining. The ratio of GUS-positive roots was used to calculate the excision efficiencies for each transgenic line. pCAMBIA1305.1 was used as a positive control, while ploxP as a negative one. WT also used as a negative control here. All data is presented as mean of three replicates with error bars indicating ± SD.

Figure 4. The recombinant splite-Cre excises DNA fragment between two Loxp sties *in vivo*. Vallidation of the non-excision (A) and excision (B) of DNA fragment in hairy roots. The amplified fragments of non-excision was 862 bp and 664bp for pNCre, pnNCre and pCCre, pnCCre, respectively. The amplified fragments of post-excision was 369 bp, while the pre-excision was 862 bp and 664 bp for pCCre-nNCre, pnCCre-nNCre and pCre, pnCre, respectively. M: DL2000 Marker; P, pre-excision signal; E, post-excision signal. pCambia and ploxP were used as control. **C: Schematic illustration of deletion in pCCre-nNCre and the sequencing result after deletion of DNA fragment.**

recombination efficiency was comparable to that of intact Cre recombinase (Fig. 5). However, no Cre recombination activity was detected when either NCre or CCre gene was expressed alone (Fig. 5).

The Cre/loxP system contains two elements: the Cre recombinase and two consensus sequences (loxP sites) [39]. Previous

studies have reported that the C-terminal domain of the Cre recombinase harbors the active site, consisting of the conserved catalytic residues Arg173, His289, Arg292 and Trp315 [19,40]. Furthermore, using chimeras of the Flp and Cre recombinases, Shaikh and Sadowski (2000) [41] demonstrated that the C-terminal domain of the Cre recombinase determined their mode of cleavage. In the present study, *in vitro* assays showed that NCre protein (aa 1-59) used alone was unable to catalyze DNA recombination (Fig. 1). In contrast, surprisingly, site-specific recombination events were observed when only CCre protein (the amino acid 60–343) was expressed. Previous work has shown that although a C-terminal peptide of Cre recombinase with 25 kDa still binds the loxP sites, but it is not able to catalyze the site-specific recombination [42]. Therefore, we speculated that, when lacking the small N-terminal of Cre recombinase, the presence of C-terminal domain of Cre proteins with more than 25 kDa did not affect the *in vitro* enzymatic activity of recombination.

The Cre recombinase catalyzed the site-specific recombination at two loxP sites which were located in the genomes, therefore, it has to be imported into the nucleus. Since the 38 kDa Cre protein is smaller than the ~50 kDa upper size limit imposed by the nuclear pore on passive diffusion of macromolecules into the nucleus [43], it is hypothesized that Cre proteins enter into the nucleus by passive diffusion through the nuclear pore [44]. However, it has been reported that even small nuclear proteins of eukaryotes are more easy to gain entry to the nucleus when carrying specific nuclear localization signals (NLSs) [45]. Previous studies have demonstrated that fusing NLS sequences can effectively increase Cre recombinase activity [18,46]. In this study, we determined the effect of NLS on the recombination activity in the split-Cre proteins. An NLS sequence from the SV40 virus was fused into the N-terminals of the NCre and CCre genes (Fig. 2). The NLS-NCre and NLS-CCre had no recombination activity in transgenic plants (Fig. 5). However, the *in vivo* recombination of the NLS-NCre and NLS-CCre proteins had higher activity compared to the wild-type Cre.

Concluding Remarks

This study provides an alternative strategy for regulation of gene expression by site-specific recombination using the split-Cre recombinase complementation approach in plants. This system has wide application prospects in plant functional genomics and genetic engineering. In general, most plant genes are expressed in different tissues and developmental stages. The split-Cre recombinase system allows spatial and temporal regulation of recombination through cell-specific gene genetic targeting by the simultaneous activity of two promoters in plants. In addition, a potential application of the split-Cre recombinase system is to control transgenes (i.e. selectable markers and novel trait genes) activation or removal them from transgenic plants, producing trait- or marker transgene-free transgenic crops.

Supporting Information

Figure S1　Prokaryotic expression and purification of split- and full-length Cre protein. M, Protein marker. Lane

1:Induced NCre; Lane 2: Non-induced NCre; Lane 3: Induced CCre; Lane 4: Non-induced CCre; Lanes 5-6: Induced and Non-induced MBP protein as control; Lanes 7-9: Purified protein of NCre, CCre and MBP; Lanes 10–11: Induced and Non-induced Cre; Lane 12: Purified Cre.

Acknowledgments

The authors thank Professor Guoliang Wang (Hunan Agricultural University, Changsha 410128, China) for providing the plant binary vector pCXSN.

Author Contributions

Conceived and designed the experiments: MW KL. Performed the experiments: MW LW LR YG JL. Analyzed the data: MW KL. Contributed reagents/materials/analysis tools: LR YG. Wrote the paper: MW KL.

References

1. Sauer B, Henderson N (1988) Site-Specific DNA Recombination in Mammalian-Cells by the Cre Recombinase of Bacteriophage-P1. Proceedings of the National Academy of Sciences of the United States of America 85: 5166–5170.
2. Sternberg N (1981) Bacteriophage-P1 Site-Specific Recombination.3. Strand Exchange during Recombination at Lox Sites. Journal of Molecular Biology 150: 603–608.
3. Delneri D, Tomlin GC, Wixon JL, Hutter A, Sefton M, et al. (2000) Exploring redundancy in the yeast genome: an improved strategy for use of the cre-loxP system. Gene 252: 127–135.
4. Gueldener U, Heinisch J, Koehler GJ, Voss D, Hegemann JH (2002) A second set of loxP marker cassettes for Cre-mediated multiple gene knockouts in budding yeast. Nucleic Acids Research 30.
5. Srivastava V, Ow DW (2003) Rare instances of Cre-mediated deletion product maintained in transgenic wheat. Plant Molecular Biology 52: 661–668.
6. Sreekala C, Wu L, Gu K, Wang D, Tian D, et al. (2005) Excision of a selectable marker in transgenic rice (Oryza sativa L.) using a chemically regulated Cre/loxP system. Plant Cell Reports 24: 86–94.
7. Luo KM, Duan H, Zhao DG, Zheng XL, Deng W, et al. (2007) 'GM-gene-deletor': fused loxP-FRT recognition sequences dramatically improve the efficiency of FLP or CRE recombinase on transgene excision from pollen and seed of tobacco plants. Plant Biotechnology Journal 5: 263–274.
8. Bai XQ, Wang QY, Chu CC (2008) Excision of a selective marker in transgenic rice using a novel Cre/loxP system controlled by a floral specific promoter. Transgenic Research 17: 1317–1326.
9. Schwenk F, Baron U, Rajewsky K (1995) A cre-transgenic mouse strain for the ubiquitous deletion of loxP-flanked gene segments including deletion in germ cells. Nucleic Acids Research 23: 5080–5081.
10. Hayashi S, McMahon AP (2002) Efficient recombination in diverse tissues by a tamoxifen-inducible form of Cre: A tool for temporally regulated gene activation/inactivation in the mouse. Developmental Biology 244: 305–318.
11. Pan XF, Wan HY, Chia W, Tong Y, Gong ZY (2005) Demonstration of site-directed recombination in transgenic zebrafish using the Cre/loxP system. Transgenic Research 14: 217–223.
12. Miller AJ, Dudley SD, Tsao JL, Shibata D, Liskay RM (2008) Tractable Cre-lox system for stochastic alteration of genes in mice. Nature Methods 5: 227–229.
13. Mantamadiotis T, Lemberger T, Bleckmann SC, Kern H, Kretz O, et al. (2002) Disruption of CREB function in brain leads to neurodegeneration. Nature Genetics 31: 1035–1043.
14. Kellendonk C, Opherk C, Anlag K, Schutz G, Tronche F (2000) Hepatocyte-specific expression of Cre recombinase. Genesis 26: 151–153.
15. Tsien JZ, Chen DF, Gerber D, Tom C, Mercer EH, et al. (1996) Subregion- and cell type-restricted gene knockout in mouse brain. Cell 87: 1317–1326.
16. Han XZ, Han FY, Ren XS, Si J, Li CQ, et al. (2013) Ssp DnaE split-intein mediated split-Cre reconstitution in tobacco. Plant Cell Tissue and Organ Culture 113: 529–542.
17. Gu W, Schneider JW, Condorelli G, Kaushal S, Mahdavi V, et al. (1993) Interaction of Myogenic Factors and the Retinoblastoma Protein Mediates Muscle-Cell Commitment and Differentiation. Cell 72: 309–324.
18. Kellendonk C, Tronche F, Monaghan AP, Angrand PO, Stewart F, et al. (1996) Regulation of cre recombinase activity by the synthetic steroid RU 486. Nucleic Acids Research 24: 1404–1411.
19. Guo F, Gopaul DN, VanDuyne GD (1997) Structure of Cre recombinase complexed with DNA in a site-specific recombination synapse. Nature 389: 40–46.

20. Hirrlinger J, Scheller A, Hirrlinger PG, Kellert B, Tang WN, et al. (2009) Split-Cre Complementation Indicates Coincident Activity of Different Genes In Vivo. Plos One 4.
21. Hirrlinger J, Requardt RP, Winkler U, Wilhelm F, Schulze C, et al. (2009) Split-CreERT2: temporal control of DNA recombination mediated by split-Cre protein fragment complementation. Plos One 4: e8354.
22. Jullien N, Sampieri F, Enjalbert A, Herman JP (2003) Regulation of Cre recombinase by ligand-induced complementation of inactive fragments. Nucleic Acids Research 31.
23. Xu YW, Xu G, Liu BD, Gu GQ (2007) Cre reconstitution allows for DNA recombination selectively in dual-marker-expressing cells in transgenic mice. Nucleic Acids Research 35.
24. Casanova E, Lemberger T, Fehsenfeld S, Mantamadiotis T, Schutz G (2003) alpha complementation in the Cre recombinase enzyme. Genesis 37: 25–29.
25. Seidi A, Mie M, Kobatake E (2007) Novel recombination system using Cre recombinase alpha complementation. Biotechnology Letters 29: 1315–1322.
26. Maruo T, Ebihara T, Satou E, Kondo S, Okabe S (2008) Cre complementation with variable dimerizers for inducible expression in neurons. Neuroscience Research 61: S279–S279.
27. Chen SB, Songkumarn P, Liu JL, Wang GL (2009) A Versatile Zero Background T-Vector System for Gene Cloning and Functional Genomics. Plant Physiology 150: 1111–1121.
28. Chen H, Nelson RS, Sherwood JL (1994) Enhanced Recovery of Transformants of Agrobacterium-Tumefaciens after Freeze-Thaw Transformation and Drug Selection. Biotechniques16: 664–&.
29. Wu Y, Zhou Y, Song J, Hu X, Ding Y, et al. (2008) Using green and red fluorescent proteins to teach protein expression, purification, and crystallization. Biochemistry and Molecular Biology Education 36: 43–54.
30. Roodveldt C, Tawfik DS (2005) Directed evolution of phosphotriesterase from Pseudomonas diminuta for heterologous expression in Escherichia coli results in stabilization of the metal-free state. Protein Engineering Design & Selection 18: 51–58.
31. Chen C, Dale MC, Okos MR (1990) Minimal Nutritional-Requirements for Immobilized Yeast. Biotechnology and Bioengineering 36: 993–1001.
32. Murashige T, Skoog F (1962) A Revised Medium for Rapid Growth and Bio Assays with Tobacco Tissue Cultures. Physiologia Plantarum 15: 473–497.
33. Fraley RT, Rogers SG, Horsch RB (1986) Genetic-Transformation in Higher-Plants. Crc Critical Reviews in Plant Sciences 4: 1–46.
34. Jefferson RA, Kavanagh TA, Bevan MW (1987) Beta-Glucuronidase (Gus) as a Sensitive and Versatile Gene Fusion Marker in Plants. Journal of Cellular Biochemistry: 57–57.
35. Kenton A, Khashoggi A, Parokonny A, Bennett MD, Lichtenstein C (1995) Chromosomal Location of Endogenous Geminivirus-Related DNA-Sequences in Nicotiana-Tabacum-L. Chromosome Research 3: 346–350.
36. Porebski S, Bailey LG, Baum BR (1997) Modification of a CTAB DNA extraction protocol for plants containing high polysaccharide and polyphenol components. Plant Molecular Biology Reporter 15: 8–15.
37. Luo KM, Zheng XL, Chen YQ, Xiao YH, Zhao DG, et al. (2006) The maize Knotted1 gene is an effective positive selectable marker gene for Agrobacterium-mediated tobacco transformation. Plant Cell Reports 25: 403–409.
38. Yang SM, Tang F, Gao MQ, Krishnan HB, Zhu HY (2010) R gene-controlled host specificity in the legume-rhizobia symbiosis. Proceedings of the National Academy of Sciences of the United States of America 107: 18735–18740.

39. Abremski K, Hoess R, Sternberg N (1983) Studies on the Properties of P1 Site-Specific Recombination - Evidence for Topologically Unlinked Products Following Recombination. Cell 32: 1301–1311.

40. Van Duyne GD (2001) A structural view of Cre-loxP site-specific recombination. Annual Review of Biophysics and Biomolecular Structure 30: 87–104.

41. Shaikh AC, Sadowski PD (2000) Chimeras of the Flp and Cre recombinases: Tests of the mode of cleavage by Flp and Cre. Journal of Molecular Biology 302: 27–48.

42. Hoess R, Abremski K, Irwin S, Kendall M, Mack A (1990) DNA Specificity of the Cre Recombinase Resides in the 25 Kda Carboxyl Domain of the Protein. Journal of Molecular Biology 216: 873–882.

43. Peters R (1983) Nuclear-Envelope Permeability Measured by Fluorescence Microphotolysis of Single Liver-Cell Nuclei. Journal of Biological Chemistry 258: 1427–1429.

44. Sauer B (1987) Functional Expression of the Cre-Lox Site-Specific Recombination System in the Yeast Saccharomyces-Cerevisiae. Molecular and Cellular Biology 7: 2087–2096.

45. Dingwall C, Laskey RA (1991) Nuclear Targeting Sequences - a Consensus. Trends in Biochemical Sciences 16: 478–481.

46. Gu H, Zou YR, Rajewsky K (2013) Independent Control of Immunoglobulin Switch Recombination at Individual Switch Regions Evidenced through Cre-loxP-Mediated Gene Targeting. Journal of Immunology 191: 7–16.

17

Structural Determinants of *Arabidopsis thaliana* Hyponastic Leaves 1 Function *In Vivo*

Paula Burdisso[1,2], Fernando Milia[1], Arnaldo L. Schapire[1,3], Nicolás G. Bologna[1,4], Javier F. Palatnik[1], Rodolfo M. Rasia[1,2]*

1 Instituto de Biología Molecular y Celular de Rosario, Rosario, Argentina, **2** Área Biofísica, Facultad de Ciencias Bioquímicas y Farmacéuticas, Universidad Nacional de Rosario, Rosario, Argentina, **3** Center for Research in Agricultural Genomics CRAG (CSIC-IRTA-UAB-UB), Edifici CRAG-Campus UAB, Bellaterra (Cerdanyola del Vallés), Barcelona, Spain, **4** Swiss Federal Institute of Technology (ETH), Zurich, Switzerland

Abstract

MicroRNAs have turned out to be important regulators of gene expression. These molecules originate from longer transcripts that are processed by ribonuclease III (RNAse III) enzymes. Dicer proteins are essential RNAse III enzymes that are involved in the generation of microRNAs (miRNAs) and other small RNAs. The correct function of Dicer relies on the participation of accessory dsRNA binding proteins, the exact function of which is not well-understood so far. In plants, the double stranded RNA binding protein Hyponastic Leaves 1 (HYL1) helps Dicer Like protein (DCL1) to achieve an efficient and precise excision of the miRNAs from their primary precursors. Here we dissected the regions of HYL1 that are essential for its function in *Arabidopsis thaliana* plant model. We generated mutant forms of the protein that retain their structure but affect its RNA-binding properties. The mutant versions of HYL1 were studied both *in vitro* and *in vivo*, and we were able to identify essential aminoacids/residues for its activity. Remarkably, mutation and even ablation of one of the purportedly main RNA binding determinants does not give rise to any major disturbances in the function of the protein. We studied the function of the mutant forms *in vivo*, establishing a direct correlation between affinity for the pri-miRNA precursors and protein activity.

Editor: Hector Candela, Universidad Miguel Hernández de Elche, Spain

Funding: This work was supported by Grants PICT 2007-720 and 2008-318 from ANPCyT to RMR. PB and ALS are recipients of a fellowship from CONICET. JFP and RMR are members of CONICET. The funders had no role in study design, data collection and analysis, decision to publish, or preparation of the manuscript.

Competing Interests: The authors have declared that no competing interests exist.

* Email: rasia@ibr-conicet.gov.ar

Introduction

MicroRNAs are a class of post-transcriptional regulators that negatively regulate the expression of mRNAs through complementary base pairing. They originate in endogenous transcipts that fold into hairpin structures (pri-miRNA). The miRNAs are located in stem-loop structures within the pri-miRNA and are released through the action of RNAse III type enzymes. The resulting ≈21 nt mature molecules are subsequently incorporated into the effector RNA-induced silencing complex (RISC), guiding the complex to target mRNAs through base pair complementarity, resulting in translation inhibition or mRNA degradation [1].

The biogenesis of miRNA in animals proceeds in two steps separated both in time and location. The nuclear RNAse Drosha performs a first cut in the pri-miRNA, releasing the stem loop (pre-miRNA). This RNA molecule is exported to the cytoplasm where Dicer (Dcr) excises the first 22 nt of the pre-miRNA, and this final miRNA molecule is subsequently transferred to the RISC complex. MicroRNA processing in plants differs substantially, as a single RNAse III enzyme, DCL1, produces both of the staggered cuts necessary to release the miRNA from the primary miRNA transcript in the nucleus [2].

A common feature to all the RNA interference (RNAi) processing machineries characterized so far is the participation of accessory RNA binding proteins in the dicing reaction. In animals, Drosha is helped by the double stranded RNA Binding Domain (dsRBD) containing protein DGCR8 (DiGeorge syndrome critical region gene 8), whereas Dicer processing of pre-miRNA requires the presence of TRBP (human immunodeficiency virus transactivating response RNA-binding protein). In *Drosophila melanogaster* while pri-miRNAs are processed by Drosha/PASHA (DGCR8 homologous) in the nucleus and Dcr-1/Loquacious (Loqs) in the cytoplasm, the exogenous double stranded RNA (dsRNA) are generated by Dcr2/Loqs-R2D2. Loqs is required by Dcr-1 for efficient processing of certain classes of pre-miRNAs into mature miRNAs, but it is dispensable for miRNA RISC loading [3,4]. In contrast, Dcr-2 is able to process dsRNA templates in the absence of R2D2, but requires this double stranded RNA Binding Protein to form the Risc Loading Complex (RLC), which thermodynamically orientates siRNA duplexes onto Ago2 for passenger strand cleavage and active siRNA loaded RISC formation [5,6]. Dcr-2 also requires the help of the dsRBD containing protein R2D2 for the generation of small interference RNA (siRNA). But the exact molecular role of these dsRBD containing proteins in RNA processing has not yet been established.

Several accessory proteins that participate in the plant miRNA processing complex have been identified, most prominently

SERRATE (SE) and HYL1 [7,8]. The protein HYL1 has been shown to be essential for accurate digestion of the miRNA precursors both *in vivo* and *in vitro* [9,10]. It was also suggested to participate in miRNA strand selection [11]. The sequence of HYL1 contains two dsRBDs in its N-terminus (residues 1–170) followed by a long, presumably unstructured, C-terminal region (residues 171–419) containing six repeats of a 28 aminoacid sequence. The two dsRBDs of the protein are sufficient for its activity in miRNA processing [12].

HYL1 and other helper proteins have similar domain architectures, consisting of two or three dsRBDs organized in tandem. However, the RNA binding properties of these helper proteins is variable. Although R2D2 contains two dsRBDs it does not bind siRNA alone, but requires the presence of Dcr-2 [5,13,14]. In TRBP, the first two domains interact with precursor RNA, whereas the third one does not [15–17]. In HYL1 the first domain is the one that dominates RNA binding, with the second one interacting only weakly [18,19]. Substrate binding is supposed to be an essential part of the helper protein function, however it has not been established whether it is an absolute requirement for the participation of these proteins in miRNA biogenesis, and what the consequences of altering the substrate binding affinity of the proteins in the processing mechanism *in vivo* may be. In order to clarify this issue in the present work we sought to identify the RNA binding determinants of HYL1 and to assess their importance on the function of the protein *in vivo*.

Materials and Methods

Plant material and growth conditions

Hyl1-2 (SALK_064863) plants of *Arabidopsis thaliana* (Col-0 ecotype), used for all experiments, were obtained from the Arabidopsis Biological Resource Center (ABRC). Plants were grown on MS medium with 50 µg/ml kanamycin and transplanted to soil at 23°C under long days (16 h light/8 h dark) in a growth room.

Construction of mutant protein vectors

The cDNA sequence of the HYL1 gene was originally isolated from a mixed cDNA library of *Arabidopsis* by PCR and cloned in pBluescript (pBS) plasmid. A BamHI site at the 5′ end and a SalI site at the 3′ end were introduced for the construction of plant binary vectors. Overlapping PCR, using the cDNA of HYL1 as template, was used to make the five mutants versions of HYL1 (K17A/R19A, K38A, H43A/K44A, Δ40–46 and R67A/K68A. The primers used for cloning and mutagenesis are shown in Text S1. Mutations were verified by DNA sequencing, and the wild-type and mutants versions of HYL1 were introduced into the binary vector CHF5 under the control of the cauliflower mosaic virus 35S.

All binary constructs were transformed into *Agrobacterium tumefaciens*. Genetic transformation of homozygous *hyl1-2* mutant plants was performed using the floral dip method [20]. For selection of transgenic plants, seeds were grown on soil supplemented with 0.2 g/l BASTA at 23°C under long days (16 h light/8 h dark) in a growth room. To classify the phenotype of T1 lines, the size and shape of the rosette leaves were examined. We analyzed 30 to 50 T1 plants of each group.

Expression analysis

Inflorescence RNA was extracted using TRIzol reagent (Invitrogen) and 0.5 µg of total RNA was treated with RQ1 RNase-free DNAse (Promega). Next, first-strand cDNA synthesis was carried out using SuperScriptTM III Reverse Transcriptase

(Invitrogen) with the appropriate primers. PCR reactions were performed in an Mx3000P QPCR System (Stratagene) using SYBR Green I (Roche) to monitor double-stranded (ds) DNA synthesis. Quantitative (q) PCR of each gene was carried out for at least three biological replicates, with technical duplicates for each biological replicate. MiR164a, miR172a and miR396a levels were concurrently determined in each sample by stem-loop RT-qPCR [21]. The relative transcript level was determined for each sample, normalized using PROTEIN PHOSPHATASE 2A cDNA level [22]. Primer sequences are detailed in Text S1.

Protein expression and purification

Fragments corresponding to the first dsRBD of HYL1 protein (HYL1-1) and their R1, R2 and R3 mutants were amplified by PCR from the binary vectors for plant transformation using the primers shown in Text S1, cloned into the pET-TEV expression vector and sequenced [23]. The plasmids were transformed in *E. coli* BL21 (DE3) cells, which were then grown at 37°C in M9 minimal medium supplemented with 1 g/l ^{15}N-NH$_4$Cl (Cambridge Isotope Laboratories) in the case of NMR experiments and in LB medium for the stability and affinity experiments.

Protein expression was induced with 1 mM IPTG (isopropyl-β-D-thiogalactopyranoside) at OD$_{600}$ ≈0.7 and cells were grown for 4–5 hours. Cells were harvested, resuspended in a buffer containing 100 mM phosphate, 10 mM Tris, 5 mM β-mercaptoethanol, 8 M Urea pH 8 and disrupted by sonication. The denatured proteins were purified using a Ni (II) column and refolded by dyalisis in 100 volumes of 100 mM phosphate, 50 mM NaCl, 5 mM β-mercaptoethanol, 50 mM glutamate, 50 mM arginine. The refolded proteins were then digested using 1:100 mass ratio of His-tagged TEV protease, to remove the His-tag, and the protease was removed by a further passage through a Ni (II) column.

RNA synthesis

RNA samples were produced by *in vitro* transcription with T7 RNA polymerase, using annealed oligonucleotides. Briefly, a mix was prepared containing 1X transcription buffer [40 mM Tris (pH 8), 5 mM DTT, 1 mM spermidine, 0.01% Triton X-100, and 80 mg/mL PEG 8000], each rNTP at 4 mM (rA, rC, rG, and rU), 20 mM MgCl2, 40 µg/mL BSA, and 1 unit of pyrophosphatase, and the annealed template at 35 µg/mL. The reaction was started by addition of T7 RNA polymerase and allowed to proceed for 3 h at 37°C. Then, 50 units of RNase-free DNase were added, and the mix was further incubated for 30 min at 37°C. The reaction mixture was then diluted 8-fold in 20 mM Tris, 10 mM EDTA, and 8 M urea (pH 8.0) and loaded on a Q-Sepharose column equilibrated with the same buffer. The column was eluted with a gradient from 0 to 1 M NaCl in the same buffer. Fractions containing RNA, as determined by A$_{260}$, were checked via denaturing 5% polyacrylamide gel electrophoresis. The fractions with the desired transcript were pooled, dialyzed three times against 200 volumes of H$_2$O, and lyophilized for storage before being used.

Fluorescence anisotropy titrations

For fluorescence anisotropy titrations, RNA fragments were labeled with fluorescein using the 5′ EndTag Nucleic Acid End Labeling System and fluorescein maleimide-thiol reactive label from Vector Laboratories. Labeled fragments were purified by phenol extraction, precipitated with ethanol, and resuspended on 10 mM phosphate buffer (pH 7.0). The double stranded RNA binding partner used is presented in Figure S1. The fluorescence anisotropy was measured on a Varian Cary Eclipse spectrofluo-

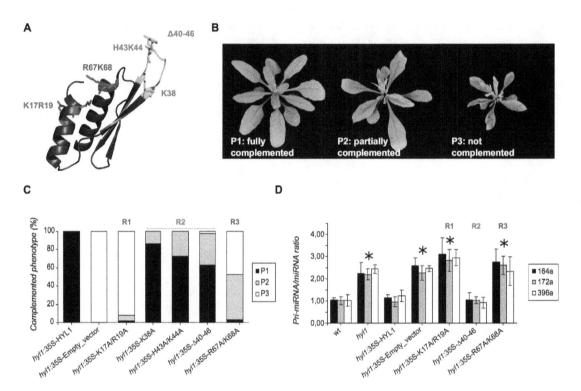

Figure 1. In vivo analysis of HYL1 dsRBD1 mutants. *A. Structure of HYL1-dsRBD1.* The residues that were mutated in the present study are highlighted in colours: region 1, red, region 2, green, region 3, blue. *B. T1 lines phenotypes.* Different phenotypes were found in T1 lines obtained after *hyl1* plants transformation. The different lines included the complete sequence for HYL1 protein with the mutations in the first dsRBD mentioned in Figure 1A. Within each set of transgenic lines, the plants were classified as fully complemented (P1), partially complemented (P2) or not complemented (P3) as function of their rosettes phenotypes. *C. Phenotype analysis of transgenic plants.* The plants were clustered according to each phenotype, as shown in B. The coloured bars represent the fraction of the T1 plants classified as P1 (black), P2 (gray) and P3 (white). *D. Functional analysis of complemented plants.* Relative transcript levels of pri-miRNA164a, pri-miRNA172a and pri-miRNA396a as well as their respective mature miRNAs were determined for each genotype, normalized using *PROTEIN PHOSPHATASE 2A* (*PP2A*, AT1G13320) and compared to WT. The graph shows the ratio of pri-miR164a/miR164a (black), pri-miR172a/miRNA172a (gray) and pri-miRNA396a/miRNA396a (white) levels for each genotype. The plants overexpressing HYL1 with the first dsRBD mutated in regions 1 and 3 (R1 and R3) presented less pri-miRNA processing efficiency compared to plants overexpressing HYL1 with the Δ40–46 deletion in region 2 (R2). Data shown are mean ± SEM of 3 biological replicates. Asterisks indicate significant differences between genotypes, as determined by ANOVA (*P<0.001).

rimeter exciting the sample at 492 nm and measuring emission at 520 nm. Anisotropy values were obtained from the average of three measurements with an integration time of 20 s. The excitation and emission slits were set to 10 nm. Labeled RNA was annealed by heating at 100°C for 5 min and chilling at 0°C in an ice-water bath. Data points were fit to the following equation:

$$r = r_0 + \left[\frac{a \times [P]}{(b + [P])} \right]$$

Where *[P]* corresponds to free protein concentration, r_0 is the anisotropy of free RNA, *a* is the amplitude of the change in

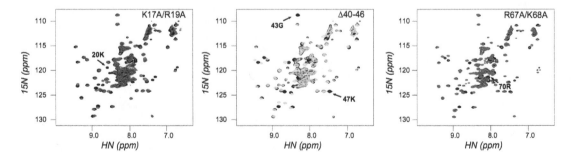

Figure 2. Folding states of HYL1-dsRBD1 mutant domains. SOFAST ^{1}H-^{15}N HMQC spectra of the mutant proteins compared to the wild type protein (black in all spectra). The mutant protein K17A/R19A is shown in red, Δ40–46 in green and R67A/K68A in blue. Signals corresponding to the mutated residues are shown with an arrow on the wild type spectra. The mutated domains in region 1, 2 and 3 retain the same folding as the wild type HYL1 dsRBD1.

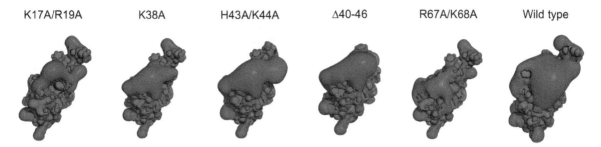

K17A/R19A   K38A   H43A/K44A   Δ40-46   R67A/K68A   Wild type

Figure 3. Electrostatic field calculations on the wild type and mutant proteins. The electrostatic field around each modeled protein and HYL1-dsRBD1 wild type was calculated using APBS [28]. The 1.5 V isocontours are shown for positive (blue) and negative (red) potential. Mutants in regions 1 and 3 display a big disruption of the positive electrostatic patch in the dsRNA-binding surface for substrate binding.

anisotropy upon binding and b is the dissociation constant. Titration curves were normalized for plotting by subtracting from each data point the value of r_0 and dividing the result by amplitude a.

Protein unfolding experiments by CD

For the stability experiments, the refolded proteins were desalted in PD-10 Desalting Columns (GE Healtcare) using 10 mM phosphate buffer pH 7. Protein concentration was measured by UV spectroscopy. HYL1-1 and HYL1-1 mutants were incubated with urea at different concentrations for three hours at room temperature to ensure equilibrium conditions and to minimize chemical modifications. Ellipticity of protein samples was evaluated using a Jasco 810 spectropolarimeter calibrated with (+) 10-camphorsulphonic acid. Far–UV CD spectra were recorded in the range between 190 and 250 nm, urea induced unfolding was monitored by changes of the ellipticity at 220 nm. Protein concentration was 10 μM, and a cell of 0.1 cm path length was used. In all cases, data were acquired at a scan speed of 20 nm min^{-1} and at least 3 scans were averaged for each sample. The signal at 250 nm was used as an internal control to correct for small fluctuations in the baseline. All measurements were done at 20°C.

Nuclear Magnetic Resonance (NMR) spectroscopy

NMR spectra were recorded at 298 K. All spectra were processed with NMRPipe [24] and analyzed with CCPNMR [25]. To evaluate the state of folding of the protein constructs, a ^{1}H-^{15}N SOFAST-HMQC spectrum [26] was acquired on a 600 MHz Bruker spectrometer.

Protein structural modeling and electrostatic field calculations

A model for the structures of the each of the mutant proteins was generated using the software Rosetta [27]. The electrostatic field of the proteins was calculated with the APBS tool [28], and represented setting 1.5 V isocontours for positive and negative potentials.

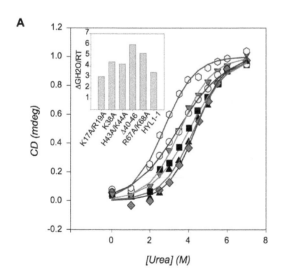

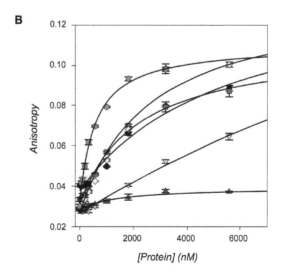

Figure 4. Biophysical characterization of HYL1-dsRBD1 mutant domains. A. *Protein stability measurements.* Induced urea unfolding of the wild type and mutant proteins were followed by circular dichroism at 220 nm. In the left, on the top, the ΔGH₂O/RT of each protein is shown. Mutants in region 2 and 3 display an increase of the stability relative to wild type, whereas the mutant in region 1 and the wild type have comparable values. B. *Substrate binding affinity of the mutant domains.* Fluorescence anisotropy of labeled substrate RNA was measured at increasing protein concentration. All the mutant proteins have lower binding affinities than the wild type domain. The mutants in region 1 and 3 have the most reduced binding affinities. In A and B symbols correspond to the following proteins: gray hexagons, wild type; black triangles, K17A/R19A; gray circles, Δ40–46; gray diamonds, K38A; black squares H43A/K44A; gray triangles R67A/K68A.

Results

Rational design of mutations in HYL1

The function of HYL1 in plants is defined by its two N-terminal double stranded RNA binding domains [12]. However, we and others have shown that both domains have different roles within the protein [18,19]. While the first domain binds tightly to substrate RNA, the second one shows little affinity for the same primary miRNA transcripts and contributes to a small extent to the overall RNA-binding affinity of the protein. Therefore, it can be concluded that the substrate recognition function of HYL1 is located in the first dsRBD. In order to understand the structural determinants of this recognition, we introduced mutations in the regions of this domain that interact with RNA. Crystal and solution structures of several dsRBDs in complex with RNA show that these domains bind dsRNA via the sidechains of three separated regions of the protein. A group of well-conserved residues in helix 1 (region 1) and the loop between strands 1 and 2 (region 2) interact with the OH moieties of backbone riboses, whereas a set of basic residues, either Arg or Lys, located at the N-terminus of helix 2 (region 3) interact with the phosphate backbone [29–33]. Our NMR characterization of the interaction between HYL1 and dsRBD1 showed a similar pattern of interactions [18]. Therefore, we decided to probe the relevance of each region to the function of the protein by the introduction of mutations based on the NMR results. With this aim we constructed a double mutant in region 1 (K17A/R19A), one point mutant, one double mutant and a deletion of the whole loop mutant in region 2 (K38A, H43A/K44A and Δ40–46) and a double mutant in region 3 (R67A/K68A) (Figure 1A). We considered that these different variants would bring new insights into the role of the different domains of HYL1 both *in vivo* and *in vitro*.

In vivo function of HYL1 variants

In order to verify the relevance of the changes introduced in HYL1-dsRBD1 for the function of the protein *in vivo* we employed a complementation experiment with the *hyl1-2* null mutant of *Arabidopsis thaliana* [34]. These plants exhibit a pleiotropic phenotype characterized by hyponastic leaves, reduced leaf size, slow growth, reduced plant height, late flowering and reduced fertility, as well as multiple lateral shoots [35]. In *hyl1-2* plants miRNA processing is impaired, leading to a decrease of mature miRNAs and an increase in the steady state levels of pri-miRNAs [7,9,10].

We transformed the mutant plants with the full-length *hyl1* genes bearing the designed mutations, under the control of the 35 S promoter of the *Cauliflower mosaic virus*. The first generation of transgenic plants show varying phenotypic features that range between strong *hyl-1* phenotype to fully complemented wild type phenotype. We classified the plants as completely rescued (P1), partially rescued (P2) or not rescued (P3), based on their phenotypic characteristics (Figure 1B). A distribution of phenotypes among the primary transgenic plants of *Arabidopsis* is expected, as the transgenes insert at random position along the genome, and are then influenced in different ways by the genomic environment. This has been previously observed for transgenes in general and also for different domains of HYL1 in particular [12]. The classification shows that the level of phenotypic rescue cluster within regions and that, unexpectedly, the highest deficiency in HYL1 function corresponds to mutants in regions 1 and 3, whereas mutants in region 2 show high levels of phenotypic rescue, indicating that the region is mostly unimportant for the function of HYL1 *in vivo* (Figure 1C). This last result was puzzling, since

several previous studies on homologous proteins have shown that this loop is essential for dsRNA recognition by dsRBDs [36,37]. With the aim of further testing this result, we produced a mutant containing a deletion of the whole loop corresponding to region 2 (Δ40–46). This mutant protein resulted functional as well, thus confirming that region 2 is dispensable for the function of HYL1 (Figure 1C).

Aside from phenotype, it is well established that the function of HYL1 is to assist DCL1 in miRNA processing [9,10]. We decided to test the function in a direct way by measuring the levels of mature miRNA and primary miRNA transcripts in the transgenic plant lines. In order to correct for possible variations of the expression levels we measured the protein expression level of different plant lines by means of western blot analysis (Figure S2) and selected lines with similar HYL1 levels for further analysis.

We quantified miR164a, miR172a and miR396a together with their corresponding pri-miRNAs in the wild type, *hyl1-2*, and *hyl1-2* mutants transformed with the different constructs (Figure 1D). In all cases, the processing activity corresponds well with the phenotypic rescue. Therefore we can confirm that the mutations in regions 1 and 3 affect the function of HYL1 during pri-miRNA processing.

Structural analysis of the mutant domains

The introduction of mutations can lead to either local or global changes in the structure of the protein. We were interested in verifying the involvement of the mutated sidechains in the function of HYL1. Therefore, with the aim of ruling out large structural rearrangements that would invalidate our results, we generated constructs for the isolated dsRBD1, labeled the proteins with ^{15}N and obtained a ^{1}H-^{15}N HMQC spectrum of each of the mutants. This kind of spectrum is exquisitely sensitive to alterations in the structure of the protein, and allows for the localization of these alterations with residue definition. The superposition of the spectra of the different mutants with that of the wild type protein shows that the structural alterations brought by the mutations are mostly restricted to the regions structurally close to the mutations themselves, not leading to global modifications (Figure 2). This is true even for the deletion mutant, where one could expect a more important rearrangement of the protein structure.

In order to find a structural basis for the alterations in HYL1 function brought about by the mutations, we obtained structural models of the mutant proteins based on the crystal structure of the wild type protein (3ADG.pdb) using the software Rosetta [27]. As expected, the point mutations do not lead to large structural rearrangements, and the deletion of the β1–β2 loop is well-tolerated by the protein fold (Figure S3). Overall, the more important difference in the HYL1 variants resides in the electrostatic potential field around the protein. The wild type protein has a large positive electrostatic potential patch around regions 1 and 3, caused by a cluster of basic residues. These residues contact the phosphate backbone of the RNA via electrostatic interactions. Removal of these basic side chains in region 1 and region 3 mutants, lead to a highly reduced electrostatic potential. In contrast, the mutants in region 2 show a mostly conserved electrostatic potential (Figure 3). Therefore, we conclude that the electrostatic attraction of the substrate by residues in regions 1 and 3 of HYL1 is essential for the correct function of the protein.

Stability and substrate binding

Having established that all the mutant proteins were correctly folded, we wondered if the differences observed on *in vivo* activities could be due to a destabilization of the protein induced by the mutations introduced. Therefore we measured the stability (ΔG°N-U) of the HYL1 dsRBD1 and the corresponding mutants. For this purpose, we used circular dichrois to monitor urea-induced denaturation. We observed that protein stability is actually increased in mutants in region 2 and 3 relative to wild type, whereas the stability of the mutant in region 1 is comparable to that of the wild type protein (Figure 4A). From these results we could conclude that mutations do not impair HYL1 dsRBD1 stability. In order to evaluate the effect of the mutations on the binding affinity for substrate RNA we resorted to a fluorescence polarization assay, using the isolated dsRBD1 constructs and the lower-stem region of pri-miR172 as binding partner. All mutant proteins showed a decrease in binding affinity. Most noticeably, the double mutant in region 1 (K17A/R19A) hardly bound RNA, even at the highest protein concentration tested, whereas the double mutant in region 3 (R67A/K68A) showed a much-reduced affinity (Figure 4B).

Quite strikingly, considering the purported importance of region 2 in binding, all mutations in this region gave only slightly diminished binding affinities. The dissociation constants of these mutants are similar, ranging between 2.5 and 5.4 µM, giving an overall coherent picture of the influence of this region in HYL1 affinity for RNA. Even when the whole loop is deleted, the protein retains significant affinity for the substrate.

Discussion

The exact function of HYL1 within the plant miRNA-processing complex is not understood at present. The protein is known to be important for the accuracy of the processing of miRNA, as it was demonstrated both *in vitro* and *in vivo* [9,10]. In fact, recent works have shown that the lack of HYL1 can be compensated with a more active form of DCL1 [38,39]. In this case, enough correctly processed miRNA can be generated, thereby limiting the effects of inaccuracy. HYL1 was also suggested to participate in miRNA strand selection and delivery to AGO1 within the RISC complex, much in the same way as TRBP in humans [11]. While there has been more effort in establishing the role of similar dsRBD-containing helper proteins in other organisms, there has been no consensus on their function so far.

Although the mechanism of HYL1 function in miRNA processing is not fully understood, binding to substrate RNA seems to be an essential part of it. RNA binding by HYL1 is dominated by dsRBD1, with marginal contribution to the binding affinity by dsRBD2 and the rest of the protein [18,19]. In the present paper we dissected the function of each of the dsRNA binding determinants within HYL1 dsRBD1.

Our characterization shows that regions 1 and 3 of the domain are the most important structural determinants of RNA binding. In most dsRBDs, region 3 contains a well-conserved KKxAK motif that recognizes the phosphodiester backbone of the dsRNA major groove. These residues form a significant electrostatic patch on the surface of the domain created by the positively charged lysine side chains (Figure 3). In HYL1, however, the motif is split between region 1 and region 3: the third lysine residue in the motif is replaced by a glutamate, but this substitution is compensated by the presence of a lysine residue in position 17, whose terminal amino group is located in a position equivalent to that of the lysine absent in region 3. The mutations that we introduced in regions 1 and 3 give rise then to a structurally similar result, that is, the disruption of this electrostatic patch. In this way we can rationalize the similar impact on both function and affinity of both mutations introduced. It was suggested that electrostatic interactions play an important role in dsRBD-RNA recognition [37,40] although these effects seem to be dissimilar between different dsRBDs [41]. The absence of alterations in stability on the mutant proteins that could hinder RNA recognition highlight that phosphate backbone recognition is essential for RNA binding by these domains.

An unexpected result in our work is that the loop $\beta1$–$\beta2$ in dsRBD1 is dispensable for HYL1 activity and has little influence on the RNA-binding affinity. This result is difficult to rationalize from a structural point of view, as the loop plays an important role in RNA binding in other reported cases, inserting in the dsRNA minor groove. It provides with a set of direct interactions with the ribose moieties and the bases, and contributes to the recognition of the of the A-form RNA double helix, as it is located exactly two turns away from the region 1 interaction position. The importance of this loop in other dsRBDs has been demonstrated through mutational analysis [36,37]. However, in HYL1 the highly conserved histidine residue at the top of the loop can be mutated and even the whole loop deleted without major changes in RNA affinity, protein stability or protein function. This shows that the importance of the loop in RNA recognition can be dissimilar among dsRBDs. In this respect, it is noteworthy that the dsRBDs of Dicer proteins have a short loop that could not in principle participate in RNA binding, or at least not if the dsRBD binds RNA in the canonical way [42,43]. The absence of the loop was suggested to hinder RNA binding by these domains [43], but it was recently shown that Dicer dsRBD do bind dsRNA and miRNA precursors [44]. This experimental evidence goes in line with our results, showing that the loop $\beta1$–$\beta2$ is dispensable for dsRNA binding by HYL1. A structural study of the complex formed by these domains or by the HYL1 deletion mutant with RNA would be necessary to understand how the binding mode of these proteins differs from that of canonical dsRBDs. The regulation of HYL1 activity by phosphorylation was also recently demonstrated [45]. Remarkably, one of the regulatory phosphorylation sites is S42, located within this loop. When this serine residue is mutated to aspartic acid, mimicking phosphorylation, the function of the protein is inhibited, whereas when the phosphorylation sites are eliminated by mutation of serines to alanines the resulting protein is fully functional. Considering such a fact, we can speculate that mutations in region 2 could hinder a natural inhibition of HYL1 activity by phosphorylation, therefore offsetting the partial loss of affinity introduced by the mutation.

In summary, we could show that there is a direct correlation between substrate binding affinity by HYL1-dsRBD1 and its *in vivo* activity. Our work also establishes that RNA binding by HYL1 dsRBD1 is essential for its function. These results contribute to understanding the participation of this protein in substrate recognition within the plant miRNA processing machinery.

Supporting Information

Figure S1   Structure of pre-miRNA 172-ls. The pre-miRNA 172 ls was used to measure the binding affinities of the wild type and mutated dsRBD1-HYL1 proteins. The double stranded RNA was labelled in the 5′ end with fluorescein (see materials and method section).

Figure S2   HYL1 expression level in inflorescences. A. The phenotype rescue of hyl1 plants do not depend to the HYL1 expression level. Transgenic plants with similar HYL1 levels are complemented in different ways and they clustered in different groups (see figure 1B). B. The expression levels of the inflorescences of 36 T1 lines were determined. The amount of recombinant HYL1 protein are highly variables. We selected 3 plants with similar protein levels to continue with further studies (e.i miRNA processing efficiency). The numbers after each label indicate the plant ID. The wells where the bands are absent indicate that protein levels are under the detection limit. The film exposure time was 2 minutes.

Figure S3 Modeled structure of HYL1-dsRBD1. Hyl-dsRBD1 Δ40–46 (left), compared to wild type HYL1-dsRBD1 (PDB 3ADG, right). The structure of the HYL1 mutant was modeled using Rosetta [27]. The final structure adopt a folding that is similar to the crystallographic structure of the wild type HYL1 dsRBD.

Text S1 Methods and primers information. Western blot method and primer sequences information.

Acknowledgments

We thank Dr. Julieta Mateos (Fundación Instituto Leloir, Buenos Aires, Argentina) for the construction of the wild type vector and for useful comments. We would like to thank María Robson, Geraldine Raimundo and Mariana de Sanctis for the language correction of the manuscript.

Author Contributions

Conceived and designed the experiments: PB NGB JFP RMR. Performed the experiments: PB FM ALS NGB. Analyzed the data: PB JFP RMR. Contributed reagents/materials/analysis tools: JFP RMR. Contributed to the writing of the manuscript: PB JFP RMR.

References

1. Filipowicz W, Bhattacharyya SN, Sonenberg N (2008) Mechanisms of post-transcriptional regulation by microRNAs: are the answers in sight? Nat Rev Genet 9: 102–114.
2. Chapman EJ, Carrington JC (2007) Specialization and evolution of endogenous small RNA pathways. Nat Rev Genet 8: 884–896.
3. Saito K, Ishizuka A, Siomi H, Siomi MC (2005) Processing of pre-microRNAs by the Dicer-1-Loquacious complex in Drosophila cells. PLoS Biol 3: e235.
4. Liu X, Park JK, Jiang F, Liu Y, McKearin D, et al. (2007) Dicer-1, but not Loquacious, is critical for assembly of miRNA-induced silencing complexes. RNA 13: 2324–2329.
5. Liu Q, Rand TA, Kalidas S, Du F, Kim H-E, et al. (2003) R2D2, a bridge between the initiation and effector steps of the Drosophila RNAi pathway. Science 301: 1921–1925.
6. Tomari Y, Matranga C, Haley B, Martinez N, Zamore PD (2004) A protein sensor for siRNA asymmetry. Science 306: 1377–1380.
7. Han M-H, Goud S, Song L, Fedoroff N (2004) The Arabidopsis double-stranded RNA-binding protein HYL1 plays a role in microRNA-mediated gene regulation. Proc Natl Acad Sci U S A 101: 1093–1098.
8. Lobbes D, Rallapalli G, Schmidt DD, Martin C, Clarke J (2006) SERRATE: a new player on the plant microRNA scene. EMBO Rep 7: 1052–1058.
9. Dong Z, Han M-H, Fedoroff N (2008) The RNA-binding proteins HYL1 and SE promote accurate in vitro processing of pri-miRNA by DCL1. Proc Natl Acad Sci U S A 105: 9970–9975.
10. Kurihara Y, Takashi Y, Watanabe Y (2006) The interaction between DCL1 and HYL1 is important for efficient and precise processing of pri-miRNA in plant microRNA biogenesis. RNA 12: 206–212.
11. Eamens AL, Smith NA, Curtin SJ, Wang M-B, Waterhouse PM (2009) The Arabidopsis thaliana double-stranded RNA binding protein DRB1 directs guide strand selection from microRNA duplexes. RNA 15: 2219–2235.
12. Wu F, Yu L, Cao W, Mao Y, Liu Z, et al. (2007) The N-terminal double-stranded RNA binding domains of Arabidopsis HYPONASTIC LEAVES1 are sufficient for pre-microRNA processing. Plant Cell 19: 914–925.
13. Pham JW, Pellino JL, Lee YS, Carthew RW, Sontheimer EJ (2004) A Dicer-2-dependent 80s complex cleaves targeted mRNAs during RNAi in Drosophila. Cell 117: 83–94.
14. Liu X, Jiang F, Kalidas S, Smith D, Liu Q (2006) Dicer-2 and R2D2 coordinately bind siRNA to promote assembly of the siRISC complexes. RNA 12: 1514–1520.
15. Lee JY, Kim H, Ryu CH, Kim JY, Choi BH, et al. (2004) Merlin, a tumor suppressor, interacts with transactivation-responsive RNA-binding protein and inhibits its oncogenic activity. J Biol Chem 279: 30265–30273.
16. Daniels SM, Melendez-Peña CE, Scarborough RJ, Daher A, Christensen HS, et al. (2009) Characterization of the TRBP domain required for dicer interaction and function in RNA interference. BMC Mol Biol 10: 38.
17. Yamashita S, Nagata T, Kawazoe M, Takemoto C, Kigawa T, et al. (2011) Structures of the first and second double-stranded RNA-binding domains of human TAR RNA-binding protein. Protein Sci 20: 118–130.
18. Rasia RM, Mateos J, Bologna NG, Burdisso P, Imbert L, et al. (2010) Structure and RNA interactions of the plant MicroRNA processing-associated protein HYL1. Biochemistry 49: 8237–8239.
19. Yang SW, Chen H-Y, Yang J, Machida S, Chua N-H, et al. (2010) Structure of Arabidopsis HYPONASTIC LEAVES1 and its molecular implications for miRNA processing. Structure 18: 594–605.
20. Clough SJ, Bent AF (1998) Floral dip: a simplified method for Agrobacterium-mediated transformation of Arabidopsis thaliana. Plant J 16: 735–743.
21. Chen C, Ridzon DA, Broomer AJ, Zhou Z, Lee DH, et al. (2005) Real-time quantification of microRNAs by stem-loop RT–PCR. Nucleic Acids Res 33: e179–e179.
22. Czechowski T, Stitt M, Altmann T, Udvardi MK, Scheible W-R (2005) Genome-Wide Identification and Testing of Superior Reference Genes for Transcript Normalization in Arabidopsis. Plant Physiol 139: 5–17.
23. Houben K, Marion D, Tarbouriech N, Ruigrok RWH, Blanchard L (2007) Interaction of the C-terminal domains of sendai virus N and P proteins: comparison of polymerase-nucleocapsid interactions within the paramyxovirus family. J Virol 81: 6807–6816.
24. Delaglio F, Grzesiek S, Vuister GW, Zhu G, Pfeifer J, et al. (1995) NMRPipe: a multidimensional spectral processing system based on UNIX pipes. J Biomol NMR 6: 277–293.
25. Vranken WF, Boucher W, Stevens TJ, Fogh RH, Pajon A, et al. (2005) The CCPN data model for NMR spectroscopy: Development of a software pipeline. Proteins Struct Funct Bioinforma 59: 687–696.
26. Schanda P, Brutscher B (2005) Very fast two-dimensional NMR spectroscopy for real-time investigation of dynamic events in proteins on the time scale of seconds. J Am Chem Soc 127: 8014–8015.
27. Rohl CA, Strauss CE, Misura KM, Baker D (2004) Protein structure prediction using Rosetta. Methods Enzymol 383: 66–93.
28. Baker NA, Sept D, Joseph S, Holst MJ, McCammon JA (2001) Electrostatics of nanosystems: application to microtubules and the ribosome. Proc Natl Acad Sci U S A 98: 10037–10041.
29. Blaszczyk J, Gan J, Tropea JE, Court DL, Waugh DS, et al. (2004) Noncatalytic assembly of ribonuclease III with double-stranded RNA. Structure 12: 457–466.
30. Masliah G, Barraud P, Allain FH-T (2013) RNA recognition by double-stranded RNA binding domains: a matter of shape and sequence. Cell Mol Life Sci 70: 1875–1895.
31. Stefl R, Oberstrass FC, Hood JL, Jourdan M, Zimmermann M, et al. (2010) The solution structure of the ADAR2 dsRBM-RNA complex reveals a sequence-specific readout of the minor groove. Cell 143: 225–237.
32. Tian B, Bevilacqua PC, Diegelman-Parente A, Mathews MB (2004) The double-stranded-RNA-binding motif: interference and much more. Nat Rev Mol Cell Biol 5: 1013–1023.
33. Wu H, Henras A, Chanfreau G, Feigon J (2004) Structural basis for recognition of the AGNN tetraloop RNA fold by the double-stranded RNA-binding domain of Rnt1p RNase III. Proc Natl Acad Sci U S A 101: 8307–8312.
34. Song L, Han M-H, Lesicka J, Fedoroff N (2007) Arabidopsis primary microRNA processing proteins HYL1 and DCL1 define a nuclear body distinct from the Cajal body. Proc Natl Acad Sci U S A 104: 5437–5442.
35. Lu C, Fedoroff N (2000) A mutation in the Arabidopsis HYL1 gene encoding a dsRNA binding protein affects responses to abscisic acid, auxin, and cytokinin. Plant Cell 12: 2351–2366.
36. Krovat BC, Jantsch MF (1996) Comparative mutational analysis of the double-stranded RNA binding domains of Xenopus laevis RNA-binding protein A. J Biol Chem 271: 28112–28119.
37. Ramos A, Grünert S, Adams J, Micklem DR, Proctor MR, et al. (2000) RNA recognition by a Staufen double-stranded RNA-binding domain. EMBO J 19: 997–1009.
38. Liu C, Axtell MJ, Fedoroff NV (2012) The Helicase and RNaseIIIa Domains of Arabidopsis Dicer-Like1 Modulate Catalytic Parameters During MicroRNA Biogenesis. Plant Physiol 159: 748–758.
39. Tagami Y, Motose H, Watanabe Y (2009) A dominant mutation in DCL1 suppresses the hyl1 mutant phenotype by promoting the processing of miRNA. RNA 15: 450–458.
40. Burdisso P, Suarez IP, Bologna NG, Palatnik JF, Bersch B, et al. (2012) Second double-stranded RNA binding domain of dicer-like ribonuclease 1: structural and biochemical characterization. Biochemistry 51: 10159–10166.
41. Bevilacqua PC, Cech TR (1996) Minor-groove recognition of double-stranded RNA by the double-stranded RNA-binding domain from the RNA-activated protein kinase PKR. Biochemistry 35: 9983–9994.

42. Du Z, Lee JK, Tjhen R, Stroud RM, James TL (2008) Structural and biochemical insights into the dicing mechanism of mouse Dicer: a conserved lysine is critical for dsRNA cleavage. Proc Natl Acad Sci U S A 105: 2391–2396.

43. Weinberg DE, Nakanishi K, Patel DJ, Bartel DP (2011) The inside-out mechanism of Dicers from budding yeasts. Cell 146: 262–276.

44. Wostenberg C, Lary JW, Sahu D, Acevedo R, Quarles KA, et al. (2012) The role of human Dicer-dsRBD in processing small regulatory RNAs. PLoS One 7: e51829.

45. Manavella PA, Hagmann J, Ott F, Laubinger S, Franz M, et al. (2012) Fast-forward genetics identifies plant CPL phosphatases as regulators of miRNA processing factor HYL1. Cell 151: 859–870.

Permissions

All chapters in this book were first published in PLOS ONE, by The Public Library of Science; hereby published with permission under the Creative Commons Attribution License or equivalent. Every chapter published in this book has been scrutinized by our experts. Their significance has been extensively debated. The topics covered herein carry significant findings which will fuel the growth of the discipline. They may even be implemented as practical applications or may be referred to as a beginning point for another development.

The contributors of this book come from diverse backgrounds, making this book a truly international effort. This book will bring forth new frontiers with its revolutionizing research information and detailed analysis of the nascent developments around the world.

We would like to thank all the contributing authors for lending their expertise to make the book truly unique. They have played a crucial role in the development of this book. Without their invaluable contributions this book wouldn't have been possible. They have made vital efforts to compile up to date information on the varied aspects of this subject to make this book a valuable addition to the collection of many professionals and students.

This book was conceptualized with the vision of imparting up-to-date information and advanced data in this field. To ensure the same, a matchless editorial board was set up. Every individual on the board went through rigorous rounds of assessment to prove their worth. After which they invested a large part of their time researching and compiling the most relevant data for our readers.

The editorial board has been involved in producing this book since its inception. They have spent rigorous hours researching and exploring the diverse topics which have resulted in the successful publishing of this book. They have passed on their knowledge of decades through this book. To expedite this challenging task, the publisher supported the team at every step. A small team of assistant editors was also appointed to further simplify the editing procedure and attain best results for the readers.

Apart from the editorial board, the designing team has also invested a significant amount of their time in understanding the subject and creating the most relevant covers. They scrutinized every image to scout for the most suitable representation of the subject and create an appropriate cover for the book.

The publishing team has been an ardent support to the editorial, designing and production team. Their endless efforts to recruit the best for this project, has resulted in the accomplishment of this book. They are a veteran in the field of academics and their pool of knowledge is as vast as their experience in printing. Their expertise and guidance has proved useful at every step. Their uncompromising quality standards have made this book an exceptional effort. Their encouragement from time to time has been an inspiration for everyone.

The publisher and the editorial board hope that this book will prove to be a valuable piece of knowledge for researchers, students, practitioners and scholars across the globe.

List of Contributors

Guimei Jiang, Peitao Lü, Jitao Liu, Junping Gao and Changqing Zhang
Department of Ornamental Horticulture, College of Agriculture and Biotechnology, China Agricultural University, Beijing, PR China

Xinqiang Jiang
College of Landscape Architecture and Forestry, Qingdao Agricultural University, Qingdao, PR China

Babatunde Bello, Xueyan Zhang, Chuanliang Liu, Zhaoen Yang, Zuoren Yang, Qianhua Wang, Ge Zhao and Fuguang Li
State Key Laboratory of Cotton Biology, Cotton Research Institute, Chinese Academy of Agricultural Sciences, Beijing, China

Deqiang Tai
Department of Plant Science and Technology, Beijing University of Agriculture, Beijing, China
College of Horticulture, Shanxi Agricultural University, Taigu, Shanxi, China

Ji Tian, Jie Zhang, Tingting Song and Yuncong Yao
Department of Plant Science and Technology, Beijing University of Agriculture, Beijing, China
Key Laboratory of New Technology in Agricultural Application of Beijing, Beijing University of Agriculture, Beijing, China

Abdul A. Waheed, Nishani D. Kuruppu, Kathryn L. Felton, Darren D'Souza'and Eric O. Freed
Virus-Cell Interaction Section, HIV Drug Resistance Program, NCI-Frederick, Frederick, Maryland, United States of America

Stephanie Jacobs, Zhenzhong Cui, Ruiben Feng and Joe Z. Tsien
Brain and Behavior Discovery Institute and Department of Neurology, Medical College of Georgia at Georgia Regents University, Augusta, Georgia, United States of America

Huimin Wang
Shanghai Institute of Functional Genomics, East China Normal University, Shanghai, China

Deheng Wang
Banna Biomedical Research Institute, Xi-Shuang-Ban-Na Prefecture, Yunnan Province, China

Qichao Zhao, Minghong Liu, Miaomiao Tan and Zhicheng Shen
State Key Laboratory of Rice Biology, Institute of Insect Sciences, Zhejiang University, Hangzhou, China

Jianhua Gao
College of Life Science, Shanxi Agricultural University, Taigu, China

Yu Han, Jiarong Meng, Jie Chen, Wanlun Cai, Yu Wang, Jing Zhao, Yueping He and Hongxia Hua
Hubei Insect Resources Utilization and Sustainable Pest Management Key Laboratory, College of Plant Science and Technology, Huazhong Agricultural University, Wuhan, P.R. China

Yanni Feng
College of Life Science and Technology, Huazhong Agricultural University, P.R. China

Dirk A. Moser
Faculty of Psychology, Genetic Psychology, Ruhr-University-Bochum, Bochum, Germany
Department of Sports Medicine, Disease Prevention and Rehabilitation, Johannes Gutenberg-University Mainz, Mainz, Germany

Luca Braga, Andrea Raso, Serena Zacchigna and Mauro Giacca
International Centre for Genetic Engineering and Biotechnology (ICGEB), Molecular Medicine, Trieste, Italy

Perikles Simon
Department of Sports Medicine, Disease Prevention and Rehabilitation, Johannes Gutenberg-University Mainz, Mainz, Germany

Joanna Kern, Silvia Leanhart, Marek Bogacz and Rafal Pacholczyk
Center for Biotechnology and Genomic Medicine, Georgia Regents University, Augusta, Georgia, United States of America

Robert Drutel
Center for Biotechnology and Genomic Medicine, Georgia Regents University, Augusta, Georgia, United States of America
Medical University of South Carolina, College of Medicine, Charleston, South Carolina, United States of America

Xiaolian Zhang, Ning Wang, Pei Chen, Mengmeng Gao, Juge Liu, Yufeng Wang, Tuanjie Zhao, Yan Li and Junyi Gai
National Key Laboratory of Crop Genetics and Germplasm Enhancement, National Center for Soybean Improvement, Key Laboratory for Biology and Genetic Improvement of Soybean (General, Ministry of Agriculture), Nanjing Agricultural University, Nanjing, Jiangsu, China

Nuri Company, Anna Nadal, Cristina Ruiz and Maria Pla
Institute for Food and Agricultural Technology, University of Girona, Girona, Spain

Irina Malinova and Joerg Fettke
Plant Physiology, University of Potsdam, Potsdam-Golm, Germany
Biopolymers analytics, University of Potsdam, Potsdam-Golm, Germany

Hans-Henning Kunz
Plant Physiology, Washington State University, Pullman, Washington, United States of America
Department of Botany II, University of Cologne, Cologne, Germany

Saleh Alseekh and Alisdair R. Fernie
Max-Planck-Institute of Molecular Plant Physiology, Potsdam-Golm, Germany

Karoline Herbst
Plant Physiology, University of Potsdam, Potsdam-Golm, Germany

Markus Gierth
Department of Botany II, University of Cologne, Cologne, Germany

Amit Kumar Chaturvedi, Manish Kumar Patel, Avinash Mishra, Vivekanand Tiwari and Bhavanath Jha
Discipline of Marine Biotechnology and Ecology, CSIR-Central Salt and Marine Chemicals Research Institute, Bhavnagar, Gujarat, India

Fang Liu, Xiaojuan Xiong, Lei Wu, Xinhua Zeng, Yinglong Cao, Yuhua Wu, Yunjing Li and Gang Wu
Key Laboratory of Oil Crop Biology of the Ministry of Agriculture, Oil Crops Research Institute, Chinese Academy of Agricultural Sciences, Wuhan, China

Donghui Fu
The Key Laboratory of Crop Physiology, Ecology and Genetic Breeding, Ministry of Education, Agronomy College, Jiangxi Agricultural University, Nanchang, China

Alice Hayward
Queensland Alliance for Agriculture and Food Innovation, The University of Queensland, Queensland, Australia

Dan Zhu
College of Life Science, Qingdao Agricultural University, Qingdao, P.R. China
Plant Bioengineering Laboratory, Northeast Agricultural University, Harbin, P.R. China

Rongtian Li
Key Laboratory of Molecular Biology, College of Heilongjiang Province, Heilongjiang University, Harbin, P.R. China

Xin Liu
College of Life Science, Qingdao Agricultural University, Qingdao, P.R. China

Mingzhe Sun, Jing Wu, Ning Zhang and Yanming Zhu
Plant Bioengineering Laboratory, Northeast Agricultural University, Harbin, P.R. China

Mengling Wen, Yuan Gao, Lijun Wang, Lingyu Ran and Jiahui Li
Key Laboratory of Eco-environments of Three Gorges Reservoir Region, Ministry of Education, Institute of Resources Botany, School of Life Sciences, Southwest University, Chongqing, China
Chongqing Key Laboratory of Transgenic Plant and Safety Control, Southwest University, Chongqing, China

Keming Luo
Key Laboratory of Eco-environments of Three Gorges Reservoir Region, Ministry of Education, Institute of Resources Botany, School of Life Sciences, Southwest University, Chongqing, China
Chongqing Key Laboratory of Transgenic Plant and

Safety Control, Southwest University, Chongqing, China
Key Laboratory of Adaptation and Evolution of Plateau Biota, Northwest Institute of Plateau Biology, Chinese Academy of Sciences, Xining, China

Paula Burdisso and Rodolfo M. Rasia
Instituto de Biología Molecular y Celular de Rosario, Rosario, Argentina
Área Biofísica, Facultad de Ciencias Bioquímicas y Farmacéuticas, Universidad Nacional de Rosario, Rosario, Argentina

Fernando Milia and Javier F. Palatnik
Instituto de Biología Molecular y Celular de Rosario, Rosario, Argentina

Arnaldo L. Schapire
Instituto de Biología Molecular y Celular de Rosario, Rosario, Argentina
Center for Research in Agricultural Genomics CRAG (CSIC-IRTA-UAB-UB), Edifici CRAG-Campus UAB, Bellaterra (Cerdanyola del Vallés), Barcelona, Spain

Nicolá s G. Bologna
Instituto de Biología Molecular y Celular de Rosario, Rosario, Argentina
Swiss Federal Institute of Technology (ETH), Zurich, Switzerland

Index

Printed in the USA
CPSIA information can be obtained
at www.ICGtesting.com
JSHW051412221024
72173JS00006B/1349